Selected Coolwater Fishes of North America

A Symposium on
Selected Coolwater Fishes of North America

Steering Committee

Arden J. Trandahl and Ward Falkner, *Cochairmen*

Bernard L. Griswold   John H. Klingbiel
Richard Sternberg   Jack R. Hammond
Shyrl E. Hood   Robert G. Martin

Special Advisors

Dale L. Henegar   G. Herbert Lawler
Richard A. Ryder   Richard H. Stroud

Sponsors

North Central Division, American Fisheries Society
Sport Fishing Institute
Fish Culture Section, American Fisheries Society
United States Fish and Wildlife Service
Freshwater Institute, Fisheries and Environment Canada

Contributors

North Central Airlines	Aquafine Corp.
Lund American, Inc.	Heath Tecna
Northern States Power Co.	Coffelt Electronics
Blue Spruce	Frigid Units
Bioproducts	Lake Aid, Inc.
UMA Engineers	Clean Flo Laboratories
Kramer, Chin & Mayo	Jungle Labs
Peterson Bros. Boat Works	Hydro Lab
Kembro	Glen L. Evans, Inc.
Peterson Trout Farm	My Buddy Products
North Star Company	Berkley and Co.
Glencoe Mills	Johnson Reels Co.
Tavolek	Normark
Aquatic Diet Technology, Inc.	E–Z Mount, Inc.
Engineered Sales Co.	Polar Craft Manufacturing Co.
Domsea Farms	Stearns Manufacturing Co.
Aquaculture Digest	Dupont
Dairyland Cooperative Power Co.	Forestry Suppliers
Ziegler Bros., Inc.	Ben Meadows Co.

Staff of the Minnesota Department of Natural Resources

Selected Coolwater Fishes of North America

Robert L. Kendall
Editor

Proceedings of a Symposium
held in St. Paul, Minnesota
March 7–9, 1978

Special Publication No. 11

American Fisheries Society
Washington, D.C.
1978

Publication of these Proceedings
was aided by grants from the

Great Lakes Fishery Commission
and the
United States Fish and Wildlife Service

Copyright 1978 by the
American Fisheries Society

Library of Congress Catalog Card Number: 78-72983

All orders should be addressed to

American Fisheries Society
5410 Grosvenor Lane
Bethesda, Maryland 21044 U.S.A.

1-15-79

Printed in the United States of America
by Lancaster Press, Inc., Lancaster, Pennsylvania

Contents

Arden Trandahl	ix	Preface
	xi	List of Fish Species

General Considerations

John J. Ney	1	A Synoptic Review of Yellow Perch and Walleye Biology
E. J. Crossman	13	Taxonomy and Distribution of North American Esocids

Ecological Considerations

Kenneth D. Carlander J. Scott Campbell Robert J. Muncy	27	Inventory of Percid and Esocid Habitat in North America
R. A. Ryder S. R. Kerr	39	The Adult Walleye in the Percid Community—A Niche Definition Based on Feeding Behaviour and Food Specificity
Donald E. Olson Dennis H. Schupp Val Macins	52	An Hypothesis of Homing Behavior of Walleyes as Related to Observed Patterns of Passive and Active Movement
Dennis H. Schupp	58	Walleye Abundance, Growth, Movement, and Yield in Disparate Environments Within a Minnesota Lake
W. B. Wrenn T. D. Forsythe	66	Effects of Temperature on Production and Yield of Juvenile Walleyes in Experimental Ecosystems
P. A. Hackney John A. Holbrook II	74	Sauger, Walleye, and Yellow Perch in the Southeastern United States
Richard B. Fitz John A. Holbrook II	82	Sauger and Walleye in Norris Reservoir, Tennessee
James P. Clugston James L. Oliver Richard Ruelle	89	Reproduction, Growth, and Standing Crops of Yellow Perch in Southern Reservoirs
Michael D. Clady	100	Structure of Fish Communities in Lakes That Contain Yellow Perch, Sauger, and Walleye Populations
Michael D. Clady Larry Nielsen	109	Diversity of a Community of Small Fishes as Related to Abundance of the Dominant Percid Fishes
John M. Casselman	114	Effects of Environmental Factors on Growth, Survival, Activity, and Exploitation of Northern Pike
E. J. Harrison W. F. Hadley	129	Ecologic Separation of Sympatric Muskellunge and Northern Pike
George B. Beyerle	135	Survival, Growth, and Vulnerability to Angling of Northern Pike and Walleyes Stocked as Fingerlings in Small Lakes with Bluegills or Minnows
Robert L. Miles	140	A Life History Study of the Muskellunge in West Virginia
John D. Minor E. J. Crossman	146	Home Range and Seasonal Movements of Muskellunge as Determined by Radiotelemetry

William R. Nelson	154	Implications of Water Management in Lake Oahe for the Spawning Success of Coolwater Fishes
James M. Engel	159	Management of Endangered Coolwater Fishes

Culture

Joseph Webster Arden Trandahl John Leonard	161	Historical Perspective of Propagation and Management of Coolwater Fishes in the United States
Leo E. Orme	167	The Status of Coolwater Fish Diets
Graden West John Leonard	172	Culture of Yellow Perch with Emphasis on Development of Eggs and Fry
Steven D. Koenig T. B. Kayes H. E. Calbert	177	Preliminary Observations on the Sperm of Yellow Perch
E. F. Schott T. B. Kayes H. E. Calbert	181	Comparative Growth of Male Versus Female Yellow Perch Fingerlings Under Controlled Environmental Conditions
John G. Nickum	187	Intensive Culture of Walleyes: The State of the Art
Delano R. Graff	195	Intensive Culture of Esocids: The Current State of the Art
Charles H. Pecor	202	Intensive Culture of Tiger Muskellunge in Michigan During 1976 and 1977
Keen Buss James Meade III Delano R. Graff	210	Reviewing the Esocid Hybrids
G. de Montalembert C. Bry R. Billard	217	Control of Reproduction in Northern Pike
Philip P. Economon	226	A Muscular Dystrophy-Like Anomaly of Walleye
R. A. Sonstegard John G. Hnath	235	Lymphosarcoma in Muskellunge and Northern Pike: Guidelines for Disease Control
Cecil L. Fox	238	The Hatchery Development Process
Harry Westers	246	Biological Considerations in Hatchery Design for Coolwater Fishes

Considerations in Percid Management

Percy W. Laarman	254	Case Histories of Stocking Walleyes in Inland Lakes, Impoundments, and the Great Lakes—100 Years with Walleyes
Michael R. Rawson Russell L. Scholl	261	Reestablishment of Sauger in Western Lake Erie
R. O. Schlick	266	Management for Walleye or Sauger, South Basin, Lake Winnipeg

Kenneth O. Paxton Frederick Stevenson	270	Food, Growth, and Exploitation of Percids in Ohio's Upground Reservoirs
Calvin L. Groen Troy A. Schroeder	278	Effects of Water Level Management on Walleye and Other Coolwater Fishes in Kansas Reservoirs

Development of Esocid Fisheries

Stephen A. Flickinger John H. Clark	284	Management Evaluation of Stocked Northern Pike in Colorado's Small Irrigation Reservoirs
Lawrence C. Belusz	292	An Evaluation of the Muskellunge Fishery of Lake Pomme de Terre and Efforts to Improve Stocking Success
Leon D. Johnson	298	Evaluation of Esocid Stocking Program in Wisconsin
Robert B. Hesser	302	Management Implications of Hybrid Esocids in Pennsylvania

Management Case Histories

E. B. Davidoff	308	The Matheson Island Sauger Fishery of Lake Winnipeg, 1972–1976
John A. Goddard Lee C. Redmond	313	Northern Pike, Tiger Muskellunge, and Walleye Populations in Stockton Lake, Missouri: A Management Evaluation
Howard E. Snow	320	Responses of Northern Pike to Exploitation in Murphy Flowage, Wisconsin
James R. Axon	328	An Evaluation of the Muskellunge Fishery in Cave Run Lake, Kentucky
Robert C. Haas	334	The Muskellunge in Lake St. Clair

General Management Concepts

Ken D. Bovee	340	The Incremental Method of Assessing Habitat Potential for Coolwater Species, with Management Implications
George F. Adams	347	An Historical Review of the Commercial Fisheries of the Boreal Lakes of Central Canada: Their Development, Management, and Potential
Dennis M. Cauvin	361	The Allocation of Resources in Fisheries: An Economic Perspective
Richard O. Anderson A. Stephen Weithman	371	The Concept of Balance for Coolwater Fish Populations
James J. Kempinger Robert F. Carline	382	Dynamics of the Northern Pike Populations and Changes that Occurred with a Minimum Size Limit in Escanaba Lake, Wisconsin
Steven L. Serns	390	Effects of a Minimum Size Limit on the Walleye Population of a Northern Wisconsin Lake
James C. Schneider	398	Selection of Minimum Size Limits for Walleye Fishing in Michigan

John A. Prentice Richard D. Clark, Jr.	408	Walleye Fishery Management Program in Texas—A Systems Approach
Robert T. Lackey	417	Fisheries Management Theory

Panels

John G. Nickum	424	Hatchery Design for Coolwater Species—A Panel
Lawrence C. Belusz	426	The Role of Private Organizations in Management of Coolwater Fisheries Resources—A Panel
Kenneth M. Muth	428	Allocating Percid Resources in the Great Lakes: Biological, Institutional, Political, Social, and Economic Ramifications—A Panel

Summary

G. H. Lawler	436	Review and a Look Ahead

Preface

This Symposium did not just happen; the idea for it emerged and caught fire nearly four years ago. It received its first official sponsorship from the North Central Division of the American Fisheries Society. Subsequently, the AFS Fish Culture Section, the Freshwater Institute, the Sport Fishing Institute, and the U.S. Fish and Wildlife Service agreed to cosponsor the symposium, and the result was an interesting and comprehensive mix of professional and governmental support. With sponsorship assured, the Steering Committee was formed, and it worked for two and a half years to bring the Coolwater Symposium to a climax.

The term "coolwater fishes" is not rigorously defined, but it refers generally to those species which are distributed by temperature preference between the "coldwater" salmonid communities to the north and the more diverse, often centrarchid-dominated "warmwater" assemblages to the south. We decided to concentrate, though not exclusively, on five species for the Symposium, hence the title, "Selected Coolwater Fishes of North America." These five—walleye, sauger, yellow perch, northern pike, and muskellunge—are of major recreational and commercial importance on this continent. Increasingly, they are cultured and stocked far beyond their native ranges in every geographical direction. Our planning, then, was international in scope. Our aim was to bring academic scientists together with agency biologists, hatchery engineers, and fisheries managers for an exchange of perspectives on the ecology, production, and enhancement of these significant resources. It is a measure of the great interest in this young field that we had no difficulty in filling such an ambitious program.

The Keynote Speaker was Richard J. Myshak, Deputy Assistant Secretary for Fish, Wildlife and Parks in the U.S. Department of the Interior. I want to offer a synopsis of his remarks, which set the Symposium's tone and context.

I find it reassuring that this is an international symposium, with U.S., Canadian, and European participants. It is a splendid example of the creative synergism that can occur when interested, dedicated people work together.

"Coolwater fishes" is a new term which has evolved in the last ten years, and which promises new quality angling experiences to many people. Advances in research, development, and management still are needed in the coolwater area. Many methods of spawning, culturing, and management being applied today are at the same stage of understanding and procedure they were 75 to 100 years ago. What advancement there has been to date has been largely the result of trial and error. Compliments are in order to those who have made significant strides, but these advances only signal what *can* be done. It remains now for us to pick up on those signals and *do* it. Perhaps 70 to 80 percent of the cultural and management concepts now used for trout and salmon can be applied in some way to the coolwater fishes, provided the research and development facilities are made available. The success of the northern pike-muskellunge hybrid is justification enough to explore other coolwater crosses. The public will continue to demand advances of this nature.

Coolwater fish scientists and managers have been a rather small group within the fishery community. You have enjoyed a comradeship that comes with this kind of smallness and closeness. This symposium says to me that you are no longer a small group, either in numbers or in knowledge. It is my hope that as you grow and gain in prominence and recognition, you also manage to hang onto the sense of togetherness that has paid off in spirited cooperation and communication—and a steadily improving fishery resource.

The Symposium met Mr. Myshak's expectations. It revealed new scientific information about coolwater fishes and important advances in their practical management and culture. Major knowledge gaps were identified, and the basis was laid for rapid progress in research and its application. The Symposium Proceedings which comprise this volume document the breadth and depth of the program, but they cannot convey the excitement, enthusiasm, and intense discussions of over 500 people during those three days in St. Paul. The prospects for coolwater fisheries now are brighter than ever before.

The Coolwater Symposium and the publication of its Proceedings could not have happened without the support of our sponsors and contributors. They are listed elsewhere in this volume, but I would like to thank them again here. The Symposium was hosted with grace and skill by the Minnesota Department of Natural Resources. Among the Minnesota staff, we are particularly grateful to Warren Scidmore, Floyd Hennagir, Charles Burrows, Henry Swanson, Howard Krosch, William Nye, Bernham Philbrook, Donald Carlson, Andrew Brewer, David Vesall, Oliver Jarvenpa, William Longley, James Schneider, Sandra Brothen, and Elaine Heinze for their contributions of time and money.

I hold a special reserve of appreciation and esteem for my Steering Committee coworkers and advisors. They worked hard and diligently over a long time to create this Symposium, and its great success reflects the quality of their efforts. Ward Falkner was the Steering Committee's Canadian Cochairman. Richard Sternberg handled all the complex local arrangements. Bernard Griswold put together the outstanding program. Shyrl Hood maintained our accounts and our solvency. Jack Hammond and Robert Martin promoted and publicized the Symposium; John Klingbiel saw that its contributions were published. Herbert Lawler, Dale Henegar, Richard Stroud, and Richard Ryder, our Special Advisors, kept our visions high but out of the clouds. It was the dedication of all these people that really made the Symposium happen. With the perogatives that stem from my initial involvement in the Symposium concept and my position as U.S. Cochairman, and with much pleasure and satisfaction, I dedicate these Proceedings to them.

ARDEN TRANDAHL

U.S. Fish and Wildlife Service
Spearfish Fisheries Center
Spearfish, South Dakota 57783

List of Fish Species

The colloquial names of most United States and Canadian fish species have been standardized in *A List of Common and Scientific Names of Fishes from the United States and Canada*, Third Edition, 1970, American Fisheries Society Special Publication No. 6. Throughout these Proceedings, species covered by the *List* are cited by their common names only. Their respective scientific names follow.

Common name	Scientific name
Alewife	*Alosa pseudoharengus*
Arctic grayling	*Thymallus arcticus*
Atlantic cod	*Gadus morhua*
Atlantic salmon	*Salmo salar*
Banded killifish	*Fundulus diaphanus*
Bigmouth buffalo	*Ictiobus cyprinellus*
Black bullhead	*Ictalurus melas*
Black crappie	*Pomoxis nigromaculatus*
Blacknose dace	*Rhinichthys atratulus*
Bluegill	*Lepomis macrochirus*
Blue pike	*Stizostedion vitreum glaucum*
Bluntnose minnow	*Pimephales notatus*
Bowfin	*Amia calva*
Brook silverside	*Labidesthes sicculus*
Brook stickleback	*Culaea inconstans*
Brook trout	*Salvelinus fontinalis*
Brown bullhead	*Ictalurus nebulosus*
Brown trout	*Salmo trutta*
Burbot	*Lota lota*
Carp	*Cyprinus carpio*
Chain pickerel	*Esox niger*
Channel catfish	*Ictalurus punctatus*
Cisco	*Coregonus artedii*
Coho salmon	*Oncorhynchus kisutch*
Colorado squawfish	*Ptychocheilus lucius*
Common shiner	*Notropis cornutus*
Creek chub	*Semotilus atromaculatus*
Cui-ui	*Chasmistes cujus*
Emerald shiner	*Notropis atherinoides*
Fallfish	*Semotilus corporalis*
Fathead minnow	*Pimephales promelas*
Flathead catfish	*Pylodictis olivaris*
Fountain darter	*Etheostoma fonticola*
Fourhorn sculpin	*Myoxocephalus quadricornis*
Freshwater drum	*Aplodinotus grunniens*
Gizzard shad	*Dorosoma cepedianum*
Golden redhorse	*Moxostoma erythrurum*
Golden shiner	*Notemigonus crysoleucas*
Goldeye	*Hiodon alosoides*
Goldfish	*Carassius auratus*
Grass pickerel	*Esox americanus vermiculatus*
Green sunfish	*Lepomis cyanellus*
Highfin carpsucker	*Carpiodes velifer*
Humpback chub	*Gila cypha*

Iowa darter	*Etheostoma exile*
Johnny darter	*Etheostoma nigrum*
Lake sturgeon	*Acipenser fulvescens*
Lake trout	*Salvelinus namaycush*
Lake whitefish	*Coregonus clupeaformis*
Largemouth bass	*Micropterus salmoides*
Logperch	*Percina caprodes*
Longear sunfish	*Lepomis megalotis*
Longjaw cisco	*Coregonus alpenae*
Longnose sucker	*Catostomus catostomus*
Maryland darter	*Etheostoma sellare*
Mooneye	*Hiodon tergisus*
Mottled sculpin	*Cottus bairdi*
Muskellunge	*Esox masquinongy*
Ninespine stickleback	*Pungitius pungitius*
Northern pike	*Esox lucius*
Pumpkinseed	*Lepomis gibbosus*
Rainbow smelt	*Osmerus mordax*
Rainbow trout	*Salmo gairdneri*
Redfin pickerel	*Esox americanus americanus*
Red shiner	*Notropis lutrensis*
River carpsucker	*Carpiodes carpio*
River redhorse	*Moxostoma carinatum*
Rock bass	*Ambloplites rupestris*
Round whitefish	*Prosopium cylindraceum*
Sand shiner	*Notropis stramineus*
Sauger	*Stizostedion canadense*
Scioto madtom	*Noturus trautmani*
Sea lamprey	*Petromyzon marinus*
Shorthead redhorse	*Moxostoma macrolepidotum*
Shortnose sturgeon	*Acipenser brevirostrum*
Silvery minnow	*Hybognathus nuchalis*
Slimy sculpin	*Cottus cognatus*
Smallmouth bass	*Micropterus dolomieui*
Smallmouth buffalo	*Ictiobus bubalus*
Sockeye salmon	*Oncorhynchus nerka*
Spotfin shiner	*Notropis spilopterus*
Spottail shiner	*Notropis hudsonius*
Spotted bass	*Micropterus punctulatus*
Spotted sucker	*Minytrema melanops*
Striped bass	*Morone saxatilis*
Tessellated darter	*Etheostoma olmstedi*
Threadfin shad	*Dorosoma petenense*
Threespine stickleback	*Gasterosteus aculeatus*
Trout-perch	*Percopsis omiscomaycus*
Walleye	*Stizostedion vitreum vitreum*
Warmouth	*Lepomis gulosus*
White bass	*Morone chrysops*
White crappie	*Pomoxis annularis*
White perch	*Morone americana*
White sucker	*Catostomus commersoni*
Yellow bass	*Morone mississippiensis*
Yellow perch	*Perca flavescens*

GENERAL CONSIDERATIONS

A Synoptic Review of Yellow Perch and Walleye Biology

JOHN J. NEY

Department of Fisheries and Wildlife Sciences
Virginia Polytechnic Institute and State University
Blacksburg, Virginia 24061

Abstract

Yellow perch and walleye systematics, reproduction, early life history, growth, food habits, mortality, annual recruitment, behavior, and interspecific relationships are briefly reviewed and compared. The species possess similar fecundity, patterns of growth and mortality, and initial food habits but differ in other major respects. Substrate preferences and disjunctive spawning periods promote reproductive segregation. Food habits diverge during the first summer as walleyes become piscivorous and yellow perch continue to consume a wide variety of planktonic and benthic invertebrates. Opportunistic feeding may partially account for the higher incidence of stunting in yellow perch populations. Where the species cohabit, synchronous fluctuations in year-class strength and inverse relations of first-year growth suggest the dependence of walleye on young-of-the-year yellow perch production. Year-class strength in each species may be controlled by environmental factors during the reproductive period where spawning habitat is limited, but complex biotic relationships probably influence recruitment in more stable systems. Scattered information indicates that yellow perch and walleye populations can be limited or displaced by certain competing species and can control the success of others.

Management of indigenous and introduced percid populations can be improved by biological research to identify stocks and ecological adaptations, analyze and predict variations in year-class strength, and document species interrelationships.

Yellow perch and walleye are the two most exploited percid species in North American fisheries. Commercial demand exceeds production, providing impetus for aquaculture development. Recreational angling pressure has stimulated management-oriented research as well as the widespread introduction of both species to new environments. Yet the current state of knowledge is generally inadequate to permit realization of sustained yield, either maximum or optimum; crucial aspects of percid biology remain for research.

In many respects, the biology of yellow perch and walleye is very similar, suggesting dependence on the same environmental circumstances for populations success. The common phenomenon of simultaneous fluctuations in year-class strengths in percid-dominated waters gives credence to that speculation, as well as to theories of species interdependence. Although yellow perch can succeed without walleye and vice versa, many of the successful fisheries within the native congruent range feature both species in abundance.

The principle objective of this paper is to provide a concise review of the literature concerning those aspects of yellow perch and walleye biology having potential relevance to management.

Taxonomic Relationships

Yellow perch and walleye, together with sauger, are the only North American Percidae which are extensively fished. Sauger and walleye are morphologically similar, coexist in many waters, and sometimes hybridize (Nelson 1968; Stroud 1949). The blue pike, generally considered a subspecies, *S. v. glaucum*, of walleye is apparently extinct in its native lakes Erie and Ontario (Zarbock 1977).

On a cosmopolitan level, both yellow perch and walleye share physical and biological features with Eurasian relatives. The European perch, *P. fluviatilis*, is almost phenotypically identical with the yellow perch (Thorpe 1977), and some authors believe them conspecific (Svetovidov 1963; Weatherley 1963). The walleye has many general and osteological similarities to the Eurasian pikeperches, which are now included in the genus *Stizostedion* (Collette and Banarescu 1977). Interbreeding of North American and Eurasian percids has not been reported.

Investigation of intraspecific variation in yellow perch and walleye has been very limited. Meristic analyses of Canadian yellow perch populations revealed little geographic trend (McPhail and Lindsey 1970). Electrophoretic studies of walleyes have shown inter-lake as well as regional genotypic variation supportive of the theory that present walleye stocks originate from either an Atlantic or a Mississippi glacial refugium (Clayton and Tretiak 1971; Uthe and Ryder 1970).

Circumstantial evidence for the existence of discrete intralacustrine subpopulations of walleye and yellow perch is strong. Life history parameters distinguish eastern and western basin walleyes of Lake Erie (Wolfert 1969) and several stocks in Lake Huron (Spangler et al. 1977). A number of tagging studies have demonstrated walleye homing to discrete spawning grounds (Crowe 1962; Forney 1963; Ryder 1968). A tagging study of yellow perch showed intralake geographic isolation (Nakashima and Legget 1975), and the existence of separate stocks within the Great Lakes and other large waters is plausible.

Reproduction

Yellow perch and walleyes are spring spawners, but timing and location of their breeding generally differ. Although both percids spawn over the same broad temperature range (3–16 C), walleye spawning has been reported to precede that of yellow perch by as much as two weeks in systems where they coexist (Maloney and Johnson 1957). In Lake Erie, walleye fry usually hatch prior to deposition of yellow perch eggs (Wolfert et al. 1975). Each species requires an extended winter chill period (minimum temperature of 10 C) for gonadal maturation (Hokanson 1977).

Habitat requirements also inhibit reproductive overlap. Both percids include lentic and lotic spawning populations, but substrate preferences differ. Walleyes select shallow (0.3–2.0 m) gravel and rubble shoals where the individual eggs are protected from dislodgement (Johnson 1961) but have been reported to spawn on flooded marsh vegetation in Wisconsin (Priegel 1970). Yellow perch spawn over submerged vegetation and brush, which anchors the characteristic ribbonlike egg mass (Muncy 1962).

Males of each species mature at earlier age than females. Onset of walleye maturity is more closely related to size than age (Table 1). Fecundity of yellow perch

TABLE 1.—*Total length (mm) and age at sexual maturity of yellow perch and walleye populations.*

Water	Male			Female		
	% Mature	Length	Age	% Mature	Length	Age
Yellow perch						
Lake Erie (Jobes 1952)	47	165	II	48	216	III
Lake Huron (el-Zarka 1959)	100	137	III	44	150	III
Severn River, Maryland (Muncy 1962)	>10	114	I	>10	173	III
Walleye						
Lake Erie (Parsons 1971)	67	368	II	100	470	III
Lake Huron (Hile 1954)	50	384	III	50	432	IV
Red Lake, Minnesota (Smith and Pycha 1961)	57	373	VI	58	424	VIII
Lac la Ronge, Saskatchewan (Rawson 1957)	50	432	VII	50	457	VIII

and walleyes shows a linear increase with weight (Tsai and Gibson 1971; Wolfert 1969). Egg production by individual yellow perch ranged from 10,000 (82 g female) to 157,000 (678 g female) in Lake Michigan (Brazo et al. 1975). Walleye egg production in the western basin of Lake Erie varied between 50,000 (315 g female) and 600,000 (5,200 g female) (Wolfert 1969).

Incubation period depends on water temperature and rate of warming and appears to be of similar duration for both species (Hokanson and Kleiner 1974; Koenst and Smith 1976). Hatch will occur in about 6 days at 20 C and 50 days at 5–6 C. At characteristic spring water temperatures (8–15 C), hatch will occur two to three weeks after spawning. A rising (0.5–1.0 C/day) thermal regime results in shorter hatching periods with few abnormalities in yellow perch (Hokanson and Kleiner 1974). Sharp (4.4 C/24 h) temperature reversals are not lethal to walleye eggs (Allbaugh and Manz 1964).

A wide variety of factors have been implicated in limiting egg deposition and hatching success in both species. Low water levels may expose walleye spawning shoals (Johnson 1961) or fail to provide flooded vegetation for yellow perch (Nelson and Walburg 1977). Heavy winds have been observed to dislodge viable eggs of both species in masses, and cast them onshore (Clady and Hutchinson 1975; Newburg 1975). Hypoxia (Oseid and Smith 1971), sediment load (Schubel et al. 1974), and wood pulp fiber (Smith and Cramer 1965) can have adverse impacts on reproductive success. Egg predation by planaria (Newburg 1975) and fish (Wolfert et al. 1975) have been cited as possible sources of substantial mortality in walleye. Less direct observations, in the form of comparisons of physical and biotic conditions during the spawning/incubation period with year-class strength, have suggested other factors as determinants in particular situations (see "Annual Recruitment," below).

Availability of suitable substrate appears to be critical for spawning success, particularly for walleyes. Johnson (1961) noted egg survival ranging from 0.6% on muck to 35.7% on gravel-rubble, generally increasing with substrate particle size. Development of artificial spawning shoals has boosted walleye reproduction in some waters where suitable spawning substrate was limited (Newburg 1975; Weber and Imler 1974).

Early Life History

Yellow perch hatch at 4.5–7.0 mm total length (TL), walleyes at 7.0–9.5 mm TL. At the prolarva stage, the species can be differentiated by myomere count, chromatophore pattern, intestine length, and morphology of teeth and jaws (Norden 1961). Yolk sac may be absorbed at 7.0 mm in yellow perch and at 9.5 mm in walleyes (Houde 1969b). Spatial segregation has been observed in Oneida Lake, New York with larval (9–11 mm) walleyes concentrated in bays while larval yellow perch of similar size were more dispersed (Houde 1969a). The swimming ability of neither species at this size is sufficient to sustain directed movement against wind-generated surface currents, but walleyes move deeper on windy days to evade dispersal. Larval yellow perch and walleyes are initially photopositive and pelagic, becoming demersal at lengths of about 25–40 mm (Forney 1966; Noble 1975).

Growth

The pattern of first-year growth in length has been reported to be sigmoid in both species, with a post-hatch accelerating phase followed by a long midsummer period of nearly linear increase which slows gradually before ceasing in early autumn (Forney 1966; Smith and Pycha 1960; Ney and Smith 1975). Daily rate of midsummer increase may be accelerated in years when initial growth is slow, reducing annual variation in mean length at age I. Explanations of this phenomenon are speculative, but may involve size-related changes in diet as a function of food availability (Ney and Smith 1975).

Factors associated with variations in first-year growth have received considerable laboratory and field study. McCormick (1976) reported maximum young-of-the-year yellow perch growth on unrestricted rations at 26–30 C. Both yellow perch and walleyes achieved optimum laboratory growth at 22 C with a 16-h photoperiod, but yellow perch growth is more photoperiod-dependent, walleye growth more temperature-dependent (Huh et al. 1976). This observation substantiates the finding of el-Zarka (1959) that first-year yellow perch growth was negatively correlated with May–June turbidity in Lake Huron. In general, attempts to explain variations in first-year growth by climatic factors have not been productive. Food availability appears to be the essential cause of growth differences, and it results from complex interactions not readily explicable by combinations of easily measured environmental variables. While growth, particularly of the more abundant yellow perch, may be space-dependent in certain situations (Graves and Kogel 1973) it more often is a function of the density of usable food. Noble (1975) stated that annual variations in first-year growth of Oneida Lake yellow perch could be explained almost entirely as a function of the density of *Daphnia* species. Ney and Smith (1975) reported better growth in the Red Lakes in years when young-of-the-year yellow perch fed principally on cladocerans rather than on benthic invertebrates. First-year walleye growth has also been related to food utilization (Smith and Pycha 1960). In waters where they coexist, young walleyes and yellow perch growth may vary inversely (Forney 1966; Kempinger and Carline 1977). Slow-growing yellow perch are available walleye forage for longer periods; their growth rate may be more important for utilization by young-of-the-year walleyes than their density.

Total first-year growth can show large differences among areas and years. Trautman (1957) reported that seasonal growth ranged from 46 to 102 mm for young-of-the year yellow perch in Ohio. Other authors have observed similar ranges in widely varied systems and situations. First-year walleye growth appears to be even more variable. While most reports cite mean total lengths of 120–180 mm at first annulus, back-calculated first-year growth was reported at 254 mm for the 1959 year class in Lake Erie (Parsons 1972) and 257 mm in Norris Reservoir, Tennessee in the years following impoundment (Stroud 1949).

Growth (Table 2) continues to slow with age after the first year. Sexual dimorphism in growth rate is characteristic of both species. Female yellow perch growth has

TABLE 2.—*Calculated total lengths (mm) of yellow perch and walleye in various waters.*

Water	Sex	Age group												
		I	II	III	IV	V	VI	VII	VIII	IX	X	XI	XII	XIII
							Yellow perch							
Lake Erie	M	91	168	213	239	257								
(Jobes 1952)	F	94	170	218	249	274	287							
Lake Huron	M	66	107	142	170	193	216	236						
(el-Zarka 1959)	F	69	110	149	190	224	259	282	297	310				
Lake Michigan	M		159	182	215	235	247	252						
(Brazo et al. 1975)	F		162	206	225	252	291	313						
Red Lakes, Minnesota	M	74	132	173	203	224	236	244	249	251	257			
(Heyerdahl and Smith 1971)	F	74	135	180	218	244	254	264	272	277	282			
							Walleye							
Lac la Ronge, Saskatchewan														
Rawson 1957)	M&F	99	224	269	312	361	404	445	483	521	554	584	615	643
Lake Oahe, South Dakota	M	163[a]	284	348	393	427	454	500	558	584				
(Nelson and Walburg 1977)	F		285	349	405	455	477	517	576	627				
Red Lakes, Minnesota	M	140	213	267	310	343	363	384	396	409				
(Smith and Pycha 1961)	F	142	211	267	310	345	378	401	424	452				
Norris Reservoir, Tennessee	M	249	406	460	485	498	505	518	564					
(Stroud 1949)	F	272	432	498	551	602	610	744	765					

[a] Males and females combined.

commonly been reported to exceed the male increment by the third year of life, but among walleyes the difference had been observed to begin as early as the first year (Stroud 1949) or as late as the eighth (Rawson 1957).

Slow growth may be a management problem in either species, but stunting is more common in yellow perch populations (el-Zarka 1959; Grice 1960; Warnick 1966), perhaps as a result of more flexible feeding habits. Extremes of walleye (and presumably yellow perch) growth are generally realized at the north-south range limits. Walleye and yellow perch production is largely limited to the summer season. Positive correlations of annual growth with temperature indicate that length of growing season may be influential (Carlander and Whitney 1961; Jobes 1952; Smith and Pycha 1961). Variations in forage abundance, including yellow perch, are sometimes responsible for annual differences in adult walleye growth (Forney 1965; Parsons 1971).

Food Habits

The young of yellow perch and walleye are opportunistic feeders, but selection for size and quality of food are evident. Both species begin feeding as prolarvae on small copepods and nauplii (Houde 1969b; Siefert 1972). Rotifers and even diatoms may be important constituents of prolarval walleye diets (Hohn 1966; Smith and Moyle 1945). Copepods and cladocerans are dominant in the diet of young-of-the-year yellow perch during the entire season and in the early food of young walleyes, and chironmids and other benthic invertebrates may also be important to both (Clady 1975; Walker and Applegate 1976). Competition between the species for particular zooplankters may be reduced by differential food selection (Bulkley et al. 1976).

Fish are rare in the young-of-the-year yellow perch diet, but walleyes are reported to each fish in even the immediate postlarval stage. Fish may become the major food of walleyes as small as 30 mm TL (Maloney and Johnson 1957), but more often are not important until a size of 75–106 mm TL is reached (Priegel 1969; Walker and Applegate 1976). Yellow perch are often the principle prey of young-of-the-year walleyes.

As adults, yellow perch consume a wider variety of organisms than do walleyes. Yellow perch feed heavily on benthic insect larvae, principally chironmids and mayflies, as well as zooplankton, amphipods, leeches, and crayfish (Clady and Hutchinson 1976; Tarby 1974; Tharatt 1960). Fish enter the summer diet of larger yellow perch and can comprise the major calorific intake (Kelso 1973). Cannibalism is intense in Oneida Lake and may have an impact on recruitment (Tarby 1974).

Adult walleyes prey primarily on fish, but mayflies and chironmids are seasonally important in some populations (Swenson 1977). Yellow perch young-of-the-year are often the predominant food, although evidence indicates that this is a function of availability rather than preference. Walleyes in Lake Michigan ignored abundant young yellow perch in favor of the larger alewife and rainbow smelt (Wagner 1972). In Lake Erie, change in preferred prey size with walleye growth was manifested in the consumption of spottail and emerald shiners, and negative selection of first-year yellow perch (Parsons 1971). In southern reservoirs, threadfin shad (Range 1973) or alewife (Boaze and Lackey 1974) may be important prey. Walleye cannibalism may be common where abundant alternate forage is lacking (Chevalier 1973) but has been reported infrequently.

Mortality

First-year mortality is extremely high in these fecund species. Loss from egg through the prolarval stage has been estimated at more than 99.5% for walleye (Forney 1976) and 82–98% for yellow perch (Clady 1976) in Oneida Lake. Most of the initial mortality occurs in the egg stage and is probably related to climatic factors, such as wind velocity and rate of water warm-up. Subsequent first-year mortality may claim more than 99% of initial survivors (Forney 1971, 1976). Trophic factors undoubtedly assume a major role with advancing age. Predation, including cannibalism, may be as important as food availability, since individuals can endure prolonged starvation (Cuff 1977; Hokanson and Kleiner 1974). A wide variety of piscivorous fish as well as leeches, diving beetles (Dytiscidae), and back-swimmers (Notonectidae) have been observed to prey on larval percids (Smith and Moyle 1945).

Annual mortality of juveniles and adults is markedly lower. Total annual mortality rates of vulnerable age classes in exploited populations range as high as 57% (Table 3), but annual natural mortality is on the order of 5–30%. In an unexploited Manitoba walleye population, total annual mortality was estimated at 63–80% (Kelso and Ward 1972), which suggests that intraspecific competition may at times promote mortality. Sexual divergence in mortality rate with increasing age has been reported for walleye but may result from the interaction of size-selective exploitation and sexual growth differential (Hile 1954). Although mature walleyes are subjected to only limited predation, adult yellow perch may be heavily utilized by esocids (Gammon and Hasler 1965; Lawler 1965). Yellow perch and walleye are afflicted by a wide variety of diseases and parasites, but reports of epidemics and mass mortality are rare. Longevity has been reported at 26 years for Lake Gogebic, Michigan walleyes (Schneider et al. 1977) and 11 years for Lake Erie yellow perch (Scott and Crossman 1973).

Annual Recruitment

Yellow perch and walleye populations show characteristically large (10- to 50-fold) annual variations in cohort recruitment. Where the two percids are codominant, year-class strengths are often synchronized between the species. Much effort has been expended to explain and then predict this variation but with only limited and local success. In certain systems, year-class strength may be set by environmental conditions at or soon after spawning. In these situations, spawning habitat is usually imperilled or unstable, and brood stock may be depleted. Walleye year-class strength in Lake Oahe is largely dependent on sufficient tributary flow to permit reproduction of the river spawning population, while yellow perch success relates to high spring lake levels and the flooding of terrestrial vegetation (Nelson and Walburg 1977). Busch et al. (1975) found walleye year-class strength in Lake Erie to vary directly with the rate of water warm-up and inversely with the frequency of strong winds during the reproductive period; they stated that abiotic factors had become critical because the population had been reduced below the level for density-dependent mechanisms to operate.

In more stable systems, the year-class strength may not be set until the second or later year of life (Forney 1971; 1976). In these waters, conditions are conducive to reproduction, and attempts to correlate

TABLE 3.—*Reported annual mortality rates of yellow perch and walleye in various waters. Notation follows Ricker (1973).*

Water	Total (A)	Conditional		Expectation of death	
		Fishing (m)	Natural (n)	Fishing (u)	Natural (v)
Yellow perch					
Red Lakes, Minnesota (Heyerdahl and Smith 1971)	0.57	0.45	0.22		
Walleye					
Oneida Lake, New York (Forney 1967)	0.11–0.54			0.10–0.47	0.01–0.07
Many Point Lake, Minnesota (Olson 1958)	0.38			0.18	0.20
West Blue Lake, Manitoba (Kelso and Ward 1972)	0.63–0.80	0.00	0.63–0.80	0.00	0.63–0.80
Missagi River, Ontario (Spangler et al. 1977)	0.564	0.324	0.355	0.266	0.298

year-class strength with physical variables during early life have generally been futile; biotic factors appear to exert the major impact. Walleye year-class strength has been suggested to depend on young-of-the-year yellow perch abundance as a buffer against cannibalism (Forney 1976; Nelson and Walburg 1977). This rationale provides an explanation for synchrony in the fluctuation of yellow perch and walleye populations, and shifts the focus of study to the former. Attempts to correlate yellow perch year-class strength with either brood stock or predator abundance have not been successful (Heyerdahl and Smith 1971; Forney 1971). In these situations, availability of food at critical life stages may merit investigation.

The practice of stocking walleye fry and fingerlings to supplement year-class production continues to receive wide application and mixed results. It has had little apparent effect in large waters with normally good natural reproduction, but has enjoyed some success in smaller lakes or where spawning conditions are unreliable (Carlander et al. 1960; Johnson 1971).

Behavior

Yellow perch and walleyes are schooling fish. Social aggregations of yellow perch are often larger than those of walleye and may be segregated by sex (Jobes 1952). Extended individual migrations have been reported for both species (Mansueti 1960; Smith et al. 1952) but walleyes may utilize limited areas as apparent home range (Ager 1976; Holt et al. 1977). Variations in experimental and angling catch and, more recently, laboratory and biotelemetry studies have provided insight into daily and seasonal patterns of activity.

Daily Activity

Juvenile and adult walleyes are photonegative (Ali and Anctil 1968; Scherer 1976) and most active at crepuscular or nocturnal light intensities. Greatest movement occurs between dawn and dusk (Kelso 1976) or when overcast (Holt et al. 1977). In contrast, adult yellow perch are day-active, with peak movement reported shortly after sunrise and before sunset (Scott 1955). Yellow perch have been observed to migrate offshore at dawn and onshore at dusk (Engel and Magnuson 1976). Walleye biotelemetry studies showed shore-parallel movement (Holt et al. 1977; Kelso 1976) which may be bottom-oriented (Ager 1976). Overlap between the species in diel activity periods may coincide with simultaneous habitat occupancy, facilitating intense crepuscular predator-prey interactions (Ali et al. 1977; Ryder and Kerr 1978, this volume).

Seasonal Behavior

Walleyes are more active in spring and fall than in summer, possibly in response to lower food availability in those seasons (Schupp 1972; Swenson and Smith 1973). Angler success is generally greater during active periods. Rate of movement and density of yellow perch schools was also found to vary directly with temperature, being much higher in summer than winter (Hergenrader and Hasler 1968). Summer depth distribution can be limited by temperature. Preferred temperature of yellow perch is reported as 20–21 C and of walleyes as 20–23 C (Coutant 1977). Walleyes have been observed to prefer deep hypoxic water to epilimnetic waters above 24 C (Fitz and Holbrook 1978, this volume). Both species feed under ice through the winter, confirming year-round activity.

Species Relationships

The interaction of percids with other resident fish species has received only limited and often tangential assessment. The northern pike is a major predator on yellow perch and walleyes, and a walleye trophic competitor over much of their mutual range (Scott and Crossman 1973). Reduction of white sucker biomass in an oligotrophic Minnesota lake was followed by a 15-fold increase in yellow perch abundance, suggesting that food competition (for burrowing mayflies) was limiting (Johnson 1977). Walleye abundance also increased, perhaps as a result of enhanced perch availability or reduced competition with white sucker for spawning substrate (Anthony and Jorgensen 1977). In the Great Lakes, yellow perch and walleye populations have declined following rapid expansion of alewife and rainbow smelt numbers. Circumstantial evidence indicates that these exotics limit percid recruitment through interference with spawning or predation on young of the year (Smith 1970; Nepszy 1977; Schneider and Leach 1977).

Percids may also have a negative impact on other species. Widespread establishment of walleye in Wisconsin lakes preceded declines in smallmouth bass populations (Kempinger and Carline 1977). Reduction of yellow perch biomass was associated with increased growth and angler harvest of smallmouth bass in a New Hampshire lake (Riel 1965). Yellow perch have also been reported to be the major predator on larval northern pike in a Michigan lake (Hunt and Carbine 1951) and to consume freshly stocked Atlantic salmon in Maine (Warner 1972).

Less direct relationships may have major impact. White perch year-class strength was positively correlated with young-of-the-year yellow perch density in Oneida Lake, which suggests that yellow perch may buffer alternate prey from walleye consumption (Forney 1974). Walleye and lake whitefish year-class strengths were positively correlated in the Red Lakes, but no explanation was evident (Peterka and Smith 1970).

Management-Related Research Needs in Percid Biology

The foregoing review of extant literature serves not only to summarize current knowledge of yellow perch and walleye biology but also to identify areas where information is insufficient for optimum resource management. I believe that the more critical aspects can be grouped in the following three categories.

Stocks and Ecological Adaptations

The existence of discrete spawning stocks of walleye within major geographically defined water bodies has been documented (Loftus 1976). Tagging studies support the concept of reproductive isolation for yellow perch and walleyes within smaller systems. It is probable that these stocks have evolved adaptations of behavior and physiological response particularly suited to the habitat they occupy; that is, stocks are not interchangeable. Support for this view is found in the occurrence of both lake and river spawning populations and in the range of reported responses of both species to experimental temperature variations (Coutant 1977).

At present, identification of individual stocks is limited to the spawning period, and the range of intraspecific ecological adaptations is virtually unknown. Yet full management of indigenous populations requires stock manipulation, and establishment of self-sustaining fisheries would be greatly facilitated by such research. Studies to identify stocks by genetic markers and to define the interaction of genetic and environmental factors are well advanced in salmonids (Suzumoto et al. 1977; Tsuyuki and Williscroft 1977). Breeding experiments to match desirable salmonid traits to the environment have shown success (Flick and Webster 1976). The technology is available for application to percid biology.

Variation in Year-Class Strength

The advantages of predicting and perhaps controlling annual recruitment are obvious. Attempts to relate fluctuations in year-class strengths of yellow perch and walleye to a host of readily measured discrete environmental and biological variables have been simplistic, widespread, and largely unsuccessful. The trophic-dynamics approach exemplified by the Oneida Lake investigations is logical, but must ultimately descend the food chain to consider limnological processes. Regardless of whether analysis begins with consideration of predation or of basic physicochemical parameters, it is apparent that recruitment is a function of complex variables which may act in a site-specific manner. Componential analysis in a systems context appears capable of providing at least first-order answers.

Species Interactions

Most studies of yellow perch or walleye biology have been directed to monospecific considerations. A few investigations have included several cohabitating species, generally in a predator-prey or food-competition context. These limited efforts indicate that walleyes and/or yellow perch interact with other species in ways which have consequences spanning the spectrum from extinction to expansion for any of the participants. MacLean and Magnuson (1977) argue that resource partitioning has evolved in percid communities to minimize interspecific conflicts; interactions with species outside the characteristic complex (such as rainbow smelt and alewife) can be expected to be more intense. Introduction of percids and other game and forage species to new habitats will continue in the effort to bolster system yield and provide diverse angling opportunity. Once established, "mistakes" are not easily or inexpensively eliminated. To obviate these high-profile errors and avoid unsuccessful introductions, some capacity to predict consequences of species interactions must be developed. An initial knowledge can be achieved through inspection of specific population trends for those systems where yellow perch or walleye have been introduced or where other species have been added to existing percid associations. More detailed understanding awaits polyspecific investigations of existing systems or experimental waters containing species groups characteristic of sites of proposed introduction. Such ecological studies are essential if percid management is to evolve predictive strategies.

References

AGER, L. L. 1976. A biotelemetry study of the movements of the walleye in Center Hill Reservoir, Tennessee. Tenn. Wildl. Res. Agency Tech. Rep. 51. 97 pp.

ALI, M. A., AND M. ANCTIL. 1968. Corrélation entre la structure rétinienne et l'habitat chez *Stizostedion vitreum vitreum* et *S. canadense*. J. Fish. Res. Board Can. 25:2001–2003.

———, R. A. RYDER, AND M. ANCTIL. 1977. Photoreceptors and visual pigments as related to behavioral responses and preferred habitats of perches (*Perca* spp.) and pikeperches (*Stizostedion* spp.). J. Fish. Res. Board Can. 34: 1475–1480.

ALLBAUGH, C. A., AND J. V. MANZ. 1964. Preliminary study of the effects of temperature fluctuations on developing walleye eggs and fry. Prog. Fish-Cult. 26:175–180.

ANTHONY, D. D., AND C. R. JORGENSEN. 1977. Factors in the declining contributions of walleye (*Stizostedion vitreum*) to the fishery of Lake Nipissing, Ontario 1960–76. J. Fish. Res. Board Can. 34:1703–1704.

BOAZE, J. L., and R. T. LACKEY. 1974. Age, growth, and utilization of the land locked alewives in Claytor Lake, Virginia. Prog. Fish-Cult. 36:163–164.

BRAZO, D. C., P. I. TACK, AND C. R. LISTON. 1975. Age, growth, and fecundity of yellow perch, *Perca flavescens*, in Lake Michigan near Ludington, Michigan. Trans. Am. Fish. Soc. 104:726–730.

BULKLEY, R. V., V. L. SPYKERMAN, AND L. E. INMON. 1976. Food of the pelagic young of walleyes and five cohabiting fish species in Clear Lake, Iowa, Trans. Am. Fish. Soc. 105:77–83.

BUSCH, W. D. N., R. L. SCHOLL, AND W. L. HARTMAN. 1975. Environmental factors affecting the strength of walleye (*Stizostedion vitreum vitreum*) year classes in western Lake Erie, 1960–1970. J. Fish. Res. Board Can. 32:1733–1743.

CARLANDER, K. D., AND R. R. WHITNEY. 1961. Age and growth of walleyes in Clear Lake, Iowa, 1935–1957. Trans. Am. Fish. Soc. 90:130–138.

———, ———, E. B. SPEAKER, AND K. MADDEN. 1960. Evaluation of walleye fry stocking in Clear Lake, Iowa, by alternate-year planting. Trans. Am. Fish. Soc. 89:249–254.

CHEVALIER, J. R. 1973. Cannibalism as a factor in first year survival of walleye in Oneida Lake. Trans. Am. Fish. Soc. 102:739–744.

CLADY, M. D. 1975. Food habits of yellow perch, smallmouth bass, and largemouth bass in two unproductive lakes in northern Michigan. Am. Midl. Nat. 91:453–459.

———. 1976. Influence of temperature and wind on the survival of early stages of yellow perch, *Perca flavescens*. J. Fish. Res. Board Can. 33: 1887–1893.

———, and B. Hutchinson. 1975. Effect of high winds on eggs of yellow perch, *Perca flavescens*, in Oneida Lake, New York. Trans. Am. Fish. Soc. 104:524–525.

———, AND ———. 1976. Food of the yellow perch, *Perca flavescens*, following a decline of the burrowing mayfly, *Hexagenia limbata*. Ohio J. Sci. 76: 133–138.

CLAYTON, J. W., AND D. N. TRETIAK. 1971. Genetics of multiple malate dehydrogenase isozymes in skeletal muscle of walleye (*Stizostedion vitreum vitreum*). J. Fish. Res. Board Can. 28:1005–1008.

COLLETTE, B. B., AND P. BANARESCU. 1977. Systematics and zoogeography of the fishes of the family Percidae. J. Fish. Res. Board Can. 34: 1450–1463.

COUTANT, C. C. 1977. Compilation of temperature

preference data. J. Fish. Res. Board Can. 34: 739–746.

CROWE, W. R. 1962. Homing behavior in walleyes. Trans. Am. Fish. Soc. 91:350–354.

CUFF, W. R. 1977. Initiation and control of cannibalism in larval walleyes. Prog. Fish-Cult. 39: 29–32.

EL-ZARKA, S. E. D. 1959. Fluctuations in the population of yellow perch, *Perca flavescens* (Mitchill) in Saginaw Bay, Lake Huron. U.S. Fish Wildl. Serv. Fish. Bull. 59:365–415.

ENGEL, S., AND J. J. MAGNUSON. 1976. Vertical and horizontal distribution of coho salmon (*Oncorhynchus kisutch*), yellow perch (*Perca flavescens*), and cisco (*Coregonus artedi*) in Pallette Lake, Wisconsin. J. Fish. Res. Board Can. 33:2710–2715.

FITZ, R. B., AND J. A. HOLBROOK. 1978. Sauger and walleye in Norris Reservoir, Tennessee. Am. Fish. Soc. Spec. Publ. 11:82–88.

FLICK, W. A., AND D. A. WEBSTER. 1976. Production of wild, domestic and interstrain hybrids of brook trout (*Salvelinus fontinalis*) in natural ponds. J. Fish. Res. Board Can. 33:1525–1539.

FORNEY, J. L. 1963. Distribution and movement of marked walleyes in Oneida Lake, New York. Trans. Am. Fish. Soc. 92:47–52.

———. 1965. Factors affecting growth and maturity in a walleye population. N.Y. Fish Game J. 12:217–232.

———. 1966. Factors affecting first-year growth of walleyes in Oneida Lake, New York. N.Y. Fish Game J. 13:147–166.

———. 1967. Estimates of biomass and mortality rates in a walleye population. N.Y. Fish Game J. 14: 177–192.

———. 1971. Development of dominant year classes in a yellow perch population. Trans. Am. Fish. Soc. 100:739–749.

———. 1974. Interactions between yellow perch abundance, walleye predation, and survival of alternate prey in Oneida Lake, New York. Trans. Am. Fish. Soc. 103:15–24.

———. 1976. Year class formation in the walleye (*Stizostedion vitreum vitreum*) population of Oneida Lake, New York, 1966–1973. J. Fish. Res. Board Can. 33:783–792.

GAMMON, J. R., AND A. D. HASLER. 1965. Predation by introduced muskellunge on perch and bass, I: years 1–5. Trans. Wis. Acad. Sci. Arts Lett. 54:249–272.

GRAVES, W. E., AND J. H. KOGEL. 1973. Effects of water volume and change of water on growth and survival of yellow perch. Prog. Fish-Cult. 35:115–117.

GRICE, F. 1960. Elasticity of growth of yellow perch, chain pickerel, and largemouth bass in some reclaimed Massachusetts waters. Trans. Am. Fish. Soc. 88:332–333.

HERGENRADER, G. L., AND A. D. HASLER. 1968. Influence of changing seasons on schooling behavior of yellow perch. J. Fish. Res. Board Can. 25:711–716.

HEYERDAHL, E. G., AND L. L. SMITH, JR. 1971. Annual catch of yellow perch from Red Lakes, Minnesota, in relation to growth rate and fishing effort. Univ. Minn. Agric. Exp. Stn. Bull. 285. 51 pp.

HILE, R. 1954. Fluctuations in growth and year-class strength of the walleye in Saginaw Bay. U.S. Fish Wildl. Serv. Fish. Bull. 54:7–59.

HOHN, M. H. 1966. Analysis of plankton ingested by *Stizostedion vitreum vitreum* (Mitchill) fry and concurrent vertical plankton tows from southwestern Lake Erie, May 1961 and May 1962. Ohio J. Sci. 66:193–199.

HOKANSON, K. E. F. 1977. Temperature requirements of some percids and adaptations to the seasonal temperature cycle. J. Fish. Res. Board Can. 34:1524–1550.

———, AND CH. F. KLEINER. 1974. Effects of constant and rising temperature on survival and development rates of embryonic and larval yellow perch, *Perca flavescens* (Mitchill). Pages 437–448 *in* J. H. S. Baxter, ed. The early life history of fish. Springer-Verlag, New York.

HOLT, C. S., G. D. S. GRANT, G. P. OBERSTAR, C. C. OAKES, AND D. W. BRADT. 1977. Movement of walleye, *Stizostedion vitreum vitreum*, in Lake Bemidji, Minnesota as determined by radiotelemetry. Trans. Am. Fish. Soc. 106(2):163–169.

HOUDE, E. D. 1969a. Distribution of larval walleyes and yellow perch in a bay of Oneida Lake and its relation to water currents and zooplankton. N.Y. Fish Game J. 16:185–205.

———. 1969b. Sustained swimming ability of larvae of walleye (*Stizostedion vitreum vitreum*) and yellow perch (*Perca flavescens*). J. Fish. Res. Board Can. 26:1647–1659.

HUH, H. T., H. E. CALBERT, AND D. A. STUIBER. 1976. Effect of temperature and light on growth of yellow perch and walleye using formulated food. Trans. Am. Fish. Soc. 105:254–258.

HUNT, B. G., AND W. F. CARBINE. 1951. Food of young pike, *Esox lucius* L., and associated species in Peterson's ditches, Houghton Lake, Michigan. Trans. Am. Fish. Soc. 80:67–83.

JOBES, F. W. 1952. Age, growth, and production of yellow perch in Lake Erie. U.S. Fish Wildl. Serv. Fish Bull. 52:205–226.

JOHNSON, F. H. 1961. Walleye egg survival during incubation on several types of bottom in Lake Winnibigosh, Minnesota, and connecting waters. Trans. Am. Fish. Soc. 90:312–322.

———. 1971. Survival of stocked walleye fingerlings in nothern Minnesota lakes as estimated from the age composition of experimental gill net catches. Minn. Dep. Nat. Resour. Invest. Rep. 314. 12 pp.

———. 1977. Responses of walleye (*Stizostedion vitreum vitreum*) and yellow perch (*Perca flavescens*) populations to removal of white sucker (*Catostomus commersoni*) from a Minnesota lake, 1966. J. Fish. Res. Board Can. 34:1633–1642.

KELSO, J. R. M. 1973. Seasonal energy changes in walleye and their diet in West Blue Lake, Manitoba. Trans. Am. Fish. Sox. 102:363–368.

———. 1976. Diel movement of walleye, *Stizostedion vitreum vitreum*, in West Blue Lake, Manitoba, as determined by ultrasonic tracking. J. Fish. Res. Board Can. 33:2070–2072.

———, AND F. J. WARD. 1972. Vital statistics, biomass, and seasonal production of an unexploited walleye (*Stizostedion vitreum vitreum*) population in West Blue Lake, Manitoba. J. Fish. Res. Board Can. 29:1043–1052.

KEMPINGER, J. J., AND R. F. CARLINE. 1977.

Dynamics of the walleye (*Stizostedion vitreum vitreum*) population in Escanaba Lake, Wisconsin, 1955-72. J. Fish. Res. Board Can. 34:1800-1811.

KOENST, W. M., AND L. L. SMITH, JR. 1976. Thermal requirements of the early life history stages of walleye, *Stizostedion vitreum vitreum*, and sauger, *S. canadense*. J. Fish. Res. Board Can. 33:1130-1138.

LAWLER, G. H. 1965. The food of the pike, *Esox lucius*, in Heming Lake, Manitoba. J. Fish. Res. Board Can. 22:1357-1377.

LOFTUS, K. H. 1976. Science for Canada's fisheries rehabilitation needs. J. Fish. Res. Board Can. 33:1822-1857.

MACLEAN, J. H., AND J. J. MAGNUSON. 1977. Inferences on species interactions in percid communities. J. Fish. Res. Board Can. 34:1941-1951.

MALONEY, J. E., AND F. H. JOHNSON. 1957. Life histories and interrelationships of walleye and yellow perch, especially during their first summer, in two Minnesota lakes. Trans. Am. Fish. Soc. 85:191-202.

MANSUETI, R. 1960. Comparison of the movements of stocked and resident yellow perch, *Perca flavescens*, in tributaries of Chesapeake Bay, Maryland. Chesapeake Sci. 1:21-25.

MCCORMICK, J. H. 1976. Temperature effects on young yellow perch, *Perca flavescens* (Mitchill). U.S. Environ. Prot. Agency Ecol. Res. Ser. 600/3-76-057, Duluth, Minn. 17 pp.

MCPHAIL, J. D., AND C. D. LINDSEY. 1970. Freshwater fishes of northwestern Canada and Alaska. Bull. Fish. Res. Board Can. 173. 381 pp.

MUNCY, R. J. 1962. Life history of the yellow perch, *Perca flavescens*, in estuarine waters of Severn River, a tributary of Chesapeake Bay, Maryland. Chesapeake Sci. 3:143-159.

NAKASHIMA, B. S., AND W. C. LEGGETT. 1975. Yellow perch (*Perca flavescens*) biomass responses to different levels of phytoplankton and benthic biomass in Lake Memphremagog, Quebec-Vermont. J. Fish. Res. Board Can. 32:1785-1797.

NELSON, W. R. 1968. Embryo and larval characteristics of sauger, walleye, and their reciprocal hybrids. Trans. Am. Fish. Soc. 97:167-174.

―, AND C. A. WALBURG. 1977. Population dynamics of yellow perch (*Perca flavescens*), sauger (*Stizostedion canadense*), and walleye (*Stizostedion vitreum vitreum*) in four main stem Missouri River reservoirs. J. Fish. Res. Board Can. 34:1748-1763.

NEPSZY, S. J. 1977. Changes in percid populations and species interactions in Lake Erie. J. Fish. Res. Board Can. 34:1861-1868.

NEWBURG, H. J. 1975. Evaluation of an improved walleye (*Stizostedion vitreum*) spawning shoal with criteria for design and placement. Minn. Dep. Nat. Resour. Div. Fish. Wildl. Invest. Rep. 340. 39 pp.

NEY, J. J., AND L. L. SMITH, JR. 1975. First-year growth of the yellow perch, *Perca flavescens*, in the Red Lakes, Minnesota. Trans. Am. Fish. Soc. 104:718-725.

NOBLE, R. L. 1975. Growth of young yellow perch (*Perca flavescens*) in relation to zooplankton populations. Trans. Am. Fish. Soc. 104:731-741.

NORDEN, C. R. 1961. The identification of larval yellow perch, *Perca flavescens*, and walleye, *Stizostedion vitreum*. Copeia 1961:282-289.

OLSON, D. E. 1958. Statistics of a walleye sport fishery in a Minnesota lake. Trans. Am. Fish. Soc. 87:57-72.

OSEID, D. M. AND L. L. SMITH, JR. 1971. Survival and hatching of walleye eggs at various dissolved oxygen levels. Prog. Fish-Cult. 33:81-85.

PARSONS, J. W. 1971. Selective food preferences of walleyes of the 1959 year class in Lake Erie. Trans. Am. Fish. Soc. 100:474-485.

―. 1972. Life history and production of walleyes of the 1959 year class in western Lake Erie, 1959-62. Trans. Am. Fish. Soc. 101:655-661.

PETERKA, J. J., AND L. L. SMITH, JR. 1970. Lake whitefish in the commercial fishery of Red Lakes, Minnesota. Trans. Am. Fish. Soc. 99:28-43.

PRIEGEL, G. R. 1969. Age and growth of the walleye in Lake Winnebago. Trans. Wis. Acad. Sci. Arts Lett. 57:121-133.

―. 1970. Reproduction and early life history of the walleye in the Lake Winnebago region. Wis. Dep. Nat. Resour. Tech. Bull. 45. 105 pp.

RANGE, J. D. 1973. Growth of five species of game fishes before and after introduction of the threadfin shad into Dale Hollow Reservoir. Proc. Annu. Conf. Southeast Assoc. Game Fish Comm. 26:510-518.

RAWSON, D. S. 1957. The life history and ecology of the yellow walleye, *Stizostedion vitreum*, in Lac la Ronge, Saskatchewan. Trans. Am. Fish. Soc. 86:15-37.

RIEL, A. D. 1965. The control of an overpopulation of yellow perch in Bow Lake, Strafford, New Hampshire. Prog. Fish-Cult. 27:37-41.

RYDER, R. A. 1968. Dynamics and exploitation of mature walleyes, *Stizostedion vitreum vitreum*, in the Nipigan Bay region of Lake Superior. J. Fish. Res. Board Can. 25:1347-1376.

―, AND S. R. KERR. 1978. The adult walleye in the percid community—a niche definition based of feeding behaviour and food specificity. Am. Fish. Soc. Spec. Publ. 11:39-51.

SCHERER, E. 1976. Overhead light intensity and vertical positioning of the walleye, *Stizostedion vitreum vitreum*. J. Fish. Res. Board Can. 33:289-292.

SCHNEIDER, J. C., P. H. ESCHMEYER, AND W. R. CROWE. 1977. Longevity, survival, and harvest of tagged walleyes in Lake Gogebic, Michigan. Trans. Am. Fish. Soc. 106:566-568.

―, AND J. H. LEACH. 1977. Walleye (*Stizostedion vitreum vitreum*) fluctuations in the Great Lakes and possible causes, 1800-1975. J. Fish. Res. Board Can. 34:1878-1889.

SCHUBEL, J. R., A. H. AULD, AND G. M. SCHMIDT. 1974. Effects of suspended sediment on the development and hatching success of yellow perch and striped bass eggs. Proc. Annu. Conf. Southeast. Assoc. Game Fish Comm. 27:689-694.

SCHUPP, D. H. 1972. The walleye sport fishery of Leech Lake, Minnesota. Minn. Dep. Nat. Resour. Fish Invest. Rep. 317. 11 pp.

SCOTT, D. C. 1955. Activity patterns of perch, *Perca flavescens*, in Rondeau Bay of Lake Erie. Ecology 36:320-327.

SCOTT, W. B., AND E. J. CROSSMAN. 1973. Freshwater fishes of Canada. Bull. Fish. Res. Board Can. 184. 966 pp.

SIEFERT, R. E. 1972. First food of larval yellow perch, white sucker, bluegill, emerald shiners, and rainbow smelt. Trans. Am. Fish. Soc. 101: 219–225.

SMITH, L. L., JR., AND R. H. CRAMER. 1965. Survival of walleye fingerlings in conifer groundwood fiber. Trans. Am. Fish. Soc. 94:402–404.

———, L. W. KREFTING, AND R. L. BUTLER. 1952. Movements of marked walleyes, *Stizostedion vitreum vitreum* (Mitchill), in the fishery of the Red Lakes, Minnesota. Trans. Am. Fish. Soc. 81:179–196.

———, AND J. B. MOYLE. 1945. Factors influencing production of yellow pike-perch, *Stizostedion vitreum vitreum*, in Minnesota rearing ponds. Trans. Am. Fish. Soc. 73:243–261.

———, AND R. L. PYCHA. 1960. First-year growth of the walleye, *Stizostedion vitreum vitreum* (Mitchill), and associated factors in Red Lakes, Minnesota. Limnol. Oceanogr. 5:281–290.

———, AND ———. 1961. Factors related to commercial production of the walleye in Red Lakes, Minnesota. Trans. Am. Fish. Soc. 90:190–217.

SMITH, S. H. 1970. Species interactions of the alewife in the Great Lakes. Trans. Am. Fish. Soc. 99:754–765.

SPANGLER, G. R., N. R. PAYNE, AND G. K. WINTERLON. 1977. Percids in the Canadian waters of Lake Huron. J. Fish. Res. Board Can. 34:1839–1848.

STROUD, R. H. 1949. Growth of Norris Reservoir walleye during the first twelve years of impoundment. J. Wildl. Manage. 13:157–177.

SUZUMOTO, B. K., C. B. SCHRECK, AND J. D. MCINTYRE. 1977. Relative resistance of three transferrin genotypes of coho salmon (*Oncorhynchus kisutch*) and their hematological response to bacterial kidney disease. J. Fish. Res. Board Can. 34:1–8.

SVETOVIDOV, A. N. 1963. The systematics, origin, and history of distribution of the Eurasiatic and North American species of *Perca*, *Lucioperca*, and *Stizostedion*. Proc. 16th Int. Congr. Zool. 1:212–219.

SWENSON, W. A. 1977. Food consumption of walleye (*Stizostedion vitreum vitreum*) and sauger (*S. canadense*) in relation to food availability and physical conditions in Lake of the Woods, Minnesota, Shagawa Lake, and western Lake Superior. J. Fish. Res. Board Can. 34:1643–1654.

———, AND L. L. SMITH, JR. 1973. Gastric digestion and food conversion efficiency in walleye. *Stizostedion vitreum vitreum*. J. Fish. Res. Board Can. 30:1327–1336.

TARBY, M. J. 1974. Characteristics of yellow perch cannibalism in Oneida Lake and the relation to first-year survival. Trans. Am. Fish. Soc. 103: 462–471.

THARATT, R. C. 1960. Food of yellow perch, *Perca flavescens* (Mitchill) in Saginaw Bay, Lake Huron. Trans. Am. Fish. Soc. 88:330–331.

THORPE, J. E. 1977. Morphology, physiology, behavior, and ecology of *Perca fluviatilis* L. and *Perca flavescens* Mitchill. J. Fish. Res. Board Can. 34:1504–1514.

TRAUTMAN, M. B. 1957. The fishes of Ohio. Ohio Univ. Press, Columbus. 683 pp.

TSAI, C., AND G. R. GIBSON, JR. 1971. Fecundity of the yellow perch, *Perca flavescens* (Mitchill), in the Patuxent River, Maryland. Chesapeake Sci. 12:270–284.

TSUYUKI, H., AND S. N. WILLISCROFT. 1977. Swimming stamina differences between genotypically distinct forms of rainbow (*Salmo gairdneri*) and steelhead trout. J. Fish. Res. Board Can. 34:996–1003.

UTHE, J. F., AND R. A. RYDER. 1970. Regional variation in muscle myogen polymorphism in walleye (*Stizostedion vitreum vitreum*) as related to morphology. J. Fish. Res. Board Can. 27:923–927.

WAGNER, W. C. 1972. Utilization of alewife by inshore piscivorous fishes in Lake Michigan. Trans. Am. Fish. Soc. 101:55–63.

WALKER, R. E., AND R. L. APPLEGATE. 1976. Growth, food, and possible ecological effects of young-of-the-year walleye in a South Dakota prairie pothole. Prog. Fish-Cult. 38:217–220.

WARNER, K. 1972. Further studies on fish predation on salmon stocked in Maine Lakes. Prog. Fish-Cult. 34:217–221.

WARNICK, D. C. 1966. Investigations in fish control. 5. Growth rates of yellow perch in two North Dakota lakes after population reduction with toxaphene. Bur. Sport Fish. Wildl. Fish Control Lab. Res. Pub. 9:1–9.

WEATHERLEY, A. H. 1963. Zoogeography of *Perca fluviatilis* (Linnaeus) and *P. flavescens* (Mitchill) with special reference to the effects of high temperature. Proc. Zool. Soc. London 141(3): 557–576.

WEBER, D. T., AND R. L. IMLER. 1974. An evaluation of artificial spawning beds for walleye. Colo. Div. Wildl. Spec. Rep. 34. 17 pp.

WOLFERT, D. R. 1969. Maturity and fecundity of walleyes from the eastern and western basins of Lake Erie. J. Fish Res. Board Can. 26:1877–1888.

———, W. D. N. BUSCH, AND C. T. BAKER. 1975. Predation by fish on walleye eggs on a spawning reef in western Lake Erie, 1969–71. Ohio J. Sci. 75:118–125.

ZARBOCK, W. M. 1977. Fish, fisheries, and water quality of the Great Lakes basin. Fisheries 2(2):2–4; 26–33.

Taxonomy and Distribution of North American Esocids

E. J. CROSSMAN

Department of Ichthyology and Herpetology
Royal Ontario Museum M5S 2C6
and Department of Zoology, University of Toronto
Toronto, Ontario M5S 1A4

Abstract

Presently there are four species of native esocids, *Esox masquinongy, E. lucius, E. niger,* and *E. americanus*, and one exotic species, *E. reicherti*, at large in North America. The four native species now include five named forms, intergrades, and natural hybrids. Taxonomic divisions within species are under study. Post-Wisconsin, natural distribution patterns have been changed by man, and man is again adjusting the distributional limits of some species by extensive introductions.

Changes in taxonomic concepts of the family and species are traced. The present distribution of each species is given in detail, and suggestions are made concerning their distributions in the past and future.

The highly specialized esocid fishes fall into a single family, and a single genus of only five species. Their closest relatives the mudminnows, *Umbra, Novumbra,* and *Dallia,* also comprise five species now considered to be best grouped in a single family. Recent studies suggest that the families Esocidae, Umbridae, and Lepidogalaxiidae make up the Suborder Esocoidei of the Order Salmoniformes. This classification places the esocids at a slightly lower level than a former one (Order Haplomi) which recognized them as distinct from, and slightly in advance of, the salmonoids and related fishes. It has been determined that many of the supposedly characteristic features (simple shoulder, etc.) which had been used to separate the esocoids are in fact shared with other salmoniforms.

The fossil history of esocid fishes is limited to the Holarctic and includes seven described froms from the Oligocene to the Pleisticene of the Old World, and seven records from the Miocene to the Pleistocene of the New World. The older Old World records are attributed to six extinct species, and the Pleistocene records to *Esox lucius*. Two New World fossil records (Pliocene and Pleistocene) were not precisely identified, and the others (Miocene to Pleistocene) were attributed to living North American species (Cavender et al. 1970; Crossman and Harington 1970). In spite of this rather extensive fossil history, or possibly as a result of it, there is disagreement regarding the place of origin (Europe or North America), and the direction and routes of dispersal (transatlantic or Bering Land Bridge) for at least some of the forms (Nelson 1972).

Systematics

A persistent enigma in the past has been the possible close relationship of the galaxioid fishes to the esocoid fishes. That relationship now seems to be most improbable (Nelson 1972). However, reexamination of the species *Lepidogalaxias salamandroides*, placed in 1961 by its describer in the family Galaxiidae, has led to the suggestion (Rosen 1974) that *Lepidogalaxias* is more closely related to esocoids than to galaxids. This idea led Rosen (1974) to a reexamination of relationships within the salmoniforms. In this he suggested that the family Lepidogalaxiidae be included in a new superfamily which, with another superfamily including Esocidae and Umbridae, should constitute the Suborder Esocoidei. Rosen also stated that although esocoid relationships are uncertain, escoids probably constitute a primitive sister group of all other salmoniforms.

This leaves us with the concept that the esocids are a group of salmonid-like fishes living in cool to warm waters, highly specialized for sprint activity and for sight predation in the main on single, relatively large individuals of other fishes. The characteristic body shape, position and size of the unpaired fins, mouth size, anterior extension of certain skull bones, and the form, number, and distribution of teeth, are all adaptations to that end.

Within the family Esocidae there are two readily recognizable groups, the pikes and the pickerels, with obvious morphological, biochemical, and genetic differences (Eckroat 1974; Beamish et al. 1971; Davisson 1972). There is however incomplete reproductive isolation between members of the two groups. Natural hybrids between certain species occur, and others can readily be produced artificially (Crossman and Buss 1965; Crossman and Meade 1977). The group of larger species, or pikes, sometimes referred to a subgenus *Esox*, includes *Esox masquinongy* Mitchill, the muskellunge, *E. lucius* Linnaeus, the pike (in Europe) or northern pike (in North America), and *E. reicherti* Dybowski, the Amur pike. The group of smaller pickerels, sometimes referred to a subgenus *Kenoza*, includes *E. niger* Lesueur, the chain pickerel, and *E. americanus* Gmelin, the redfin and grass pickerels. The pikes and pickerels are most easily separated by counting the number of sensory pores in the mandible, and the number of branchiostegal rays arising from each of the ceratohyal and epihyal bones (Crossman 1960).

Key to Esocids Presently Found in North America

1a Submandibular pores usually 5 or more on each side, rarely 4 on one side only; lateral line scales as high as 176; one or both of cheeks and opercula not fully scaled; subgenus *Esox* 2

1b Submandibular pores usually 4 on each side, rarely 3 or 5 on one side only; lateral line scales rarely exceed 135; both cheeks and opercles more or less fully scaled; subgenus *Kenoza* 4

2a Submandibular pores usually 6–9 on each side, rarely 5 or 10 on one side only; branchiostegal rays 16–19 on each side, usually 8 + 10 (ceratohyal + epihyal); neither cheeks nor opercula completely scaled .... MUSKELLUNGE, *Esox masquinongy*

2b Submandibular pores usually 5 on each side, rarely 4 or 6 on one side only; branchiostegal rays 11–16 on each side; cheeks more or less fully scaled, opercula appoximately half scaled 3

3a Branchiostegal rays 13–16 on each side, usually 7 + 8; vertebrae 57–64; color pattern with dark ground color and light bars (in young) or horizontal rows of oval yellow to white spots ....
.... NORTHERN PIKE, *Esox lucius*

3b Branchiostegal rays 11–15 on each side but usually 5 + 8 or 6 + 7; vertebrae 63–67; color pattern with silvery background and dark brown to black bars (in young) or large oval spots ....
.... AMUR PIKE, *Esox reicherti*

4a Branchiostegal rays 14–17 on each side, usually 6 + 9 ....
.... CHAIN PICKEREL, *Esox niger*

4b Branchiostegal rays 11–13 on each side 5

5a Branchiostegal rays usually 5 + 7 or 5 + 8 on each side, snout short and convex in profile; more than 5 cardiod scales in triangle between pelvic fins, more than 5 cardiod scales in oblique line from origin of anal fin to mid dorsal line .... REDFIN PICKEREL, *Esox a. americanus*

5b   Branchiostegal rays usually 4 + 7 or 4 + 8 on each side, snout longer and concave in profile; fewer than 5 cardiod scales in triangle between pelvic fins, and fewer than 5 in oblique line from origin of anal fin to mid dorsal line .... GRASS PICKEREL, *Esox americanus vermiculatus*

General Distributions

The pikes include one circumpolar species, one endemic to eastern North America, and one endemic to eastern Asia, whereas the pickerels were originally restricted to eastern and central North America.

The interest in this group of fishes is indicated by their liberation in exotic localities either for their availability as a sport fish, or in hopes of controlling the numbers of other species of fishes lower on man's artificial scale of importance. To give a few examples only: *Esox lucius, E. masquinongy,* and *E. niger* were introduced into impoundments in the Atlas Mountains of Morocco, *E. lucius* in 1935, the others in 1964 (Anonymous 1965). *E. niger* occurs in Ohio in a single location as a result of an introduction in 1935 (Armbruster 1959), and that location may have been the source for the Moroccan introduction. *E. lucius* apparently occurs in Australia but is considered a trash fish there (personal communication, B. Rickards, Cambridge, England). *E. masquinongy* and *E. a. vermiculatus* were introduced, apparently without success, into California in the early 1890's.

A second wave of artificial extension of the territory, of at least the larger esocids, began in the 1960's. This second effort, principally in the U.S., involved the spread of the northern pike and the muskellunge in the New England, the Southeastern, and the Rocky Mountain states. It also included renewed interest in the propagation and liberation of the muskellunge × northern pike hybrid, and the importation of the only species not native to North America, *E. reicherti*, the Amur pike. The Amur pike is included in the key, but will not be dealt with in detail in this discussion of North American esocids. Characteristics of this species were given by Crossman and Meade (1977).

Various aspects of these more recent introduction are treated elsewhere in these symposium proceedings.

The Pikes, subgenus *Esox*

Northern Pike

The northern pike is the circumpolar species and the only species in the family with a broad geographical and environmental range.

Although generally considered to constitute a single species over the whole of its holarctic distribution, the Eurasian and North American populations have been isolated at least since the late Pliocene. In addition there are certain morphological differences and other circumstantial distinctions (i.e. body proportions, usual maximum size, utilization of brackish water) used at times to suggest separation at least at the subspecies level. Comparisons of proteins (Crossman unpublished) of Volga River and North American specimens, and a comparison of chromosomes of English and North American specimens (Beamish et al. 1971) indicated no differences which might reflect genetic divergence. Differences in meristic characters exist but seem to be clinal and those found in Canadian populations probably result from Pleistocene isolation of different stocks in separate refugia (McPhail and Lindsey 1970). A recent study of pike in the fenlands of England (Crossman unpublished) suggests that some of the other superficial differences are basically phenotypic. It would seem unwise to suggest any taxonomic separation of the populations. We are probably dealing simply with a single, widely distributed, and highly variable species. In fact there is not enough difference to warrant even the distinctive common names presently in use. The species had long been known as *the* pike in Europe before it was discovered that it occurred in North America. Since there is no North American species presently designated as southern pike, the geographical modifier used in North America tends only to suggest a difference between populations on the two continents which seems not to exist. It may have been more appropriate in the past when the walleyes (*Stizostedion*) were referred to as pikeperches or walleyed pikes.

Hybrids occur naturally where northern pike and other esocids occur together. The hybrid with muskellunge (tiger muskellunge or norlunge) has received far more attention than any other. Hybrids can usually be distinguished on the basis of pattern, body proportions, and certain meristics, number of vertebrae in particular (Crossman and Buss 1965; Crossman and Meade 1977).

Over the whole of at least the northern portion of the distribution of the northern pike, a mutant form occurs irregularly. This form, usually called silver pike, is indistinguishable from ordinary northern pike except that it is always silvery blue or silvery green, and lacks the white to yellow spots characteristic of the species. It is different in apparently being more tenacious of life than ordinary northern pike (Lawler 1960, 1964). This mutant was at one time thought to be a form of muskellunge.

Early authorized but unrecorded introductions, and unauthorized extensions of the range of northern pike now make it difficult to reconstruct the natural range of this species. Sometime before 1850 the northern pike had been introduced in the Connecticut River and by 1864 was fairly common near Bellows Falls, Vermont. The best evidence is the knowledge of introductions in recent years, and the existence of this species in the Mississippi Refugium at least during the Wisconsin maximum of Pleistocene glaciation. These would suggest that distribution prior to introductions (see hatched area of Fig. 1) included the territory occupied today in Canada and Alaska. In the eastern U.S. the "southern" limit of distribution probably followed the lowlands west of the Green Mountains, south and west through the western parts of New York and Pennsylvania to the Ohio River, west through Indiana and Illinois, and then north in the Missouri River to include parts of Missouri, the Dakotas, and northern Montana.

The absence of northern pike today, and almost certainly in the past as well, from the eastern arctic coast of Canada, from part of Labrador, and from Cape Breton and the maritime provinces is probably due to a combination of unsuitable habitat and a failure to colonize the area. Esocid habitats exist to a limited extent in the maritime provinces but the only esocid which occurs there today (chain pickerel) was introduced from the south after 1850.

There was apparently a reduction of territory probably involving South Dakota, Nebraska, Missouri, southern Iowa, and central Illinois. This reduction may have resulted, at least in part, from man-made changes to the habitat.

More recently the development of impoundments has led to an expansion of range to the west, and in Atlantic coast states. If we include the likely portions of those states which have indicated introductions, the present distribution exceeds, or will extend, to the broken line on Figure 1.

These introductions will greatly extend the area in which northern pike and chain pickerel will be sympatric. They have overlapped recently only to a limited extent in Quebec, New York, and Pennsylvania. The two species hybridize readily, and in certain habitats northern pike may overrun chain pickerel as it appears to overrun muskellunge.

The northern pike is normally a resident of small lakes; of the shallower vegetated areas of larger lakes, marshes, and backwater sloughs; and, to a much lesser extent, of rivers. The northern pike is the only species in the family to colonize arctic environments. Consequently it should succeed elsewhere in cooler water situations where other esocids will not. Northern pike and chain pickerel or muskellunge might become established in different waters, or in

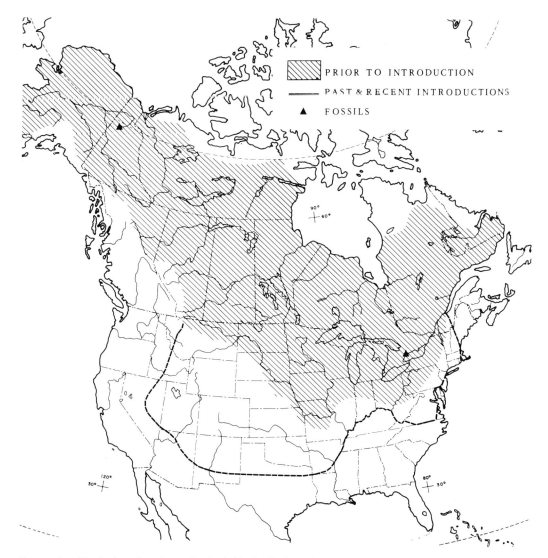

FIGURE 1.—*Distribution of northern pike. Probable distribution prior to introductions is indicated by the crosshatched area. Area within broken line includes extension of former range by past and recent introductions.*

separate areas of the same bodies of water. In the north, populations of northern pike often occur in the shallower parts of lakes which also support successful populations of one or more salmonids.

Muskellunge

This largest of the esocids has had the most chequered taxonomic history. Three basic color patterns, with intergrading variations, occur in the species. All three patterns can be found in the same habitat (Seaborne 1937) but relative frequency is such that the different patterns have been said to characterize populations in various areas. The Great Lakes and St. Lawrence River populations are most frequently spotted, populations in the Ohio River system are usually strikingly barred, and populations in Wisconsin, Minnesota, and western Ontario are often without prominent pattern or have diffuse bars or blotches. The type locality for the species is supposedly Lake Erie. When the apparent correlation between pattern and location was described, three species of muskellunge were established, *E. masquinongy* for the spotted Great Lakes form, *E. ohioensis* for the barred form in the Ohio River system, and *E. immaculatus* for the western populations (or northern as they were then called). Gradually it was realized that the distinction was seriously in doubt. A study published by Hourston (1955) suggested certain differences between western (Lake of the Woods), central (Kawartha Lakes), and eastern (St. Lawrence River) populations in Ontario and Quebec. However, neither specimens from southern parts of those areas, nor specimens from the Ohio River system entered into that study. Gradually interpretation of the distinction changed and the forms became known as *E. m. masquinongy*, *E. m. ohioensis*, and *E. m. immaculatus*. If the correlation between water system and subspecies is adhered to it implies differences between populations in northern and eastern Wisconsin (Great Lakes) and those in southern or western Wisconsin (western). It would also require separation of the populations in the Kawartha Lakes of Ontario from the Great Lakes type although the waters are contiguous. The tendency in recent years has been to consider the forms all part of a single species *E. masquinongy* but as recently as 1975 the subspecies designation was still being used for at least populations in the Ohio River system (Riddle 1975). A study (Crossman unpublished) involving meristics, larval development of color pattern, and proteins, and involving individuals from all three "areas," suggests as much variability within groups as between, and corroborates the concept of a single variable species, in which pattern seems determined by environmental factors. Any solution of this subspecies or racial problem grows steadily more difficult as the Ohio River form has been liberated in Pennsylvania tributaries of Lake Erie. Both the Ohio form and an inland Ontario form (Kawartha Lakes: faint bars to immaculate) have been introduced into various Quebec waters tributary to the St. Lawrence River. In early fish-cultural practise, if male muskellunge were not readily available, male northern pike were used to fertilize muskellunge eggs and the resulting young were liberated as "muskellunge." In addition there is concern in certain states that different genetic strains of muskellunge with differing inherent growth limitations occur. This problem would also seem to be phenotypic rather than genotypic, but is under study.

In the 1800's the distinction between northern pike and muskellunge was not clear and led to many nomenclatorial problems (see below). This situation was further clouded by Godfrey's (1945) description of a supposed new species *Esox amentus* from Vermillion Lake, Ontario. The animal was designated the "True Tiger." Bishop (1946) brought attention to the unorthodox new species and suggested that the specimen resembled "supposed hybrids between the Chautauqua and St. Lawrence muskellunge." A later study (Cameron 1948) of specimens from the same location determined that the animal represented naturally occuring northern pike × muskellunge hybrids, and the name was dropped.

For many years the color mutant of northern pike, now referred to as the silver pike, was thought to be a variant of muskellunge and was called the silver muskellunge.

The following discussion of distribution (Fig. 2) ignores the subareas of the "races." The recent natural distribution can be assumed to include the following territory, starting with the northeastern limit: from the St. Lawrence River immediately downstream of the mouth of the Chaudiere River; south in the Hudson River valley; south, west of the Appalachian Mountains, from New York to eastern Tennessee (including a small part of the Little Tennessee and French Broad systems in North Carolina [Menhinick et al. 1974] and probably in the past the Tennessee system in extreme northern Alabama); north through Kentucky, extreme southern Indiana, eastern Ohio, Michigan (but only peripheral to the Great Lakes), extreme northeastern Illinois, and throughout Wisconsin; south in the main stem of the Mississippi River to southern Iowa (rare); north through eastern Minnesota to the Winnipeg River in Manitoba (recent natural westward expansion); east and south in Ontario through the systems of the English and Rainy rivers;

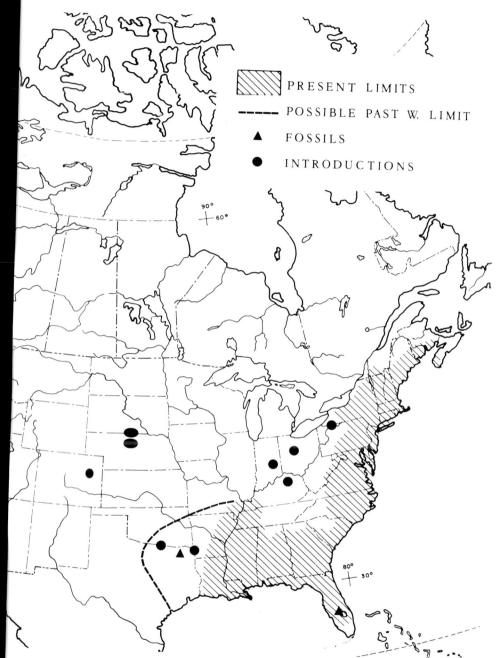

—*Distribution of chain pickerel. The crosshatched area indicates present limits. The broken line indicates western limit in the past.*

it is said to inhabit larger mountains. Where well established it is part of a community involving some tion of ictalurids, centrarchids, catostomids, and several cyprinids.

It can withstand temperatures probably to 35 C, levels of pH near lethal for most other fishes (Wich and Mullen 1958), and salinity as high as 22‰ (Schwartz 1964). This last tolerance has probably allowed it

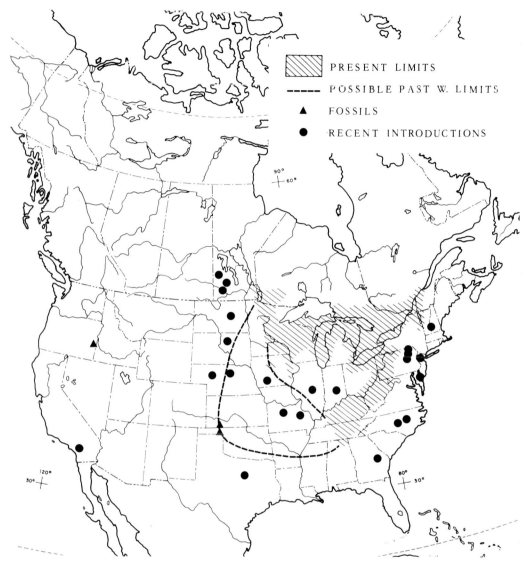

FIGURE 2.—*Distribution of muskellunge. The crosshatched area indicates the present limits. The two broken lines indicate possible western limits in the past.*

south to follow the south shore of Lake Superior; north in Ontario to the Chapleau region; and east from that point, along an irregular line through Ontario and Québec, to the northeastern limit.

Isolated occurrences in Iowa, and known historical losses of territory in Illinois and Indiana, suggest that the western boundary may once have been the Mississippi River from western Tennessee, including at least eastern Iowa, and northward including more of western Minnesota.

Fossil records (Smith 1954; Smith 1963) tentatively attributed to the muskellunge, and a possible relict population in Nebraska could be considered evidence that the territory in the Pleistocene extended to Kansas and Oklahoma. Another fossil tentatively attributed to this species (Cavender et al. 1970) would place the muskellunge in Oregon in the late Miocene.

More recent introductions of muskellunge have been made in Atlantic coastal, midwest, and plains states; in Texas; and in Manitoba. In many of these areas populations may have to be maintained by constant introduction. If the populations are maintained they will greatly expand the North American distribution of this species.

Historical considerations of the spread or constriction of the territory of the muskellunge is complicated by confusion which existed in the 1800's. The distinction between northern pike and muskellunge was not clear to many authors and the two species were often reversed. One must be very cautious with early distribution records published under such names as *Esox lucius, E. nobilior, E. nobilis,* and *E. estor,* or the same names in the genus *Lucius* (Crossman and Goodchild 1978).

Muskellunge usually exhibit the affinity for vegetation characteristic of most members of the family. Very large adults, however, are often captured in moderately deep water along rocky shorelines, where emergent or visible vegetation of any nature is sparse to absent.

Throughout its range the muskellunge occurs more frequently (at times other than spawning periods) than northern pike in slow meandering streams, and in the quieter portions of large, moderately swift rivers. It may in fact be more of a river species than a lake species. Its presence in the south is limited to rivers and it doubtless waited out at least the Wisconsin glaciation in these southern river environments. The only species of esocid not native to North America, *Esox reicherti,* is the only other silvery esocid with a pattern of dark spots like some muskellunge. *E. reicherti* is said to lack an affiliation with vegetation and to occur in the rocky, central portions of moderately flowing streams.

In some drainage systems in North America the muskellunge occurs or occurred exclusive of other esocids. If at a later time the northern pike colonizes, or is introduced into these systems, the muskellunge population seems invariably to decline (Johnson 1975; Threinen and Oehmcke 1950). See the work of Harrison and Hadley (1978, this volume) regarding ecological separation. In addition, the two species hybridize naturally and in certain places possibly very frequently. Of 69 specimens examined by Cameron (1948) from Vermillion Lake, Ontario, six were hybrids. The hybrid exhibits characteristics very attractive to an angler but the hybrid has seriously limited potential for natural reproduction. Possibly the muskellunge would be more successful if introduced into areas where it would be sympatric with only one of the smaller pickerels. In this way competition and the harmful erosion of interbreeding would possibly be less damaging to muskellunge populations.

The Pickerels, Subgenus *Kenoza*

The common name pickerel (the diminutive of the word pike), now restricted to the smaller esocids, has caused considerable confusion. The names pickerel and pike, with various modifiers, have been used interchangeably in different places for several species of *Esox,* including northern pike. This problem is compounded, at least in Ontario, because the walleye has been traditionally referred to as pickerel or yellow pickerel. The walleye, a very popular sportfish, is widespread and well known. The only true pickerel in Ontario, the grass pickerel, is much less abundant and almost unknown to anglers. When seen, the grass pickerel is usually thought to be the young of northern pike.

Chain Pickerel

As a result of its abundance in the New England states, the chain pickerel has been relatively well known probably even prior to its formal description in 1818. One must, however, be cautious of early published records. Specimens now thought to have been *E. niger* were published, starting in 1814, under the following names: *E. lucius; E. reticulatus; E. niger; E. tredecem-lineatus; E. tredecem-radiatus; E. tridecemlineatus; E. crassus;* and *E. affinis.* Many of these names were also used with the genera *Picorellus* and *Lucius.* In addition, 31 common names involving the terms pike, pickerel, and jack with various modifiers, have been applied to this species in different localities (Crossman and Lewis 1973).

In general, the chain pickerel has been considered to constitute a single species over the whole of its range. As early as 1854 a pickerel from the Tennessee River in Alabama was described as a separate species, *E. crassus* (Agassiz 1854) and vague means of separating this form from *E. reticulatus* (=*E. niger*) were given. Later all previously used names were synonymized with *E. niger.* Weed (1927) claimed there were differences in head proportions in "northern" and "southern" populations, and there have been suggestions through the years that the coastal populations and the Mississippi Valley populations were in some way different (Hubbs and Lagler 1964). Tarzwell (1942) alluded to the chain pickerel in Tennessee as *E. niger crassus* in a table, but only as *E. niger* in the text. Some distinctive characteristics may exist, possibly as a result of the sort of isolation postulated for *E. americanus* (Crossman 1966). An examination of morphometric characters is underway to test this possibility.

Another form of chain pickerel, reported from the Finger Lakes region of New York (Menzel and Green 1972), is a silvery color mutant lacking the chain-like pattern characteristic of the species. This is a counterpart to the situation for northern pike, but the New York record is apparently the only report of this mutation in chain pickerel.

Often one sees repeated statements of the supposed original distribution of the chain pickerel. Combined, those statements involved a range from southwestern Maine, southward east of the Green Mountains to Crooked Lake, Orange County, Florida, and west to Mammoth Springs and other tributaries of the White River in Arkansas.

More recently the general distribution (Fig. 3) could be said to extend from restricted localities in Yarmouth and Digby counties in Nova Scotia, through southern New Brunswick and southern Maine, west to the eastern townships of Quebec, south from New York (including tributaries of Lake Ontario) through northeastern West Virginia to at least the Everglades canals and Lake Okeechobee in Florida (Hutt 1967), west through most of each of the Gulf states to the [...] and north throug[h ...] southeastern Miss[...] Kentucky. Populat[ions ...] homa border outs[ide ...] resulted from intr[...]

As early as the [...] had been extende[d ...] natural expansio[n ...] man-made conne[ctions ...] (Weed 1927). Th[...] now occupied by [...] Maine, New Bru[nswick ...] man 1959), Verm[ont ...] New York is at le[ast ...] those early extensi[...] tions result from i[...] 1900's and appar[...]

The chain pick[erel ...] duced outside thi[s ...] Erie system of N[...] Ohio (Armbruste[r ...] northern Kentu[cky ...] Nebraska, and C[...] McCabe 1958).

The fossil rec[ord ...] only possibly attri[...] These include a [...] central Texas (U[...] Pliocene record [...] 1962). If the foss[il ...] cies they suggest [...] the Texas-Oklah[oma ...] populations.

Over much of [...] chain pickerel o[...] as one of the for[...] adjacent waters. [...] more often in la[...] and lakes. Where [...] occurred togethe[r ...] Carolina the abu[ndance ...] of chain pickere[l ...] streams (Cross[man ...] pickerel is symp[atric ...] pickerels extensi[vely ...]

The chain pick[erel ...] of habitat types [...] has the typical [...] shallow, heavily [...] it known in deep[...] or no vegetatio[n ...] duced outside i[...]

FIGURE 3[...]
a possibl[e ...]

Missour[i ...]
tain str[eams ...]
usually [...]
combina[tion ...]
percids,[...]

to utilize brackish estuaries of the Atlantic and Gulf coasts to move from one river system to another. These broad environmental tolerances make the chain pickerel highly suitable for introduction into waters no longer suitable for some other species. Its presence in trout and salmon water, and far worse its introduction into salmonid waters, is a serious matter to anglers. The result of contact of chain pickerel with salmonids varies from place to place. Barr (1963) said that when chain pickerel gets into a reclaimed pond stocked with brook trout "within a few years the pond becomes worthless for trout fishing." In contrast, as early as 1908 (Kendall and Goldsborough 1908) it was suggested that reduction of salmonids by chain pickerel occurred in streams "where the [trout] were declining for other reasons."

Redfin and Grass Pickerels

Esox americanus is now generally considered to be divisable into two subspecies: *E. a. americanus* Gmelin—redfin pickerel; and *Esox a. vermiculatus* Lesueur—grass pickerel. Weed (1927) considered this complex to consist of three subspecies: *E. a. americanus* on the east coast; *E. a. vermiculatus* in the Mississippi River system; and *E. a. umbrosus* in the Great Lakes. This distinction was not retained, however, and the redfin and grass pickerels were long considered separate species. Usage changed (Hubbs and Lagler 1947, 1958) from separate species to subspecies following Legendre (1952). A study (Crossman 1966) of material from localities over the whole distribution suggested that two forms intergrade in the south to such an extent that populations in those parts of Georgia, Florida, Alabama, and Mississippi drained by Gulf of Mexico tributaries from the Suwanee River to the Biloxi River could not be assigned to either form. It suggested also that the part of Florida from the St. John River to Lake Okeechobee is part of the zone of overlap, as opposed to being the southern limit of the Atlantic coastal form. The study postulated that the fusion in the zone of overlap represents secondary mixing of two stocks which had previously diverged in response to isolation on opposite sides of the Appalachian Mountains. The two subspecies show only slight divergence and are distinguishable on the basis of snout length and profile, and the numbers of specialized (cardiod) scales on various parts of the body.

The present distribution of redfin pickerel (Fig. 4) starts in Quebec in the region of Lac St. Pierre (an expansion of the St. Lawrence River), continues south in the Hudson River system to coastal New York, and northward through the New England states to southern Massachusetts and southeastern New Hampshire. It occurs generally in at least the eastern portions of the states from Pennsylvania to Georgia. It is most abundant below the fall line in northern states but penetrates almost to the headwaters of streams in Georgia. The most southern populations are in the St. Mary's River system of southeastern Georgia and northeastern Florida. The territory occupied in north central New York and Quebec is a natural extension resulting from the interconnection of the Hudson and Richelieu rivers. The species was first reported from Quebec as recently as 1944 (Cuerrier 1947).

Populations clearly recognizable as grass pickerel extend from the Pearl River in Louisiana westward to the Brazos River system in Texas, north through the eastern portions of Texas and Oklahoma, southern and eastern Missouri, the eastern border of Iowa, to southeastern Wisconsin (an isolated distribution in Villas and Oneida counties in northern Wisconsin is probably the result of an introduction in the 1940's: Kleinert and Mraz 1966), east across southern Michigan, to the north shore of Lake Erie, the north shore of Lake Ontario (Niagara Peninsula; Bay of Quinte eastward), and east to the St. Lawrence and its tributaries as far as the mouths of the Chateauguay and Ottawa rivers in Quebec.

The eastern boundary follows the lowlands west of the Appalachian Mountains from the Finger Lakes district of New York, south through extreme northwest Pennsylvania, and southwestward through the western portions of the states from West Virginia to Mississippi, including the Ten-

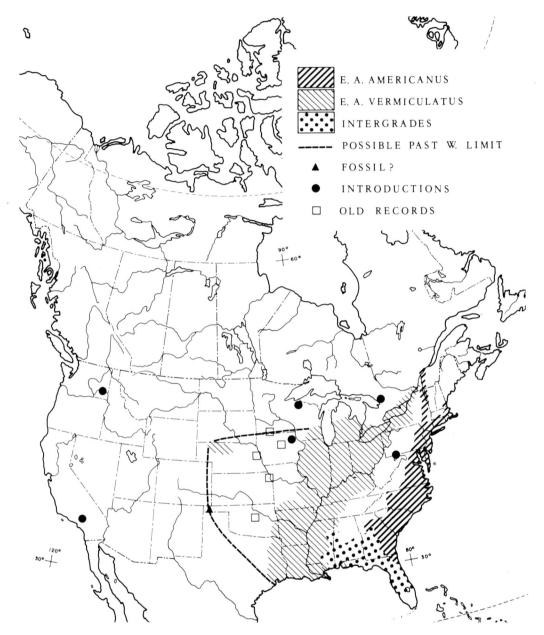

FIGURE 4.—*Distribution of* Esox americanus. *The present area of the two subspecies and of probable intergrades is indicated. The broken line represents a possible western limit in the past.*

nessee River in northwestern Alabama. Grass pickerel also occur in the Sandhill Lakes of northcentral Nebraska.

Populations of grass pickerel in Nebraska, old records in Iowa and Oklahoma west of the present territory, and possible fossil records in Oklahoma (Pliocene: Smith 1962) and Texas (Pleistocene: Uyeno 1963) could indicate a distribution in the past extending west to the area indicated by the dashed line on Figure 4.

Interest in this species is very low as a result of its size. Such lack of interest usually precludes artificial extension of the distribution of a fish. However, the redfin pickerel has been introduced west of the mountains in Deep Creek, Maryland (Schwartz 1960). The grass pickerel has been introduced in California (unsuccessfully), Colorado, Washington, northern Wisconsin, and central Ontario (successfully).

Both forms are typical of small quiet, vegetated bodies of water such as streams, ponds, borrow pits, drainage ditches, and weedy bays of small lakes. The redfin pickerel often occurs in acid, dark waters, whereas the grass pickerel is usually in neutral to basic, silted water.

Each of the two forms frequently occurs sympatrically with another esocid, and natural hybridization takes place. Often one of the pickerels is replaced in larger but typical esocid habitats, by chain pickerel on the Atlantic and Gulf coasts, and by northern pike in the Mississippi and Great Lakes systems.

As is the case with the other esocids, these two pickerel normally occur in predictable fish communities containing centrarchids, catostomids, ictalurids, percids, and cyprinids.

Acknowledgments

The many people who so kindly contributed to the preparation of the annotated bibliographies on northern pike and muskellunge made the preparation of this paper much easier. Mr. F. Sherwood assisted with the preparation of the figures. Various projects which contributed to the information contained here were supported by grants from the National Research Council of Canada (A-1705), and from the Canadian National Sportsmen's Show. Dr. R. J. Muncy and Mr. E. Harrison kindly provided access to data in preparation for this symposium.

References

ANONYMOUS. 1965. Pike in Morocco. Fishing Club du Moyen—Atlas, Annu. Rev. 1965:22–23.
AGASSIZ, L. 1854. Notice of a collection of fishes from the southern bend of the Tennessee River, in the State of Alabama. Am. J. Sci. Arts, Sec. Ser. 17(50):297–308.
ARMBRUSTER, D. C. 1959. Observations on the natural history of the chain pickerel (*Esox niger*). Ohio J. Sci. 59(1):55–58.
BARR, L. M. 1963. The chain pickerel. Maine Fish Game, Fall 1963:13–15.
BEAMISH, R. J., M. J. MERRILEES, AND E. J. CROSSMAN. 1971. Karyotypes and DNA values for members of the suborder Esocoidei (Osteichthyes: Salmoniformes). Chromosoma (Berl.) 34:436–447.
BISHOP, S. C. 1946. An alleged new species of muskellunge. Copeia 1946:43.
CAMERON, G. S. 1948. An unusual maskinonge from Little Vermillion Lake, Ontario. Roy. Ont. Mus. Zool. Contrib. 31:223–229.
CAVENDER, T. M., J. G. LUNDBERG, AND R. L. WILSON. 1970. Two new fossil records of the genus *Esox* (Teleostei, Salmoniformes) in North America. Northwest Sci. 44(3):176–183.
CROSSMAN, E. J. 1960. Variation in number and assymetry in branchiostegal rays in the family Esocidae. Can. J. Zool. 38(2):363–375.
———. 1962. The redfin pickerel, *Esox a. americanus* in North Carolina. Copeia 1962:114–123.
———. 1966. A taxonomic study of *Esox americanus* and its subspecies in eastern North America. Copeia 1966:1–20.
———, AND K. BUSS. 1965. Hybridization in the family Esocidae. J. Fish. Res. Board Can. 22(5): 1261–1292.
———, AND C. D. GOODCHILD. 1978. An annotated bibliography of the muskellunge, *Esox masquinongy* (Osteichthyes: Salmoniformes). Roy. Ont. Mus. Life Sci. Misc. Publ. 131 pp.
———, AND C. R. HARINGTON. 1970. Pleistocene pike, *Esox lucius*, and *Esox* sp., from the Yukon Territory and Ontario. Can. J. Earth Sci. 7(4):1130–1138.
———, AND G. E. LEWIS. 1973. An annotated bibliography of the chain pickerel, *Esox niger* (Osteichthyes: Salmoniformes). Roy. Ont. Mus. Misc. Publ. 81 pp.
———, AND J. W. MEADE. 1977. Artificial hybrids between Amur pike *Esox reicherti*, and North American esocids. J. Fish. Res. Board Can. 34(12):2338–2343.
CUERRIER, J. P. 1947. *Esox americanus* in Quebec. Copeia 1947:62.
DAVISSON, M. T. 1972. Karyotypes of the teleost family Esocidae. J. Fish. Res. Board Can. 29(5): 579–582.
ECKROAT, L. R. 1974. Interspecific comparisons of the lens proteins of Esocidae. Copeia 1974: 977–978.
GODFREY, J. JR. 1945. Muskies unlimited. Outdoors, Oct. 1945. pp 11, 36–37.
HARRISON, E. J., AND W. F. HADLEY. 1978. Ecological separation of sympatric muskellunge and northern pike. Am. Fish. Soc. Spec. Publ. 11:129–134.
HOURSTON, A. S. 1955. A study of variations in the maskinonge from three regions in Canada. Roy. Ont. Mus. Zool. Palaeontol. Contrib. 40:13 pp.
HUBBS, C. L., AND K. F. LAGLER. 1947. Fishes of the Great Lakes region. Cranbrook Inst. Sci. Bull. 26:186 pp.
———, AND ———. 1958. Fishes of the Great Lakes Region. Univ. Mich. Press, Ann Arbor. 213 pp.
———, AND ———. 1964. Fishes of the Great Lakes region. Univ. Mich. Press, Ann Arbor. 213 pp.
HUTT, A. 1967. The chain pickerel. Fla. Wildl. 20(11): 16–19.
JOHNSON, L. D. 1975. Statewide fisheries research.

Population levels of natural muskellunge populations. Wis. Dep. Nat. Resour. D. J. Proj. F-83-R-9, Study 211(1973). 7 pp.

KENDALL, W. C., AND E. L. GOLDSBOROUGH. 1908. The fishes of the Connecticut lakes and neighbouring waters, with notes on the plankton environment. U.S. Bur. Fish., Rep. Comm. Fish, Doc. 633. 77 pp.

KLEINERT, S. J., AND D. MRAZ. 1966. Life history of the grass pickerel (*Esox americanus vermiculatus*) in southeastern Wisconsin. Wis. Conserv. Dep. Tech. Bull. 37. 40 pp.

LAWLER, G. H. 1960. A mutant pike, *Esox lucius*. J. Fish. Res. Board Can. 17(5):647–654.

———. 1964. Further evidence of hardiness of "silver" pike. J. Fish. Res. Board Can. 21(3):651–652.

LEGENDRE, V. 1952. Clef des poissons de pêche sportive et commerciale de la Province de Québec. Soc. Can. d'Ecol. Montreal. 84 pp.

MCCABE, B. C. 1958. *Esox niger* LeSueur. Tabular treatment of the life history and ecology or the chain pickerel. Nat. Acad. Sci. Comm. Handb. Biol. Data. 45 pp.

MCPHAIL, J. D., AND C. C. LINDSEY. 1970. Freshwater fishes of northwestern Canada and Alaska. Bull. Fish. Res. Board Can. 173. 381 pp.

MEHINICK, E. F., T. M. BURTON, AND J. R. BAILEY. 1974. An annotated checklist of the freshwater fishes of North Carolina. J. Elisha Mitchell Sci. Soc. 90(1):24–50.

MENZEL, B. W., AND D. M. GREEN, JR. 1972. A color mutant of the chain pickerel, *Esox niger* LeSueur. Trans. Am. Fish. Soc. 101(2):370–372.

NELSON, G. J. 1972. Cephalic sensory canals, pitlines, and the classification of esocid fishes, with notes on galaxiids and other teleosts. Am. Mus. Novit. 2492. 49 pp.

RIDDLE, J. W. 1975. Status of the native muskellunge, *Esox masquinongy ohioensis*, of the Cumberland plateau, Tennessee. M.S. Thesis. Tenn. Tech. Univ. 70 pp.

ROSEN, D. E. 1974. Phylogeny and zoogeography of salmoniform fishes and relationships of *Lepidogalaxias salamandroides*. Bull. Am. Mus. Nat. Hist. 153(2):267–325.

SCHWARTZ, F. J. 1960. The pickerels. Md. Conserv. 37(4):23–26.

———. 1964. Natural salinity tolerances of some freshwater fishes. Underwater Nat. 2(2):13–15.

SCOTT, W. B., AND E. J. CROSSMAN. 1959. The freshwater fishes of New Brunswick: a checklist with distributional notes. Roy. Ont. Mus. Div. Zool. Palaeontol. Contrib. 51. 37 pp.

SEABORNE, E. 1937. Variations in the maskinonge in the Sauble River, Ontario. Copeia 1937:237–238.

SMITH, C. L. 1954. Pleistocene fishes of the Berends fauna of Beaver County, Oklahoma. Copeia 1954:282–289.

———. 1962. Some Pliocene fishes from Kansas, Oklahoma, and Nebraska. Copeia 1962:505–520.

SMITH, G. R. 1963. A late Illinois fish fauna from southwestern Kansas and its climatic significance. Copeia 1963:278–285.

TARZWELL, C. M. 1942. The fish population of Wheeler Reservoir. Trans. Am. Fish. Soc. 71(1941):201–214.

THREINEN, C. W., AND A. OEHMCKE. 1950. The northern invades the musky's domain. Wis. Conserv. Bull. 15(9):10–12.

UYENO, T. 1963. Late Pleistocene fishes of the Clear Creek and Ben Franklin local fauna of Texas. South. Methodist Univ., J. Grad. Res. Cent. 31(3):168–171.

WEED, A. C. 1927. Pike, pickerel and muskalonge. Field Mus. Nat. Hist. Zool. Leafl. 9:153–204.

WICH, K., AND J. W. MULLAN. 1958. A compendium of the life history and ecology of the chain pickerel *Esox niger* (LeSueur). Mass. Div. Fish. Game. Fish. Bull. 22. 22 pp.

ECOLOGICAL CONSIDERATIONS

Inventory of Percid and Esocid Habitat in North America[1]

KENNETH D. CARLANDER

Department of Animal Ecology
Iowa State University, Ames, Iowa 50011

J. SCOTT CAMPBELL

Department of Fisheries and Environment
Fisheries and Marine Service, Freshwater Institute
Winnipeg, Manitoba R3T 2N6

ROBERT J. MUNCY

Iowa Cooperative Fishery Research Unit[2]
Iowa State University, Ames, Iowa 50011

Abstract

A questionnaire sent to the chief fishery staff member of each of the U.S. states and Canadian provinces provided the basic data for estimating habitat occupied by walleyes, saugers, yellow perch, northern pike, and muskellunge in North America. Species habitat areas, grouped by size categories within lakes, impoundments, and streams, were listed by major drainage patterns in 4 east to west and 15 north to south regional groupings.

The present distributional patterns of each species resulting from introduced, as well as native, populations revealed major concentrations of these species around the Great Lakes. The areas occupied by percids and esocids as a percentage of total freshwater area in North America were 54% for northern pike, 1% for muskellunge, 32% for walleyes, 10% for saugers, and 26% for yellow perch. Areas occupied by the five species, as percentage of total freshwater areas, varied widely in different regions.

Although the coolwater species of fish, that are the subject of this symposium, support significant fisheries and have received considerable study, we know of no previous estimates of the magnitude of their habitat in North America. The geographical ranges of the species are known, but not the amount of water which is inhabited by, or could be managed for, the species. Estimates of habitat provided in this paper are based on information received from the fishery staff in each state and province.

Survey

Collection of Data

A questionnaire was mailed in April 1977 to the chief fishery staff member of each U.S. state and Canadian province requesting a summary of the numbers and surface area of waters—by four size groups for lakes and impoundments and by three stream widths—that contained native or introduced populations of northern pike, muskellunge, sauger, walleye, or yellow perch. Potential habitat for future introductions was identified separately. In addition, information was requested on current habitat management by four categories: (1) unmanaged; (2) stocked only; (3) managed; and (4) managed primarily for other species. Additional information was requested by correspondence and telephone. In a few instances, we made preliminary rough estimates which the staff member of the state or province concerned revised for specific species or habitat types. All preliminary state or provincial totals were resubmitted to respondents for review of our interpretation of their original data and classifications as well as methods of conversion, expansion, and summation.

---

[1] Joint contribution: Journal Paper No. J-9047 of the Iowa Agriculture and Home Economics Experiment Station, Ames, Iowa. Project 2002.

[2] The Iowa Cooperative Fishery Research Unit is jointly sponsored by Iowa State University of Science and Technology, Iowa Conservation Commission and U.S. Fish and Wildlife Service.

Our own estimates were used for Alaska, Maine, Michigan, Northwest Territories, Yukon, Labrador, and Quebec.

Surface area estimates were more easily obtained for natural lakes than for impoundments or streams. States and provinces with limited habitats for the species in question usually were able to give more reliable estimates than were those with widespread habitat. For some of the more northerly states and provinces recorded data on percid or esocid habitat were incomplete, as were surveys of total water areas; consequently, the agencies were unable to classify the waters into the various habitat types and size categories. Several reporting units had computerized data on water areas by size categories and species; many others, however, had to compile detailed listings of habitats for each species. Some of the more northern states and provinces provided reports of total water areas or standing water areas and miles of streams, from which we estimated the percentage area occupied by each species to obtain hectares without detailed estimates for different size categories of lakes or impoundments. Sometimes, it was not possible to estimate the numbers of smaller lakes in each area especially in northern Canada with resulting estimates tending to be conservative. Most respondents stated that they did not report as habitat those waters containing small populations or stray occurrences.

Habitat estimates for the Great Lakes (including Lake St. Clair) were obtained from the U.S. Fish and Wildlife Service Great Lakes Fishery Laboratory, Ann Arbor, Michigan, and were based on areas historically occupied by adult populations, but adjusted downward for losses because of habitat degradation. These estimates were circulated to other researchers in surrounding states or provinces for suggestions or revisions.

Computation and Summarization of Data

After the receipt of reports from the majority of the states and provinces, we compiled the data by watershed and latitude regions. This allowed us to follow up instances of differences in conflicting data on shared waters from adjacent reporting units. Regional summaries also gave preliminary indications of the reliability of, and relative agreement among, the independent reports by each state or province.

The Canadian Atlantic Region was composed of Quebec, Newfoundland and Labrador, Nova Scotia, New Brunswick, and Prince Edward Island. Northeastern Atlantic states (6) extended from Maine and Vermont southward through Connecticut. Central Atlantic states (7) included New York and Pennsylvania southward through North Carolina. South Carolina, Georgia, and Florida comprised the Southeastern Atlantic states. Ontario was the Canadian Mid-North America province adjoining the Great Lakes (including Lake St. Clair). Michigan, Wisconsin, and Minnesota comprised Lake states. The eight Midwest states ranged from Iowa east into West Virginia, south into Tennessee, and west into Missouri. Gulf states were comprised of Alabama, Mississippi, Louisiana, Arkansas, and Texas. Northwest Territories, Alberta, Saskatchewan, and Manitoba comprised the Canadian Plains. The five United States Plains states extended from North Dakota south through Oklahoma. Eight Rocky Mountain states extended from Idaho and Montana southward into Arizona and New Mexico. The Pacific Coast states were Washington, Oregon, and California with Alaska considered as a separate region. British Columbia and the Yukon comprised western-most Canada. Detailed reports and worksheets for the individual states and provinces are on file at three locations: the Denver Public Library, Fish and Wildlife Reference Service; Iowa State University Library, Reference Section; and the Library of the Freshwater Institute in Winnipeg.

Stream length estimates were requested in three width groupings: (1) over 100 m; (2) 25–100 m; and (3) less than 25 m. Unless actual average width estimates were given in original reports or suggested as better estimates in follow-up evaluation by respondents, we multiplied arbitrary widths of 150 m, 50 m, and 20 m by stream lengths in km in the three width categories to obtain surface areas in hectares for streams. Size categories for lakes and impoundments were over 5,000, 1,000–5,000, 500–1,000, and under 500 hectares.

In compiling the regional summary tables, if one state or province in the group furnished combined categories or did not furnish a detailed breakdown of numbers or types of waters for a species, we combined the waters in that category or omitted the category from tables containing that regional grouping. The accuracy of the estimates and regional summaries may be related directly to the value that fishermen in the areas concerned place on each species, and inversely to species distribution. Therefore, yellow perch and northern pike estimates probably are low, but area estimates for muskellunge are more accurate. We hope that this initial summary of habitat will stimulate improved future estimates and comparisons with detailed data on file in the three libraries mentioned above.

Northern Pike

Distribution

Northern pike habitat was reported extensive from Pennsylvania westward to Montana and northward to the Arctic Ocean in Alaska and the Northwest Territories of Canada (Table 1). Expansion of northern pike habitat in U.S. Plains and Rocky Mountain states has followed construction of large impoundments and foothill reservoirs not totally suitable for coldwater species, especially in the Missouri River systems of the Dakotas, Montana,

TABLE 1.—*Northern pike habitat (thousands of hectares) in North America presented in groupings of provinces and states from east to west and north to south by major drainage patterns. Question marks indicate numbers are unknown; long dashes indicate combined area or width categories.*

Regions	Agencies reporting[a]	Lakes		Impoundments		Streams			Total hectares (1,000's)
		No. by area groups[b]	Hectares (1,000's)	No. by area groups[b]	Hectares (1,000's)	No. by width groups[c]	No. km (1,000's)	Hectares (1,000's)	
Atlantic									
Canadian	2/5	320-640-1,005-?	12,960	—	—	—	—	—	12,960
U.S. Northeast	5/6	1-0-2-40	33	0-3-0-4	2	3-5-7	1	8	43
Central	5/7	4-15-7-114	1,090	4-10-14-108	79	21-66-39	5	46	1,215
Southeast	0/3								0
Total			14,083		81			54	14,218
Mid-North America									
Ontario	1/1	220-450-690-?	8,900	—	—	—	—	—	8,900
Great Lakes	6/6	6-0-0-0	1,290						1,290
U.S. Lake states	3/3	—4,862+—	1,381	—	—	657+	28	110	1,491
Midwest	5/8	0-5-6-108	19	1-10-7-57	35	8-4-2	3	46	100
Gulf	2/5		0	4-0-1-18	66			0	66
Total			11,590		101			156	11,847
Plains									
Canadian	4/4	879-767-2,970-?	40,402	2-3-3-8	28	?	?	74	40,504
U.S.	5/5	1-5-35-122	50	14-14-9-223	500	7-12-23	9	50	600
Total			40,452		528			124	41,104
Rocky Mountain & Pacific									
Alaska	1/1	?	501	—	—	?	?	500	1,001
Yukon & British Col.	2/2	24-31-122-?	638	—	—	—	—	—	638
U.S. Rocky Mountain	8/8	1-0-0-16	5	3-17-4-59	148	3-7-1	1	8	161
Pacific	0/3								0
Total			1,144		148			508	1,800
Grand total									68,969

[a] States or provinces reporting habitat in each region.
[b] Numbers of lakes or impoundments by size groupings: number at left 5,000+ hectares; 2nd no. 1,000–5,000 hectares; 3rd no. 500–1,000 hectares; 4th no. <500 hectares.
[c] Numbers of streams with average width over 100 m, 25–100 m, and <25 m width.

Nebraska, and Colorado. Except for Texas, the southern states seem to be stocking fewer waters with northern pike than with muskellunge or walleyes.

The widespread distribution (Table 1), and the tendency for northern pike to use shallow weedy areas of lakes and marches, and backwater sloughs of streams, created problems in estimating occupied habitat in the northern states, Canada, and Alaska. The question of summer occupancy of shallow spawning and nursery areas subject to winterkill influenced habitat estimates, but summer habitat areas are nevertheless included as important to the evaluation of the northern pike resource. In most cases, a percentage calculated from known or estimated water areas for lakes and streams probably represents conservative estimates in Great Lake states, Canada, and Alaska.

California, Missouri (Goddard and Redmond 1978, this volume), and Georgia reported unsuccessful or unsatisfactory results from the stocking of northern pike. Virginia, Missouri, and Iowa are stocking muskellunge in waters previously stocked with northern pike. Texas stocking records for 1954–1973 indicate equal areas stocked with northern pike and muskellunge, but in 1974–1977, some reservoirs stocked with northern pike also received muskellung or muskellunge × northern pike hybrids. No states or provinces reported northern pike as a species scheduled for expanded distribution. However, Federal Aid reports from Alaska suggested that northern pike can provide increased sport fishing there and may become economically important in the future.

Management

The loss of aquatic vegetation at spawning sites after several years of reservoir impoundment, as well as draining and filling of lake marshes and backwater areas along streams, may have reduced habitat and influenced the declining interest in introducing and managing northern pike outside its major natural range. Stocking and establishment of northern pike in irrigation reservoirs and drainage canals in Plains and Rocky Mountain states and in wildlife management lakes and marshes managed primarily for wildlife seems to be expanding the occupied areas.

Legal size regulations for northern pike were not listed by most states. Vermont listed a 30.5-cm minimum legal length for northern pike, muskellunge, and pickerel, but no size limit for northern pike in Lake Champlain. Iowa is considering commercial fishing for northern pike in upper Mississippi River pools. In Canada, the northern pike is considered both a sport and commercial species, being protected by angling limits and commercial quotas in some provinces. During 1955–1976, the marketed catch in Canada fluctuated from 3,045 metric tons (t) to 4,500 t and averaged 3,500 t (Fisheries and Marine Service 1977). In 1976, the total commercial catch in Manitoba amounted to almost 60% of the total Canadian catch for that year, and substantial percentages of the remainder were taken in Saskatchewan and Ontario. Alt (1970) reported that northern pike were used for dog food in Alaska. The species was not listed in 1973 U.S. commercial fishery statistics from Alaska (NMFS 1976). The U.S. 1973 commercial fisheries catches were listed by regions (1976) as follows: Great Lakes, 14.1 t for U.S. portion and 23.6 t for Canada; Lake of the Woods, 27.7 t for the U.S., and 154.2 t for Canada; and in Minnesota 7.1 t from Rainy Lake and 10 t from the Mississippi River and its tributaries. Although northern pike was not identified in the catches in U.S. freshwater fisheries, the total in 1973 would have been 59 t compared with over 3,270 t from Canada.

Muskellunge

Distribution

Because the muskellunge is highly prized as a trophy sport fish, we believe that the distribution was rather accurately reported by all respondents (Table 2). Maryland reported an occasional muskellunge taken in fish traps at the Susquehanna River dams, but considered these fish to have

TABLE 2.—Muskellunge, including hybrid, habitat (thousands of hectares) in North America presented in grouping of provinces and states from east to west and north to south by major drainage patterns. Question marks indicate numbers unknown; long dashes indicate combined area or width categories.

Regions	Agencies reporting[a]	Lakes		Impoundments		Streams			Total hectares (1,000's)
		No. by area groups[b]	Hectares (1,000's)	No. by area groups[b]	Hectares (1,000's)	No. by width groups[c]	No. km (1,000's)	Hectares (1,000's)	
Atlantic									
Canadian	1/5	5-8-10-?	39			?	1	90	129
U.S. Northeast	1/6	1-0-0-0	3			0-1-0	Tr.	Tr.	3
Central	6/7	2-3-3-18	53	2-12-9-108	54	19-44-13	5	70	177
Southeast	1/3		0	0-1-0-0	1		3	0	1
Total			95		55			160	310
Mid-North America									
Ontario	1/1	14-25-18-46	151						151
Great Lakes	3/6	3-0-0-0	62						62
U.S. Lake states	3/3	545	453	—		9-42-11	2	19	472
Midwest	8/8	0-3-0-2	5	3-7-10-29	68	3-19-85	3	11	84
Gulf	1/5		0	1-0-2-6	28				28
Total			671		96			30	797
Plains									
Canadian	1/3	1-1-1-4	125	0-0-1-1	1				126
U.S.	3/5	1-0-0-6	19	0-2-2-4	5	0-2-0	1	5	29
Total			144		6			5	155
Rocky Mountain & Pacific									
Alaska	0/1								0
Yukon & British Col.	0/2								0
U.S. Rocky Mountain	0/8								0
Pacific	0/3								0
Total									
Grand total									1,262

[a] States or provinces reporting habitat in each region.
[b] Numbers of lakes or impoundments by size groupings: number at left 5,000+ hectares; 2nd no. 1,000–5,000 hectares; 3rd no. 500–1,000 hectares; 4th no. <500 hectares.
[c] Numbers of streams with average width over 100 m, 25–100 m, and <25 m width.

strayed downstream after stocking in Pennsylvania. As expected from the natural range of muskellunge (Crossman 1978, this volume), the major habitat was in the Great Lakes states of Minnesota, Wisconsin, and Michigan; Lakes St. Clair (Haas 1978, this volume), Ontario, and Erie; and inland waters of Pennsylvania and New York (Table 2).

Vermont reported that muskellunge populations were expanding in the Missisquoi Bay portion of Lake Champlain. Virginia, North Carolina, New Jersey, and Delaware reported introduced muskellunge populations in reservoirs or streams. Georgia reported that limited numbers of introduced muskellunge remained in Blue Ridge Reservoir. Texas stocking records for 1974–1977 indicated an expanding stocking program for muskellunge and hybrid muskellunge in reservoirs. limited but unmeasured populations were being found above and below these reservoirs. Among the eight Midwest states, Tennessee, Missouri, and Ohio led in reported muskellunge habitat, mainly in impoundments; only West Virginia, Kentucky, Ohio, and Tennessee reported stream habitat for muskellunge; and Iowa and Illinois were the only states reporting the introduction of muskellunge into natural lakes.

For the U.S. Plains states, North Dakota reported 93% of habitat, whereas Kansas and Oklahoma reported none. Manitoba reported all of the 125,000 hectares in Canadian Plains region (Table 2). Five states (Illinois, Indiana, Wisconsin, North Dakota, Nebraska) indicated potential stocking in 37 small lakes totaling 10,069 hectares and 27 impoundments totaling 40,223 hectares. California reported unsuccessful past stocking of muskellunge.

Management

Most states imposed creel limits of one fish and minimum size limits of 61–66 or 76.2 cm. In Ontario and Quebec, this species generally can be taken only from July to October, in numbers no greater than two daily per angler, and the fish must exceed 71.1 cm in length. Provincial authorities have estimated that 34.2 t of muskellunge were captured in 1955 by anglers fishing in Ontario waters alone (Scott and Crossman 1973). Insofar as is known, no commercial fishing is permitted for muskellunge.

Hybrid muskellunge are being stocked in streams by Virginia because the state has only limited muskellunge brood stocks; also the hybrids grow faster and are considered to be better predators on suckers and sunfishes. Some states indicated that management of native or introduced populations of muskellunge involved restrictions on stocking of northern pike, to reduce possible angler creeling of small muskellunge as well as to maintain separate population stocks. Some waters, therefore, could be considered as being managed for muskellunge.

Walleye

Distribution

Walleye native habitats were reported (Table 3) to overlap those of the sauger; however, walleyes were found throughout the Great Lakes and over a wider area in Canada. Walleyes have been introduced into Atlantic drainage rivers throughout the eastern United States, Gulf drainages of Mississippi and Texas, and Mississippi River drainage impoundments on the eastern ranges of the Rocky Mountain

TABLE 3.—*Walleye habitat (thousands of hectares) in North America presented in grouping of provinces and states from east to west and north to south by major drainage patterns. Question marks indicate numbers are unknown; long dashes indicate combined area or width categories.*

Regions	Agencies reporting[a]	Lakes		Impoundments		Streams			Total hectares (1,000's)
		No. by area groups[b]	Hectares (1,000's)	No. by area groups[b]	Hectares (1,000's)	No. by width groups[c]	No. km (1,000's)	Hectares (1,000's)	
Atlantic									
Canadian	1/5	47-97-150-?	1,728						1,728
U.S. Northeast	3/6	1-1-0-4	44	0-3-0-0	2	1-11-0	1	5	51
Central	7/7	2-10-6-59	100	4-17-12-130	87	33-77-15	7	71	258
Southeast	2/3		0	5-5-0-2	77			0	77
Total			1,872		166			76	2,114
Mid-North America									
Ontario	1/1	220-450-690-7,000	8,000			?			8,000
Great Lakes	6/6	6-0-0-0	3,220						3,220
U.S. Lake states	3/3	—2,312—	1,117			?		75	1,192
Midwest	8/8	0-5-5-17	14	12-22-10-84	226	17-17-11	9	148	388
Gulf	4/5		0	38-26-6-14	565	1-7-11	2	10	575
Total			12,351		791		11	233	13,375
Plains									
Canadian	4/4	534-1,052-1,652-?	24,203	1-1-2-2	10	?	10	74	24,287
U.S.	5/5	1-7-9-48	44	20-24-13-111	588	8-18-20	9	55	687
Total			24,247		598			129	24,974
Rocky Mountain & Pacific									
Alaska	0/1								0
Yukon & British Col.	1/2	0-2-10-?	20			?		150	170
U.S. Rocky Mountain	8/8	1-0-0-0	40	12-25-12-41	337	5-11-0	2	17	394
Pacific	2/3		0	5-1-0-0	94	1-0-0	Tr.	Tr.	94
Total			60		431			167	658
Grand total									41,121

[a] States or provinces reporting habitat in each region.
[b] Numbers of lakes or impoundments by size groupings: number at left 5,000+ hectares; 2nd no. 1,000–5,000 hectares; 3rd no. 500–1,000 hectares; 4th no. <500 hectares.
[c] Numbers of streams with average width over 100 m, 25–100 m, and <25 m width.

States. Washington, Oregon, and British Columbia reported that the walleye range is expanding in the Columbia River system. California reported unsuccessful attempts to introduce walleyes. Prentice et al. (1977) reported walleyes in California but not in Washington or Oregon.

Management

Some states manage large impoundments and lakes as primary walleye waters, with seasons and laws to regulate walleye fishing, whereas other states and provinces maintain year-round fishing. Creel limits were the most commonly listed management tool although only Vermont reported a minimum-length (38.1 cm) regulation; other states may have similar size regulations, but not all state and provincial regulations were available to us. Prentice et al. (1977) discussed walleye stocking and sport fishery management in the United States.

In Canada, commercial fisheries for this species are regulated by quota, minimum size, and minimum mesh size regulations. U.S. statistics for 1973 (NMFS 1976) reported commercial catches at 221 t of which 192 t were from international lakes and 3 t from the Mississippi River and its tributaries in Minnesota. In years 1955–1976, the commercial catch of walleyes from Canadian waters ranged from 9,091 t to 2,727 t and averaged 5,319 t. A total commercial catch of about 4,545 t was taken in 1976 in Canada of which 2,545 t were from Manitoba, 636 t from Saskatchewan, and 636 t from Ontario (Fisheries and Marine Service 1977).

Sauger

Distribution

Sauger habitats (Table 4) were reported for southwestern Quebec, southern Ontario, Manitoba, Saskatchewan, and Alberta in Canada; southward to the Red and Arkansas rivers of Texas, Oklahoma, and Arkansas; and eastward into Tennessee and Ohio rivers. Saugers were reported in Lakes Huron and Erie; Minnesota and Wisconsin reported sauger habitat in the Mississippi drainage. Michigan reported saugers in two lakes totaling 5,000 hectares. Ontario reported that 57 surveyed lakes with a combined freshwater area of approximately 327,000 hectares contained saugers. This is a conservative estimate and probably the number of lakes containing saugers will approach 1,000 after detailed survey data for the whole province become available. New York and Vermont reported sauger populations in Lake Champlain. Virginia reported 113 km of sauger habitat in headwaters of the Tennessee River drainage. Limited populations were reported in Clark Hill Reservoir (shared by Georgia and South Carolina) on the Savannah River. Mississippi has introduced saugers from the Tennessee River reservoir shared with Alabama. Arkansas, Mississippi, and Louisiana noted occasional sauger in the lower Mississippi River but did not think this section of the river should be considered major sauger habitat.

In the northern portion of the sauger's range, sauger habitat was reported for large lakes (e.g., Lake Winnipeg), although Alberta indicated that sauger distribution is basically limited to rivers in that province, whereas in the southern portion of the range, only Nebraska reported a lake as large as 101 hectares containing saugers. The major area category in U.S. Midwest, Plains, and Gulf states was larger impoundments on large streams containing native populations. Only Ohio and Mississippi reported introductions of sauger into new waters. Ohio reported restocking saugers in Lake Erie (Rawson and Scholl 1978, this volume) and establishing populations in two small impoundments. Georgia's report of saugers in Clark Hill Reservoir was not explained as to source of fish. Montana and Wyoming reported sauger habitat in seven larger impoundments and eight large streams, and Wyoming reported that saugers co-existed with trout. No other western state or province with a Pacific drainage reported presence of saugers or previous attempts at introduction. Ohio reported an interest in additional stocking of saugers but indicated

TABLE 4.—*Sauger habitat (thousands of hectares) in North America presented in grouping of provinces and states from east to west and north to south by major drainage patterns. Question marks indicate numbers are unknown; long dashes indicate combined area or width categories.*

Regions	Agencies reporting[a]	Lakes		Impoundments		Streams			Total hectares (1,000's)
		No. by area groups[b]	Hectares (1,000's)	No. by area groups[b]	Hectares (1,000's)	No. by width groups[c]	No. km (1,000's)	Hectares (1,000's)	
Atlantic									
Canadian	1/5	11-22-35-?	399	—	—	—	—	—	399
U.S. Northeast	1/6	1-0-0-0	70						70
Central	2/7	1-0-0-0	39			0-1-0	Tr.	1	40
Southeast	2/3		0	2-0-0-0	28				28
Total			508		28			1	537
Mid-North America									
Ontario	1/1	140-300-400-?	4,800	—	—	—	—	—	4,800
Great Lakes	2/6	2-0-0-0	700						700
U.S. Lake states	3/3	5-6-1-11	296	5-3-0-0	25	6-8-11	3	26	347
Midwest	8/8		0	11-4-3-1	184	11-7-5	8	143	327
Gulf	4/5		0	6-0-0-0	109	0-2-0	1	5	114
Total			5,796		318			174	6,288
Plains									
Canadian	2/4	58-85-136-?	6,000				1	20	6,020
U.S.	5/5	0-0-0-1	0	7-0-0-0	372	8-10-0	4	39	411
Total			6,000		372			59	6,431
Rocky Mountain & Pacific									
Alaska	0/1		0		0			0	0
Yukon & British Col.	0/2								0
U.S. Rocky Mountain	2/8		0	4-2-1-0	116	4-4-0	1	14	130
Pacific	0/3								0
Total					116			14	130
Grand total									13,386

[a] States or provinces reporting habitat in each region.
[b] Numbers of lakes or impoundments by size groupings: number at left 5,000+ hectares; 2nd no. 1,000–5,000 hectares; 3rd no. 500–1,000 hectares; 4th no. <500 hectares.
[c] Numbers of streams with average width over 100 m, 25–100 m, and <25 m width.

that a source of hatchery eggs was a major obstacle.

Management

No respondents indicated that they were managing waters specifically for saugers, and from almost all reports it appeared that saugers were coexisting with walleyes, especially in larger river systems. In the unaltered environment, the range of the sauger, in general, extends further to the south than that of walleyes. With the creation of other water conditions in large impoundments, and the introduction of walleye, the range of walleyes at many points now extends further south than that of sauger. Further information on this aspect of distribution is provided by Hackney and Holbrook (1978, this volume).

Since 1956, the annual commercial catch of saugers in Canada has averaged over 1,747 t. In 1976, 1,727 t were caught of which approximately 98% came from Manitoba, and the rest from Ontario (Fisheries and Marine Service 1977).

Yellow Perch

Distribution

The distribution and abundance of habitat within the individual states and provinces proved to be more difficult to determine or estimate for yellow perch than for the other four species. Three aspects contributing to the difficulties were the yellow perch's lower esteem as a sport fish, its continuing range expansion, and its frequent occurrence in small water bodies. In addition to overlapping ranges with walleye in lakes, impoundments, and streams of the upper Midwest, Plains, and Lake states of the United States, and of Canada (Table 5), the yellow perch is native to the Atlantic drainage of the Northeast and Central Atlantic states where its range extends into estuarine waters with salinities up to 5‰. Within the last 20 years, the yellow perch range has been expanding in the northeastern states, especially northward in Maine. Many of the yellow perch populations in Nova Scotia and New Brunswick may have resulted from accidental introductions. Dams of impoundments on New England streams have tended to hinder expansion by blocking upstream movements but also have favored expansion by warming the water and providing additional habitat for residual yellow perch populations. In northern states, the yellow perch was reported as not being abundant in small streams.

Georgia reported that yellow perch populations were expanding in up to 60% of its rivers and absent in only the Blue Ridge Reservoir, one of its 27 reservoirs. Clugston et al. (1978, this volume) discussed yellow perch range extension in southeastern reservoirs. Yellow perch habitat extensions into Gulf states were reported only for the Tennessee River and limited portions of the Rio Grande in Texas. In states bordering the Mississippi River, the yellow perch was reported to be rare in northern Missouri and absent from Oklahoma southward. Arizona, New Mexico, and Nevada reported scattered introduced populations in one lake, seven impoundments, and two streams. California reported yellow perch in three impoundments and two streams of the Klamath River (Coots 1956). Oregon and Washington reported that the yellow perch

TABLE 5.—*Yellow perch habitat (thousands of hectares) in North America presented in grouping of provinces and states from east to west and north to south by major drainage patterns. Question marks indicate numbers are unknown; long dashes indicate combined area or width categories.*

Regions	Agencies reporting[a]	Lakes			Impoundments		Streams			Total hectares (1,000's)
		No. by area groups[b]		Hectares (1,000's)	No. by area groups[b]	Hectares (1,000's)	No. by width groups[c]	No. km (1,000's)	Hectares (1,000's)	
Atlantic										
Canadian	3/5	45-101-153-?		1,694	1-6-10-11	26	?	15	37	1,757
U.S. Northeast	6/6	6,866+	—			485	?	?	142	627
Central	7/7	4,000+	—	303	1-11-13-739	58	?	?	409	770
Southeast	2/3			0	9-14-6-7	130	?	?	39	169
Total				1,997		699			627	3,323
Mid-North America										
Ontario	1/1	145-290-455-?		5,600						5,600
Great Lakes	6/6	6-0-0-0		4,240						4,240
U.S. Lake states	3/3	6,317+	—	1,485					68	1,553
Midwest	5/8	502+	—			82	5-13-0	1	18	100
Gulf	2/5			0	3-0-0-0	60	0-1-0	Tr.	Tr.	60
Total				11,325		142			86	11,553
Plains										
Canadian	4/4	362-646-1,164-?		17,752		3				17,755
U.S.	4/5	2-1-33-153		67	7-10-10-66	231	2-4-20	5	25	323
Total				17,819		234			25	18,078
Rocky Mountain & Pacific										
Alaska	0/1									0
Yukon & British Col.	1/2	1-6-1-14		56				Tr.	3	59
U.S. Rocky Mountain	8/8	2-1-3-50		57		275	6-10-5	2	20	352
Pacific	3/3	3,707+		195			3-5-8	2	17	212
Total				308		275		4	40	623
Grand total										33,577

[a] States or provinces reporting habitat in each region.
[b] Numbers of lakes or impoundments by size groupings: number at left 5,000+ hectares; 2nd no. 1,000–5,000 hectares; 3rd no. 500–1,000 hectares; 4th no. <500 hectares.
[c] Numbers of streams with average width over 100 m, 25–100 m, and <25 m width.

was well established and expanding its range. The species has spread northward into British Columbia from introductions in Washington.

No state or province suggested stocking yellow perch. As previously stated, respondents from Northeast and Pacific states expressed a desire to limit or remove yellow perch populations.

Management

Management of yellow perch ranged from none to the imposition of creel and commercial size limits. Most western states reported attempts to prevent further extension of its range into salmonid waters. Northeastern states, such as Maine, reported no management of yellow perch. Slow growth and small size of the yellow perch in northeastern United States and eastern Canada were suggested as reasons for its low esteem, along with its competition with salmonids. Midwestern and Central Atlantic states and central Canada hold the yellow perch in higher regard because of its value as a major species in winter, late fall, and spring fisheries. In the early 1900's, Atlantic coastal states, such as Maryland, operated egg-taking hatchery operations to improve commercial catches of yellow perch, which peaked in 1934 at 7,401 t (NMFS 1976). Later evaluations indicated that the loss of yellow perch habitat was a major cause of declining stocks and that the hatching and stocking of sac-fry did little to increase yellow perch populations. Recent attempts toward confined culture of yellow perch may renew interest in culturing this species (West and Leonard 1978, this volume).

The commercial catch of yellow perch during 1975 was listed as 2,000 t for the United States and 6,600 t for Canada (FAOUN 1976). Annual Canadian commercial catches of yellow perch fluctuated between 3,318 and 14,136 t from 1955 to 1976. In 1974, 6,409 t were caught, of which 6,091 t were landed in Ontario—largely from Lake Erie (Fisheries and Marine Service 1977). Atlantic tidal streams of states from New York to North Carolina provided a commercial catch in 1973 to 22 t, mainly from Maryland's Chesapeake Bay (17 t). The U.S. catch was 1,371 t from the Great Lakes and 21 t from Mississippi River and tributaries in Minnesota (NMFS 1976).

Coolwater Habitat in North America

Comparisons of the relative amounts of occupied habitat for five species (Tables 1–5) as percentages of total freshwater in the United States (Martin 1977) and Canada (Anonymous 1974) in Table 6 do not take into account the differences in abundance or standing stocks in different reporting units within regions. Therefore, the percentage area only indicates that the species occur in habitat areas but may not denote the relative importance in different fisheries. Yellow perch habitat in the Central Atlantic states included extensive estuarine tidal areas in Delaware and Maryland resulting in a higher percentage (53%) than if brackish waters were excluded. The percentage given is the total area occupied, freshwater and estuarine, divided by the total area of freshwater. Table 6 reveals major concentrations of coolwater species habitat around the Great Lakes but not necessarily in the Great Lakes.

Discussion

The reported occupied habitats of the five species were the combined result of the original natural distributions plus successful introduction and the spread of some species into new waters that would sustain adult fish. Although some introduced populations now are naturalized, highly prized sport species such as muskellunge, walleye, and, to a lesser extent, northern pike in the United States are maintained far outside their former native range through the stocking of eggs, fry, and fingerlings in waters lacking spawning habitat. In Canada, the natural range of the percid and esocid species has not changed substantially, nearly all of the five species being able to sustain themselves except in a few instances where an intensive commer-

TABLE 6.—*Percentage of reported esocid and percid habitat in North America (Tables 1–5) within reported total freshwater habitat in the United States (Martin 1977) and Canada (Anonymous 1974).*

Regions	No. agencies	Freshwater hectares (1,000's)	Percentage				
			Northern pike	Muskellunge	Walleye	Sauger	Yellow perch
Atlantic							
Canadian	5	22,192	58.4	0.6	7.8	1.8	7.9
U.S. Northeast	6	885	5	0.4	6	8	71
Central	7	1,453	84	12	18	3	53[a]
Southeast	3	1,427	0	0.1	5	2	12
Mid-North America							
Ontario	1	8,900	100	1.7	90.6	54.4	63.4
Great Lakes	6	24,249	5	0.2	13	3	17
U.S. Lake states	3	1,893	79	25	63	18	82
Midwest	8	1,490	7	5.5	26	22	7
Gulf	5	2,697	2	1	21	4	2
Plains							
Canadian	4	52,115	77.7	0.2	46.6	11.5	34.1
U.S.	5	1,278	5	2.2	54	32	25
Rocky Mountain & Pacific							
Alaska	1	5,006	20	0	0	0	0
Yukon & British Col.	2	2,254	28.3	0	7.5	0	2.6
U.S. Rocky Mountain	8	1,369	12	0	29	9	26
Pacific	3	1,172	0	0	8	0	18
North America		128,380	53.7	1.0	32.0	10.4	26.2[a]

[a] The percentage given is the total area occupied, freshwater and estuarine, divided by the total freshwater area.

cial fishery has developed. Over the years, attempts have been made in Ontario and Manitoba to enhance some heavily exploited walleye and muskellunge populations through the planting of eggs and fry but there is little evidence that this technique has been effective. For muskellunge × northern pike hybrids, continued restocking is necessary, and judging by recent results in Missouri and Virginia, it is evidently justified. The range of the less prized yellow perch continues to increase by natural expansion and by unplanned introductions.

State and provincial data sheets indicate that Minnesota seems to serve as the hub for the five percid and esocid species, their present distribution radiating mainly northward for northern pike, eastward for mukellunge, eastward and northwestward for yellow perch, southwestward and northwestward for walleyes, and southeastward and westward for saugers (Table 6). Minnesota contained the largest freshwater areas for all five species in the United States and Ontario seems to contain the largest freshwater area for these species in Canada. In the United States, the northern Mississippi River and Ohio River drainages along with the Great Lakes are the major natural habitats of muskellunge, northern pike, walleye; and sauger; in Canada, northern pike and walleye are naturally distributed throughout the Great Lakes, the Hudson Bay and Mackenzie River drainages having spread northward from the upper Mississippi system during deglaciation. The natural habitats of yellow perch, sauger, and especially muskellunge are not nearly as extensive in Canada as those of northern pike and walleye. The St. Lawrence River system has not served effectively for establishing muskellunge, walleye, and sauger in the Canadian Atlantic provinces, although northern pike are extensively distributed throughout Quebec and the inland drainages of Labrador, and yellow perch are found in Nova Scotia and New Brunswick. Walleyes were reported by Virginia and North Carolina as native populations in the Roanoke River drainage, which is noted for records of stream capture from New River of the Ohio River drainage. The Savannah river drainage between Georgia and South Carolina represents another southern extension of percid populations. The Tennessee River, flowing into Alabama and Mississippi, provides native walleye, sauger, and yellow perch habitat in the Gulf states region. Florida was the only continental state and Prince Edward Island the only province reporting none of the five species present.

Western and northeastern states, Atlantic provinces, and British Columbia indicated little or adverse interest in the expansion of coolwater species, especially if at the expense of coldwater salmonids. Central Atlantic states, such as Pennsylvania, suggested that additional coolwater stream habitat could be available with improved pollution control. Large impoundments seem to be the major waters that recently have been and in the future will be considered as coolwater habitat. Most states reported only limited data and management goals for stream habitat.

In Canada, populations of walleye, sauger, yellow perch, and northern pike have been commercially exploited for some time, large quantities having been taken from the Great Lakes and the large inland lakes of Manitoba. At the same time, some of these populations, especially of walleye have been exploited for sport. For the most part, saugers are caught only by commercial fishermen because anglers do not normally fish for that species in Ontario and the Prairie provinces. The regulation of the sport and commercial fisheries and the management of these populations have been difficult and at times unsuccessful, with the result that a number of walleye populations have been overexploited and eventually lost. This has been especially true for some Great Lakes walleye populations when the stocks were exposed to overexploitation, pollution, and interaction with foreign species (Schneider and Leach 1977). Provincial management of commercial percid populations now includes closed areas and seasons, catch quotas, minimum mesh-size regulations, and limited entry into the fishery as measures to ensure the maintenance of the remaining stocks.

References

ALT, K. T. 1970. The northern pike in Alaska. Alaska Dep. Fish Game Wildl. Notebook Ser. Fish. 5. 2 pp.

ANONYMOUS. 1974. Canada year book. Information Division, Statistics Canada, Ottawa. 914 pp.

CLUGSTON, J. P., J. L. OLIVER, AND R. RUELLE. 1978., Reproduction, growth, and standing crops of yellow perch in southern reservoirs. Am. Fish. Soc. Spec. Publ. 11:89–99.

COOTS, M. 1956. The yellow perch, *Perca flavescens* (Mitchill), in the Klamath River. Calif. Fish Game 42(7):219–228.

CROSSMAN, E. J. 1978. Taxonomy and distribution of North American esocids. Am. Fish. Soc. Spec. Publ. 11:13–26.

FAOUN (FOOD AND AGRICULTURE ORGANIZATION OF THE UNITED NATIONS). Yearbook of fishery statistics—1975. Vol. 40. F.A.O.U.N., Rome, Italy.

FISHERIES AND MARINE SERVICE. 1977. Annual statistical review of Canadian fisheries: 1955–1976. Intelligence Services Div., Marketing Service Branch, Fisheries and Environment Canada, Ottawa. Vol. 9. 200 pp.

GODDARD, J. A., AND L. C. REDMOND. 1978. Northern pike, tiger muskellunge, and walleye populations in Stockton Lake, Missouri: a management evaluation. Am. Fish. Soc. Spec. Publ. 11:313–319.

HAAS, R. C. 1978. The muskellunge in Lake St. Clair. Am. Fish. Soc. Spec. Publ. 11:334–339.

HACKNEY, P. A., AND J. A. HOLBROOK II. 1978. Sauger, walleye, and yellow perch in the southeastern United States. Am. Fish. Soc. Spec. Publ. 11:74–81.

MARTIN, R. G. 1977. Trends in angling pressure and license sales. Fisheries 2(4):30–31.

NMFS (NATIONAL MARINE FISHERIES SERVICE). 1976. Fishery statistics of the United States—1973. U.S. Nat. Mar. Fish. Serv. Stat. Dig. 67. 458 pp.

PRENTICE, J. A., R. D. CLARK, JR., AND N. E. CARTER. 1977. Walleye acceptance—a national view. Fisheries 2(5):15, 17.

RAWSON, M. R., AND R. L. SCHOLL. 1978. Reestablishment of sauger in western Lake Erie. Am. Fish. Soc. Spec. Publ. 11:261–265.

SCHNEIDER, J. C., AND J. H. LEACH. 1977. Walleye (*Stizostedion vitreum vitreum*) fluctuations in the Great Lakes and possible causes, 1800–1975. J. Fish. Res. Board Can. 34(10):1878–1889.

SCOTT, W. B., AND E. J. CROSSMAN. 1973. Freshwater fishes of Canada. Bull. Fish. Res. Board Can. 184. 966 pp.

WEST, G., AND J. LEONARD. 1978. Culture of yellow perch with emphasis on development of eggs and fry. Am. Fish. Soc. Spec. Publ. 11:172–176.

The Adult Walleye in the Percid Community—A Niche Definition Based on Feeding Behaviour and Food Specificity[1,2]

R. A. RYDER

Fisheries Branch
Ontario Ministry of Natural Resources,
Thunder Bay, Ontario P7B 5E7

S. R. KERR

Marine Ecology Laboratory
Bedford Institute of Oceanography
Dartmouth, Nova Scotia B2Y 4A2

Abstract

A methodology is proposed for niche contouring of a single species, the walleye, as well as of the percid community according to the Hutchinson-Fry niche concepts. A percid "harmonic" community is defined as one consisting of four basic species components which contribute to both persistence of the community and its continuing integrity. Mesotrophy provides fundamental environmental conditions for harmonic percid communities which are aggregated about a median of central tendency representing an optimum combination of abiotic factors. Astatic fish aggregations occur remote from the optimum in either direction, and lack continuing integrity. The walleye niche is described on the basis of the species' food habits, feeding behaviour and interspecific ethology within the percid community. The walleye is an opportunistic piscivore, which partitions the available food resources in time with other piscivores by feeding during twilight periods or nocturnally. Therefore, species niche is characterized in terms of individual metabolic capacities while community niche is defined on the basis of emergent system properties.

Niche, as defined by Hutchinson (1957, 1965), consists of an abstract hypervolume that is generated by assigning a single dimension to each factor that affects an organism's survival. Application to this concept of the views of Fry (1947, 1971) allows niche hyperspace description in terms of scope for metabolic activity at both the autecological and synecological levels of organization as elucidated in a previous paper (Kerr and Ryder 1977).

We here restrict our consideration to certain fundamental niche dimensions, namely, food, spawning, and interspecific ethology. Environmental heterogeneity (and the abiotic factors contributing to it) will also be considered as one of the major niche determinants. Feeding behaviour and food specificity are emphasized, not only because of the wealth of documentary evidence pertaining to this aspect in the primary scientific literature (e.g. Eschmeyer 1950; Price 1963; Fedoruk 1966; Swenson and Smith 1976) but also because of the evidence for niche partitioning with respect to food resources (Schoener 1974). However, the niche description we propose for the walleye is based, in addition, on more than 20 years of original data including over 1,000 hours of underwater observations while scuba diving in lakes and rivers of the boreal forest.

Succinctly stated, we provide a theoretical framework that describes the principal components and properties of a fish community—in this instance, the percid community. In recognition of the walleye as a major component of this community, we demonstrate how commonly measured dimensions of the walleye niche may be placed in perspective with other community components as a contribution to the understanding of total community interactions and the expression of community emergent properties. For illustrative purposes, we describe but a single dimension

---

[1] Contribution No. 78-8 of the Ontario Ministry of Natural Resources, Research Section, Fisheries Branch, Box 50, Maple, Ontario L0J 1E0.
[2] Bedford Institute of Oceanography Contribution.

of walleye niche (food and feeding) with the expectation that other investigators will, in the future, outline the other niche dimensions for both the walleye and other percid community species.

The Percid Community Concept

Fish communities as here defined are assemblages of co-existing fish species aggregated symbiotically (broad sense) so that the association demonstrates certain identifiable internal properties as well as external emergent properties (Kerr 1974). The community persists and maintains its integrity through interactions among its component fish species such as predation, competition, mutualism, and parasitism, as well as between them and other biotic and abiotic elements of the ecosystem. The species interactions internal to the community may be thought of as "community physiology" while "community morphology" may be considered to be the observable, emergent properties such as standing stocks (expressed as numbers or ichthyomass), community mortality or system decomposability (Simon 1969). Together they constitute dimensions of the community niche within the ecosystem.

The idealized percid community may be represented by the median of central tendency of the community curve representing mesotrophic waters (Fig. 1). (A proposed morphogenesis for this suite of curves has been described by Kerr [1977], but the precise derivation is not essential to our basic consideration, that is, community morphology.) At the median, the community shows maximum persistence with substantial resilience to exogenous stresses and consequently maintains a high level of integrity. As the ends of the curve are approached, a departure from ideal abiotic conditions occurs and increased interactions (direct competition and predation, for example) take place with component species of other communities, resulting in depressed persistence and low levels of integrity. Where the curves for

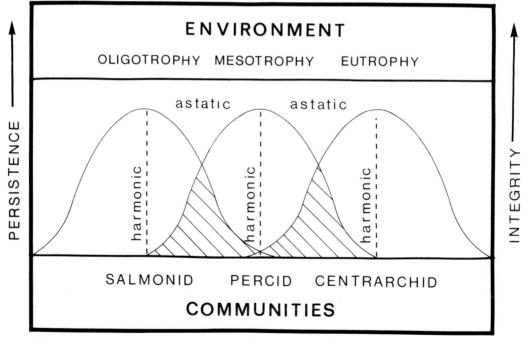

FIGURE 1.—*Schematic representations of three fish communities of the boreal forest zone showing the medians of central tendency (----) producing harmonic species associations complementary to prevailing environmental conditions. This, in turn, results in high levels of community persistence and integrity. The crosshatched areas indicate astatic species associations of low persistence and variable levels of community integrity.*

1969) dimensions. Niche separation of saugers from walleyes and yellow perch may be predicated on size differences of the fishes or their food items (Wilson 1975; Kerr and Ryder 1977), or more subtly through differences in light perception (Ali and Anctil 1977; Ali, et al. 1977; Ryder 1977).

Mutually Exclusive Species

The mutually exclusive fishes (Table 1-E) do not normally occur with the walleye in the same waters although they may be indigenous to the same watershed. The brook trout may coexist with the walleye in larger lakes and rivers, but in these instances they do not normally occupy the same space concurrently. Brook trout in Lake Nipigon and Lake Superior, for example, occupy shallow littoral zones in clear, cold bays, while walleyes are virtually non-existent in these bays except near the more turbid and warmer deltaic areas of larger inflowing rivers. Conversely, walleyes are abundant in the shallow, weedy, or relatively turbid bays of the same large lakes quite discrete from brook trout habitat. In small streams and lakes, walleyes and brook trout do not usually coexist whereas in the larger streams, the latter species is most often relegated to the colder headwaters while the walleye occurs in the warmer and more turbid downstream regions.

Walleyes are even less prone to co-occur with the brook stickleback, the two dace (*Phoxinus* spp.) found in small boreal forest lakes, and many other cyprinid species. Brook stickleback and dace are common occupants of small bog lakes or streams, habitat not likely conducive to walleye reproductive success. Predation by both walleyes and northern pike may preclude these species from the larger waters to which they have contiguous access.

Food and Feeding Habits

The food habits of adult walleyes have been rather thoroughly documented in the primary scientific literature. A general review of this literature (Eschmeyer 1950; Colby et al. 1978) revealed that the walleye is highly piscivorous but opportunistic in both its selection of forage fishes as well as invertebrate foods. Expediency in food selection is aptly demonstrated when one considers the predatory role played by the walleye in the various fish communities in which it occurs (Table 2). Thus within a centrarchid community inhabiting a eutrophic lake, walleyes tend to feed on the predominant available forage species, namely, centrarchids (Priegel 1963). In oligotrophic lakes dominated primarily by a salmonid complex and associated species, walleyes feed, for the most part, on the ninespine stickleback, sculpins, *Cottus* spp., coregonines, and two species of suckers, *Catostomus* spp. (Clemens et al. 1923; Rawson 1951). The invertebrate food also reflects the oligotrophic nature of the system as *Mysis relicta* is of considerable importance as a food item besides the usually ingested Ephemeroptera nymphs and *Gammarus*. In general, however, walleyes occurring in oligotrophic lakes tend to occupy the more eutrophic littoral or deltaic zones where food items may be representative of mesotrophic conditions (Table 2).

The more characteristic circumstances we wish to describe herein involves the food habits of adult walleyes as part of a harmonic percid community inhabiting lakes in the mesotrophic range. This essentially covers the midrange of the species and undoubtedly includes most of the prime natural walleye waters in North America (e.g., Regier et al. 1969). These waters may be classified on a trophic basis by the morphoedaphic index (MEI): total dissolved solids/mean depth (Ryder et al. 1974). The MEI provides a ready means for determining the relative closeness of the study lakes to the mesotrophic condition. The four lakes under immediate consideration, Gogebic (46°30′N, 89°33′W), Falcon (49°51′N, 95°18′W), Shebandowan (48°42′N, 90°15′W), and Lake of the Woods (49°30′N, 94°60′W), occupy the mesotrophic range as indicated by their metric morphoedaphic indices— i.e. 9.7, 5.2, 4.6, and 10.5, respectively. An MEI of about 6.6 (metric) denotes a lake with classical mesotrophic characteristics (Ryder 1965) and the range bracketed by the four lakes indicates a relatively narrow

TABLE 2.—*Trophic range of lakes zoogeographically available to percid communities and walleyes. Optimum abiotic and biotic conditions for thriving percid communities fall in the mesotrophic range.*

Lake trophic level	Harmonic community	Secondary components	Walleye standing stocks (biomass)	Walleye production	Principal food items of walleyes
Ultra-oligotrophic	salmonine (*Salvelinus*)	Nil			
Oligotrophic	salmonid	percid	low	low	coregonines, ninespine stickleback, *Catostomus*, burbot, *Cottus*, *Mysis*
Mesotrophic	percid	salmonid, centrarchid	high	moderate	percids, coregonines, darters, *Catostomus*, trout-perch, darters, Ephemeroptera, *Gammarus*
Eutrophic	centrarchid	percid, cyprinid	low to high	high	centrarchids, cyprinids
Ultra-eutrophic	cyprinid (*Cyprinus*)	Nil			

spread of conditions from the most oligotrophic in nature (Shebandowan) to the most nearly eutrophic (Lake of the Woods).

As indicated from stomach samples in each lake, fishes constituted substantial (Falcon Lake—58%) to major (Lake of the Woods—87%) percentages of the total food items ingested. As the remaining food items were chiefly arthropods and, in particular, Ephemeroptera, the proportional contribution of fish food on a volumetric, mass, or energetic basis (e.g., Kelso 1973) to the nutrition of walleyes must be substantial. Table 3 ranks the major species of fishes consumed by walleyes in these four mesotrophic lakes based on frequency of occurrence.

Fish Food Items

YELLOW PERCH

No single fish species predominated as a walleye food item in all four lakes. The yellow perch was the most prevalent food item in walleye stomachs on a frequency of occurrence basis in three of the lakes but second to *Coregonus* sp. in Shebandowan Lake (Table 3). It should be noted that Shebandowan Lake is the most nearly oligotrophic of the four lakes (MEI = 4.6) and the cold-water habitat (deep enough for the occurrence of dimictic thermal stratification) occupies about one-half the total lake area. Yellow perch are abundant in some of the shallower bays but are generally scarce throughout the main lake basin. Even so, the crepuscular forays of walleyes take them into the shallower bays where the perch are abundant and more vulnerable to predation with decreasing light intensities (Ryder 1977). Hasler and Villemonte (1953) have documented diel movements of schools of yellow perch from greater to lesser depths just prior to sundown. Upon reaching the littoral, the fish settled onto the substrate supported by their pectoral fins. A reverse movement to deeper and limnetic waters took place at daybreak. One of us (RAR) has collected yellow perch at night in the shallows while scuba diving in Shebandowan Lake merely

TABLE 3.—*Fish food items from walleye stomachs ranked according to frequency of occurrence and including only those items common to all four lakes. These data were abstracted from Eschmeyer (1950), Fedoruk (1966), present study, and Swenson and Smith (1976), respectively, for Gogebic Lake, Falcon Lake, Shebandowan Lake, and Lake of the Woods.*

Taxon	Gogebic Lake	Falcon Lake	Shebandowan Lake	Lake of the Woods	Overall rank
Yellow perch	1	1	2	1	I
Coregonus sp.[a]	5	0[b]	1	2	II
Ninespine stickleback	2	5	2	0	III, IV
Stizostedion sp.[c]	8	2	4	5	III, IV
Cyprinidae	3	4	0	3	V
Trout-perch	6	0	3	4	VI
Burbot	0	0	3	6	VII
White sucker	8	2	0	0	VIII
Etheostomatinae	8	0	0	5	IX, X
Smallmouth bass	0	3	0	0	IX, X
Northern pike	7	0	0	0	XI
No. sampled	196	288	258	1,270	

[a] *Coregonus* sp. was chiefly cisco although lake whitefish were occasionally encountered in stomach samples.
[b] 0 denotes the occurrence of the species in the lake but not in stomach samples of walleyes.
[c] *Stizostedion* sp. is walleye except in Lake of the Woods where sauger occurred as a food item as well.

by enveloping them in his hand as they rested in contact with the substrate or on submerged macrophytes (see also Emery 1973). Cruising, and presumably feeding, walleyes were observed in the same area, which suggests extreme vulnerability of yellow perch in this situation. Walleyes captured in the same vicinity with gill nets had substantial numbers of young-of-the-year perch in their stomachs. Walleyes have a subretinal tapetum lucidum which provides them with an ability to see and feed at low light intensities (Scherer 1961; Ali et al. 1977; Ryder 1977), plus their recurved canine teeth (Scott and Crossman 1973) make them admirably adapted to searching for and capturing resting yellow perch under low levels of subsurface illumination. The greatest feeding rates occur during the twilight periods of dawn and sunset (Ryder 1977), when the concurrent movements of yellow perch and walleyes onto shoal areas, and decreased activity of the yellow perch and increased activity of walleyes suggest a predator-prey relationship close to optimum efficiency—i.e. a maximum mass of food obtained per unit expenditure of energy over a relatively short time span.

CISCO

The food item of second average importance for walleyes in the four lakes was *Coregonus* spp., but more specifically ciscos comprised the bulk of this food item (Table 3). As ciscos are generally known to be limnetic plankton feeders (Pritchard 1931), restricted to offshore waters at a depth consistent with their relatively cool thermal requirements, then it must be suspected that adult walleyes feed rather deeply into the limnetic zone, at least intermittently. Underwater observations in Shebandowan Lake revealed that adult walleyes for the most part occupied littoral areas down to the upper strata of the metalimnion. Twenty-four hour gill-net sets, however, showed that at least occasional forays are made even into the hypolimnion following vernal thermal stratification. This factor coupled with the knowledge that ciscos migrate into the hypolimnion when surface water temperatures rise above 20 C (Fry 1937) would suggest a concurrent but intermittent diurnal regulation of spatial occupancy during most of the summer months for both adult walleyes and ciscos, a condition that subjects the latter species to walleye predation. Emery (1973) noted that juvenile ciscos occurred commonly at night in shallow waters but adults were rare. Walleyes are also taken abundantly with ciscos in spring, early summer, and again in autumn in Lake Nipissing, and Fry (1937) noted large walleyes feeding on fingerling ciscos in August, just above the thermocline. Walleyes and ciscos, therefore, may coexist at least intermittently in the same space for much of the year. During the summer months, however, after thermal separation of these species occurs, the eurythermal characteristic of the walleye enables it to invade the hypolimnion habitat of the cisco, which in turn may sporadically occupy a portion of the walleye habitat at the top of the metalimnion. Thermal stratification, consequently, does not effectively isolate predator from prey during summer months in this instance.

NINESPINE STICKLEBACK

The ninespine stickleback is the food item of third average importance for the walleye in the four lakes (Table 3). However, for two of the lakes, Gogebic and Shebandowan, it is of secondary importance but only ranks tenth in Lake of the Woods. Diving observations in Shebandowan have shown ninespine sticklebacks to be near bottom in the deeper waters in general, and below the top of the metalimnion (7 m) during the late summer thermal stratification period. Ninespine sticklebacks were usually highly visible to divers against a light sandy bottom where they were, no doubt, vulnerable to walleye predation. Little is known of the crepuscular or nocturnal behaviour or habitat of ninespine sticklebacks except that Emery (1973) noted they could be closely approached at night despite the fact they tended to be flighty during daylight hours. Later in the summer, large aggregations of ninespine sticklebacks were often observed 10 to 15 m off bottom in open water. The ninespine stickleback, therefore, is vulnerable to walleye predation at almost any time, although effectively only when the walleye penetrates the hypolimnion or occupies the open limnetic zone.

WALLEYE

Young-of-the year, yearling, and subadult walleyes contributed substantially to all food items ingested (Table 3). Many observers have noted young-of-the-year walleyes schooling with young-of-the-year yellow perch (e.g., Maloney and Johnson 1957), the principal food of both fingerling and adult walleyes. The vulnerability of young-of-the-year walleyes to predation from adults undoubtedly is increased by their shoaling association with the primary food item, yellow perch. As walleyes demonstrate a change from positive to negative phototaxis during their first year of life (Moore 1944), they move into deeper water (Raney and Lachner 1942) where they become still more exposed to the feeding forays of adult walleyes, an occurrence due to both coincidental co-occupation of the same space and because of morphological changes in the retina in the young walleyes. The diel activity pattern of the young walleyes converges with that of the adult making the former more vulnerable to predation in both the time and space dimensions.

Cannibalism among walleyes has been previously documented (Chevalier 1973) and dates from almost first feeding throughout its life. Diving observations (RAR) have shown that active schools of walleyes in Shebandowan Lake were constituted of sizes ranging from yearlings to the largest of adults. On occasion the peduncle and caudal fin of a yearling walleye was seen protruding from the mouth of a large adult. It would seem that coexistence in the same school of walleyes does not exempt a smaller fish from becoming prey to a larger fish. Yet if this were of frequent occurrence it would be expected that such schools would consist only of large walleyes. However, many yearling walleyes during daylight hours do not join the schools of larger adults but rather secrete themselves on the substrate under convenient boulders or sunken trees, no doubt safe from predation in this behaviour mode. It must be expected, then, that most yearling walleyes are cannibalized during crepuscular feeding forays when body vibrations may attract larger fish during periods of subsurface light attenuation.

Forney (1974) has proposed a rationale for walleye cannibalism. He noted that young yellow perch were proportionately most abundant in Oneida Lake as well as in walleye stomach samples where they "were consistently selected by walleyes." Cannibalism by older walleyes increased during periods of low levels of yellow perch abundance—a suggestion that yellow perch act as a buffer species controlling the degree of cannibalism. Hence, indirectly, the abundance of young yellow perch controls the size of future year classes of walleyes through the negative feedback mechanism of cannibalism.

CYPRINIDAE

Minnows of many species are often common fishes of north-temperate boreal lakes and sometimes provide a substantial proportion of the forage base (Scott and Crossman 1973). They are abundant in Lake of the Woods but extremely scarce in Shebandowan and consequently are the food item of third importance in the former lake but were not encountered in the stomach samples of walleyes from Shebandowan Lake (Table 3). North American cyprinid ethology is not well documented but there are temperate species that may be active diurnally, nocturnally, or during dusk hours (Emery 1973) and consequently at least some species may be vulnerable to walleye predation through coincidental occupation of space. Forage cyprinids of the north-temperate lakes are usually small, schooling fishes that feed on diatoms, algae, small zooplankters, or insects (Scott and Crossman 1973), and occupy a diversity of habitat usually above the metalimnion. Their affinity for ecotones, either shoreline, substrate, underwater structures, or water surface, concentrate schooling cyprinids in an extremely vulnerable position with respect to predation. It is likely that their relatively low levels of abundance in boreal forest lakes when compared to young-of-the-year yellow perch or cisco, accounts for the fifth-place ranking of cyprinids among all fish food items encountered in walleye stomachs.

TROUT-PERCH

Trout-perch likely possess a retinal tapetum lucidum, and are believed to be crepuscular and nocturnal feeders (Scott and Crossman 1973). They form an obvious portion of the inshore fauna at night (Hubbs and Lagler 1947), and McPhail and Lindsey (1970) have documented large inshore movements which would coincide to some degree with the diel activity pattern of the walleye, the principal predator of the trout-perch in the Red Lakes, Minnesota, according to Magnuson and Smith (1963). The preferred food of the trout-perch is benthos, particularly chironomid larvae, ephemeropterans, and amphipods (Clemens et al. 1923; Price 1963), which, together with the proclivity of trout-perch for inshore movements at dusk suggests a coincidental occupation of space with foraging walleyes. Trout-perch, however, constituted only a small proportion of walleye food in the four study lakes. This condition may be attributed to a reduced activity on the part of both the predator and the prey. Trout-perch are known to make post-dawn movements to deeper waters in transparent lakes (Scott and Crossman 1973) or a general retreat to a boulder or sunken log shelter in shallower waters (RAR underwater observation). Both these stratagems tends to protect the trout-perch from high levels of subsurface illumination. Like the walleye, the possession of a tapetum in the trout-perch likely gives it nocturnal advantage in avoiding predators but becomes a handicap in clear water lakes necessitating retreat from high daytime light intensities.

BURBOT

Burbot formed an insignificant overall portion of the food items ingested by adult walleyes in the four study lakes although it was of tertiary importance in Shebandowan Lake, where 100% of those ingested were advanced young-of-the-year and yearling burbot. Extensive underwater observations on burbot in Shebandowan Lake (Ryder and Varmo 1978) revealed that advanced young-of-the-year and yearling burbot were secreted in the shallows (0.5–4 m) beneath stones, bark, sunken trees, or other substrate structure by day and became active only at night. Nocturnal forays consisted of food searches from boulder to boulder on the bottom and often covered several metres of barren sand between physical structures. In the latter instance young burbot would be particularly vulnerable to walleye forays as both species occupy the same space simultaneously. Indeed, adult walleyes were often seen at night foraging in the vicinity of young burbot, the latter feeding in a heads-down position near a stone or sunken log. The food of young burbot (late young of the year and yearlings) is almost entirely invertebrate and benthic in nature (Robins and Deubler 1955; Ryder and Varmo 1978). Where sunken structures were in close proximity as on a boulder shoal, it is unlikely that young burbot would be vulnerable to predation by cruising walleyes because of proximate shelter. The burbot's tendency to be nocturnally efficient would provide it with survival value not only for feeding efficiency but also for the avoidance of predation.

WHITE SUCKER

The white sucker, despite its general occurrence in all four study lakes, was not a frequently ingested food item except in Falcon Lake where it was second only to the yellow perch. Divers in Shebandowan Lake have rarely encountered adult white suckers, except during spawning, which suggests the species possesses an "early warning device" perhaps through its acoustico-lateralis system or Weberian apparatus, the latter particularly well-developed in ostariophysids (Lagler et al. 1962).

Fish Feeding Behaviour

In reconsideration of the two fish foods most frequently ingested by walleyes in this study, namely the yellow perch and cisco, it is of importance to note that during summer stratification the spatial segregation of these two species was distinct. It has been suggested by Engel and Magnuson (1973) that spatial segregation of yellow perch and cisco in northern lakes of the United States offered a unique opportunity to avoid direct interaction and to further divide the available resources of the lake. The physiological capabilities of the walleye have allowed it to forage in all three thermal zones of stratified lakes and consequently to feed opportunistically on both the perch and cisco and indeed on all fish species of appropriate size for ingestion in mesotrophic lakes. Those species most effectively escaping walleye predation are the shelter-seekers, especially the etheostomatines, sculpins, and young-of-the-year northern pike and smallmouth bass. Trout-perch and young burbots tend to use large boulders or sunken logs for diurnal shelter also but their nocturnal forays no doubt expose them to a modicum of walleye predation. In summary, it would seem that adult walleyes will prey fortuitously on any species of fish of suitable size with which it co-occurs on a time-space basis. Ambient temperature and oxygen variability in north-temperate mesotrophic lakes does not seem to inhibit this opportunistic feeding behaviour but subsurface illumination constraints are a primary consideration in both the time and space dimensions (Ryder 1977). The occurrence of the walleye, therefore, is complementary to those of the mainly diurnal piscivores such as the northern pike and lake trout because it partitions food resources temporally in such a way as to provide for optimum utilization and hence maximum production from the system (e.g., Paloheimo and Dickie 1970).

Invertebrate Foods

Almost all food studies of adult walleyes list invertebrate foods as of major importance. In the Shebandowan Lake study, invertebrates occurred in about one-third of the stomachs examined and of these the sub-imagoes of Ephemeroptera were by far the most abundant. These latter constitute a seasonal food during June and July of almost universal importance for adult walleyes in boreal forest lakes. The catastrophic decline of walleyes in western Lake Erie has, on occasion, been attributed to the equally drastic decline of the formerly abundant *Hexagenia*, at one time an important food item of the walleye there, although this supposition has never been formally proven (Regier et al. 1969).

Crustaceans occurred in about 7% of the stomachs examined and *Mysis* was the most frequent of these. Individual stomachs were occasionally packed with up to 50 or more *Mysis* following autumnal overturn in October. Summer samples contained but few *Mysis* in individual stomachs, a suggestion that walleyes made somewhat rapid and brief feeding forays through the metalimnion into the colder waters of the hypolimnion where *Mysis* and ninespine sticklebacks were preyed upon.

Crayfish were abundant in Shebandowan Lake, but were not often ingested, although their habit of emerging from shelter at night (Crocker and Barr 1968) made them particularly vulnerable to walleye predation. It must be assumed, however, that because of some behavioural anomaly of the prey (or perhaps the predator), crayfish cannot normally be considered as a prime forage organism for walleyes. Conversely, this condition may have evolved from niche partitioning (e.g., Schoener 1974) with the adult smallmouth bass and burbot, both of which prey heavily on adult crayfish in Shebandowan Lake. Even here, a time-and-space partitioning occurs, with the adult burbot preying on crayfish almost exclusively at night on deeper shoals, and the smallmouth bass ingesting crayfish normally during daylight or crepuscular periods in the shallower littoral zone.

In defining the food dimension of a niche, it is informative to consider items of food that are abundant but not foraged (e.g., Ivlev 1961). Four rather large invertebrate organisms in Shebandowan fall into this category. Each of these organisms in turn constitutes a substantial portion of the total animal biomass in the littoral areas of the lake. Three of these organisms are molluscs—2 large gastropods (shell length >4 cm) and 1 pelycopod (shell length ca. 8 cm). None of these has been observed in a walleye stomach in Shebandowan Lake despite their dense concentrations on the littoral substrate. It may be concluded that the behavioural characteristics of either predator or prey are not conducive to predation even when they occupy the same space simultaneously, or conversely, that either the calcareous shells or pheromones of these molluscs offer suitable protection from the walleye, and render them inappropriate as a food item.

Finally, the most obviously abundant invertebrate foods occupying the substrate and limnetic zones of Shebandowan Lake, which clearly are potentially vulnerable to fish predation, particularly when swimming with their slow undulating movements, are the leeches (Hirudinea). These were considered to be the least desirable walleye food relative to their abundance and conspicuousness. Only three leeches (unidentified to genus) were taken from one walleye stomach out of a total of 258 stomachs examined. Large horse leeches, *Haemopsis*, were particularly abundant in the lake and were often in excess of 20 cm in length. However, horse leeches were not found in the stomachs of smallmouth bass either. Divers who were able to feed crayfish to adult smallmouth bass by hand were unable to entice the same fish to taste a *Haemopsis*. Therefore, despite its potential vulnerability to predation, *Haemopsis* is not a selected food in Shebandowan Lake for walleyes, presumably because of an abundance of food of greater preference or perhaps as the result of an avoidance reaction to *Haemopsis* secretions. Other species of leeches are often consumed in numbers by both smallmouth bass and walleyes elsewhere, however (Eschmeyer 1950).

Coherence of Percid Communities

Clearly, regular and definitive patterns of organization exist in percid communities, as assessed in the present instance relative to the walleye. The patterns we have emphasized, specifically the niche dimensions pertaining to food selection, behaviour, and environmental heterogeneity, in general act to limit the range of interactions observable within percid communities. We have elsewhere raised the view that such limits are essential to the persistence and integrity of natural communities, in the sense that system connectance is reduced to stable levels by the provision of such niche refugia (Kerr and Ryder 1977).

The cline of lake trophic status (Fig. 1) represents an aggregated index of a number of individual niche parameters. Relative to this cline, we find that the persistent communities in the harmonic region are characterized by clear mechanisms allowing interactions among species in both time and space, but limiting the interactions between communities. The extreme limbs of the percid mesotrophic cline, on the contrary, are characterized by less typical organizational patterns leading, in our view, to community structures of reduced persistence and integrity.

Our assessment has been qualitative in large part. That is, we have concentrated on the mechanisms which serve to maintain community connectance at stable levels. It is implicit, however (Kerr and Ryder 1977), that within the cline of mesotrophic variables, the individual metabolic capacities of the organisms making up percid communities reach their maxima in the harmonic region. In essence, intermediate mesotrophy optimizes both the individual niche capacities of percid community components and the structural organization of the community itself.

In this sense we have recognized both an individual niche, characterized in terms of individual metabolic capacities, and a community niche defined in terms of the emergent properties of integral communities. The harmonic region of the percid distribution superimposed on lake trophic status represents one such realization of a community niche (Fig. 1). Niche analysis at the community level, in the present terms, seems to us potentially profitable either in terms of ordinary analysis of variance or, perhaps most appropriately, by spectral analysis (Platt and Denman 1975), which partitions community variability into its natural harmonics (seasonal, diel, etc.) and its associated astatic components.

Acknowledgments

We appreciate the scientific review provided by A. H. Lawrie, J. H. MacLean, and H. A. Regier. We are indebted to J. M.

Casselman for presenting our contribution to the Symposium on Selected Coolwater Fishes of North America.

References

ADAMS, G. F., AND C. H. OLVER. 1977. Yield properties and structure of boreal percid communities in Ontario. J. Fish. Res. Board Can. 34:1613–1625.

ALI, M. A., AND M. ANCTIL. 1977. Retinal structure and function in the walleye (*Stizostedion vitreum*) and sauger (*S. canadense*). J. Fish. Res. Board Can. 34:1467–1474.

———, R. A. RYDER, AND M. ANCTIL. 1977. Photoreceptors and visual pigments as related to behavioral responses and preferred habitats of perches (*Perca* spp.) and pikeperches (*Stizostedion* spp.). J. Fish. Res. Board Can. 34:1475–1480.

CHEVALIER, J. R. 1973. Cannibalism as a factor in first year survival of walleye in Oneida Lake. Trans. Am. Fish. Soc. 102:739–744.

CLEMENS, W. A., J. R. DYMOND, N. K. BIGELOW, F. B. ADAMSTONE, AND W. J. K. HARKNESS. 1923. The food of Lake Nipigon fishes. Publ. Ont. Fish. Res. Lab. 16:173–188.

COLBY, P. J., R. E. McNICOL, AND R. A. RYDER. 1978. Synopsis of biological data for the walleye, *Stizostedion vitreum vitreum*. FAO (F.A.O.U.N.) Fish. Synop. (In prep.)

COLLETTE, B. B., M. A. ALI, K. E. F. HOKANSON, M. NAGIEC, S. A. SMIRNOV, J. E. THORPE, A. H. WEATHERLEY, AND J. WILLEMSEN. 1977. Biology of the percids. J. Fish. Res. Board Can. 34:1890–1899.

CROCKER, D. W., AND D. W. BARR. 1968. Handbook of the crayfishes of Ontario. Roy. Ont. Mus., Univ. Toronto Press, Toronto. 158 pp.

EMERY, A. R. 1973. Preliminary comparisons of day and night habits of freshwater fish in Ontario lakes. J. Fish. Res. Board Can. 30:761–774.

ENGEL, S., AND J. J. MAGNUSON. 1973. Spatial interactions and physico-chemical relationships among fishes in Pallette Lake, Wisconsin. Limnol. Lab. Univ. Wis. 34 pp. (Unpublished.)

ESCHMEYER, P. H. 1950. The life history of the walleye, *Stizostedion vitreum vitreum* (Mitchill), in Michigan. Mich. Inst. Fish. Res. Bull. 3. 99 pp.

FEDORUK, A. N. 1966. Feeding relationship of walleye and smallmouth bass. J. Fish. Board. Can. 23:941–943.

FORNEY, J. L. 1974. Interactions between yellow perch abundance, walleye predation, and survival of alternate prey in Oneida Lake, New York. Trans. Am. Fish. Soc. 103:15–24.

FRY, F. E. J. 1937. The summer migration of the cisco, *Leucichthys artedi* (Le Sueur), in Lake Nipissing, Ontario, Publ. Ont. Fish. Res. Lab. 55. 91 pp.

———. 1947. Effects of the environment on animal activity. Univ. Toronto Stud. Biol. Ser. 55, Publ. Ont. Fish. Res. Lab. 68:1–62.

———. 1971. The effect of environmental factors on the physiology of fish. Pages 1–98 *in* W. S. Hoar and D. J. Randall, eds. Fish physiology, vol. VI. Academic Press, New York.

HASLER, A. D., AND JAMES R. VILLEMONTE. 1953. Observations on the daily movements of fishes. Science 118 (3064):321–322.

HUBBS, C. L., AND K. F. LAGLER. 1947. Fishes of the Great Lakes Region. Univ. Mich. Press, Ann Arbor. 213 pp.

HUTCHINSON, G. E. 1957. Concluding remarks. Cold Spring Harbor Symp. Quant. Biol. 22:415–427.

———. 1965. The ecological theater and the evolutionary play. Yale Univ. Press, New York. 139 pp.

IVLEV, V. S. 1961. Experimental ecology of the feeding of fishes. Yale Univ. Press, New Haven. 302 pp.

JOHNSON, F. H., AND J. G. HALE. 1977. Interrelations between walleye (*Stizostedion vitreum vitreum*) and smallmouth bass (*Micropterus dolomieui*) in four northeastern Minnesota lakes, 1948–69. J. Fish. Res. Board Can. 34:1626–1632.

KELSO, J. R. M. 1973. Seasonal energy changes in walleye and their diet in West Blue Lake, Manitoba. Trans. Am. Fish. Soc. 102(2):363–368.

KERR, S. R. 1974. Structural analysis of aquatic communities. Proc. 1st. Int. Cong. Ecol.: 69–74.

———. 1977. Structure and transformation of fish production systems. J. Fish. Res. Board Can. 34:1989–1993.

———, AND R. A. RYDER. 1977. Niche theory and percid community structure. J. Fish. Res. Board Can. 34:1952–1958.

LAGLER, K. F., J. E. BARDACH, AND R. R. MILLER. 1962. Ichthyology. John Wiley and Sons, New York. 545 pp.

MAGNUSON, J. J., AND L. L. SMITH JR. 1963. Some phases of the life history of the trout-perch. Ecology 44(1):83–95.

MALONEY, J. E., AND F. H. JOHNSON. 1957. Life histories and inter-relationships of walleye and yellow perch, especially during their first summer in two Minnesota lakes. Trans. Am. Fish. Soc. 85: 191–202.

McPHAIL, J. D., AND C. C. LINDSEY. 1970. Freshwater fishes of northwestern Canada and Alaska. Bull. Fish. Res. Board Can. 173. 381 pp.

MOORE, G. A. 1944. The retinae of two North American teleosts, with special reference to their tapeta lucida. J. Comp. Neurol. 80(3):369–379.

PALOHEIMO, J. E., AND L. M. DICKIE. 1970. Production and food supply. Pages 499–527 *in* J. H. Steele, ed. Marine food chains. Oliver and Boyd, Edinburgh.

PLATT, T., AND K. L. DENMAN. 1975. Spectral analysis in ecology. Annu. Rev. Ecol. Syst. 6:189–210.

PRICE, J. W. 1963. A study of the food habits of some Lake Erie fish. Bull. Ohio Biol. Surv. 11(1):1–89.

PRIEGEL, G. R. 1963. Food of walleye and sauger in Lake Winnebago, Wisconsin. Trans. Am. Fish. Soc. 92:312–313.

———. 1969. The Lake Winnebago sauger. Age, growth, reproduction, food habits and early life history. Wis. Dep. Nat. Resour. Tech. Bull. 43. 63 pp.

PRITCHARD, A. L. 1931. Taxonomic and life history studies of the ciscoes of Lake Ontario. Publ. Ont. Fish. Res. Lab. 41. 78 pp.

RANEY, E. C., AND E. A. LACHNER. 1942. Studies of the summer food, growth, and movements of young yellow pike-perch, *Stizostedion v. vitreum*, in Oneida Lake, New York. J. Wildl. Manage. 6(1):1–16.

RAWSON, D. S. 1951. Studies of the fish of Great Slave Lake, J. Fish. Res. Board. Can. 8:207–240.

REGIER, H. A., V. C. APPLEGATE, AND R. A. RYDER. 1969. The ecology and management of the walleye in western Lake Erie. Great Lakes Fish. Comm. Tech. Rep. 15:101 pp.

ROBINS, C. R., AND E. E. DEUBLER JR. 1955. The life history and systematic status of the burbot, *Lota lota lacustris* (Walbaum), in the Susquehanna River system. N.Y. State Mus. Sci. Serv. Circ. 39. 49 pp.

RYDER, R. A. 1965. A method for estimating the potential fish production of north-temperate lakes. Trans. Am. Fish. Soc. 94(3):214–218.

———. 1972. The limnology and fishes of oligotrophic glacial lakes in North America (about 1800 A.D.). J. Fish. Res. Board Can. 29:617–628.

———. 1977. Effects of ambient light variations on behavior of yearling, subadult, and adult walleyes (*Stizostedion vitreum vitreum*). J. Fish. Res. Board Can. 34:1481–1491.

———, S. R. KERR, K. H. LOFTUS, AND H. A. REGIER. 1974. The morphoedaphic index, a fish yield estimator—review and evaluation. J. Fish. Res. Board Can. 31:663–688.

———, AND R. VARMO. 1978. Behavior, growth and food habits of young-of-the-year burbot (*Lota lota*) in a boreal forest lake. (In prep.)

SCHERER, E. 1971. Effects of oxygen depletion and of carbon dioxide buildup on the photic behavior of the walleye (*Stizostedion vitreum vitreum*). J. Fish. Res. Board Can. 28:1303–1307.

SCHOENER, T. W. 1974. Resource partitioning in ecological communities. Science. 185:27–39.

SCOTT, W. B., AND E. J. CROSSMAN. 1973. Freshwater fishes of Canada. Bull. Fish. Res. Board Can. 184. 966 pp.

SIMON, H. A. 1969. The sciences of the artificial. M.I.T. Press, Cambridge, Mass. 123 pp.

SWENSON, W. A. 1977. Food consumption of walleye (*Stizostedion vitreum vitreum*) and sauger (*S. canadense*) in relation to food availability and physical conditions in Lake of the Woods, Minnesota, Shagawa Lake and western Lake Superior. J. Fish. Res. Board Can. 34:1643–1654.

———, AND L. L. SMITH JR. 1976. Influence of food competition, predation, and cannibalism on walleye (*Stizostedion vitreum vitreum*) and sauger (*S. canadense*) populations in Lake of the Woods, Minnesota J. Fish. Res. Board Can. 33:1946–1954.

WELCH, H. 1967. Energy flow through the major macroscopic components of an aquatic ecosystem. Ph.D. Thesis. Univ. Ga., Athens.

WILSON, D. S. 1975. The adequacy of body size as a niche difference. Am. Nat. 109:769–784.

An Hypothesis of Homing Behavior of Walleyes as Related to Observed Patterns of Passive and Active Movement

Donald E. Olson

Minnesota Department of Natural Resources
P.O. Box 823, Detroit Lakes, Minnesota 56501

Dennis H. Schupp

Minnesota Department of Natural Resources
Brainerd, Minnesota 56401

Val Macins

Ontario Ministry of Natural Resources
Kenora, Ontario P9N 3X9

Abstract

An hypothesis that walleye homing is an adult-learned behavior rather than a natal-imprinted response is presented. Marked adult walleyes tend to home to spawning areas. Individual walleyes tend to return to the same open-water feeding areas in successive years. Movement of immature walleyes often differs from that of adults in the same waters. Intensity of walleye homing varies in separate waters and appears to be influenced by physical characteristics of the environment and strengthened by repeated migrations. River and wind currents commonly move walleye eggs and fry great distances from the site of egg deposition before fry are sufficiently developed to commence feeding. This makes natal conditioning to spawning areas unlikely.

Review of literature pertaining to walleye movement shows that marked spawning walleyes are far more likely to be recaptured in subsequent years at the same spawning site than at other suitable spawning areas (Smith et al. 1952; Crowe 1962; Forney 1963; Olson and Scidmore 1962; Priegel 1968; Rawson 1957). Also, adult walleyes marked and released during open-water seasons (postspawning) are frequently recaptured near their release site during the year of marking and also in subsequent years (Forney 1963; Magnin and Beaulieu 1968; Schupp 1972). These observations suggest that some adult walleyes in certain populations migrate between home spawning and home feeding areas.

Walleye movement patterns in separate waters are not presented here in detail. Rather, we have synthesized an hypothesis for the development of walleye homing behavior which is consistent with observations of passive and active walleye movements.

Mixing of genetically discrete stocks, the result of decades of walleye propagation and redistribution, has not likely significantly modified primitive movement patterns. Transplantation studies of fingerling salmon have demonstrated that the homing mechanism is inherited, but not the memory of the characteristics of the homestream (Cooper et al. 1976).

Homing as used here for walleyes is defined by Gerking (1959) as "the choice that a fish makes between returning to a place formerly occupied instead of going to other equally probable places." It is not implied in this definition that homing is a return to either a natal area or to a home feeding area formerly occupied by the parent fish, though both unquestionably occur frequently in walleye populations.

Home spawning areas are the chosen locations for spawning. Home feeding areas are the chosen locations for feeding during open-water seasons and perhaps also during months of ice cover.

Home spawning areas generally are river inlets but include shoal areas in lakes where spawning is concentrated in successive years.

Home feeding areas may vary greatly in size, and those of individual fish may overlap, but they are in preference to other potentially suitable sites.

Homing to Spawning Areas

Evidence of homing among spawning walleyes has been obtained primarily by marking and releasing adult walleyes at spawning sites and then comparing the distribution of recaptured fish among these sites during subsequent spawning seasons. Such investigations have been carried on at widely separate geographic locations in North America and within waters which range in size from the Great Lakes to inland lakes of less than 8 km^2.

All marking studies reviewed which contained sufficient data for an analysis of spawning migrations reported tendencies for homing among spawning walleyes. Most tagged walleyes recovered in spawning runs in Upper and Lower Red Lakes, Minnesota, were fish originally marked at those locations (Smith et al. 1952). However, some walleyes tagged at two widely separated tributaries were recaptured in a third stream, which suggests that homing instinct may be weak and that fish may often go to the closest available spawning stream.

Crowe (1962) reported homing tendencies at spawning among all walleye populations investigated in Michigan. The Newaygo dam is a spawning site for walleyes of the Muskegon River system and the eastern shore of Lake Michigan. Of 7,155 marked walleyes released in the Muskegon River system during spawning, 1947–1961, 156 were recaptured at the dam site in subsequent spawning migrations. Marked walleyes were recaptured at widely dispersed locations in other seasons but none were reported at other sites during spawning seasons. In the inland waterways of Michigan's lower peninsula, 1,932 walleyes were tagged and released during spawning seasons, 1948–1961. During marking and in subsequent spawning seasons, 27 were recaptured at "home" spawning sites and 4 recaptured at other than their release site. In Bay De Noc, Lake Michigan, 5,155 walleyes were tagged during spawning, 1958–1961. Of 137 walleyes recaptured in nets after they had been free for a year or longer, 130 were recaptured at the point of release.

During six spawning seasons, 1955–1960, walleyes were marked and remarked with year-identifying fin clips at the Ottertail River inlet to Many Point Lake, Minnesota (Olson and Scidmore 1962). Walleye spawning runs at the Ottertail River inlet had a higher ratio of marked to unmarked fish than was found among adult walleyes captured during summer dispersal in Many Point Lake. The proportion of walleyes in spawning runs which had originally been marked at this site was 1.5 to 1.8 times greater than among walleyes over 46 cm, total length, in angler catches of the preceeding summer. There was a greater tendency to return for spawning walleyes which had used the site more often in previous years suggesting that repeated trips to the spawning area reinforced homing behavior or that walleyes with feeding areas nearer the spawning site returned more frequently. Total numbers of spawners captured in the inlet were greater in springs of higher water flow but the proportion of marked fish in spawning runs was greatest at low water flows. This suggests that those fish which had previously spawned at the inlet were better able to locate the area when water flow was low.

Walleyes were tagged during spawning in waters of the Fox and Wolf Rivers which join in Big Lake Butte des Morts 16 km above Lake Winnebago, Wisconsin (Priegal 1968). Twenty four percent of 322 walleyes tagged in the Fox River were recaptured in Lake Winnebago and the Fox River and 8 percent of 235 walleyes tagged in the Wolf River were recaptured from the Wolf River and downstream lakes but there was no known interchange between the Wolf and Fox river systems.

Only a small percentage of marked spawning walleyes are known to have strayed to other spawning areas but some interchange between spawning areas appears to be characteristic of walleye behavior. A change of spawning grounds within a single season was reported for two male walleyes in Oneida Lake (Forney 1963). Whitney (1958) reported extensive intermingling of spawners in the same season at two sites located 2.8 km apart in Clear Lake, Iowa.

Wide dispersal during summer months of walleyes marked in spawning runs is frequently reported. This indicates that fish from separate home feeding areas use a common spawning area. However, it is unlikely that walleyes which use a common spawning area are of genetically discrete stocks. This would require that fish from separate stocks recognize and mate only with others from their particular subpopulation, or that the separate stocks occupy spawning areas at different times in the spawning period. Such behavior is inconsistent with the observed differences in spawning migration patterns of males and females. Males remain on spawning areas for extended periods but females move out shortly after spawning (Eschmeyer 1950; Whitney 1958; Forney 1963; Rawson 1957).

Rawson (1957) suggested that discrete stocks spawned in two separate tributaries to Lac La Ronge. Adult walleyes in spawning runs of the Montreal River were considerably smaller in size than those in the Potatoe River. Recaptures of walleyes tagged in the two river systems indicated that little intermingling occurred during summer months. Most postspawning recaptures were taken within 3.2 km of their respective spawning site.

Home Feeding Areas

Though some adult walleyes range great distances to and from spawning areas, movement following postspawning migrations does not appear to be extensive and some walleyes appear to select the same general location for feeding in successive years.

During trapping and electrofishing along a 4-km section of shoreline in Oneida Lake, fall 1957, Forney (1963) noted a steady increase in recaptures as fin-clipping progressed which would not be expected to occur in a 207-km^2 lake had the walleyes moved extensively soon after marking. When the entire shoreline area of the lake was sampled again in fall 1958, 44 fin-clipped walleyes were recaptured of which 34 were taken in an area roughly corresponding to the release site. Fin-clipped walleyes from this group were also recaptured at two distant spawning sites during the interim spring; it was assumed that many had traveled great distances to spawn and then returned to a "home" area.

Oneida Lake anglers reported the locations of recapture of 63 of 301 walleyes tagged at four separate locations during fall 1957. Most recoveries were made in the area of the lake where they had been marked. Many were assumed to have migrated to spawning sites during the spring between marking and recapture.

At 450-km^2 Leech Lake, Minnesota, walleyes were tagged and released at widely scattered trawling stations during postspawning seasons of 1966 and 1967 (Schupp 1972). The distribution of walleyes recaptured in Leech Lake was reexamined in greater detail for this report. Of 2,917 walleyes marked with numbered tags, 617 returned by angling or trawling were identified as to their locations of recapture. Walleyes marked at separate stations intermingled but recaptures in the year of marking and in subsequent years tended to be near the locations at which they had been marked. Sixty-eight percent of 148 fish recaptured in the year of marking were taken less than 3.2 km from their marking site. The percentages of walleyes recaptured less than 3.2 km from their marking site in years subsequent to marking were 63% after 1 year, 48% after 2 years, and 50% recaptured 3 or more years after tagging.

There was greater dispersal from their marking sites among walleyes recaptured during May and June than among those recaptured during July through October. Greater dispersal during the earlier period is likely attributable to the recapture of some fish which had not yet completed postspawning migrations. Fifty-three percent of 348 walleyes recaptured during May and June were within 3.2 km of the trawling station at which they had been marked as compared with 68 percent of 269 fish recaptured during July through October.

That Leech Lake walleyes tend to occupy the same feeding areas in successive summers is evident when recaptures by trawling are considered separately from angler returns. A total of 58 tagged walleyes were recaptured in repeated trawling of marking stations and all were at the same stations where they were first captured and released. Twenty-four had been at large during one interim spawning season and two had been out for two springs before they were recaptured.

Little tendency for long distance movement was shown for walleyes tagged in the Saint Lawrence River from 1954 to 1958 (Magnin and Beaulieu 1968). Of 479 walleyes tagged, 124 were recaptured and 89 were taken less than 16 km from the tagging point.

Drift of Eggs and Fry

Unlike salmon which deposit their eggs in gravel redds in streams, spawning walleyes broadcast their eggs in river currents or on windswept lake shoals. Consequently, walleye eggs may be transported great distances from the site of deposition during the period of embryo development. Emerging fry, though motile, are also subject to transport by wind and river currents and typically commence feeding in the upper portion of the limnetic zone frequently far from the site at which they were spawned.

Houde and Forney (1970) demonstrated that newly hatched fry (prolarvae) drift passively for about four days in bottom currents of Oneida Lake. Postlarvae were able to maintain a position in the upper 3 m of the water column and were transported by surface wind currents. Fry were not capable of regulating their distribution until one to two weeks after hatching. Stocked walleye fry were concentrated in bays of the western part of the lake and along the south shore eight to ten days after stocking where they remained until their transition to a demersal existence at 25–30 mm total length.

Movements of Juveniles

Movement patterns of fingerling and immature walleyes are as yet poorly defined, but frequently differ from those of adults. Walleye fingerlings in Lake Winnibigoshish and connected waters occupied shallow water, 0.3 to 1.2 m depth, during July and early August, while older walleyes were most abundant in water between 1.2 and 3.7 m deep. In late August and September, walleyes of all ages tended to mix at depths between 1.5 and 4.6 m. During hot weather in August, age I and II fish remained mostly within the 4.6-m contour but older walleyes were taken in deeper water (Johnson 1969).

Movement patterns of marked immature walleyes in Great Lake investigations (Ferguson and Derksen 1971; Wolfert 1963) suggest a rapid dispersal from markings sites. A clearly divergent movement between juveniles and adults during spring (postspawning) and summer was indicated in Lake St. Clair and western Lake Erie. Juveniles tended to remain in Lake St. Clair or move downstream into western Lake Erie while adults tended to move upstream into the St. Clair River and south-Lake Huron. Western Lake Erie and its south shore appear to be a mixing area for juveniles from Lake Erie and Lake St. Clair while adults move easterly along Lake Erie's north shore during summer, but tend to return to the western basin for spawning.

There was no discernable summer homing pattern shown in the locations of recapture by anglers of 260 walleyes from 1,703 mostly immature walleyes marked in trawl hauls in Cass Lake, Minnesota, 1971–1973 (Robert Strand, Minnesota Department of Natural Resources, personal communication).

Tagged walleyes, which were mostly immature fish, recaptured by anglers in Lake Winnibigoshish, Minnesota, were also widely dispersed from the trawling locations at which they had been tagged (Thomas Osborn, Minnesota Department of Natural Resources, personal communication). At Leech Lake, however, tagged immature walleyes showed a pattern of recapture similar to that of adults which indicated that they also tended to remain near a home feeding area during summer months (Schupp 1972).

Development of Walleye Homing

Walleye homing behavior appears to be weakly developed compared with that of Pacific salmon. We propose that walleye homing is an adult-learned behavior which is more strongly displayed by fish with home feeding areas near a particular spawning site and/or which is reinforced by repeated migrations. Conditions for the short-lived Pacific salmon which likely provided an evolutionary advantage for a precise homing behavior are not the same for walleyes. That is, homing of salmon to natal streams likely evolved because of the small area of suitable spawning habitat relative to the immense marine area available for population expansion. Suitable spawning habitat is not so restricted for the longer-lived walleye which broadcasts its eggs and utilizes both river and lake-shoal environments. It is further argued that natal imprinting of walleyes is unlikely considering the early period in larval development at which they must be conditioned to the spawning area. The homing behavior of salmon, which hatch and commence feeding in their natal stream, appears to become established shortly before they migrate to sea. Fingerling salmon reared in one stream but transplanted to a second stream shortly before migration homed to the second stream to spawn (Donaldson and Allen 1958; Peck 1970; Jensen and Duncan 1971).

Adult walleye migrations to home feeding areas is also considered to be learned. Walleyes occupying separate home feeding areas are not necessarily discrete genetic stocks though discrete stocks may exist where distance or barriers serve to isolate both spawning and feeding areas. Spawning migrations from home feeding areas are likely to the nearest detectable spawning site in most instances, even though migrations of great distances apparently occur.

The passive movement of walleye eggs and fry is largely determined by river and wind currents and the hydrographic features of the lake. Juvenile movement is also likely influenced by currents and partial barriers. The destination of a walleye's initial spawning migration may well be determined by its somewhat chance location at maturity from which it is guided to a spawning site by currents and the movement of adult walleyes with established movement patterns. Dispersal to summer feeding areas following initial spawning may also be guided by established migration patterns of older walleyes.

It is to be expected that movement patterns and the intensity of homing behavior will vary between populations in separate waters. Lake size, shape, bottom contours, the numbers and locations of suitable lake and river spawning areas, and the age structure of the population are all probable determinants of movement patterns. Thermal gradients influence both spawning migrations and summer distributions. In Georgian Bay of Lake Huron, walleye stocks of the Shawanaga River appear to be thermally restricted to the lower Shawanaga basin and only during extremely hot summers do they move into the colder waters of Shawanaga Bay (Spangler et al. 1977).

Warmer water temperatures in streams preceding and during ice breakup is considered to be an important factor for initiating spawning runs (Rawson 1957). Whether or not distinctive stream odors also play a part in homing of spawning walleyes has not been determined. Olfactory homing to lake-shoal spawning sites or to home feeding areas could be confused by changing patterns of internal currents and would appear to require particularly sensitive olfactory perception. Restricted ranging area in some small lakes might completely mask walleye homing to feeding areas. Conversely, where suitable spawning habitat is severely restricted, homing of spawning walleyes might appear to be unusually strong.

Home feeding areas may be lacking or poorly defined in regular-shaped, thermally unstratified lakes with expansive areas of homogeneous habitat. That partial barriers somewhat restrict walleye movement was reported by Holt et al (1977). Walleyes bearing sonic tags were tracked in Lake Bemidji, Minnesota. The test fish moved within well-defined depth ranges (1.6 to 5.0 m) and did not readily move around submerged bars. When tagged fish approached a bar, they either turned back upon their previous path or followed bathymetric contours around the submerged structure.

Recaptures of walleye tagged in four separate basins of Lake of the Woods, Canada, were at progressively closer average distances to their marking site where the basin shoreline was of greater irregularity (Macins, unpublished data).

If, indeed, homing behavior is strengthened by repeated migrations, populations comprised of a large proportion of older fish would be expected to display stronger homing tendencies than heavily-exploited populations of a younger age structure.

The hypothesis that walleye homing is an adult-learned behavior rather than a natal conditioning appears to be in accord with the species' early life history and with observations which indicate that the apparent intensity for homing may be associated with various physical features of the environment and strengthened by repeated migrations.

If walleye homing is an adult-learned behavior, the reestablishment of spawning runs in once-polluted but rehabilitated rivers might be speeded by transplanting adult walleyes from other waters during spawning migration.

References

COOPER, J. C., A. T. SCHOLZ, R. M. HORRALL, A. D. HASLER, AND D. M. MADISON. 1976. Experimental confirmation of the olfactory hypothesis with homing, artificially imprinted coho salmon (*Oncorhynchus kisutch*). J. Fish. Res. Board Can. 33:703–710.

CROWE, W. R. 1962. Homing behavior in walleyes. Trans. Am. Fish. Soc. 91(4):350–354.

DONALDSON, R., AND G. H. ALLEN. 1958. Return of silver salmon *Oncorhynchus kisutch* (Walbaum) to point of release. Trans. Am. Fish. Soc. 87:13–22.

ESCHMEYER, P. H. 1950. The life history of the walleye *Stizostedion vitreum vitreum* (Mitchell) in Michigan. Mich. Dep. Conserv. Inst. Fish. Res. Bull. 3. 99 pp.

FERGUSON, R. G., AND A. J. DERKSEN. 1971. Migration of adult and juvenile walleyes (*Stizostedion vitreum vitreum*) in southern Lake Huron, Lake St. Clair, Lake Erie, and connecting waters. J. Fish. Res. Board Can. 28:1133–1142.

FORNEY, J. L. 1963. Distribution and movement of marked walleyes in Oneida Lake, New York. Trans. Am. Fish. Soc. 92:47–52.

GERKING, S. D. 1959. The restricted movement of fish populations. Biol. Rev. Cambridge Phil. Soc. 34:221–242.

HOLT, C. S., G. D. S. GRANT, G. P. ABERSTAR, C. C. OAKES, AND D. W. BRADT. 1977. Movement of walleye, *Stizostedion vitreum*, in Lake Bemidji, Minnesota as determined by radio-biotelemetry. Trans. Am. Fish Soc. 106(2):163–169.

HOUDE, E. D., AND J. L. FORNEY. 1970. Effects of water currents on distribution of walleye larvae in Oneida Lake, New York. J. Fish. Res. Board Can. 27:445–456.

JENSON, A., AND R. DUNCAN. 1971. Homing of transplanted coho salmon. Prog. Fish-Cult. 33:216–218.

JOHNSON, F. H. 1969. Environmental and species associations of the walleye in Lake Winnibigoshish and connected waters, including observations on food habits and predator-prey relationships. Minn. Fish. Invest. 5:5–36.

MAGNIN, E., AND G. BEAULIEU. 1968. Deplacements de dore jaune *Stizostedion vitreum* (Mitchill) du fleuve Saint-Laurent d'apre's les donnees du marquage. Nat. Can. 95:897–905.

OLSON, D. E., AND W. J. SCIDMORE. 1962. Homing behavior of spawning walleyes. Trans. Am. Fish. Soc. 91(4):355–361.

PECK, J. W. 1970. Straying and reproduction of coho salmon, *Oncorhynchus kisutch*, planted in a Lake Superior tributary. Trans. Am. Fish. Soc. 99(3):591–595.

PRIEGEL, G. R. 1968. The movement, rate of exploitation and homing behavior of walleyes in Lake Winnebago and connecting waters, Wisconsin, as determined by tagging. Trans. Wis. Acad. Sci. Arts Lett. 56:207–223.

RAWSON, D. S. 1957. The life history and ecology of the yellow walleye *Stizostedion vitreum*, in Lac la Ronge, Saskatchewan. Trans. Am. Fish. Soc. 86:15–37.

SCHUPP, D. H. 1972. The walleye fishery of Leech Lake, Minnesota. Minn. Dep. Nat. Resour., Invest. Rep. 317.

SMITH, L. L. JR, L. W. KREFTING, R. L. BUTLER. 1952. Movements of marked walleyes, *Stizostedion vitreum vitreum* (Mitchell) in the fishery of the Red Lakes, Minnesota. Trans. Am. Fish. Soc. 81:179–196.

SPANGLER, G. R., N. R. PAYNE, AND G. K. WINTERTON. 1977. Percids in the Canadian waters of Lake Huron. J. Fish. Res. Board Can. 34:1839–1848.

WHITNEY, R. R. 1958. Numbers of mature walleyes in Clear Lake, Iowa, 1952–3, as estimated by tagging. Iowa State J. Sci. 33(1):55–79.

WOLFERT, D. R. 1963. The movements of walleyes tagged as yearlings in Lake Erie. Trans. Am. Fish. Soc. 92:414–420.

Walleye Abundance, Growth, Movement, and Yield in Disparate Environments Within a Minnesota Lake

DENNIS H. SCHUPP

Minnesota Department of Natural Resources
Division of Fish and Wildlife
Brainerd, Minnesota 56401

Abstract

Walleye abundance and yield was highest and growth was most rapid in mesotrophic areas of a large lake with disparate environments ranging from eutrophic to morphometrically oligotrophic. Movement of tagged walleyes varied among the different environments. The following factors appeared to influence these parameters most: (1) basin morphometry and its effect on turbidity, temperature, and productivity; (2) availability of spawning habitat; and (3) abundance and growth of young yellow perch, the principal prey of walleyes.

Walleyes are influenced by a variety of environmental factors that affect reproduction, abundance, behavior, growth, and survival. They are tolerant of a great range of environmental systems, but appear to reach greatest abundance in large, shallow, turbid lakes (Scott and Crossman 1973). Other environmental factors that appear to influence walleyes are availability and quality of spawning habitat (Johnson 1961), water clarity (Ryder 1977), temperature (Hokanson 1977), and availability of food (Chevalier 1973; Forney 1966, 1976; Johnson 1977; Swenson 1977). The relative effect of each of these factors is difficult to assess in any lake system since they can combine to provide an infinite range of habitats in which walleyes may thrive.

The purpose of this paper is to describe the response of walleyes to different environmental conditions within a large Minnesota lake. Walleye abundance, growth, movement, and yields characteristic of these environments are used to synthesize conclusions on the combined influence of environmental factors and to evaluate potential limiting factors within each habitat.

Lake Description

Leech Lake, in north central Minnesota, is an irregularly shaped lake of 45,134 hectares (Fig. 1). The lake was formed by a moraine dam, part of a system of terminal moraines (Zumberge 1952). Since 1884, the lake has been part of a system of reservoirs retaining water for navigation on the upper Mississippi River. Four major tributaries and many smaller streams drain a watershed of 301,000 hectares. Water levels are controlled by a dam at the outlet. The original operating limits allowed water fluctuations up to 1.75 m, but the ordinary range of annual fluctuations under current operation is 0.75 m, thus conditions are lakelike.

The 306 km of shoreline are about equally divided between high or intermediate shore bordered by upland cover of mixed conifers and hardwoods and low shore bordered by muskeg and sedges interspersed with stands of tamarack or aspen. The lake bottom along the high or intermediate shorelines to a depth of 5 m consists mainly of rock and rubble overlying gravel. The lake bottom along the low shorelines is mainly sand to a depth of 4–5 m. Most of the low shoreline occurs at the head of the numerous bays. Rooted aquatic plants are common only in sheltered bays or on shorelines protected from westerly winds.

Trophic conditions within the lake range from eutrophic to morphometrically oligotrophic but the main basin is mesotrophic. Pertinent physical and chemical data are listed in Table 1.

The fish community is dominated by percids and esocids. Walleyes, northern pike, and muskellunge are the principal piscivores. Yellow perch are abundant throughout the lake. White suckers, cisco, lake whitefish, and rock bass are common in the mesotrophic areas. These species, with the exception of the salmonids, are also present in the weedy, eutrophic bays. In addition, these bays support populations of brown and black bullheads, black crappies, pumpkinseeds, and bluegills.

Study Areas

Four areas studied which are representative of the range of habitats available are Walker Bay, Steamboat Bay, and two areas in the main basin, Whipholt and Pelican Island (Fig. 1).

The morphometry of Walker Bay is similar to that of many oligotrophic lakes on the Precambrian Shield, but the chemical fertility of the water is high (Table 1) and thus the basin is classified as mesotrophic. The morphoedaphic index (MEI: Ryder 1965) is 16.5 and the theoretical fish yield is 3.9 kg/hectare. Based on Figure 2 of Ryder et al. (1974), Walker Bay is in that class of waters where primary production may be reduced in deeper strata. The maximum surface water temperature recorded was 24.4 C. A thermocline forms at about 10 m during mid-summer. Oxygen is present below the thermocline to a depth of at least 20 m. The bay is ice-covered from early December to early May in most years. Walleyes spawn along most of the rocky east and south shores of the bay.

Steamboat Bay is a very shallow, eutrophic area. The maximum depth (4.6 m) occurs in an inundated river channel paralleling the north and east sides of the bay. Most of the bay is less than 2 m deep and dense growths of rooted vegetation occur throughout. The theoretical fish yield, based on the MEI of 81.8, is 8.7 kg/hectare. The bay falls in the class of waters where

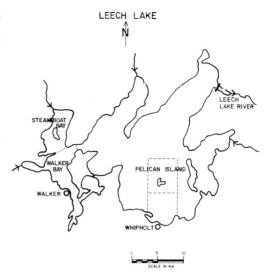

FIGURE 1.—Map of Leech Lake and study areas within the lake.

winterkill might restrict yield (Ryder et al. 1974). Prior to closure of the dam, most of the bay was shallow marsh interspersed with small areas of open water. Water temperatures change rapidly and are usually a few degrees below the mean air temperature at any time. The bay is ice-covered from early November to mid-April most years. The rock and rubble spawning substrate preferred by walleyes is limited and organic debris is common on the substrate.

The mesotrophic main basin areas differ mainly in the relative amounts of substrate materials and the regularity of bottom contours. The Whipholt area on the south shore of the main basin is a windswept bay characterized by gently sloping sand beaches largely barren of vegetation. Rock and gravel are present mainly on points

TABLE 1.—*Physical and chemical characteristics of Leech Lake and study areas within the lake.*

Measurement	Leech Lake	Walker Bay	Whipholt	Pelican Island	Steamboat Bay
Water area, hectares	45,134	3,333	3,001	2,770	2,374
Mean depth, m	5.6	10.4	5.8	5.4	1.8
Maximum depth, m	54.1	54.1	11.5	8.2	4.6
Transparency, m (Secchi disc)	1.5–4.1	3.0–3.7	1.5–4.1	1.5–4.1	2.0
Morphoedaphic index	31.4	16.5	30.4	32.7	81.8
Total dissolved solids, mg/liter	172.9	172.0	176.5	176.5	147.2
Total alkalinity, mg/liter	136.5	142.5	139.0	139.0	110.0

TABLE 2.—Mean number of fish captured per hour of trawling in Leech Lake study areas. Hours of trawling are in parentheses beneath the sampling locality. TL is total length.

Species	Walker Bay (20.4 h)	Whipholt (6.9 h)	Pelican Island (6.4 h)	Steamboat Bay (10.6 h)
Walleyes	49.4	94.1	86.9	36.4
Young of year	19.0	15.1	29.1	1.0
<254 mm TL	16.0	51.7	15.2	20.0
>254 mm TL	14.4	27.3	42.6	15.4
Northern pike	1.2	0.8	3.5	9.4
Yellow perch	1,954.7	7,826.7	4,732.4	1,300.5
Young of year	1,250.3	7,747.7	3,841.8	83.8
Age I+	704.4	79.0	890.6	1,216.7
Minnows[a]	66.7	905.9	322.1	137.1
Darters[b]	76.1	247.7	211.5	0.0
Trout-perch	6.5	652.3	4.1	127.6

[a] Spottail shiner, emerald shiner, bluntnose minnow.
[b] Logperch, johnny darter, Iowa darter.

and about the one small island in the area. The rocky areas are the only locations where walleye eggs have been found. The area about Pelican Island in the main basin is characterized by rocky reefs to depths of 3–4 m radiating in all directions from the island. Walleye eggs were found on all shoals about the island and the deeper shoals may also be used for spawning. The MEI of 30.4 and 32.7 for the Whipholt and Pelican Island areas, respectively, are nearly the same as that calculated for Leech Lake as a whole. These values are near the optimum for north temperate waters as postulated by Ryder et al. (1974). The theoretical yield for these areas is 5.4 kg/hectare. Maximum surface water temperatures were 24.4 C. Neither of the main basin areas stratify thermally. The main basin is ice-covered from late November to late April each year.

Methods

Data used in this report were collected during a study of the sport fishing yield of walleyes (Schupp 1972). A modified grid pattern was used to delineate sampling areas and all data were identified with these areas. Indices of fish abundance were determined from otter-trawl catches from mid-May through September, 1966–1968. Walleye growth rates were back-calculated from scales of fish caught by trawling for the 1962 to 1966 year classes. The mean annual increments were summed for age classes I–IV to estimate average growth. Sex was not determined, so growth data beyond age IV (earliest age of maturity) are not included. Walleye movements were determined from trawl-caught fish larger than 254 mm total length (TL) in 1966 and 1967 tagged with numbered disc-dangler tags attached through the musculature posterior of the dorsal fin with stainless steel surgical wire.

Indices of Fish Abundance

The abundances of walleyes and associated prey species were generally highest in the main basin areas (Table 2). Catches of walleyes of all sizes were about twice as high at Whipholt and Pelican Island as in Walker Bay and in Steamboat Bay.

The size distribution of walleyes caught varied markedly among the four areas. Catches of young of the year (hereafter referred to as young) were highest at Pelican Island. Walleyes less than 254 mm TL (excluding young) predominated in the catches from Whipholt. These fish were mainly ages I–III. Fish larger than 254 mm TL made up 49% of the catches at Pelican Island, nearly three times as great as from Walker and Steamboat Bays and 1.5 times greater than at Whipholt.

Catches of prey fishes were also highest in the main basin areas. Yellow perch young dominated the catches in all areas but Steamboat Bay. The mean catch of young yellow perch at Whipholt ranged

TABLE 3.—Mean annual increment (mm) in total length and coefficient of variation (C.V.) of walleyes from the 1962–1966 year classes captured by trawling in Leech Lake study areas.

Year of growth	Walker Bay (n = 381)	Whipholt (n = 354)	Pelican Island (n = 243)	Steamboat Bay (n = 162)
1	140	143	151	140
C.V.	0.096	0.094	0.080	0.070
2	69	77	85	79
C.V.	0.212	0.172	0.168	0.192
3	61	71	66	65
C.V.	0.214	0.198	0.203	0.189
4	48	53	50	51
C.V.	0.263	0.192	0.196	0.244

TABLE 4.—Values of t and probabilities P of differences in mean annual increments of walleyes in the first four years of life occurring by chance among Leech Lake study areas.

	Year of growth	Steamboat Bay		Whipholt		Pelican Island	
		t	P	t	P	t	P
Walker Bay	1	−0.30	0.383	−1.82	0.027	−8.70	<0.001
	2	−5.65	<0.001	−5.13	<0.001	−11.58	<0.001
	3	−2.18	0.015	−4.63	<0.001	−3.15	0.001
	4	−1.51	0.069	−2.42	0.008	−1.04	0.149
Steamboat Bay	1			−1.32	0.095	−2.41	0.009
	2			0.84	0.202	−3.45	<0.001
	3			−2.48	0.007	−0.43	0.334
	4			−0.70	0.242	0.74	0.231
Whipholt	1					−5.48	<0.001
	2					−4.72	<0.001
	3					2.37	0.010
	4					1.65	0.051

from 2.0 to 92.5 times as great as in other areas. Yellow perch age I+ were most abundant in Steamboat Bay, followed in order by Pelican Island, Walker Bay, and Whipholt. The catch of older yellow perch in Steamboat Bay was 15 times that at Whipholt. The catch of other small prey fishes (minnows, darters, trout-perch) was also highest at Whipholt, ranging from 3.4 to 12.1 times as great as in the other areas. Catches of these groups were lowest in Walker Bay.

Northern pike are abundant and may compete with or prey on walleyes. Catches of northern pike were highest in Steamboat Bay and lowest at Whipholt and in Walker Bay. Abundance at Pelican Island was between these extremes.

Walleye Growth

The growth of walleyes was more rapid in the main basin areas. The most rapid growth was observed for fish from Pelican Island while the slowest growth was for fish from Walker Bay (Table 3). Mean growth increments for Walker Bay fish were significantly lower ($P < 0.05$) than main basin fish for each of the first three years of life and significantly lower than Steamboat Bay fish in the second and third year (Tables 3 and 4). Fish from Steamboat Bay grew more slowly than fish from Whipholt, though the difference was significant only in the third year, and slower than fish from Pelican Island the first two years of life. Fish from Pelican Island grew significantly faster than those from Whipholt the first two years of life but significantly slower the following two years. Growth in the first two years of life accounted for most of the differences in TL at age IV among the four areas.

The slower growth of walleyes in Walker and Steamboat Bays was associated with greater variation in growth within different age classes. These differences appear to be related to the abundance and growth of young yellow perch, the principal prey. Young yellow perch made up more than 90% of all prey fish consumed during the growing season by walleye age classes 0 and I and more than 70% for older fish (unpublished data). The lower abundance of young yellow perch in Walker and Steamboat Bays was associated with more rapid yellow perch growth. Young yellow perch in Walker Bay averaged 68 mm TL and 3.3 g by late September, 1966, while those at Whipholt averaged 45 mm and 0.9 g. Young walleyes sampled in late September at Whipholt were feeding almost entirely on yellow perch young, but in Walker Bay only the larger young walleyes were eating yellow perch. Slower growing walleyes in Walker Bay contained invertebrates or small minnows. The ratio of TL of young perch to TL of young walleyes in Walker Bay exceeded 0.45 throughout the 1966 growing season while at Whipholt, the ratio had declined to less than this by mid-July and by late September was about 0.30 (Fig. 2). The ratio at Whipholt throughout the summer was similar to the length ratio se-

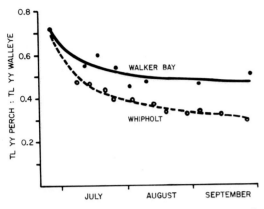

FIGURE 2.—Ratio of TL (total length, mm) of young-of-the-year yellow perch to TL (mm) of young-of-the-year walleye caught by trawling in Walker Bay and in Whipholt area during 1966, Leech Lake.

lected by similar sized walleyes in Lake Erie (Parsons 1971). The lengths of young yellow perch available in Walker Bay were selected primarily by walleyes 200–250 mm TL in Lake Erie (Parsons 1971), a size reached in the second growing season in Leech Lake. Condition factor (K) for walleye young at Whipholt varied little over the whole size range (130–185 mm TL), averaging 1.38. In Walker Bay, the size ranged from 107–168 mm TL and condition factor ranged from 1.24 for the smallest sizes to 1.34 for the largest.

There is also some evidence that growth compensation in the first four years of growth is more likely to occur in the main basin areas. Negative correlations between annual increments were more common in main-basin growth calculations than in those for Walker and Steamboat Bays. Negative correlations were observed in 7 of 11 comparisons from Pelican Island and in 5 of 9 comparisons from Whipholt. In contrast, negative correlations were observed in only 1 of 8 comparisons from Steamboat Bay and 3 of 12 comparisons from Walker Bay. The change from significantly more rapid growth for Pelican Island fish compared to Whipholt fish at ages I and II to significantly more rapid growth for Whipholt fish at ages III and IV (Tables 3 and 4) tends to support the evidence for growth compensation in the main basin.

Walleye Movements

The movement of tagged walleyes differed among the four areas and these behavioral differences appear to be related to the different environments. Walleyes tagged in Walker Bay appeared to stray least from the study area (Table 5). Thirteen fish tagged in other areas were reported from Walker Bay and eight of these were from Steamboat Bay. No seasonal pattern of movements between Walker Bay and other areas was detected.

Walleyes tagged in Steamboat Bay appeared to have a distinctive seasonal movement pattern. The percentages of returns in the bay was highest in fall and spring and lowest in July and August (Table 5). No walleyes tagged in Steamboat Bay were reported caught there between July 15 and August 21. The seasonal movements suggest that the behavior may be a response to changing water temperatures, but the low rate of return in mid-summer also coincided with the low abundance of demersal young perch in Steamboat Bay and high abundance in other areas. Walleyes tagged in Steamboat bay and caught elsewhere were usually in contiguous areas (8 in Walker Bay and 4 in the western portion of the main basin). Only six walleyes tagged in other areas were reported from Steam-

TABLE 5.—Monthly returns of tagged walleyes caught by anglers that were both tagged and caught in each Leech Lake Study area.

Study area	Total tagged	Total returns	% of returns in study area (n)						
			Total	May	June	July	August	September	October
Walker Bay	230	39	89.7	0.0 (2)	100 (11)	100 (10)	90.9 (11)	80.0 (5)	
Whipholt	154	47	59.6	33.3 (12)	57.1 (14)	60.0 (5)	83.3 (6)	80.0 (10)	
Pelican Island	260	100	75.0	64.3 (28)	73.1 (26)	70.0 (10)	91.7 (12)	81.8 (22)	100 (2)
Steamboat Bay	181	45	68.9	100 (1)	63.6 (11)	44.9 (9)	55.6 (9)	93.3 (15)	

boat Bay and four of these were from the western portion of the main basin.

Walleyes tagged at Whipholt appeared to stray most from the study area, and there was a seasonal pattern to the recaptures (Table 5). The limited spawning area available at Whipholt may explain the low returns in spring and the abundance of prey (Table 2) may have attracted fish to return in mid-summer and fall. No distinctive spatial movement pattern to or from the area could be detected.

Walleyes tagged at Pelican Island appeared to stray less than those at Whipholt and Steamboat Bay, but more than those from Walker Bay (Table 5). Twenty of the recaptures reported from elsewhere were caught in contiguous areas in the eastern portion of the main basin. Nineteen walleyes tagged in other areas were caught at Pelican Island and 15 (79%) of these were caught in May and early June. All but two came from contiguous areas in the eastern part of the main basin. This movement pattern suggests that these fish may have spawned at Pelican Island.

Sport Fishing Success and Yield

The sport fishing yield (kg/hectare) and angling success (kg/fishing hour) for walleyes also appeared to be related to the trophic status of the study areas. The highest yield of all species was from Steamboat Bay (Table 6). The walleye yield was above the lakewide average as were yields of all other species. The yield of northern pike was more than twice the lakewide average. Walleye fishing success was near the lakewide average.

The highest walleye yield was at Pelican Island and walleye fishing success was also best there. The yield was 1.6 times the lakewide average and fishing success 1.4 times the lakewide average. Fishing success and yields for other species were below average.

The lowest total yield and walleye yield was at Whipholt. However, fishing success for all species was similar to the lakewide averages. Fishing pressure at Whipholt is relatively low compared to the rest of Leech Lake.

Yields of walleyes and total yields in

TABLE 6.—Sport fishing yield (kg/hectare) and fishing success (kg/fishing hour) in Leech Lake study areas.

	Leech Lake	Walker Bay	Whipholt	Pelican Island	Steamboat Bay
Total yield	4.48	3.79	1.77	4.86	8.38
Walleye	2.14	1.54	0.73	3.40	2.52
Yellow perch	0.79	0.80	0.30	0.33	1.09
Northern pike	1.40	1.20	0.52	0.98	3.19
Other[a]	0.15	0.25	0.22	0.15	1.58
Fishing success	0.25	0.13	0.26	0.25	0.33
Walleye	0.12	0.05	0.11	0.17	0.10
Yellow perch	0.04	0.03	0.04	0.02	0.04
Northern pike	0.08	0.04	0.08	0.05	0.13
Other[a]	0.01	0.01	0.03	0.01	0.06

[a] Muskellunge, brown bullhead, centrarchids.

Walker Bay were below the lakewide averages. Fishing pressure was relatively high but fishing success for walleye was lower than average and less than one-third the rate at Pelican Island.

Discussion

The response of walleyes to environmental influences in Leech Lake appeared to be most favorable in the mesotrophic main basin areas. Optimum combinations of water clarity (Ryder 1977) and preferred temperatures (Hokanson 1977) are more likely to occur there. Prey abundance was greater in these areas. The observations reported here support the inferences of Leach et al. (1977) and Kitchell et al. (1977) regarding optimum conditions for walleyes in the trophic continuum.

The higher quality of the main basin areas for walleyes was reflected in their greater abundance and more rapid growth. Pelican Island and Whipholt differ mainly in the quality and quantity of spawning area available. This difference apparently influenced the seasonal movements of walleyes tagged in these areas and may be related to the differences in angling yields observed. About one-third of the annual Leech Lake walleye harvest occurred in May when walleyes were seasonally less abundant at Whipholt.

Availability of spawning areas also appeared to influence walleyes inhabiting Steamboat and Walker Bays. The abundance of walleyes age I+ in these bays was similar; however, young walleyes were common in Walker Bay but not in Steamboat Bay. Growth in the first year was similar in these bays and was apparently influenced by the relatively low abundance and large size of young yellow perch. The subsequent differences in growth were probably the result of behavioral differences among older fish. Fish using Steamboat Bay appeared to be more likely to move to main basin areas where prey abundance was greater. This movement was most evident in July and August, the months of most rapid growth and highest prey abundance (Schupp 1972).

The biotic factors favoring walleyes most in the main basin areas appeared to be the greater abundance and slower growth of young yellow perch. This is consistent with the findings of Forney (1966; 1976) in Oneida Lake, New York where growth of young walleyes varied inversely with growth of young yellow perch and the abundance of yearling and young yellow perch accounted for much of the variability in relative survival of young walleyes. A higher incidence of cannibalism on young walleyes in Oneida Lake was observed when the density of young yellow perch was low (Chevalier 1973). Early growth compensation might also be less likely to occur where low abundance and large size of preferred prey confers an advantage on faster growing individuals of a year class.

The study areas within the lake encompass much of the range of habitats in which walleyes are found in Minnesota. Walker Bay is morphometrically similar to many lakes in northeastern Minnesota that have natural walleye populations limited mainly by the morphometry of the basin. Walleye yields in waters of this type may increase if more of the system's energy can be channeled through a preferred prey species. Intensive removal of white suckers in Wilson Lake, Minnesota was followed by a marked increase in yellow perch abundance and increased yields of walleyes (Johnson 1977).

Basin morphometry also influences conditions in Steamboat Bay. The depletion of oxygen through the decay of organic matter in winter would cause winterkill were the bay a closed system. Fish yields can be high if winterkills are not annual. Minnesota lakes with winterkills on the average of once every 3 to 5 years provide good walleye fisheries following stocking of walleye fry (Scidmore 1970).

The dearth of good spawning area appeared to be the main factor influencing walleyes at Whipholt. The addition of gravel and rubble to marginal walleye spawning areas was followed by increased egg deposition and survival (Johnson 1961; Newburg 1975), but walleye fisheries in most lakes where limited by reproduction are maintained by stocking walleye fry or fingerlings. The area around Pelican Island is typical of the state's best walleye waters. Lakes with these characteristics provide high walleye yields under intensive fishing pressure. Management in these prime waters is limited to catch regulations (closed seasons and bag limits).

Angling yields from the study areas, with the exception of Whipholt, were near the theoretical yields estimated from the MEI and are characteristic of other Minnesota waters with similar habitats studied over the last 25 years. Creel surveys on two large walleye lakes, repeated after 15- to 20-year intervals, indicate that walleye yields have not changed despite significant increases in fishing pressure. This suggests that the walleye yields observed were near their upper limits.

References

CHEVALIER, J. R. 1973. Cannibalism as a factor in first year survival of walleye in Onieda Lake. Trans. Am. Fish. Soc. 102(4):739–744.

FORNEY, J. L. 1966. Factors affecting first-year growth of walleyes in Oneida Lake, New York N. Y. Fish Game J. 13:146–167.

———. 1976. Year-class formation in the walleye (*Stizostedion vitreum vitreum*) population of Oneida Lake, New York, 1966–73. J. Fish. Res. Board Can. 33:783–792.

HOKANSON, K. E. F. 1977. Temperature requirements of some percids and adaptations to the seasonal temperature cycle. J. Fish. Res. Board Can. 34:1524–1550.

JOHNSON, F. H. 1961. Walleye egg survival during incubation on several types of bottom in Lake Winnibigoshish, Minnesota, and connecting waters. Trans. Am. Fish. Soc. 90(3):312–322.

———. 1977. Responses of walleye (*Stizostedion vitreum vitreum*) and yellow perch (*Perca flavescens*) populations to removal of white sucker (*Catostomus commersoni*) from a Minnesota lake, 1966. J. Fish Res. Board Can. 34:1633–1642.

KITCHELL, J. F., M. G. JOHNSON, C. K. MINNS, K. H. LOFTUS, L. GREIG, AND C. H. OLVER. 1977. Percid habitat: The river analogy. J. Fish. Res. Board Can. 34:1936–1940.

LEACH, J. H., M. G. JOHNSON, J. R. M. KELSO, J. HARTMANN, W. NÜMANN, AND B. ENTZ. 1977. Responses of percid fishes and their habitats to eutrophication. J. Fish. Res. Board Can. 34:1964–1971.

NEWBURG, H. J. 1975. Evaluation of an improved walleye (*Stizostedion vitreum*) spawning shoal with criteria for design and placement. Minn. Dep. Nat. Resour. Invest. Rep. 340. 39 pp.

PARSONS, J. W. 1971. Selective food preferences of walleyes of the 1959 year class in Lake Erie. Trans. Am. Fish. Soc. 100(3):474–485.

RYDER, R. A. 1965. A method for estimating the potential fish production of north-temperate lakes. Trans. Am. Fish. Soc. 94(3):214–218.

———. 1977. Effects of ambient light variations on behavior of yearling, subadult, and adult walleyes (*Stizostedion vitreum vitreum*). J. Fish. Res. Board Can. 34:1481–1491.

———, S. R. KERR, K. H. LOFTUS, AND H. A. REGIER. 1974. The morphoedaphic index, a fish yield estimator—review and evaluation. J. Fish. Res. Board Can. 31:663–688.

SCHUPP, D. H. 1972. The walleye fishery of Leech Lake, Minnesota. Minn. Dep. Nat. Resour. Invest. Rep. 317. 11 pp.

SCIDMORE, W. J. 1970. Using winterkill to advantage. Pages 47–52 *in* E. Schneberger, ed. A symposium on the management of midwestern winterkill lakes. Am. Fish. Soc. North Central Div. Spec. Publ.

SCOTT, W. B., AND E. J. CROSSMAN. 1973. Freshwater fishes of Canada. Bull. Fish. Res. Board Can. 184. 965 pp.

SWENSON, W. A. 1977. Food consumption of walleye (*Stizostedion vitreum vitreum*) and sauger (*S. canadense*) in relation to food availability and physical environmental conditions in Lake of the Woods, Minnesota, Shagawa Lake, and western Lake Superior. J. Fish. Res. Board Can. 34:1643–1654.

ZUMBERGE, J. H. 1952. The lakes of Minnesota, their origin and classification. Minn. Geol. Surv. Bull. 35. 99 pp.

Effects of Temperature on Production and Yield of Juvenile Walleyes in Experimental Ecosystems[1]

W. B. WRENN AND T. D. FORSYTHE

Tennessee Valley Authority
Decatur, Alabama 35602

Abstract

Yield (net biomass change) and total production were determined for populations of juvenile walleyes held in 12 outdoor experimental channels for 118 days (May–September) at four thermal regimes. Average yield ranged from 36 kg/hectare at the ambient temperatures of Wheeler Reservoir (Tennessee River) to 0 kg/hectare at temperatures 6 C above ambient. Total mortality apparently occurred in the +6 C regimen when temperatures exceeded 34 C in August. Ambient temperatures exceeded 28 C for 78 days. At the next highest regimen (+4 C), mean yield (18 kg/hectare) was not significantly different from that in the ambient regimen, despite temperatures of 32 to 33 C for 75 days. With the exception of the highest, lethal, temperature regimen, the specific mortality rate of walleye was similar for all temperature levels, 0.55%/day–0.60%/day. Total production ranged from 5.5 g/m² at ambient temperature to 2.8 g/m² in the highest regimen (+6 C). Walleyes in this study were more temperature tolerant than previously reported in laboratory studies.

Temperature requirements of walleyes in the Tennessee River drainage as well as in other waters are not well established. Since temperature is recognized as an important physical factor in determining fish distribution and since walleyes are more commonly associated with northern lakes and streams, the general assumption has been that walleyes are excluded from most southern waters due to higher summer temperatures. To some extent this assumption appears valid for the Tennessee River drainage where walleyes are abundant in many of the relatively cooler upland storage impoundments such as Norris Reservoir (Fitz and Holbrook 1978, this volume) and are uncommon in warmer mainstream reservoirs which at the southernmost latitudes extend across northern Alabama. However, in their review of percid populations in the southeast, Hackney and Holbrook (1978, this volume) reported that walleye distribution and abundance were relatively independent of temperature. Also, the complexity of establishing thermal requirements for coolwater species was placed in proper perspective by Hokanson (1977), who reported that water temperatures in percid habitat vary seasonally from less than 4 C for more than eight months in the northernmost latitudes to about 32 C in the southernmost extremities of the percid range.

Although records of natural distribution have application in determining thermal requirements of a fish species, apparently the most sensitive function for establishing an upper temperature limit for long-term exposure is growth rate (Brungs and Jones 1977; National Academy of Science and National Academy of Engineering 1973; McCormick et al. 1971). Temperature requirements for growth have customarily been determined from laboratory studies. In order to evaluate growth at the population level, the purpose of this experiment was to determine the effect of elevated water temperatures on production and yield of populations of juvenile walleyes held in large outdoor channels. In cooperation with the U. S. Environmental Protection Agency, this study was designed to present a new and more realistic approach for the assessment of thermal requirements.

Methods

Experimental Channels

Twelve outdoor channels were located adjacent to the Browns Ferry Nuclear Plant on Wheeler Reservoir (Tennessee River) in northern Alabama. Each of the

[1] The work upon which this publication is based was performed pursuant to an interagency agreement between the Tennessee Valley Authority and the U.S. Environmental Protection Agency.

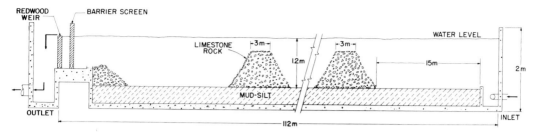

FIGURE 1.—*Longitudinal section of one of the 12 outdoor experimental channels.*

112-m concrete channels had a water surface area of 480 m². Substrates were provided which consisted of six alternate rock (5–15 cm diam) and six mud-silt zones (Fig. 1). The water supply, 0.66 m³/min per channel, was pumped from the reservoir which provided colonization of diverse communities of algae, zooplankton, and macroinvertebrates (Armitage et al. 1978). This system was designed so that four temperature regimens, with three channels per regimen, could be established. The ambient temperature of water pumped from the reservoir provided the lowest regimen or treatment. Three elevated temperature regimens, superimposed on the seasonal and diurnal temperature cycle of the reservoir, were established by passing the inflowing ambient water through stainless-steel heat exchangers before it entered the channels. The heat source for the exchangers was the condenser cooling water pumped from the nuclear plant. Flow rates and the physical dimensions of the channels were such that the elevated increments established at the inlet were expected to be maintained at "constant" levels (±1 C) only in the upper 44 m of the channels. Downstream from this point the temperature normally declined as it began to approach ambient temperatures. In order to confine fish populations to the upper "constant" sections or to the highest temperatures in some cases, the channels could be divided by fine mesh (2 mm) screen barriers into upper and lower sections of 190 and 290 m², respectively. In the present study two of the three channels for each regimen were divided and the third undivided; therefore, there was a total of 20 experimental walleye populations to be evaluated.

The nominal elevated temperature regimens were 2, 4, and 6 C above the ambient reservoir temperature. Actual temperature regimens obtained are presented in Figure 2. These regimens were based on the mean temperature at 1 m depth which represented the average minimum temperature accessible to the walleye. In the upper sections temperature was measured at the inlet and 38 m down-channel, and in the lower sections at 70 m and 102 m down-channel. Although temperature was monitored automatically (hourly) at these four locations in each channel, the temperature curves in Figure 2 were based on the mean of temperatures measured at 0700 and 1530 h which were taken with a calibrated (ASTM) electrical thermometer. Abrupt declines in the elevated regimens were caused by valve failures or the loss of heat (unit outage or load reduction) from the nuclear plant.

Dissolved oxygen (maximum–minimum) ranged from 10.8–3.9 mg/liter in the upper sections of the ambient channels to 16.0–2.0 mg/liter in the upper sections of the +6 C regimen; minimum dissolved oxygen levels in the upper sections of the +6 C regimen usually did not drop below 4 mg/liter. Dissolved nitrogen saturation in the heated channels did not exceed 109%. Mean total hardness (as $CaCO_3$) was 47 mg/liter, and it was the same in all channels.

Fish

Walleyes (mean total length, 42 mm; mean weight, 0.5 g) were stocked in the channels on 13 May 1977. They were provided by the Carbon Hill National Hatchery (Alabama) which reared them for a month after obtaining the stock initially as sac-fry

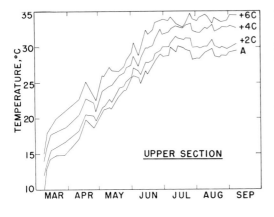

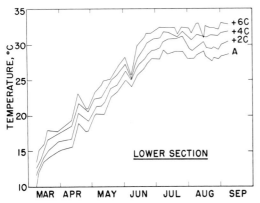

FIGURE 2.—*Average minimum daily temperatures in experimental channels, 1977; A = ambient temperature regimen of Wheeler Reservoir (Tennessee River). Upper section included 40% of the area of each channel.*

from the Senecaville National Hatchery (Ohio). Origin of the broodstock was Linesville, Pennsylvania. Numbers stocked in the upper and lower sections of divided channels were 12 (632/hectare) and 20 (690/hectare), respectively. The undivided channel in each treatment (regimen) received 32 walleyes (667/hectare). Adult bluegills, expected to spawn in the channels to provide prey for the walleye, were stocked on 8 March at 150 per channel or 60 and 90 in the upper and lower sections, respectively. Stocking densities for bluegills and walleyes were derived, on the basis of maximum walleye growth and survival, from previous evaluation of stocking rates with both species under ambient temperature conditions in the channels (Forsythe 1977). To determine growth each walleye population was sampled monthly by electrofishing (pulsed DC; 20 cycles/s; 150 V AC input; output ca. 2 A). Decreases in population numbers were estimated from daily visual inspection of the channels for dead fish. After 118 days the channels were drawn down, and the walleyes and bluegills were recovered with rotenone.

Production in this study is defined as the total incremental growth of fish tissue throughout the duration of the study, including that formed by individuals which did not survive to the end of the experiment (Ivlev 1966 as cited by Chapman 1971). Total production, as wet weight, was estimated numerically according to the Ricker formula (1946), on the basis of four monthly sampling intervals between 13 May and 9 September. For each sampling interval, production was calculated as $P = G \bar{B}$; Where P is production in kg, G is the instantaneous rate of increase in weight, and $\bar{B}$ is the mean biomass. Exponential models of growth and mortality were applied to the calculations of G and $\bar{B}$ (Chapman 1971). Estimates of the mean individual weights and population numbers were used in the calculation of production for the first three intervals, whereas actual mean weights and numbers of walleye recovered by rotenone were used for the last interval. We defined yield as the net biomass change of each population, and it was calculated as the difference between the biomass of the population stocked initially and the biomass of the population recovered at the end of 118 days.

Results

After 118 days, mean yield (net biomass change) of walleye from the upper channel sections ranged from 36 kg/hectare in the ambient regimen to 0 kg/hectare in the highest regimen (+6 C), in which none of the walleye survived to the end of the test (Table 1). Ambient temperature conditions exceeded 28 C for 78 days, and the maximum was 30 C for about 5 days (Fig. 2). Yield from the upper sections of the intermediate temperature regimens, +2 C and +4 C, decreased to 29 and 18 kg/hectare, respectively. With the exception of the +6 C regimen (treatment), the yields among treatments in the upper sections were not significantly different (analysis of variance, $P \leq 0.05$). Also, there were no significant differences in yield among treatments for the lower sections. With the exception of the +6 C treatment, there were no significant differences in yield from upper and lower sections within treatments. Yields from the undivided channels, similar to those for the lower sections, ranged from 26 kg/hectare in the +2 C and +4 C regimens to 20 kg/hectare in the +6 C regimen. Maximum temperatures at which walleye survived and continued to grow (Table 2) were 32–33 C, which occurred for about 75 days in the upper sections of the +4 C regimen (Fig. 2); similar thermal conditions also occurred in the lower sections of the +6 C regimen.

Total production estimates followed the same trend as that for yield in the upper sections in that production was highest (5.5 g/m^2) in the ambient temperature regimen, intermediate (3.9 g/m^2) in the +4 C regimen, and lowest (2.8 g/m^2) in the +6 C regimen (Table 2). In the lower sections and undivided channels production was not indicative of yield in every case; e.g., production in the undivided channels was highest in the ambient regimen (5.1 g/m^2 or 51 kg/hectare), but the yield (21 kg/hectare) was next to the lowest for the four temperature regimens.

Production at the end of the first month (13 May–8 June) was progressively higher as the temperature regimen increased (Table 2). In the upper sections production in the +6 C regimen was 1.1 g/m^2 which was more than twice that for the ambient regimen (0.4 g/m^2). However, at the end of the third month (8 August), production estimates for the ambient regimen (upper section) (2.3 g/m^2) exceeded that in the +6 C regimen by 1.4 g/m^2. Any change in production for the +6 C treatment (upper section) after 8 August was not accounted for since no walleye were recovered from this treatment at the end of the test. Walleye production was also greater initially in the other elevated temperature regimens, but production in the ambient channels eventually equaled or exceeded this early gain.

Based on the actual numbers of walleye recovered after 118 days, survival in all treatments was overestimated (Table 2) for the first three production intervals, which would cause production to be overestimated also. Although we believe that the unaccounted mortalities (or possibly escapement) occurred when the walleye were very small (40–50 mm total length), no attempt was made to distribute these missing fish in estimating production.

Mean survival of walleye after 118 days ranged from 63% in the upper sections of the ambient regimen to no survival in the upper sections of the +6 C regimen (Table 1). The last observed mortality in the +6 C regimen was 15 August; four additional mortalities occurred in this treatment from 22 July to 15 August. No moribund or distressed walleye were observed prior to finding dead fish. Average minimum temperatures during this interval ranged from 33.6 to 34.7 C. Mean survival for the five walleye populations in the ambient regimen was 50% which was essentially the same as that (52%) in all elevated regimens, except the +6 C treatment. Mean individual weight gain of walleye surviving for the duration of this test ranged from 91 g in the ambient regimens (upper sections) to 57 g in the undivided channel at the +6 C regimen, while the respective specific growth rates (%/day) ranged from 4.40 to 4.01 (Table 1). The biomass of young-of-the-year bluegills recovered at the end of the test, mean of 200 kg/hectare, indicated that prey was abundant in all temperature regimens.

Discussion

The ambient temperature regimen in the channels (Fig. 2), typical of summer temperatures in mainstream reservoirs of the Tennessee River, had no apparent adverse effect on the final biomass of experimental populations of juvenile walleyes. Carlander (1977) reported that the biomass of walleyes in three lakes under 100 hectares averaged 26.8 kg/hectare; mean biomass of walleye in 23 North American lakes was 16 kg/hectare. Mean biomass of the five walleye populations held at ambient tem-

TABLE 1.—Mean growth, survival, and yield (net biomass) of walleye in 12 experimental channels for 118 days (May–September). Water temperatures in channels were ambient (Wheeler Reservoir, Tennessee River) or elevated +2 C, +4 C, or +6 C above ambient in the upper channel sections. Within each temperature regimen, two channels were divided into upper (isothermal) and lower (cooling) sections, and one channel was undivided.

Parameter	Upper section				Lower section			
	A	+2 C	+4 C	+6 C	A	+2 C	+4 C	+6 C
(1) Length increase[a] (mm)	175	175	141	147[b]	178	167	153	146
(2) Weight increase[a] (g)	90	88	61		88	77	66	65
(3) Specific growth rate (wt.) (%/day)	4.40	4.37	4.07	[c]	4.38	4.27	4.14	4.13
(4) Specific mortality rate (%/day)	0.41	0.52	0.66		0.63	0.54	0.63	0.59
(5) Survival (%)	63	54	46	0	48	53	48	50
(6) Yield (kg/hectare) (±1 SD)	36 (8.4)	29 (11.3)	18 (6.6)	0	25 (2.9)	28 (2.8)	21 (1.6)	22 (0.6)

[a] Initial mean total length (TL) – 42 mm; mean wt – 0.5 g.
[b] Mean total length of 3 dead walleye recovered in mid-August.
[c] Not computed, no walleye surviving at end of test.

peratures in the channels was 28 kg/hectare. Mean individual growth (total length > 200 mm) in the ambient regimen was equal to the rapid growth reported for walleyes in Norris Reservoir and other southern waters (Stroud 1949; Hackney and Holbrook 1978, this volume). The mean specific mortality rate (0.60%/day) was within the range of natural mortality rates determined for juvenile walleye in Oneida Lake (Forney 1971).

The upper channel sections of the elevated temperature regimens were considered to simulate conditions that would occur if the temperature increased (either naturally or man-induced) in an entire

TABLE 2.—Mean production of walleye in experimental channels. Water temperatures in channels were ambient (Wheeler Reservoir, Tennessee River) or elevated +2 C, +4 C, +6 C above ambient in the upper channel sections. Within each temperature regimen, two channels were divided into upper (isothermal) and lower (cooling) sections, and one channel was undivided.

Time interval	Individual size (g)				Instantaneous growth rate				Population numbers[a]			
	A	+2 C	+4 C	+6 C	A	+2 C	+4 C	+6 C	A	+2 C	+4 C	+6 C
					Upper Section							
13 May	0.5	0.5	0.5	0.5					12	12	12	12
13 May–8 Jun	5	9	10	12	2.66	3.25	3.38	3.66	12	12	12	11
9 Jun–10 Jul	24	34	26	24	1.47	1.31	0.92	0.63	12	12	12	11
11 Jul–8 Aug	62	58	53	42	0.95	0.52	0.73	0.56	12	12	11	10
9 Aug–8 Sept	91	87	62	0	0.38	0.42	0.15	0	7	7	6	0
Total												
					Lower Section							
13 May	0.5	0.5	0.5	0.5					20	20	20	20
13 May–8 Jun	5	9	8	13	2.66	3.27	3.10	3.71	20	20	20	20
9 Jun–10 Jul	28	32	26	35	1.57	1.23	1.14	0.95	20	20	19	19
11 Jul–8 Aug	61	58	41	59	0.81	0.61	0.48	0.54	20	20	19	19
9 Aug–8 Sept	82	77	60	65	0.28	0.28	0.38	0.10	10	11	10	10
Total												
					Undivided							
13 May	0.5	0.5	0.5	0.5					32	32	32	32
13 May–8 Jun	7	7	13	14	3.05	3.05	3.76	3.84	32	32	32	32
9 Jun–10 Jul	28	27	32	36	1.30	1.27	0.84	0.89	32	32	31	31
11 Jul–8 Aug	59	57	40	58	0.75	0.75	0.22	0.48	32	32	31	29
9 Aug–8 Sept	78	70	68	57	0.28	0.21	0.53	−0.02	13	18	18	17
Total												

[a] Mean of 2 replicates in upper and lower sections rounded to whole fish.

TABLE 1.—Continued.

Parameter	Undivided			
	A	+2 C	+4 C	+6 C
(1) Length increase[a] (mm)	168	162	151	132
(2) Weight increase[a] (g)	78	70	68	57
(3) Specific growth rate (wt.) (%/day)	4.27	4.18	4.16	4.01
(4) Specific mortality rate (%/day)	0.76	0.49	0.49	0.54
(5) Survival (%)	41	56	56	23
(6) Yield (kg/hectare) (±1 SD)	21	26	26	20

water body such that cooler refuge areas would not be available to the walleye. After 118 days mean individual size of walleyes in the +2 C and +4 C regimens was less than that for walleyes in the ambient regimen, and no walleye survived to the end of the test in the +6 C regimen (Table 1). Obviously the total mortality in the +6 C regimen was ecologically significant; however, the significance of the reduced size of walleyes in the +4 C regimen was not clear since the size attained by walleyes in this treatment was not abnormal (mean total length 183 mm and weight 62 g). Under conditions similar to those in the channels (abundant forage and no competing fishes), Walker and Applegate (1976) reported that walleyes attained a mean length of 167 mm and weight of 41 g in a 12.5-hectare pothole in South Dakota (13 June–13 September).

Effects of temperature on walleye growth and survival in the upper channel sections contrasted with results obtained by Smith and Koenst (1975) in a laboratory study. They reported that the optimum temperature for growth of juvenile walleyes in a 28-day experiment was 22 C and that the incipient lethal temperature (50% mortality) was 31.6 C; the temperature for zero net growth was 29 C as extrapolated by Hokanson (1977). Ambient temperatures in our channels were near or above 29 C for the last 75 days of the experiment, and mini-

TABLE 2.—Continued.

Time interval	Biomass (g)				Production (g/m²)			
	A	+2 C	+4 C	+6 C	A	+2 C	+4 C	+6 C
				Upper Section				
13 May								
13 May–8 Jun	30	46	51	61	0.42	0.79	0.91	1.17
9 Jun–10 Jul	174	255	210	204	1.35	1.76	1.02	0.68
11 Jul–8 Aug	479	544	439	337	2.39	1.49	1.68	0.99
9 Aug–8 Sept	708	610	454	0	1.42	1.35	0.36	0
Total					5.58	5.38	3.96	2.84
				Lower Section				
13 May								
13 May–8 Jun	50	77	70	108	0.46	0.87	0.75	1.38
9 Jun–10 Jul	330	400	323	463	1.79	2.17	1.27	1.52
11 Jul–8 Aug	858	869	633	890	2.40	1.83	1.05	1.66
9 Aug–8 Sept	951	974	677	857	0.92	0.94	0.89	0.30
Total					5.57	5.34	3.95	4.87
				Undivided				
13 May								
13 May–8 Jun	105	105	178	190	0.66	0.66	1.40	1.53
9 Jun–10 Jul	560	544	704	782	1.52	1.43	1.24	1.44
11 Jul–8 Aug	1,392	1,285	1,111	1,380	2.16	2.00	0.52	1.37
9 Aug–8 Sept	1,406	1,525	1,231	1,292	0.82	0.65	1.36	−0.05
Total					5.16	4.75	4.51	4.29

mum temperature in the +4 C regimen exceeded 29 C for 95 days. According to the interval production statistics (Table 2), walleyes in both the ambient and +4 C regimen were growing during these periods. The upper lethal temperature for walleyes in the channels (+6 C regimen, upper section) was about 34 C although this was not determined on the basis of 50% mortality. Differences in the results between the laboratory and channel studies probably were due to the low acclimation temperature (26 C) and the direct transfer technique of handling the fish that were employed in the laboratory study (Hokanson 1977).

Thermal conditions in the undivided channels were considered to simulate thermal plumes such as those associated with heated discharges from power plants. In this case the highest temperature in the plume would occur in 40% of the mixed water body (channel), but the walleyes were not prevented from seeking cooler zones along a thermal gradient or vice versa. Although the mean size of walleyes attained after 118 days in the elevated regimens was less (8 to 21 g) than that in the ambient channel, the yield, which reflected both growth and survival, was about equal to or higher than that in the undivided ambient channel (Table 1). Apparently the thermoregulatory behavior of walleyes (Ferguson 1958; Hokanson 1977) enabled the population in the undivided channel of the +6 C regimen to avoid the temperatures that were lethal (ca. 34 C) for walleye confined to the upper channel sections in this treatment. Average minimum temperatures in the lower section of the +6 C regimen ranged from 32 to 33 C during the period (22 July–15 August) when lethal conditions were reached in the upper section. Based on the interval production statistics between 10 July and 8 August (Table 2), growth occurred during this interval in both the lower divided sections and the undivided channel in the +6 C regimen.

Although the juvenile stage is perhaps the most thermally tolerant stage in the life cycle of percids, with walleye being the least tolerant species (Hokanson 1977), results from the present channel experiment indicated that walleyes were more thermally tolerant than previously considered. Thermal tolerance of walleyes in this study was more similar to that determined for young yellow perch (McCormick 1976). It has been suggested that there may be differences in the thermal tolerance of different genetic strains; however, there was no intent to test this hypothesis although a northern stock of walleye was used in the present study. Along with the need for additional information on temperature requirements of the adult walleye stage, the genetic question remains for future research.

References

ARMITAGE, B. J., T. D. FORSYTHE, E. B. RODGERS, AND W. B. WRENN. 1978. Colonization by periphyton, zooplankton, and macroinvertebrates. Browns Ferry Biothermal Res. Ser. I. U.S. Environ. Prot. Agency, Ecol. Res. Ser. (In press.)

BRUNGS, W. A., AND B. R. JONES. 1977. Temperature criteria for freshwater fish: protocol and procedures. U.S. Environ. Prot. Agency, Duluth, Minn. EPA-600/3-77-061. 130 pp.

CARLANDER, K. D. 1977. Biomass, production, and yields of walleye (*Stizostedion vitreum vitreum*) and yellow perch (*Perca flavescens*) in North American lakes. J. Fish. Res. Board Can. 34(10): 1602–1612.

CHAPMAN, D. W. 1971. Production. Pages 199–214 *In* W. E. Ricker, ed. Methods for assessment of fish production in freshwaters. IBP Handbook No. 3, 2nd ed. Blackwell, Oxford and Edinburg.

FERGUSON, R. G. 1958. The preferred temperature of fish and their midsummer distribution in temperate lakes and streams. J. Fish. Res. Board Can. 15:607–624.

FITZ, R. B., AND J. A. HOLBROOK II. 1978. Walleye and sauger in Norris Reservoir, Tennessee. Am. Fish. Soc. Spec. Publ. 11:82–88.

FORNEY, J. L. 1971. Analysis of year class formation in a walleye population. Fed. Aid Proj. F-17-R, Rep. Job I-a, N.Y. State Dep. Environ. Conserv.

FORSYTHE, T. D. 1977. Predator-prey interactions among crustacean plankton, young bluegill (*Lepomis macrochirus*) and walleye (*Stizostedion vitreum vitreum*) in experimental ecosystems. Ph.D. Thesis. Mich. State Univ., East Lansing. 104 pp.

HACKNEY, P. A., AND J. A. HOLBROOK II. 1978. Sauger, walleye, and yellow perch in the southeastern United States. Am. Fish. Soc. Spec. Publ. 11: 00–00.

HOKANSON, K. E. F. 1977. Temperature requirements of some percids and adaptations to the seasonal temperature cycle. J. Fish. Res. Board Can. 34(10):1524–1550.

MCCORMICK, J. H. 1976. Temperature effects on young yellow perch, *Perca flavescens* (Mitchill). U.S. Environ. Prot. Agency, Ecol. Res. Ser. EPA-600/3-76-057. 18 pp.

———, B. R. JONES, AND R. F. SYRETT. 1971. Temperature requirements for growth and survival for larval ciscos (*Coregonus artedii*). J. Fish. Res. Board Can. 28:924–927.

NATIONAL ACADEMY OF SCIENCES AND NATIONAL ACADEMY OF ENGINEERING. 1973. Water quality criteria 1972. U.S. Gov. Print. Off., Washington, D.C.

RICKER, W. E. 1946. Production and utilization of fish populations. Ecol. Monogr. 16:373–391.

SMITH, L. L., JR., AND W. M. KOENST. 1975. Temperature effects on eggs and fry of percoid fishes. U.S. Environ. Prot. Agency, Duluth, Minn. EPA-660/3-75-017. 91 pp.

STROUD, R. H. 1949. Growth of Norris Reservoir walleye during the first 12 years of impoundment. J. Wildl. Manage. 13(2):157–177.

WALKER, R. E., AND R. L. APPLEGATE. 1976. Growth, food, and possible ecological effects of young-of-the-year walleyes in a South Dakota prairie pothole. Prog. Fish-Cult. 38(4):217–220.

Sauger, Walleye, and Yellow Perch in The Southeastern United States

P. A. HACKNEY AND JOHN A. HOLBROOK II

Tennessee Valley Authority
Norris, Tennessee 37828

Abstract

Sauger, walleye, and yellow perch are all native to the southeastern United States, the southerly limit of their natural range. Important fisheries exist only for sauger in low-gradient mainstream reservoirs and walleye in upland tributary reservoirs. Mortality of sauger is very high and year-class strength quite variable. Growth is rapid for both sauger and walleye, and fisheries for trophy walleye formerly existed in many reservoirs for several years following impoundment. Walleye fisheries declined in all reservoirs, but were restored from stockings of northern strains. Self-sustaining populations of walleye now exist although trophy fish are no longer taken. The factors limiting percid abundance and distribution in the southeastern United States are unclear. Water temperature does not appear to be the limiting factor in most instances.

The southeastern United States is the natural southerly limit of the three percid species of angling interest: sauger; walleye; and yellow perch. These species were formerly confined to streams or brackish waters in the south. However, recent construction of numerous impoundments and other developmental activities have led to substantial modification of original habitats and distributions in this area. Stockings outside their original ranges have expanded the distribution of these species and have provided further information on biology and fisheries characteristics.

This work describes the distribution, "races," growth, cycles of abundance, and fisheries of these species in the southeastern United States. Of additional interest are those factors which appear to limit the distribution of these species in the southernmost portions of their ranges.

Distributions

Sauger

Saugers are native to the Mississippi River drainage and fishable populations are found in the Ohio River and all mainstream reservoirs on the Tennessee and Cumberland Rivers (Fig. 1). In addition, sauger are abundant in Norris and Douglas Reservoirs—deep impoundments located on major tributaries to the Tennessee River. Saugers (and walleyes; Netsch and Turner 1964) were present in Cherokee Reservoir for many years following impoundment but have disappeared. Outside the Mississippi River drainage, sauger (limited numbers) have been successfully introduced to Clark Hill Reservoir, Georgia (K. W. Primmer, personal communication); and the Appalachicola River, Florida (J. M. Barkuloo, personal communication). Stocked sauger failed to establish populations in the Chattahoochee River, Georgia; Lake Norman, North Carolina (H. M. Ratledge, personal communication); and Lewis Smith Lake, Alabama (S. L. Spencer, personal communication).

Walleye

Walleyes are also native to the Mississippi River drainage (Fig. 2). Walleyes are rare in mainstream reservoirs of the Tennessee and Cumberland Rivers but sizable populations occur in deep tributary reservoirs. Outside the Mississippi River drainage, walleye fisheries have been established in Lakes Hartwell and Burton, Georgia, and Lakes James, Kerr, and Gaston, North Carolina.

The Mississippi River drainage stock of walleye was widespread in the southeastern United States. Before the advent of impoundments it was restricted to streams and ranged over all of Kentucky, almost all of Tennessee, parts of Virginia, North Carolina, and Georgia, and much of Mississippi. The current angling record of 11.3 kg was taken from the Cumberland River drainage and was probably an individual of this stock.

Two other naturally occurring, apparently disjunct populations of walleye occur in the southeast outside the Mississippi River drainage—a mid-Atlantic population and a Gulf Coast population. The mid-Atlantic walleyes were found in North Carolina in the Pasquotank River, Edenton Bay, the Roanoke River (Smith 1907) and at the fall line in the Nottaway River system of Virginia (R. G. Martin, personal communication). This stock may have produced the walleye fisheries found in Kerr (Buggs Island) and Gaston Reservoirs on the Roanoke River (R. G. Martin and W. B. Smith, personal communications).

The Gulf Coast stock of walleye (Brown 1962; Cook 1959; Smith-Vaniz 1968) is more of biological than angling interest as specimens are usually taken only incidently by fishermen. This stock is known from brackish waters in Mobile Bay and occurs from the extreme western portion of the Florida panhandle to the Pearl River, Mississippi. The largest populations appear in the cooler streams, particularly below cool hydroelectric discharges. This race is unusual in that it does not establish in reser-

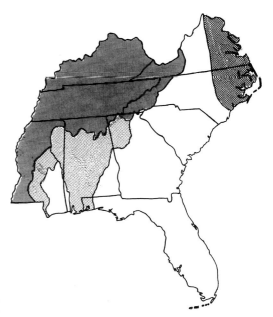

FIGURE 2.—*Distribution of walleye in the southeastern United States showing the Mississippi River, Gulf Coast, and Atlantic Coast "races."*

voirs nor are individuals encountered there. This is surprising since many of the reservoirs appear to be suitable habitat and walleye populations persist in streams above and below these reservoirs following impoundment. The angling record is a fish of 3.75 kg from Alabama.

The current walleye populations in Center Hill and Dale Hollow Reservoirs apparently developed from New York State stocks thought to have included both yellow (*S. vitreum vitreum*) and blue walleyes (*S. v. glaucum*). Individuals with large eyes and a bluish cast have been observed but could not be positively identified as blue walleye (Libbey 1969; Muench 1966). Scott (1976) did not find any blue walleyes during extensive investigations of Center Hill and concluded "that the present stock is the yellow walleye, *Stizostedion vitreum vitreum*."

Yellow Perch

The yellow perch originally occurred along the Atlantic Coast south to North Carolina (Jordan and Evermann 1902) and in Gulf Coast drainage streams of Alabama (Smith-Vaniz 1968) where it is apparently native though rare. The species range has

FIGURE 1.—*Distribution of sauger in the southeastern United States.*

now been extended through introductions to include portions of the Mississippi Valley, South Atlantic, and Florida Gulf Coast streams. In North Carolina the yellow perch seems to be "moving" upstream into mountain impoundments (W. M. Ratledge, personal communication), and in Virginia it has been stocked in the Ohio River drainage (R. G. Martin, personal communication). The yellow perch is common in most Georgia streams (and their impoundments), except the Coosa (K. W. Primmer, personal communication) and, by way of the Chattahoochee River, is found in the Appalachicola in Florida (J. M. Barkuloo, personal communication).

The species now is extending its range rapidly within the Tennessee River drainage (Timmons 1975). Since its discovery in 1953 in Chatuge Reservoir, North Carolina and Georgia, the species has spread downstream in the Hiwassee River and become established in Hiwassee and Apalachia Reservoirs. Upon reaching the Tennessee River it has become established in Chickamauga Reservoir and has radiated downstream through Nickajack, Guntersville, and Wheeler Reservoirs, and upstream through Watts Bar and Melton Hill Reservoirs (unpublished TVA data). It seems likely that yellow perch will eventually colonize the entire mainstream Tennessee Reservoir system and probably the Cumberland River system as well. However, it is surprising to us that a species which was unable to extend its range southward in the Mississippi and Ohio Rivers should suddenly be extending its range northward from a southern introduction.

Percid Fisheries Characteristics

The largest percid fishery in the southeastern United States is for sauger in the mainstream reservoirs of the Cumberland and Tennessee Rivers. Sauger fisheries are also found in two Tennessee tributary impoundments, Norris and Douglas. No significant sauger fisheries are present in the southeastern impoundments outside of the Tennessee and Cumberland River systems. A number of reservoirs contain walleye fisheries but with the exception of Claytor Lake, Virginia, Lakes Burton and Hartwell, Georgia, and Lake James and Kerr and Gaston Reservoirs in North Carolina, most are also found in the Cumberland or Tennessee drainages. Only in Norris Reservoir, Tennessee, do both sauger and walleye fisheries coexist.

Both sauger and walleye fisheries are typically seasonal, at least in the Tennessee Valley. Usually saugers congregate in the tailwaters of the next upstream dam beginning in late fall and remain there through the spring. This concentration provides the bulk of the sauger fishery. During years of very low density, even this "concentrated" fishery may be too poor to attract many anglers.

The fish are scattered throughout the reservoir the remainder of the year and are relatively unavailable to fishermen. We suspect this unavailability results both because of the relatively low density of the dispersed stocks and because relatively few fishermen know how to catch sauger during summer. In years of peak abundance, like 1977, saugers are taken in good numbers in all areas of the reservoirs throughout the year and may even be the dominant fishery in some impoundments.

In most reservoirs, walleye fisheries are similar except that the fish only congregate when they are on the actual spawning run and hence are vulnerable for a shorter period of time. Some troll fisheries do exist for both sauger and walleye during periods they are not concentrated. However, these fisheries are largely seasonal components of the overall sport fishery in the Tennessee Valley. Relatively few fishermen specifically seek these species except when they are concentrated—usually either in tailwaters, in steam plant discharge basins (Wrenn 1975), or on spawning runs in rivers. In addition, the size of the populations supporting the fishery varies and is usually dependent on two- and/or three-year-old fish.

Yellow perch fisheries in the southeast are unknown to us. They are taken incidently by anglers and specimens in excess of 0.5 kg are occasionally observed though most are too small to be of interest to fisherman.

Cycles of Abundance

Saugers experience considerable fluctuations in abundance in the mainstream reservoirs of the Tennessee River system (Table 1). When sauger stocks are high they are generally high in all reservoirs, and when low are low in all reservoirs. Saugers are migratory and pass through locks of mainstream Tennessee Dams. Many fish apparently migrate into this system from the Ohio River (Cobb 1960).

The factors responsible for producing large sauger stocks in some years have not been identified but are likely abiotic effects on reproductive success. Walleye fisheries are not known to undergo such large fluctuations, possibly because their greater longevity tends to damp out variations in reproductive success.

Mortality and Growth

Mortality

Mortality of sauger is exceptionally high in the Tennessee and Cumberland River impoundments. The estimated mortality for 1,123 age II and older individuals collected in gill nets (37 mm bar mesh) from seven reservoirs was 87.7% annually ($r^2 = 0.986$; instantaneous mortality coefficient, $Z = 2.097$). Although these data may be biased by mesh selectivity, it is doubtful if serious error exists due to the relatively small length differences among age I to V saugers (301–465 mm). Only two individuals were age V and only 15 were age IV in the sample of 1,123 sauger.

The reason(s) for this high mortality rate is (are) unknown. Excessive exploitation is a possible cause although this seems unlikely to us. However, it is obvious that the maximum biomass of a year class is attained sometime before age II, after which attrition is extremely rapid.

Our data are insufficient for mortality studies of walleye but the mortality rate in the southeast is apparently greater than in more northerly waters.

Growth

Numerous authors (Stroud 1949; Libbey 1969; Scott 1976) have reported that walleyes in the southeast grow faster and mature earlier but do not live as long as fish in more northerly cooler waters. By the end of the second year, walleyes in most southern waters range from 300 to over 400 mm total length (Table 2), they are recruited to the fishery, and both sexes are mature. Females typically grow faster and live longer; larger (over 4 kg) walleyes are invariably females.

Southeastern saugers do not grow as fast as walleye. Like the walleyes, however, southern saugers grow faster than northern fish in the early years (Table 3) and females grow faster, live longer, but mature somewhat later than males (Hassler 1957). Few fish attain age III and practically none survive to age IV or V. More northerly saugers live longer and thus are able to attain the same ultimate size as southern saugers (Table 3).

Post-Impoundment Changes in Walleye Fisheries

Reproductive Success

Walleye fisheries in many reservoirs in the southeastern U.S. expanded immediately following impoundment but later de-

TABLE 1.—*Sauger density (numbers/hectare) in cove rotenone samples of mainstream Tennessee River impoundments. Annual averages are for all reservoirs.*

Year	Density
1949	5.2
1950	4.1
1951	19.1
1952	10.9
1953	14.3
1954	22.8
1955	21.0
1956	8.7
1957	8.3
1958	18.6
1959	18.4
1960	14.9
1961	16.7
1962	4.9
1963	12.7
1964	7.4
No samples	
1969	0.2
1970	3.0
1971	5.3
1972	7.3
1973	7.5
1974	6.9
1975	9.4
1976	14.9

TABLE 2.—Mean total calculated lengths (mm) at annulus for walleye (sexes combined) in southern reservoirs and the Mississippi River.

Location	Annulus									
	1	2	3	4	5	6	7	8	9	10
Norris Reservoir, Tenn. (Stroud 1949)	264	416	474	505	528	533	558	632		
Norris Reservoir, Tenn. (TVA, unpublished data)	247	404	475	524						
Melton Hill Reservoir, Tenn. (Fitz 1968)	248	330	368							
Watauga Reservoir, Tenn. (TVA, unpublished data, 1972)	228	340								
Fontana Reservoir, N.C. (Louder 1965)	167	287	388	452	513	568				
Mississippi River (Vasey 1967)	175	307	414	492	541					
Canton Reservoir, Okla. (Lewis 1970)	309	426	495	553	607	650	704			
Center Hill Reservoir, Tenn. (Scott 1976)	267	421	480	508	533	559	592	664	709	716
Center Hill Reservoir, Tenn. (Muench 1966)	248	415	497	512	583	610	683	715		
Center Hill Reservoir, Tenn. (Netsch and Turner 1964)	202	414	541	617	688	665	778			
Dale Hollow Reservoir, Tenn. (Libbey 1969)	264	409	488	546	602	638	645			
Dale Hollow Reservoir, Tenn. (Netsch and Turner 1964)	198	432	569	660	711	742	800	808	800	
Claytor Lake, Va. (Roseberry 1950)	251	386	503	580	663	701	759	818		

clined. Following impoundment in 1949, the walleye population in Center Hill Reservoir expanded, peaked in 1954, and then declined to virtual extinction by 1957 (Muench 1966). Growth of walleyes during this period was extremely rapid (Table 2) as compared both to other southeastern reservoir walleye populations and to wall-

TABLE 3.—Mean total calculated lengths (mm) at annulus for sauger (sexes combined) in Tennessee and Cumberland River impoundments and two other populations.

Location	Annulus						
	1	2	3	4	5	6	7
Norris Reservoir, Tenn. (Hassler 1957)	213	337	396	436	472	497	515
Norris Reservoir, Tenn. (TVA, unpublished data)	202	355	412	457	491		
Old Hickory Reservoir, Tenn. (TVA, unpublished data, 1976)	191	312	408	517			
Melton Hill Reservoir, Tenn. (Fitz 1968)	228	338	368				
Douglas Reservoir, Tenn. (TVA, unpublished data, 1973)	216	348	410				
Kentucky Lake, Ky. and Tenn. (TVA, unpublished data, 1964)	149	314	386	502			
Kentucky Lake, Ky. and Tenn. (TVA, unpublished data, 1976)	208	322	402				
Barkley Reservoir, Ky. (TVA, unpublished data, 1977)	197	301	350	375			
Guntersville Reservoir, Ala. (TVA, unpublished data, 1975)	245	356	416				
Pickwick Reservoir, Ala., Miss., Tenn. (TVA, unpublished data, 1975)	223	338	385				
Watts Bar Reservoir, Tenn. (TVA, unpublished data, 1976)	223	343	420	459	465		
Mississippi River (Vasey 1967)	144	269	355	414	450	480	513
Lake Sakakawea, N.D. (Wahtola et al. 1972)	131	225	296	357	423	475	481

eye growth rates in Center Hill in later years. Muench believed this was due to the expanded area and abundant prey available to the walleye following impoundment but offered no data to support this conclusion. The fishery was restored by fry stocking (New York eggs) during 1954–1960 (Scott 1976) and Center Hill now contains a good walleye fishery.

Less information is available for Dale Hollow Reservoir, Tennessee and Kentucky, but a similar sequence of events apparently occurred. The excellent fishery in the early 1950's declined and was similarly restored by stocking fry (New York source) during the 1954–1965 period (Libbey 1969).

Other reservoir walleye fisheries which have been restored by stocking include Blue Ridge, Georgia (K. W. Primmer, personal communication) and Lake Cumberland, Kentucky, (P. W. Pfieffer, personal communication). Most Tennessee River impoundments which currently contain good walleye fisheries (Norris, Watauga, Hiwassee, Fontana, Chatuge, Nantahala, Santeetalah) were either stocked with walleyes from northern sources or historic records are insufficient to determine if naturally reproducing native populations occurred prior to stocking. Two impoundments on the Roanoke River, Kerr (Buggs Island) and Gaston, apparently had naturally reproducing native stocks which expanded immediately following impoundment but have since declined.

These findings all follow a pattern: naturally occurring "southern" stocks of walleye, at least in the Tennessee, Cumberland, and Roanoke Rivers (and perhaps the New River), were trapped behind new impoundments and apparently reproduced during the first few years following impoundment. They were evidently unable to sustain their populations and eventually declined. Only if a stocking program was initiated did the fishery revive. Since the fisheries are now self-sustaining there must be a difference between the stocked fish and the original "southern" stocks. Available data reveal that most of the fish which were (and are now being) stocked in southeastern impoundments are of New York (Cayuga Lake) or Ohio (Lake Erie) origin. This "northern" stock now spawns in rivers and in the lake proper whereas the "southern" stock probably was an obligate river spawner, perhaps even having specific well-defined spawning ground which, if eliminated, doomed the population. We suspect that most reservoir fisheries in the southeast are now entirely of "northern" fish.

Growth Differences

Besides the Center Hill data (Table 2), other data suggest that the "southern" stock had a faster growth rate, at least at older ages, than present stocks. As shown by Table 2 and discussed by Libbey (1969), growth of younger fish in Dale Hollow was similar for 1953 (Netsch and Turner 1964) and 1968–1969. However, fish of age IV–VII (all females) had grown faster in 1953 than did individuals collected in 1968–1969. These older fish (1953 data) would have been spawned 3–6 years after impoundment of Dale Hollow. Similarly, the growth of age I–IV Claytor Lake walleye in 1948–1949 was comparable to growth in other waters (Table 2) but older fish grew extremely fast (Roseberry 1950). However, because Claytor Lake was impounded and stocked in 1939, it is not possible to determine whether the stocked or native fish were the fast growers. While rapid growth due to expanded area and/or increased prey could occur for the first one or two year classes following impoundment, this does not explain why a year class produced five years after impoundment would have better growth throughout its older (but not its younger) age classes than a year class produced 10 or 15 years after impoundment.

Exceptionally large (over 8 kg) walleyes are no longer taken from southeastern impoundments. We reject the argument that increased harvest has prevented walleyes from attaining large size because (1) some walleye populations are lightly exploited, (2) we have seen no evidence that exploitation of reservoir stocks is greater now than for the same stocks in rivers prior to impoundment, and (3) we doubt that exploitation is now greater for reservoir stocks than it was 15–20 years ago when fewer walleye fisheries existed and interest in some fisheries was very high because trophy fish were present. We conclude that the "southern" stock has the capability to grow larger than the "northern" stock, at least at older ages.

Also, the "northern" stock may not live as long as the "southern" stock, at least in the southeast.

If our ideas on the "southern" and "northern" races of walleye are correct, the stocking of the "northern" race with its lake spawning habits is probably responsible for the establishment of walleye fisheries in southeastern impoundments. In establishing self-reproducing stocks we may have sacrificed the ability of these reservoirs to produce trophy-size walleye. Managers confronted with put-grow-take walleye fisheries may want to consider the "southern" stock because of its rapid growth and/or trophy size. However, additional research will be required to confirm or deny these suppositions.

Limiting Factors

In the Tennessee River impoundments saugers are moderately abundant below elevations of about 250 m above sea level. Walleyes appear sporadically below 300 m but are abundant at greater elevations. We suspect that the apparent relationship with increasing elevation is really one of the lake type. Lakes in the higher elevations tend to be cool, deep, and clear with much rock outcropping. Those at lower elevations tend to be warm, shallow, and turbid with mud-silt bottoms.

Norris Reservoir, Tennessee, is unusual in that walleyes and saugers were both present in good numbers (Fitz and Holbrook 1978, this volume). Saugers are currently relatively scarce in this reservoir. Both species also formerly occurred in Cherokee Reservoir (Netsch and Turner 1964) where saugers were dominant. Both are now absent from this lake, apparently because of a dam precluding spawning migration. The two species seldom coexist in reservoirs of the southeastern United States except in instances where one (usually walleye) is rare.

The headwater characteristics may determine the distribution of sauger in Tennessee River Valley impoundments. Those reservoirs containing sauger fisheries all have large rivers flowing into them. In the mainstream impoundments, these "rivers" are actually the tailwaters of the dams immediately upstream. The large congregations of sauger found below these dams a few months prior to and during spawning time in both the Tennessee Valley and in Lewis and Clark Lake and the Missouri River (Nelson 1968) and the occurrence of most Norris Reservoir saugers in spawning condition in the headwater rivers (Fitz and Holbrook 1978, this volume) suggest that a large river habitat is required for reproduction and/or early survival. With the exception of Norris and Douglas Reservoirs (and formerly Cherokee Reservoir), these environments are not present above any of the Tennessee Valley tributary impoundments.

The avoidance of temperatures above 24 C observed for Norris walleyes (Dendy 1945; Fitz and Holbrook 1978, this volume) has been suggested as the reason for their absence in mainstream Tennessee River reservoirs. In the tributary reservoirs containing walleyes, minimum temperatures in the epilimnion exceed 24 C for a mean annual period of 50 days whereas in the mainstream impoundments (which are typically not thermally stratified) the period averages 86 days at corresponding depths. However, work in heated channels (Wrenn and Forsythe 1978, this volume) and the success of walleye introduction into Oklahoma and Texas suggests that water temperature is probably not the factor limiting distribution of walleye.

The factors that limit the southerly distribution of percids are unclear, and do not appear to include temperature, at least in the sense of being lethal. It even seems improper to label these as "coolwater" species since walleye and yellow perch, at least, are found as far south as the brackish waters of the Gulf of Mexico. Although "heat-tolerant" percid races in the southeastern United States are a possibility, it seems more likely to us that other factors are involved.

References

BROWN, B. E. 1962. Occurrence of the walleye, *Stizostedion vitreum*, in Alabama south of the Tennessee Valley. Copeia 1962(2):469–471.

COBB, E. S. 1960. The sauger fishery in the lower Tennessee River. Dingell-Johnson Proj. F-12-R, Large impoundment investigations, Final Rep., Tenn. Game Fish Comm. 16 pp.

COOK, F. A. 1959. Freshwater fishes in Mississippi. Miss. Game Fish Comm., Jackson. 239 pp.

DENDY, J. S. 1945. Fish distribution, Norris Reservoir, Tennessee, 1943. II. Depth distribution of fish in relation to environment (factor), Norris Reservoir. J. Tenn. Acad. Sci. 20(1):114–135.

FITZ, R. B. 1968. Fish habitat and population changes resulting from impoundment of Clinch River by Melton Hill Dam. J. Tenn. Acad. Sci. 43(1):7–15.

———, AND J. A. HOLBROOK II. 1978. Sauger and walleye in Norris Reservoir, Tennessee. Am. Fish. Soc. Spec. Publ. 11:82–88.

HASSLER, W. W. 1957. Age and growth of the sauger (*Stizostedion canadense canadense* (Smith)) in Norris Reservoir, Tennessee. J. Tenn. Acad. Sci. 32(1):55–76.

JORDAN, D. S., AND B. W. EVERMANN. 1902. American food and game fishes. Doubleday, Page and Co. 574 pp.

LEWIS, S. A. 1970. Age and growth of walleye, (*Stizostedion vitreum vitreum* (Mitchill)), in Canton Reservoir, Oklahoma. Proc. Okla. Acad. Sci. 50:84–86.

LIBBEY, J. E. 1969. Certain aspects of the life history of the walleye, *Stizostedion vitreum vitreum* (Mitchill), in Dale Hollow Reservoir, Tennessee and Kentucky, with emphasis on spawning. M.S. Thesis. Tenn. Tech. Univ., Cookeville. 55 pp.

LOUDER, D. E. 1965. The biology of Fontana Reservoir fishes. Pages 3–9 *in* Dingell-Johnson Proj. F-16-R-1, Job III A, Work Plan III: Power reservoir investigations, Annu. Prog. Rep., N. C. Wildl. Resour. Comm.

MUENCH, K. A. 1966. Certain aspects of the life history of the walleye, *Stizostedion vitreum vitreum* (Mitchill), in Center Hill Reservoir, Tennessee. M.S. Thesis. Tenn. Tech. Univ., Cookeville. 66 pp.

NELSON, W. R. 1968. Reproduction and early life history of sauger, *Stizostedion canadense*, in Lewis and Clark Lake. Trans. Am. Fish. Soc. 97(2):159–166.

NETSCH, N. F., AND W. L. TURNER. 1964. Creel census and population studies. Dingell-Johnson Proj. F-2-R, Final Rep., Tenn. Game Fish Comm. 40 pp.

ROSEBERRY, D. A. 1950. Fishery management of Claytor Lake, an impoundment on the New River in Virginia. Trans. Am. Fish. Soc. 80:194–209.

SCOTT, E. M., JR. 1976. Dynamics of the Center Hill walleye population. Tenn. Wildl. Resour. Agency Tech. Rep. No. 76-55, Nashville, Tenn. 86 pp.

SMITH, H. M. 1907. The fishes of North Carolina. II. Pages 248–252 *in* North Carolina geological and economic survey. E. M. Uzzell and Co., Raleigh, N.C.

SMITH-VANIZ, W. F. 1968. Freshwater fishes of Alabama. Paragon Press, Montgomery, Ala. 211 pp.

STROUD, R. H. 1949. Growth of Norris Reservoir walleye during the first 12 years of impoundment. J. Wildl. Manage. 13(2):157–177.

TIMMONS, T. J. 1975. Range extension of the yellow perch, (*Perca flavescens* (Mitchill)), in Tennessee. J. Tenn. Acad. Sci. 50(3):101–102.

VASEY, F. W. 1967. Age and growth of walleye and sauger in pool II of the Mississippi River. Iowa State J. Sci. 41(4):447–466.

WAHTOLA, C. H., D. E. MILLER, AND J. B. OWEN. 1972. The age and growth of walleye (*Stizostedion vitreum*) and sauger (*Stizostedion canadense*) in Lake Sakakawea, North Dakota. Proc. N. D. Acad. Sci. 25(2):72–83.

WRENN, W. B. 1975. Seasonal occurrence and diversity of fish in a heated discharge channel, Tennessee River. Proc. Annu. Conf. Southeast Assoc. Game Fish Comm. 29:235–247.

———, AND T. D. FORSYTHE. 1978. Effects of temperature on production and yield of juvenile walleye in experimental ecosystems. Am. Fish. Soc. Spec. Publ. 11:66–73.

Sauger and Walleye in Norris Reservoir, Tennessee

RICHARD B. FITZ AND JOHN A. HOLBROOK II

Division of Forestry, Fisheries, and Wildlife Development
Tennessee Valley Authority
Norris, Tennessee 37828

Abstract

Norris Reservoir, Tennessee, contains populations of both walleye and sauger. Following impoundment in 1936 saugers were an important member of the sport fishery but have since declined to 1% of the catch, although they are still fairly abundant. Since 1938 first-year growth of saugers has declined but growth of older age classes has increased. Walleye growth in the first four age classes has shown no trends. In both species, heaviest, longest, oldest, and fastest growing specimens are typically females. Norris populations also grow faster than more northern populations. In contrast to walleyes, saugers are seldom caught in vertical gill nets, apparently spawn over large distances of tributary rivers, and do not appear to spawn in the reservoir proper. Walleyes caught in vertical gill nets appear to stay close to the thermocline when summer stratification is most intense even if it means occupying water with oxygen concentration of 1–2 mg/liter. In cooler months walleyes also apparently move toward the surface at night.

Both sauger and walleye naturally occurred in the unimpounded Tennessee River system, and following impoundment in 1936, fisheries for both species developed in Norris Reservoir, Tennessee. In addition to native stocks, walleyes from Lake Erie were introduced to the reservoir during its early years. Norris is now the only impoundment in the southeastern United States containing significant populations of both species.

Norris Dam is located in eastern Tennessee on the Clinch River, 128 km from its confluence with the Tennessee River. The dam creates a reservoir (Fig. 1) of 13,840 surface hectares at normal full pool (310 m above mean sea level). Two major rivers (Clinch and Powell) enter the reservoir from primarily forested lands and are not influenced by large metropolitan effluents. Mining for coal on the northern rim of the basin creates local problems with acid drainage, but the acid is neutralized naturally before entering the reservoir. From midsummer through the early fall months, Norris Reservoir stratifies thermally, and dissolved oxygen concentrations of less than 1.0 mg/liter occur below depths of 10 m.

Fisheries

Standing stocks of fishes in Norris Reservoir have been estimated by cove rotenone sampling over the past 15 years. They have averaged 162 kg/hectare in 1960–1970; however, studies in August 1975 showed an increase to 268 kg/hectare (Table 1). This increase was reflected in virtually all species; however, threadfin shad contributed 47% of the gain. Standing stock biomass (Table 1) of Norris walleyes has been higher than that of saugers since 1960, although values for both species are probably underestimates since neither species would be expected in coves during warmer months when samples are taken.

During 1974–1975 saugers and walleyes represented 4.9 and 16.8%, respectively, of all fish netted in the shoreline areas and 0.4 and 20.4% of the fish caught in vertical nets. In contrast, in a 1943 netting study Cady (1945) found 5.4% walleye, but sauger was the dominant species (46.6%) in Norris.

Growth

Sauger

Sauger growth in Norris Reservoir was calculated for a 13-year period (1937–1949) by Hassler (1957). For that period, mean total length attained at the end of years one to seven was 213, 337, 396, 436, 472, 497, and 515 mm, respectively, and growth was rapid compared to other populations. Subsequent data confirm Hassler's work.

Yearly growth increments for sauger age classes (Fig. 2) have been variable as previously noted by Hassler (1957). Regressions

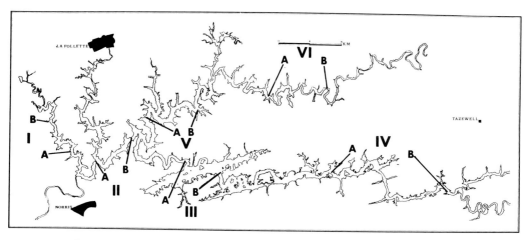

FIGURE 1.—Gill net stations for Norris Reservoir fish distribution studies, 1974–1975.

of growth increment versus time show a significant ($P < 0.05$) decline in first-year growth since Norris was impounded. However, growth at older ages has increased and yearling growth increased significantly ($P < 0.01$). Because of the inverse relationship between first- and second-year growth, length attained at the end of the second year has remained relatively constant through time while third-year length has increased.

Hassler (1955) found that sauger growth was significantly ($P < 0.01$) and positively correlated with the number of days between the first days of 15.4 C water temperatures in both spring and fall (measured at 0.15 m depth). In addition, a negative relationship was shown between attained length and the number of days required to warm Norris Reservoir from 9.9 to 15.4 C. Hassler concluded that an extended period of these temperatures in the spring resulted in prolonged feeding by young saugers on plankton (rather than young shad) which diminished growth.

Females grew faster than males in all age classes but significantly so only in ages two to five (Hassler 1957). Of 5,500 sauger examined by Hassler, the heaviest (1.87 kg) and longest (553 mm) were females. The heaviest and longest males were 1.36 kg and 532 mm, respectively.

Average condition factors (k) for Norris sauger are 1.43 for females and 1.42 for males (Hassler 1957). Standard length (L, mm)–weight (W, kg) relationships are:

Males, $\log W = -5.071 + 3.088 \log L$;

Females,

$$\log W = -5.4738 + 3.2458 \log L;$$

Combined,

$$\log W = -5.4180 + 3.2259 \log L.$$

Walleye

Growth of Norris walleyes in 1936–1940 was discussed by Eschmeyer and Jones (1941), and Stroud (1949) summarized data for the first 12 years (1936–1947) following impoundment. During the period 1936–1947, Norris walleyes grew to an average of 264 and 416 mm (total length) at the end of the first and second years of

TABLE 1.—Standing stock (kg/hectare) of fishes in Norris Reservoir, August 1975 and 1960–70.

Species	1975 (12 samples)	1960–1970 (13 samples)
Gizzard shad	111.7	99.4
Threadfin shad	53.6	8.4
Bluegill	22.3	10.7
Carp	19.6	17.3
Freshwater drum	11.6	2.8
Smallmouth bass	6.4	4.9
Largemouth bass	5.6	4.2
Channel catfish	3.7	3.6
Flathead catfish	3.7	0.8
Striped bass	2.9	0
Spotted bass	2.5	1.3
White bass	1.5	1.5
Walleye	2.5	0.8
Sauger	0.7	0.3

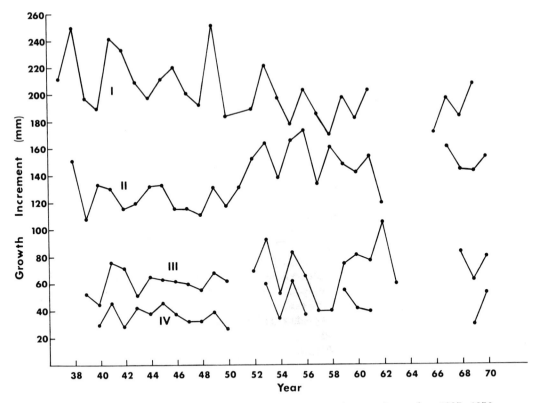

FIGURE 2.—Mean calculated incremental growth of Norris Reservoir sauger by age class, 1937–1970.

growth, respectively (Stroud 1949). The minimum legal size during these years was 381 mm; therefore, most Norris walleyes were harvestable by the end of the second year of growth. Mean lengths attained by the end of the third through eighth years of growth were calculated to be 474, 505, 528, 533, 558, and 632 mm, respectively. Stroud (1949) found that the growth of Norris walleyes was very rapid compared with that in other waters. Walleye growth has been quite variable since 1943 (Fig. 3), but unlike the case for sauger, there have been no long-term growth trends for age classes I–IV.

Stroud (1949) also found that during early spring, growth of one-year-old fish is relatively slow but rapid in older fishes. Growth of all age classes of walleye slows from mid-June to mid-August, but increases in early fall and is complete by the end of November although guts are generally well filled in winter (Dendy 1946).

During their first year females outgrew males; however, the differences were most pronounced after three years of age. Of the 528 males examined during Stroud's (1949) studies, the oldest was eight years, the longest was 589 mm, and the heaviest was 1.9 kg. Of the 514 females, the oldest female was also eight years, the longest was 805 mm, and the heaviest was 5.4 kg. Larger walleyes (9 kg) used to be taken from Norris Reservoir and were undoubtedly females.

Condition factors of Norris walleyes (Stroud 1949) were 1.38 for age I fish and 1.51 for older fish, sexes combined. The standard length (L, mm), weight (W, kg) relationship for all fish is $\text{Log } W = -5.069 + 3.079 \log L$.

Harvest

During the late 1940's and early 1950's sauger comprised an important portion of the sportfish harvest; however, since that time it has been of relatively minor importance (Table 2).

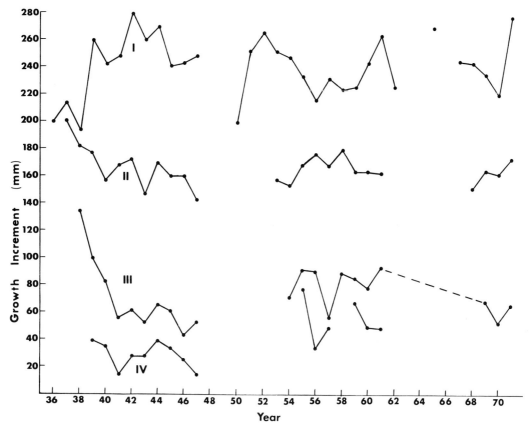

FIGURE 3.—*Mean calculated incremental growth of Norris Reservoir walleye by age class, 1936–1971.*

The average size of sauger caught over the years has been variable. Prior to 1950, there was a 381-mm minimum length limit on both saugers and walleyes. Saugers caught by sport fishermen during these years averaged 407 mm and represented age II. Following removal of the length restrictions, mean length decreased to 376 mm during the 1950's, to 344 mm in 1963, and recovered to 352 mm in 1970. Since 1970, the mean length of saugers in the creel steadily increased to 434 mm by 1975.

Walleye harvest increased during the late 1940's, but unlike that of sauger, it has held relatively steady since then and averages close to 12% of the total sport harvest (Table 2). The size of walleyes caught over the years has followed the same pattern shown for saugers. During the years of length limit restrictions, walleyes caught averaged 496 mm total length. Following removal of these restrictions, mean length dropped to 434 mm during the 1950's and to 393 mm in 1963. During the 1970's the mean length increased and by 1974 was 474 mm.

Exploitation

Exploitation rates for saugers and walleyes were determined over an 18-year period (Chance 1958). The data reveal mean annual rates of 22.1% (SD = 6.4) for saugers and 23.2% (SD = 7.7) for walleyes. The individual annual rates show a rather consistent harvest from year to year for both species; however, returns from a 1975 tagging study were only 8.6% for saugers and 5.6% for walleyes.

Distribution

During March 1974–January 1975, a netting project was conducted on Norris

TABLE 2.—*Percentage composition of sport species harvested from Norris Reservoir, 1938–1975.*

Year[a]	Sauger	Wall-eye	Large-mouth bass[b]	Small-mouth bass	Spotted bass[b]	Black crappie	White crappie[c]	White bass[d]	Blue-gill	Striped bass[e]	Others
1938		3	39	24					30		4
1939	2	9	56	24		1			6		2
1940	3	9	41	29		8			8		3
1944	4	3	36	13		43			1		1
1945	3	3	8	4		80			1		1
1946	4	8	12	7		65			2		1
1947	8	6	17	9		50			8		3
1948	16	14	10	8		45			4		5
1949	6	12	9	11		58			3		4
1950	14	16	21	21		26			2		3
1951	4	9	16	29		40		1	2		1
1952	1	8	12	26		41		5	3		3
1953	1	14	4	15		48		4	4		3
1954		3	9	17		61		5	2		3
1963	3	15	6	15	Tr[f]	1	32	8	6		14
1970	2	17	5	18	Tr	1	31	4	12	Tr	10
1971	1	8	6	7	1	1	49	8	13		6
1972	1	18	4	19	Tr	9	33	4	10		2
1973	Tr	16	6	12	1	24	28	3	6		4
1974	Tr	11	14	18	1	10	26	1	14	Tr	5
1975	Tr	4	7	14	Tr	5	56	6	4	Tr	4

[a] 1938–1954 data from Chance (1958).
[b] Largemouth bass records included spotted bass through 1954.
[c] White crappie introduced in 1953.
[d] White bass introduced in 1949.
[e] Striped bass introduced in 1965.
[f] Trace, <1%.

Reservoir to determine the spatial, vertical, and temporal distributions of walleyes and saugers as related to water temperature and dissolved oxygen. Six littoral areas each containing two subareas (Fig. 1) were sampled with standard sinking gill nets (46 m × 2.4 m × 38 mm bar mesh), and three pelagic areas (main lake—areas I and II; Clinch—areas III and IV; and Powell—areas V and VI) were each fished with two floating vertical gill nets (7.3 m × 22.9 m × 38 mm and 51 mm bar mesh). All netting stations were fished one week in each month sampled. A similar study was conducted in 1943 (Cady 1945; Dendy 1945) but was confined to the Clinch River section of the reservoir and only littoral areas were sampled.

Monthly catch rates (Fig. 4) indicate the saugers were evenly distributed over the six littoral areas during the 10-month period and that catch rates in most stations sharply declined after April, increased during the fall, and declined from December to January. The similarity in patterns of monthly catch rates among stations also suggests no mass movements from one area of the lake to another. Large numbers of saugers in spawning condition were found in the Clinch and Powell Rivers during April and May of 1975. These fish were found farther upstream and were distributed over a much greater river distance than walleyes. Few saugers in spawning condition were found in the reservoir itself during the period. Few saugers were caught in the vertical nets.

As in the case of saugers, there was no significant difference in spatial distribution of walleyes over the six stations for the netting period (Fig. 5). Variations in catch per hour within each netting station show spring and fall peaks, a summer depression, and a winter depression which is more consistent than that seen for saugers. Comparison with sauger catches (Fig. 4) indicates walleyes were caught about twice as often in littoral nets.

Electrofishing samples during April and May 1975 revealed many walleyes in spawning condition just above the headwaters of the reservoir in the Clinch and Powell Rivers. These fish were taken over shoals in water temperatures of 8–12 C. Large numbers of apparently spawning walleyes were also taken over extensive gravel areas in the reservoir itself during this period.

The vertical gill net data (Table 3) indicate that as the thermocline increased in median depth during the period of stratification (May–August), there was a corresponding increase in walleye depth. Walleyes appeared to avoid temperature above 24 C even to the point of entering relatively deoxygenated water (1 to 2 mg/liter) below the thermocline. No walleyes were captured in water with less than 1 mg/liter dissolved oxygen.

In March, May, June, and July 1974, the vertical gill nets were lifted both at dawn and at dusk. Eighty-four percent of the walleyes were caught during the night hours at a mean depth of 8.8 m less than fish caught during the day. This demonstrates the influence of light intensity on percids as previously reported by Ali and Anctil (1968). As water temperatures increased, the diel differences in depth decreased.

Food Habits

The stomach contents of Norris Reservoir saugers and walleyes during 1943 and 1944 were qualitatively determined by Dendy (1946). Saugers ate primarily gizzard shad (62% occurrence) and young black crappies (15%). Less than 10% of the stomachs contained bass. Many sauger stomachs contained small gizzard shad during January and February, a period when gizzard shad tend to be lethargic because of cold water. Walleyes also utilized small gizzard shad predominantly (71% occurrence) fol-

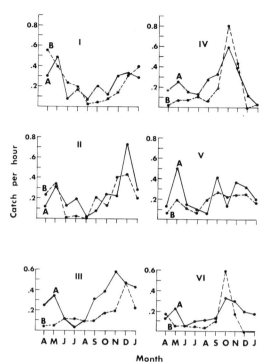

FIGURE 5.—Sinking gill catch (numbers) per hour of walleye by reservoir area and month, 1974–1975.

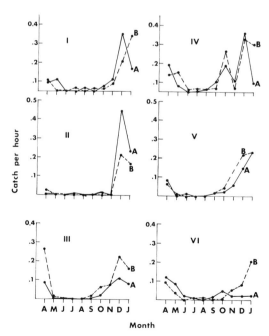

FIGURE 4.—Sinking gill net catch (numbers) per hour of sauger by reservoir area and month, 1974–1975.

TABLE 3.—Mean depth (m) of walleye and median depth of thermocline during March 1974–January 1975. Temperature stratification was absent during months without thermocline data.

Month	Main lake		Clinch arm		Powell arm	
	Walleye	Thermocline	Walleye	Thermocline	Walleye	Thermocline
March	4.1		2.6		2.8	
May	7.2	6.0	5.4	6.0	5.6	6.5
June	4.1	8.0	5.4	8.0	5.0	8.2
July	10.8	11.0	10.2	10.2	11.6	9.0
August	11.5	9.0	7.5		10.4	9.0
September	12.7		9.4		10.9	
October	7.1		7.3		12.6	
November	8.5		7.1		6.3	
December	10.2		9.0		7.8	
January	12.0		9.9		13.8	

lowed by black crappies (11%). Stroud (1949) reported that walleyes less than 508 mm long probably cannot eat adult (203+ mm) gizzard shad. Since male walleyes seldom exceeded this length, he concluded that only the female walleye utilized the adult gizzard shad which undoubtedly increased the differential in growth between the sexes.

Dendy (1946) reported no relation between depth of capture and food eaten for walleyes and also suggested that, because of the large numbers of gizzard shad in Norris and their wide vertical distribution, they are present at all depths occupied by Norris predator fishes (including three bass species). He therefore concluded that serious food competition among largemouth bass, walleyes, and saugers did not exist because each species tended to be concentrated at different depths.

Since 1944, threadfin shad have replaced gizzard shad as the primary food source for the two percids, and in 1974 and 1975 no crappies were found in any of 500 stomachs examined.

Discussion

Early data from Norris showed an exploitation rate of 23% per year for saugers and walleyes; however, age-frequency data revealed only a few fish older than three years, indicating a significant natural mortality occurred in older fish. These observations suggested to Eschmeyer (1944) that fish populations in Norris were underexploited and, together with other factors, ultimately led to the abandonment of closed seasons on Norris Reservoir (and eventually on other waters in the U.S.). Information from 1975 indicates that exploitation of both saugers and walleyes has not increased since the 1940's and, in fact, may have decreased. The percentage contribution of saugers to the creel has been relatively insignificant since the late 1940's, and we speculate that this reflects a decline in sauger numbers following the characteristic "boom" years of a new reservoir. However, because current cove and littoral netting samples show little difference in abundances of the two species, we suspect that saugers are relatively more difficult to catch (probably because of their bottom-dwelling habits), and compared to the walleye, must be extremely abundant to support significant fisheries.

References

ALI, M. A., AND M. ANCTIL. 1968. Correlation entre la structure retinienne et l 'habitat chez (*Stizostedion vitreum vitreum* et *S. canadense*). J. Fish. Res. Board Can. 25(9):2001–2003.

CADY, E. R. 1945. Fish distribution, Norris Reservoir, Tennessee, 1943. I. Depth distribution of fish in Norris Reservoir. J. Tenn. Acad. Sci. 20(1): 103–114.

CHANCE, C. J. 1958. History of fish and fishing in Norris- a TVA tributary reservoir. Proc. Annu. Conf. Southeast. Assoc. Game Fish Comm.: 116–127.

DENDY, J. S. 1945. Fish distribution, Norris Reservoir, Tennessee, 1943. II. Depth distribution of fish in relation to environmental factors, Norris Reservoir. J. Tenn. Acad. Sci. 20(1):114–135.

———. 1946. Food of several species of fish, Norris Reservoir, Tennessee. J. Tenn. Acad. Sci. 21(1): 105–127.

ESCHMEYER, R. W. 1944. Norris Lake fishing-1944. Tenn. Dep. Conserv. Bull. 18 pp.

———, AND A. M. JONES. 1941. The growth of game fishes in Norris Reservoir during the first five years of impoundment. Trans. 6th North Am. Wild. Conf. 222–240.

HASSLER, W. W. 1955. The influence of certain environmental factors on growth of Norris Reservoir sauger (*Stizostedion canadense canadense* (Smith)). Proc. Annu. Conf. Southeast. Assoc. Game Fish Comm.:111–119.

———. 1957. Age and growth of the sauger (*Stizostedion canadense canadense* (Smith)) in Norris Reservoir, Tennessee. J. Tenn. Acad. Sci. 32(1):55-76.

STROUD, R. H. 1949. Growth of Norris Reservoir walleye during the first 12 years of impoundment. J. Wildl. Manage. 13(2):157–177.

Reproduction, Growth, and Standing Crops of Yellow Perch in Southern Reservoirs

JAMES P. CLUGSTON, JAMES L. OLIVER, AND RICHARD RUELLE

U.S. Fish and Wildlife Service, Southeast Reservoir Investigations
Clemson, South Carolina 29631

Abstract

Although water temperatures required for spawning of yellow perch were met in 17 southern reservoirs, prespawning water temperatures were in the range considered to impair reproductive success (during most years, 8–10 C was the lowest temperature measured). Yellow perch spawned about 2 months earlier in South Carolina than in northern states, but at about the same water temperature (10 C). Maturity indexes for gravid females, the number of eggs produced by a female of a given length, and the age at maturity were similar at northern and southern latitudes. Although yellow perch in the South appeared to have a longer growing season than fish in the North, southern fish did not grow significantly faster. Young of the year grew from a length of 7 mm in early March to about 40 mm in early June. Average lengths of yellow perch at the end of each of the first 6 years of life in Keowee Reservoir, South Carolina, were 77, 132, 184, 217, 250, and 252 mm. Standing crops of yellow perch varied from less than 0.1 to more than 14 kg/hectare in southern reservoirs—considerably lower than the standing crops in many northern small ponds and lakes that have been studied. In general, the yellow perch is relatively unimportant to the southern angler.

The range of yellow perch has been extended southward over the years, through intentional and unintentional transplants. It is now found in many reservoirs, cool tailwaters below reservoirs, and in other cool streams of South Carolina, Georgia, and Alabama. It has not been reported from Mississippi (Barry O. Freeman, personal communication) or Louisiana (Kenneth C. Smith, personal communication). The southern limits appear to be the Apalachicola River below Jim Woodruff Dam (Lake Seminole), Gadsden County, Florida (Kilby et al. 1959) and Gunnison Creek, Mobile County, Alabama (Spencer 1965). Both sites are near latitude 30°45′N.

We have studied the fish communities in Keowee Reservoir, South Carolina, since 1972, with the overall objective of evaluating the effects of the heated effluent from a nuclear power plant on the aquatic life in this reservoir. Since changes in growth and reproduction of fishes usually follow temperature changes or other habitat modifications, we have included studies of the life histories of fishes in Keowee Reservoir. The research covers a period before and after the nuclear plant began producing electricity. In this paper we compare reproduction, growth, and standing crop of yellow perch from Keowee Reservoir with the same characteristics of this species in its natural range and in other southern reservoirs.

Keowee Reservoir is a 7,435-hectare impoundment in the upper Savannah River Basin of northwestern South Carolina (latitude 34°50′N). The reservoir filled between 1968 and 1971 and has a mean depth of 15.8 m at full pool elevation of 243.8 m above mean sea level. Keowee Reservoir serves as a cooling reservoir for the 2,658-MWe Oconee Nuclear Station, supplies water for a 140-MWe hydroelectric power plant, and is the lower pool for a 610-MWe pumped-storage hydroelectric plant. Jocassee Reservoir (3,063 hectares) is the upper pool for the pumped storage system.

Full scale production of electricity by the Oconee Station did not begin until late 1974. The station draws water from depths of 19.8 to 27.4 m for condenser cooling. The water temperature is raised about 10 C before the water is returned to Keowee Reservoir. The addition of the heated water has changed the temperature profiles throughout the year (Fig. 1). Before the plant began operation, a well defined thermocline (upper limit near a depth of 4 m) was present throughout the summer. A thermocline was lacking or difficult to discern after the discharge of heated water

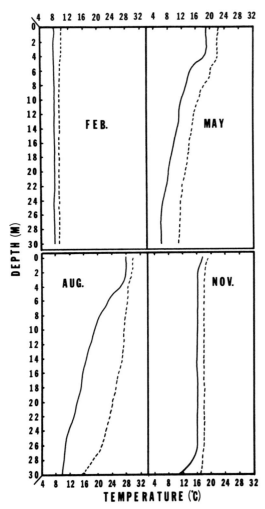

FIGURE 1.—*Water temperature profiles from Keowee Reservoir, South Carolina, 1973 (solid lines) and 1975 (broken lines). Profiles were taken about 11 km from the heated discharge.*

began. Generally, summer water temperatures at a depth of 6 to 26 m ranged from 3 C to 10 C higher in 1975 than in 1973.

Reproduction

Age and Size at Maturity

Gonads from 130 male and 139 female yellow perch collected by gill netting and electrofishing December–June, 1973–1976, were examined and classified as mature or immature. Fish were categorized by length (all length measurements are of total length) and age (see later section on growth). Males matured at a younger age and at a smaller size than females. We found seven mature age 1 male yellow perch longer than 92 mm, and all of 23 age 2 males longer than 130 mm were mature. The smallest (130 mm) and youngest mature female was age 2. It was uncommon, however, for fish of this size to be mature; other age 2 females 132 to 166 mm in length were immature. Some age 3 females in the 130–135 mm length range were immature, but all age 3 females longer than 145 mm were mature. About 75% of the female spawning population was composed of age groups 3 and 4.

Similar findings have been reported from other latitudes. Males began maturing at age 1 (102 mm) in Maryland (Muncy 1962), and at age 1 (157 mm) in Wisconsin (Herman et al. 1959). Females first matured at age 3 in Lake Ontario (Sheri and Power 1969) and Lake Michigan (Brazo et al. 1975). Clady (1976) found that 72.3% of the age 3 and all age 4 female yellow perch in Oneida Lake, New York, were mature.

Maturity Indexes

Gonads were excised and weighed, and maturity indexes, expressed as the percentage of body weight contributed by ovary weight (LeCren 1951), were calculated for adult female perch collected January–June, 1973–1974. Maturity indexes ranged from 9.8% to 20.0% in January and from 10.2% to 26.1% in February. Spent females were first collected in March. Indexes during March ranged from 20.0% to 27.4% for unspent fish and from 0.5% to 2.7% for spent fish. Most females collected from April through June had spawned. Large fish had higher maturity indexes than did small fish during the 2 months preceeding spawning. A similar relation also was found in Michigan (Brazo et al. 1975). Maturity indexes for female yellow perch collected during January and February before spawning in Keowee Reservoir could be predicted from the least squares equation: maturity index (%) = $-9.74 + 0.11197$ (total fish length); $r = 0.82$; $\alpha \leq 0.01$; $N = 15$.

Fecundity

We estimated the number of ova in 25 adult females collected from January through April, 1973–1974. Each pair of ovaries was weighed, and four cross sections from different locations within the ovaries were removed and weighed separately. The ova in each section were counted and the average number per gram for the four sections was used to estimate the total number of ova in the ovaries. We distributed ova from a random sample of each section evenly in a circular plankton counting cell and measured the diameters of the first 50 ova, using a microscope equipped with an ocular micrometer.

Numbers of mature ova ranged from 2,764 for a fish 130 mm total length to 48,272 for a fish 283 mm long. Fecundity was an exponential function of fish length and was described by the least squares equation: Log of fecundity = −4.21565 + 3.58816 (log of fish length); $r = 0.97$; $\alpha \leq 0.01$. Fecundity estimates for fish 130 to 280 mm long, calculated at 10 mm length increments, were similar to those for yellow perch in other areas of the country (Brazo et al. 1975; Sheri and Power 1969; Tsai and Gibson 1971).

Diameters of mature ova increased from about 0.80 to 1.42 mm in mid-January, to 0.95 to 1.81 mm in mid-February, and to 1.11 to 2.05 mm in early March. Mansueti (1964) reported that before water hardening, fertilized ova ranged in diameter from 1.62 to 2.09 mm.

Ova from the largest (oldest) fish had the largest diameters during January and February. The maximum diameter of ova from a fish 130 mm long was 1.13 mm, whereas the maximum diameter of ova from a fish 270 mm long was 1.75 mm. The relation between the mean diameter of mature ova and fish length was linear as expressed by the equation: ovum diameter = 0.3411 + 0.00385 (fish length); $r = 0.73$; $\alpha \leq 0.01$; $N = 15$.

Spawning Season

Periods of yellow perch reproduction in Keowee Reservoir were estimated primarily by collecting limnetic larval and juvenile fish with a frame trawl (1.3-m square aluminum frame with a 6.1-m long, 0.8-mm mesh net) as described by Ruelle et al. (1977). Samples were collected weekly at four locations at different distances from the heated discharge (the nearest was adjacent to the heated discharge and the farthest was 11 km from the discharge) from March through September, 1973–1977, at depths of 1 and 5.5 m. In 1973 and 1976 samples were also collected every two weeks at 11 and 16 m. Temperature profiles were taken when samples were collected. Sampling was begun each year before yellow perch were available to the trawl, and continued through mid-September to verify that spawning was completed and to collect larvae of species that spawned later.

Larval yellow perch were first captured in early March, when water temperatures were 9.2–13.3 C (Table 1). Each year the first specimens captured were 5–7 mm long and many had not completely absorbed their yolk sac. During most years, some newly hatched fish 5–7 mm long were captured every week through April. Catches per haul were usually highest in early April, when the water had warmed to 12–16 C and most of the fish were 8–11 mm long. The highest temperature

TABLE 1.—Mean numbers of young-of-the-year yellow perch caught per frame trawl haul at depths of 1 and 5.5 m from Keowee Reservoir 1973–1977, and dates of, and water temperatures (C) during, earliest, highest, and latest catches.

Year and depth (m)	Number per haul	First catch Date	First catch Temp.	Temp. range during highest catch	Last catch Date	Last catch Temp.
1973						
1.0	26.2	3/13	9.6	12.0–13.0	6/11	26.7
5.5	12.1	3/13	9.5	11.5–13.0	6/11	19.5
1974						
1.0	11.8	3/4	11.9	13.0–15.0	6/12	26.0
5.5	4.3	3/4	11.5	13.0–14.0	6/12	24.8
1975						
1.0	1.1	3/1	9.8	13.5–15.5	5/28	26.9
5.5	0.5	3/4	9.2	12.0–14.0	5/28	24.0
1976						
1.0	0.8	3/15	12.5	14.8–17.1	5/18	21.0
5.5	0.2	3/22	12.8	14.2–16.0	5/18	21.0
1977						
1.0	0.9	3/23	11.0	12.8–17.0	5/27	23.0
5.5	0.7	3/15	13.3	12.6–16.0	5/27	23.7

at which larval perch were collected was 26.9 C (in May 1975). On the basis of time of capture of larvae, the maturity indexes, sizes of the ova, and time of collection of spent females, it appears that yellow perch begin spawning at the end of February or in early March in Keowee Reservoir, when the temperature is about 10 C. They apparently continue to spawn intermittently through April.

Water temperature criteria for yellow perch spawning have been described for northern latitudes. Hokanson (1977) showed that a chill period of 4 C for about 185 days produced the most viable embryos among prespawning conditions tested. Reproductive success was reduced in yellow perch exposed to winter temperatures higher than 4 C or to durations of the 4 C temperature for less than, or greatly exceeding, 185 days. The lowest percentages of viable embryos per female (0 to 12%) were produced by a chill temperature of 10 C, the highest temperature tested.

Hokanson (1977) reported spawning in water at temperatures from 3.9 to 18.6 C. The highest percentage (93%) of viable embryos was measured at a medium spawning temperature of 11.9 C (with a chill temperature of 4 C for 185 days). Spawning usually occurs in Wisconsin in April or early May when water temperatures are 7.2 to 11.1 (Herman et al. 1959), and in California at temperatures of 7.2 to 12.8 C (Coots 1966).

Young of the Year

Young-of-the-year yellow perch were more abundant at 1 m than at 5.5 m at all sites during 1973–1977 (Table 1). Annual average catches in 1976 at depths of 11 and 16 m (not shown) ranged only from 0.07 to 0.09 fish per haul.

Differences in catch rates between the four sampling sites in 1973–1976 were described by Ruelle et al. (1977). Numbers of larval yellow perch were lowest near the heated effluent. The scarcity there was attributed to possible avoidance of the area by adult spawners because of the elevated temperatures, or the flushing and diluting effect of the larva-free water discharged by the nuclear plant.

TABLE 2.—*Mean length (mm) of young-of-the-year yellow perch from Keowee Reservoir 1973–1977. Fish captured by frame trawl, mid-water trawl, seine, and rotenone. Numbers of fish are shown in parentheses.*

Collecting method and year	March		April		May		June		July	
	1–15	16–31	1–15	16–30	1–15	16–31	1–15	16–30	1–15	16–31
Trawl										
1973	6.1 (9)	6.8 (566)	7.9 (1,362)	11.2 (1,644)	15.7 (694)	19.9 (431)	23.8 (231)	(0)		
1974	7.0 (19)	9.0 (1,402)	a	14.4 (420)	15.8 (151)	21.5 (6)	41.4 (41)	(0)		
1975	7.4 (5)	7.2 (81)	9.1 (276)	13.7 (71)	11.1 (12)	25.8 (35)	(0)	(0)		
1976	7.7 (4)	8.7 (89)	9.9 (307)	10.8 (126)	13.2 (37)	(0)	43.5 (4)	(0)		
1977	7.5 (2)	8.3 (15)	8.9 (117)	11.0 (28)	9.3 (15)	(0)	(0)	(0)		
Seine or rotenone										
1973							40.8 (13)	46.9 (12)	51.2 (21)	54.9 (27)
1974				(0)	31.0 (9)	37.2 (63)	41.2 (33)	(0)	56.4 (17)	65.3 (11)
1975							(0)	(0)	61.6 (20)	63.4 (8)
1976							(0)	66.0 (3)	(0)	65.7 (6)
1977							49.1 (18)	(0)	(0)	(0)

a No sample taken.

Temperatures of the discharged water during spring were not in the range considered lethal for yellow perch. Fewer young-of-the-year yellow perch were captured near the heated effluent than at the other three sites in 1977. Water temperatures at the sampling station nearest the heated discharge usually were about 4 to 5 C higher than those at the other stations.

An earlier attempt was made to correlate the relative abundance of young-of-the-year yellow perch and winter water temperatures in Keowee Reservoir (Ruelle et al. 1977). During the winters of 1973 through 1976 the water temperatures in Keowee Reservoir were lowest in February or early March. The water cooled to 8 C in 1973, 10 C in 1974, and 9.3 C in 1975 and 1976. However, during the exceptionally harsh winter of 1977 the water temperatures reached 10 C in December (1976), 5.8 C in January, and 5.5 C in February. The water remained in the 5.5–5.8 C range for about 40 days. The catch during 1973–1976, as shown in Table 1, was inversely related to winter water temperature ($r = 0.81$; $\alpha \leq 0.01$). However, even though the water cooled to 5.5–5.8 C in 1977, compared with a lowest measurement of 8 C in 1973, the catch of young yellow perch in spring was about the same in 1977 as in 1975 and 1976.

Growth

Young of the Year

We compared growth rates for the first 7 months of life from semimonthly length-frequency distributions of yellow perch captured by trawl, seine, and the application of rotenone in coves (see later section on standing crop). Since young yellow perch 30 to 45 mm long inhabit both inshore and offshore areas and their migration patterns are unknown, the data were separated to show the limnetic (trawl) collections and the littoral (seine and rotenone) collections (Table 2). The Wilcoxon signed ranks test (Conover 1971) was used to test for differences ($\alpha = 0.05$) in semimonthly fish lengths between years. Since the sampling methods were size selective and sampled different habitats, the limnetic and littoral data were tested independently.

Young yellow perch were first captured by trawl during the first week of March when they averaged 6.1 to 7.7 mm in total length (Table 2). They grew to 11–16 mm by the first week of May every year except 1977 when they were smaller. By early June, limnetic fish averaged about 40 mm in 1974 and 1976 and 24 mm in 1973 (none were collected in June 1975 and 1977).

Growth rates between years for yellow perch captured in trawls rank as follows (year of slowest growth to left and fastest at right):

1973	1975	1976	1977	1974
	———	———	———	

There were no statistically significant differences ($\alpha = 0.05$) in growth rates between years underlined with the same line. Slowest growth (1973) and fastest growth (1974) both occurred before full-scale operation of the Oconee Station.

Young-of-the-year yellow perch seined were 40–50 mm long by early June—about the same length as those in the

TABLE 2.—Continued.

Collecting method and year	August		September	
	1–15	16–31	1–15	16–30
Trawl				
1973				
1974				
1975				
1976				
1977				
Seine or rotenone				
1973		63.6 (152)	72.6 (25)	80.0 (2)
1974	59.9 (204)	57.2 (4)	68.0 (3)	(0)
1975	53.5 (108)	65.0 (7)	60.0 (8)	61.3 (12)
1976	64.2 (117)	64.6 (34)	64.4 (7)	(0)
1977	59.0 (114)	68.0 (78)		

TABLE 3.—*Average calculated total length at end of each year of life for each year class of yellow perch (sexes combined) captured in Keowee Reservoir, 1972–1977. Numbers of fish are shown in parentheses.*

Year class	Age (years)						
	1	2	3	4	5	6	7
1967	92 (1)	131 (1)	220 (1)	237 (1)	253 (1)		
1968	104 (6)	145 (6)	204 (6)	243 (6)	278 (6)		
1969	92 (51)	141 (51)	186 (50)	221 (28)	235 (6)	231 (2)	
1970	97 (162)	161 (161)	190 (85)	216 (20)	265 (4)	241 (2)	250 (1)
1971	96 (118)	152 (101)	192 (42)	227 (12)	252 (4)	286 (2)	
1972	85 (102)	121 (52)	161 (28)	191 (15)	221 (4)		
1973	72 (162)	86 (74)	141 (12)	219 (4)			
1974	58 (131)	83 (61)					
1975	54 (152)						
1976	69 (8)	92 (2)					
Weighted average length	77 (893)	132 (509)	184 (224)	217 (86)	250 (24)	252 (6)	250 (1)
Average length increment	77	55	52	33	33	2	

June trawl samples of 1974 and 1976 (Table 2). At the end of August, average total lengths were 57 to 68 mm. Sample sizes were small for the rest of the sampling period; however, two fish averaged 80 mm by the end of September. A slow growth rate or no growth is suggested by the constant mean lengths from late July to September in 1974, 1975, and 1976. This constancy of length may have resulted from the migration of the larger fish to deep areas of the reservoir, where they were not sampled. There were no significant differences in yellow perch growth rates for 1973 through 1976 for fish captured by seine and rotenone. Too few fish were captured in 1977 to test for differences.

Adults

We examined the scales from 1,236 yellow perch collected with gill nets, electroshocker, or rotenone from May 1972 through August 1977 to determine age and estimate growth. We removed scale samples from below the lateral line immediately behind the pectoral fin. Scales were formed on yellow perch in this area when the fish were about 23 mm long. We back calculated lengths at each annulus, using a modification of the direct-proportion formula, which corrects the calculated length for the approximate fish size at the time of scale formation (Ricker 1971). Sex was determined for most fish collected in 1972 and 1973, and for a few collected in later years. Analysis of variance indicated no significant differences in length between sexes at each age through age 4 ($\alpha \leq 0.005$). (Sample sizes were too small to support this comparison for older fish.) Therefore we combine the sexes in the discussion of yellow perch growth (Table 3).

Subsamples of scales from 668 yellow perch in age groups 1 through 7 were examined chronologically to determine time of annulus formation. Annulus formation depended on age and maturity. As in many other species, immature perch began growing and forming annuli at about the time mature fish started spawning (Nelson 1969; Ruelle 1971). In age 1 fish, annuli were visible on the scales of 50% of the fish collected 16–28 February and of all fish collected by 1–15 April. Some age 2 yellow perch formed annuli by 1–15 March and all had formed annuli by 1–15 May. Yellow perch in all age groups had completed annulus formation by 1 June.

The average calculated lengths of yellow perch at the time of annulus formation for the first six years of life in Keowee Reservoir were 77, 132, 184, 217, 250, and 252 mm (Table 3). First-year growth was apparently completed by September, since calculated lengths at time of first annulus formation corresponded well with empirical measurements after 7 months of life (Table 2). Growth during the first year of life generally has decreased since 1968. The length of age 2 fish increased between 1967 and 1970 and then declined through 1974. The growth rate of age 3 fish declined steadily between 1967 and 1973. Trends were not apparent for age 4 and older fish.

TABLE 4.—*Comparison of mean total lengths (mm) of yellow perch at the end of each year of life from four northern waters and Keowee Reservoir, South Carolina. Keowee estimates are not corrected to compensate for fish length at time of scale formation (as in Table 3) so data between states are comparable.*

State and source	Age (years)								
	1	2	3	4	5	6	7	8	9
Michigan									
Hile and Jobes (1941)			171	217	247	273	293	315	
South Carolina									
Present study	65	126	182	219	255	258			
Minnesota									
Carlander (1949)	97	137	178	204	230	251	262	278	280
North Dakota									
Wahtola et al. (1971)	81	126	155	181					
Pennsylvania									
Miller and Buss (1961)	50	102	145	183	208	229	249	259	

The growth increment demonstrated by yellow perch during the third growing season was inversely related to the mean length of the fish at the end of the second growing season ($r = 0.90$; $\alpha \leq 0.05$). In 1972, age 2 fish had a mean length of 161 mm and an incremental increase in the third year of 29 mm. In 1975, age 2 fish had a mean length of 86 mm at the end of the second growing season, but had an incremental increase of 55 mm during the third growing season. Slow growth in the first 2 years was compensated for by increased growth in the third growing season, and all fish were about the same length at the end of the fourth year of growth.

Yellow perch spawned earlier in the spring in Keowee Reservoir than in more northern waters. If we assume that the summer water temperatures did not exceed ranges higher than that required for optimum growth and that sufficient food was available, yellow perch had a longer growing season in southern than in northern waters. However, growth rates varied considerably throughout the yellow perch's range and it is difficult to evaluate differences in growth rates between the two areas of the country. In Keowee Reservoir, the weighted mean total length at the end of the first growing season for 10 years was 77 mm (Table 3). The weighted calculated mean length at the end of the first growing season (corrected for fish size at time of scale formation) over 11 years for yellow perch from Lower Red Lake, Minnesota, was 64 mm (Pycha and Smith 1955) and an average for 11 years in Oneida Lake, New York, was 65 mm (Forney 1971). Much of the literature contains growth information not adjusted for fish length at time of scale formation (Table 4). First year growth better than that in Keowee Reservoir was reported from Minnesota (Carlander 1949) and North Dakota (Wahtola et al. 1971). Lengths attained in later years of life were relatively close in most areas. At age 6, the largest fish were from Saginaw Bay, Michigan (Hile and Jobes 1941). Overall, slowest growth was reported from Pennsylvania (Miller and Buss 1961).

In Keowee Reservoir, few fish 5 years old or older were captured (Table 3); however, one yellow perch 7 years old was caught. Yellow perch attained an age of 6 years in most waters (Table 4), and the oldest fish reported in the literature was 13 years old (Miller and Buss 1961).

Standing Crop

The standing crop of fishes in Keowee Reservoir was estimated by the "cove rotenone" technique. This method has been used extensively in the southeastern United States. Procedures were basically those recommended by Surber (1959). Three coves (1.13, 1.25, and 1.42 hectares) were sampled each year in 1972–1976 and a fourth cove (0.77 hectare) was included in 1976. Before application of 1 mg/liter of 5% emulsified rotenone, the mouth of each cove was blocked with a 16-mm mesh net. Fish were recovered for

TABLE 5.—*Estimated total fish standing crops (kg/hectare) and yellow perch standing crops from cove rotenone samples in Keowee Reservoir and Jocassee Reservoir, South Carolina, 1972–1976.*

Year	Keowee Total	Keowee Yellow perch	Jocassee Total	Jocassee Yellow perch
1972	98.3	0.7	152.6	0.3
1973	44.7	1.1	83.9	0.1
1974	79.5	3.1	19.8	0.1
1975	56.7	1.0	28.4	1.3
1976	49.1	2.2	41.2	0.8

2 days after the rotenone was applied, sorted to species in 25-mm length groups, counted, and weighed. Similar samples were collected from three coves in Jocassee Reservoir (on the headwaters of Keowee Reservoir) and these data are provided for comparison.

Thirty-two species of fish from Keowee Reservoir and 26 species from Jocassee Reservoir were collected in the cove rotenone samples. Total standing crops ranged from 45 to 95 kg/hectare in Keowee Reservoir and 28 to 153 kg/hectare in Jocassee Reservoir (Table 5). A general decline in total standing crop is common in new reservoirs and is believed to result from a stabilization of environmental conditions and fish biomass levels (Jenkins 1977). Although the total fish standing crop declined, yellow perch standing crops changed little from year to year in either reservoir. However, the age and size composition of yellow perch in Keowee Reservoir rotenone samples from 1972–1974 differed from those in 1975–1976. In 1972–1974 about 98% of the yellow perch standing crop was composed of fish age 1 and older, whereas in 1975–1976 this percentage was only about 60. The low numbers of fish over 2 years old collected since 1974 appear in Table 3. Length of the largest fish captured each year in the rotenone samples declined from 271 mm in 1972 to 145 mm in 1976. Fifty yellow perch over 150 mm long were collected in 1972–1974, but only 9 in 1975–1976.

It is difficult to compare our observations with those in northern waters because cove rotenone samples are seldom collected north of the Mason-Dixon line and we found no reports of standing crops in large northern reservoirs or lakes. Carlander (1977) summarized yellow perch standing crops from a number of small lakes and ponds north of latitude 42°N. Estimates resulted from complete kills of the fish population by the use of chemicals and mark-and-recapture experiments. Estimates of yellow perch standing crops were: lakes in which yellow perch were less than 20% of the total standing crop, 1 to 36 kg/hectare; lakes in which yellow perch were at least 20% of the total standing crop, 4.5 to 65 kg/hectare; and lakes with only yellow perch, 39 to 215 kg/hectare. Mean standing crops of yellow perch were 0.5 kg/hectare in Jocassee Reservoir and 1.6 kg/hectare in Keowee Reservoir.

Many southern reservoirs have been sampled by the cove rotenone method. The National Reservoir Research Programs, U.S. Fish and Wildlife Service, has compiled physical, chemical, and biological information from U.S. reservoirs since 1963. We summarized pertinent information on yellow perch standing crop and sport fish harvest information from this work (R. M. Jenkins, personal communication) and recently canvassed the southern states for additional standing crop data, creel data, and temperature profiles from these reservoirs (Table 6). Monthly water temperature information was available from a portion of the listed southern reservoirs containing yellow perch; however, for most it covered only 1 or 2 years. Complete monthly temperature profiles for several years were available from Jocassee, Keowee, Hartwell, and Par Reservoirs.

Creel information on yellow perch was available from 6 of the 17 reservoirs. Catches ranged from less than 0.1 kg/hectare to 0.1 kg/hectare per year. Creel surveys were conducted on many of the other reservoirs but yellow perch catches were insignificant and included in the species category as "other." Yellow perch caught by fishermen in Keowee and Jocassee Reservoirs before 1976 were placed in this category. During the first 6 months of 1976, 0.03 kg/hectare yellow perch were caught in Keowee Reservoir.

The mean standing crops of yellow perch in 17 Georgia and South Carolina reservoirs ranged from less than 0.1 kg/hectare to 14.2 kg/hectare. The highest estimate, from Lake Jackson, Georgia, included one sample of 50.1 kg/hectare. The lowest estimates were from reservoirs farthest north (Jocassee and Keowee) and those farthest south (Blackshear, George, and Seminole). Recently, the two most northern reservoirs contained the smallest total standing crop: 40–50 kg/hectare (Table 5). The remaining reservoirs generally were older than Keowee and Jocassee Reservoirs, were constructed in more fertile watersheds, and supported total standing crops between 75 and 225 kg/hectare. The small number of yellow perch in Keowee and Jocassee Reservoirs may result partly from low basic fertility and overall low fish production. Jenkins (1977) found that total dissolved solids was the most important independent variable for predicting fish standing crops. Total dissolved solids were 18 mg/liter in Jocassee Reservoir and 20 mg/liter in Keowee Reservoir. All other southern reservoirs listed in Table 6 had total dissolved solids from 30 to 52 mg/liter. The relatively low standing crop in the three most southern reservoirs may be related to water temperature or other ecological factors.

In general, it appears that yellow perch populations should be small or not exist in much of the South if the water temperature spawning criteria reviewed by Hokanson (1977) are considered. Chill temperatures of 4–6 C were rare, and when they occurred, they lasted for only a few days. Although temperature data were limited, we found temperature records in this range for four reservoirs: Lanier (1966); Sinclair (1967); Hartwell (1970); and Keowee (1977). Annual winter water temperatures varied between reservoirs because of latitude, elevation, and meterological conditions. As examples, Lake Murray cooled to 8.2 C in 1973 and to 12.8 C in 1974. Keowee Reservoir cooled to 8.0 C in 1973 and 10.1 C in 1974. During most years, the lowest temperatures measured in the 17 reservoirs were 8–10 C, usually in February. The headwaters of Lake

TABLE 6.—*Mean standing crop of yellow perch (kg/hectare) in southern reservoirs. Latitude indicates the approximate midpoint of each reservoir and elevation is meters above mean sea level; "t" indicates winter water temperature data were available and "c" indicates creel information was available for selected years.*

Reservoir	Lat. (N)	Elevation above mean sea level (m)	Number years sampled	Standing crop (kg/hectare)
Jocasse, S.C.	35°00'	338.3	5 t,c	0.5
Keowee, S.C.	34°50'	243.8	9 t,c	1.6
Hartwell, S.C., Ga.	34°30'	201.2	8 t,c	5.0
Wateree, S.C.	34°25'	68.9	8	2.1
Secession, S.C.	34°20'	167.6	2	11.6
Lanier, Ga.	34°15'	326.1	7 t,c	3.1
Greenwood, S.C.	34°15'	134.1	7 t,c	5.3
Murray, S.C.	34°05'	111.3	10 t	8.5
Clark Hill, S.C., Ga.	33°50'	100.6	12 t	4.4
Marion, S.C.	33°30'	22.9	11	5.1
Jackson, Ga.	33°25'	152.4	6 t,c	14.2
Moultrie, S.C.	33°20'	22.9	10 t	6.3
Par, S.C.	33°14'	60.7	3 t	4.2
Sinclair, Ga.	33°10'	103.6	5 t	0.8
Blackshear, Ga.	31°55'	72.2	7	0.6
W.F. George, Ga., Ala.	31°45'	30.5	2	<0.1
Seminole, Ga., Fla.	30°46'	23.5	11 t	<0.1

Seminole (the most southern reservoir) had February temperatures of 8.0, 9.0, and 10.5 C during 1973, 1971, and 1972, respectively (only data available). Water temperature measurements from the main portion of this reservoir were not available but the coldest water discharged through the penstocks during February 1973 was 12.3 C. As pointed out earlier, yellow perch live in Lake Seminole's tailrace (Kilby et al. 1959).

Summer water temperatures of 28–30 C were common in the upper strata of many southern reservoirs and surface temperatures of 32 C, the lethal threshold described by Hart (1952), were measured in many of them. However, summer stratification usually provided deep layers of water at or near the adult yellow perch's preferred temperatures of 18–21 C and the juveniles' preferred temperatures of 20–24 C (Ferguson 1958; McCauly and Read 1973). Since cove rotenone samples usually are taken in August and generally sample the shallowest areas, it is possible that the standing crops of yellow perch reported from southern reservoirs are underestimates. Many yellow perch probably were in the deep, cool water, away from the coves sampled. During the collection of rotenone samples, water temperatures of

18–21 C were available in the Jocassee Reservoir coves, 1972–1976. However, adult preferred temperatures were at depths of the cove samples in Keowee Reservoir only during 1973. Related data from the other reservoirs were not available.

Despite the uncertainties regarding the true abundance of yellow perch in southern reservoirs, the creel data point out the insignificance of this species as a sport fish in the South. It is rarely fished for, and is usually caught by accident. Since the reservoirs described (Table 6) are south of this species' natural range, it is reasonable to expect only small populations. Water temperatures suitable for spawning were measured in the southern reservoirs, but prespawning temperatures were in the range Hokanson (1977) found resulted in impairment of reproductive success. Since similar data were not found in the literature, we cannot compare the relative abundance of larval yellow perch in South Carolina with abundances in the North, and we cannot determine if reproduction was impaired or if population sizes were adjusted during some other stage of the life cycle. The presence of yellow perch in the South emphasizes the need for better defining yellow perch temperature requirements in warm regions.

It is known that the success of an organism depends on the completeness of a complex of conditions—both physical requirements and biological interrelations (Odum 1971). We discussed the importance of water temperature as a factor controlling yellow perch reproduction and distribution. Few additional environmental characteristics have been evaluated in the field or laboratory, and little is understood about the interrelations between yellow perch and other species. Possibly environmental criteria for all life processes can be better understood, tested, and isolated if the study of habitats marginal for a particular species—such as the South for yellow perch—are stressed.

Acknowledgments

The authors gratefully acknowledge Otho D. May Jr., Miller G. White III, Lawrence J. Tilly, Russ Ober, Sam L. Spencer and William L. Shelton for providing unpublished standing crop, creel, and water temperature information.

References

Brazo, D. C., P. I. Tack, and C. R. Liston. 1975. Age, growth, and fecundity of yellow perch, *Perca flavescens*, in Lake Michigan near Ludington, Michigan. Trans. Am. Fish. Soc. 104(4): 726–730.

Carlander, K. D. 1949. Growth rate studies of saugers, *Stizostedion canadense canadense* (Smith) and yellow perch, *Perca flavescens* (Mitchill) from Lake of the Woods, Minnesota. Trans. Am. Fish. Soc. 79:30–42.

———. 1977. Biomass, production, and yields of walleye (*Stizostedion vitreum vitreum*) and yellow perch (*Perca flavescens*) in North American lakes. J. Fish. Res. Board Can. 34(10):1602–1612.

Clady, M. D. 1976. Influence of temperature and wind on the survival of early stages of yellow perch, *Perca flavescens*. J. Fish. Res. Board Can. 33(9):1887–1893.

Conover, W. J. 1971. Practical nonparametric statistics. John Wiley and Sons, New York. 462 pp.

Coots, M. 1966. Yellow perch. Pages 426–430 *in* Alex Calhoun, ed. Inland fisheries management. Calif. Dep. Fish Game, Sacramento.

Ferguson, R. G. 1958. The preferred temperature of fish and their midsummer distribution in temperate lakes and streams. J. Fish. Res. Board Can. 15:607–624.

Forney, J. L. 1971. Development of dominant year classes in a yellow perch population. Trans. Am. Fish. Soc. 100(4):739–749.

Hart, J. S. 1952. Geographic variations of some physiological and morphological characters in certain freshwater fish. Univ. Toronto Stud. Biol. Ser. 60; Publ. Ont. Fish. Res. Lab. 62. 79 pp.

Herman, E., W. Wisby, L. Wiegert, and M. Burdick. 1959. The yellow perch: its life history, ecology, and management. Wis. Conserv. Dep. Publ. 228. 14 pp.

Hile, R., and F. W. Jobes. 1941. Age, growth, and production of the yellow perch, *Perca flavescens* (Mitchill), of Saginaw Bay. Trans. Am. Fish. Soc. 70:102–122.

Hokanson, K. E. F. 1977. Temperature requirements of some percids and adaptations to the seasonal temperature cycle. J. Fish. Res. Board Can. 34(10):1524–1550.

Jenkins, R. M. 1977. Prediction of fish biomass, harvest, and prey-predator relations in reservoirs. Pages 282–293 *in* W. Van Winkle, ed. Proceedings of the conference on assessing the effects of power-plant-induced mortality on fish populations. Pergamon Press, New York.

Kilby, J. D., E. Crittenden, and L. E. Williams. 1959. Several fishes new to Florida freshwaters. Copeia 1959(1):77–78.

LeCren, E. D. 1951. The length-weight relationship and seasonal cycle in gonad weight and condition in the perch (*Perca fluviatilis*). J. Anim. Ecol. 20, 201–219.

Mansueti, A. J. 1964. Early development of the

yellow perch *Perca flavescens*. Chesapeake Sci. 5:46–66.

McCauley, R. W., and L. A. A. Read. 1973. Temperature selections by juvenile and adult yellow perch (*Perca flavescens*) acclimated to 24 C. J. Fish. Res. Board Can. 30:1253–1255.

Miller, J., and K. Buss. 1961. The age and growth of fishes in Pennsylvania. Penn. Fish Comm., Harrisburg. 26 pp.

Muncy, R. J. 1962. Life history of the yellow perch, *Perca flavescens*, in estuarine waters of Severn River, a tributary of Chesapeake Bay, Maryland. Chesapeake Sci. 3:134–159.

Nelson, W. R. 1969. Biological characteristics of the sauger population in Lewis and Clark Lake. U.S. Fish. Wildl. Serv. Tech. Pap. 21. 11 pp.

Odum, E. P. 1971. Fundamentals of ecology. W. B. Saunders, Philadelphia. 574 pp.

Pycha, R. L., and L. L. Smith, Jr. 1955. Early life history of the yellow perch, *Perca flavescens* (Mitchill), in the Red Lakes, Minnesota. Trans. Am. Fish. Soc. 84:249–260.

Ricker, W. E. 1971. Methods for assessment of fish production in fresh waters. Int. Biol. Programme Handbook No. 3. Blackwell Scientific Publications, Oxford and Edinburgh. 348 pp.

Ruelle, R. 1971. Factors influencing growth of white bass in Lewis and Clark Lake. Am. Fish. Soc. Spec. Publ. 8:411–423.

———, W. Lorenzen, and J. Oliver. 1977. Population dynamics of young-of-the-year fish in a reservoir receiving heated effluent. Pages 46–67 *in* W. Van Winkle, ed. Proceedings of the conference on assessing the effects of power-plant-induced mortality on fish populations. Pergamon Press, New York.

Sheri, A. N., and G. Power. 1969. Fecundity of the yellow perch, *Perca flavescens* (Mitchill), in the Bay of Quinte, Lake Ontario. Can. J. Zool. 47:55–58.

Spencer, S. L. 1965. Yellow perch, native of north, harvested in Mobile County. Ala. Conserv. (Oct.–Nov.):3.

Surber, E. W. 1959. Suggested standard methods of reporting fish population data for reservoirs. Proc. Annu. Conf. Southeast. Assoc. Game Fish Comm. 13:313–325.

Tsai, C., and G. R. Gibson, Jr. 1971. Fecundity of yellow perch, *Perca flavescens* Mitchill, in the Patuxent River, Maryland. Chesapeake Sci. 12(4):270–284.

Wahtola, C. H., B.L. Evenhuis, and J. B. Owen. 1971. The age and growth of the yellow perch *Perca flavescens* (Mitchill) in the Little Missouri Arm of Lake Sakakawea, North Dakota, 1968. Proc. N. D. Acad. Sci. 24:39–44.

Structure of Fish Communities in Lakes That Contain Yellow Perch, Sauger, and Walleye Populations

MICHAEL D. CLADY[1]

Department of Natural Resources
Cornell University, Ithaca, New York 14853

Abstract

Yellow perch and walleye populations appear to thrive in the same lakes most often, and the presence of sauger usually indicates that the portion of the community made up of yellow perch and walleyes is less. Species diversity is correlated negatively with the relative importance of yellow perch in the fish community and positively with the proportion of the community made up of sauger, but apparently is not related to relative abundance of walleye. Fish communities containing different combinations of large percids tend to be characterized by non-percid species belonging to different reproductive guilds.

Populations of yellow perch, sauger, and walleye in lakes have been extensively and intensively studied. Although some of these studies were strictly species-specific, many also dealt with other fishes living in the lake, particularly with those that were known competitors or prey of the large percids. These latter studies have also been largely at the population level, however, and usually emphasize only the few most abundant fishes interacting with yellow perch, sauger, and walleye. Few investigations have been made of the entire fish community in lakes containing established populations of large percids (Clady 1976a; Clady and Nielsen 1978, this volume), although more or less typical community types, particularly those in lakes containing well established yellow perch-walleye populations, have long been recognized (Maloney and Johnson 1957; Moyle et al. 1950).

The relative abundances of the species comprising the fish communities in several percid lakes have been monitored for several years—20 years in Oneida Lake, New York. The purpose of this paper is to compare this species composition with that in selected lakes and reservoirs containing large percids to determine if the types of dominant fishes and fish community diversity are related to relative abundance of yellow perch, sauger, and walleye. Common species of fish occurring in lakes containing large percids were classified ecologically according to the reproductive guilds of Balon (1975). Use of Balon's classification does not imply complete acceptance of the hypothesis that spawning strategy interaction or related factors limit the number of species that can successfully coexist. Rather, since percid populations are present in all lakes examined, it is assumed that the reproductive guilds of non-percid species, as well as species diversity of the community, are similar in several lakes and thus may suggest the existence of community types. Such types which can be defined ecologically may be more useful than standard taxonomic classifications as a basis for discussion.

Sources of Data

Catches in trap and gill nets, seines, fry samplers, and trawls were used to determine the relative abundance of both large percids and the remaining fish species in Oneida Lake, New York. Data from studies in other percid lakes that also determined relative catch of all fish species in the same and similar gears were also examined. Some intensively studied percid lakes (Escanaba, Red) are not included in this analysis because (1) different methods of capture were used, (2) the entire fish community was not sampled, or (3) through oversight I did not locate the appropriate data in the literature.

Space limitations do not allow a detailed description of the data base from all lakes,

---

[1] Present address: Oklahoma Cooperative Fishery Research Unit, Oklahoma State University, Stillwater, Oklahoma 74074.

but the following description of sampling techniques in Oneida Lake suggest the general type of data used. During April and May of 1969, 1971, 1973, and 1975 about five trap nets of the type described by Forney (1961) were set to catch spawning yellow perch and walleyes, but all fish were counted (unpublished data). Effort averaged about 85 net nights per year. During May and June of 1965–1976, several estimates of populations of larval yellow perch (Noble 1968) and walleye (Forney 1975a) were made with Miller high-speed fry samplers and the catch of all species was recorded. Total number of samples in a typical year was 800 (200 5- to 7.5-min hauls with 4 samplers). Larval densities used in this paper are unpublished and from Clady (1976b, 1976c, 1976d) and Forney (1975b, 1976a). Two 91-m experimental gill nets were set overnight at 1 of 15 sites at weekly intervals from early June to mid-September of 1958–1973 and 1975 (unpublished data; Clady 1974; Clady and Hutchinson 1976). All species were also enumerated in catches of 5.9-m otter trawls fished weekly for 5 min at 4–10 stations from late July through October of 1959–1975 (Forney 1971, 1974). Catches used in this paper are unpublished, from Clady (1976b, 1977), and from Forney (1976a, 1976b). Also, 9 sites were seined at least weekly in June through August of 1959–1970 and 1975 with a 23-m bag seine (Forney 1971; Clady 1976a). Seine catches are unpublished and from Clady (1974). With the exception of larval yellow perch and walleyes captured in high-speed nets (unweighted mean), overall structure of the fish community in Oneida Lake was derived from a mean of yearly catches of all species in each gear, weighted to account for differences in effort. For data from other lakes also, means were weighted accordingly in all cases where relative effort was given for each sampling interval (the great majority). Otherwise unweighted means are used to describe community structure. Diversity of fish communities was measured with the Shannon-Weaver index ($\bar{d}$):

$$\bar{d} = -\sum_{i=1}^{s} (n_i/n) \log_2(n_i/n);$$

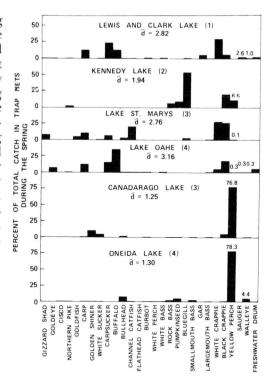

FIGURE 1.—The species composition (%) and diversity ($\bar{d}$) of spring catches in trap nets. (See Table 2 for sources of data. Number of years in parentheses.)

where n_i is the number of fish in the ith species; n is the total number of fish; and s is the total number of species (Wilhm 1976).

Results

The relative abundances and diversities of fishes taken by various gear are given in Figures 1–6. Since numbers of uncommon and rare species were occasionally not reported, diversities of these communities are shown as approximations, albeit close ones because these rare species contribute little to the diversity index.

The relative abundance of yellow perch, sauger and walleye did not appear to be strongly interdependent. When considered by method of capture, only the percentages of yellow perch and walleye taken in trap nets during the summer ($r_s = 0.94$; $P < 0.01$) and the percentages of walleyes and sauger taken in trawls ($r_s = 0.83$; $P < 0.05$) were significantly correlated (r_s

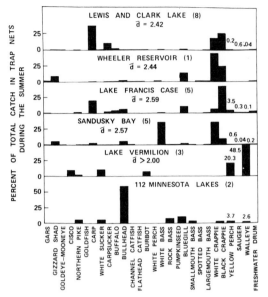

FIGURE 2.—*The species composition (%) and diversity ($\bar{d}$) of summer catches in trap nets. (See Table 2 for sources of data. Number of years in parentheses.)*

= Spearman rank correlation; Siegel 1956). Considered by capture method, relative abundance was always positively related for yellow perch and walleye ($r_s = 0.15$ to 0.50), usually negatively related for yellow perch and sauger ($r_s = 0.04$ to -0.49), and variable for sauger and walleye ($r_s = -0.31$ to 0.53).

Diversity of the entire fish community (Table 1), however, was correlated negatively with relative importance of yellow perch ($r_s = -0.65$; $P < 0.01$; Fig. 7), but positively with the proportion of the community made up of sauger ($r_s = 0.66$; $P < 0.01$). For all gear combined, diversity indices of communities containing over 20% yellow perch and communities containing saugers (Table 1), were significantly lower ($t = 3.35$, df = 32, $P < 0.01$) and higher ($t = 4.10$, df = 32, $P < 0.01$), respectively, than other communities. The proportion of the community constituted of walleye was not related to diversity ($r_s = -0.04$).

The reproductive guilds of fishes common in one or more gears appeared to be related to the presence of yellow perch (a phyto-lithophil) and sauger and walleye (both lithophils, Table 2). Lakes in which the yellow perch was the only large percid present were characterized by the presence of large numbers of a lithophil (bluegill), various phytophils (including several centrarchids), and a polyphil (pumpkinseed). No pelagophils, litho-pelagophils, or phyto-lithophils other than yellow perch were common. Lakes and reservoirs containing only sauger or walleye, but no other large percids, commonly contained large numbers of a litho-pelagophil (gizzard

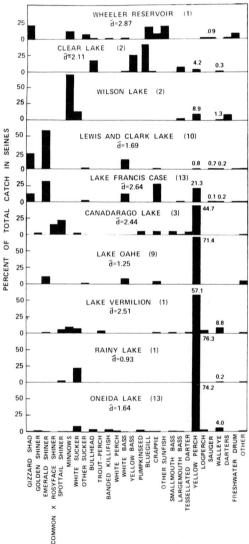

FIGURE 3.—*The species composition (%) and diversity ($\bar{d}$) of catches in seines. (See Table 2 for Sources of data. Number of years in parentheses.)*

shad), a lithophil (bluegill) and various phytophils. Phyto-lithophils, psammophils, speleophils, and polyphils were rare or absent.

Unlike the previous three examples of monotypic percid communities, the reproductive environment required by fishes common in lakes containing both yellow perch and walleye are all closely related to a firm substrate (rock, gravel, sand, and holes). Pelagophils and phytophils were uncommon or absent, but litho-pelagophils, psammophils, speleophils and phyto-lithophils other than yellow perch (various members of the genus *Morone*) are common.

Lakes and reservoirs with yellow perch-sauger-walleye populations have a fish community characterized by pelagophils (the only species combination to do so) and

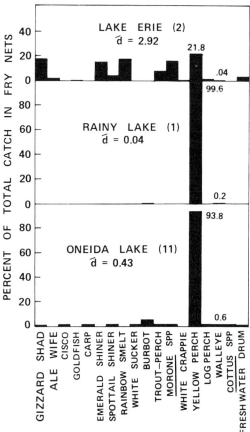

FIGURE 5.—*The species composition (%) and diversity ($\bar{d}$) of catches in fry nets. (See Table 2 for sources of data. Number of years in parentheses.)*

all other guilds except polyphils. Phytophils are abundant, as in yellow perch-only and sauger-only situations, in contrast to lakes with yellow perch-walleye communities.

Discussion

There are several obvious weaknesses in this presentation on community structure in percid lakes. The scope is narrow, all diversity indices are not equally precise, and even though comparisons were confined largely to catches in the same gear, differences exist in vulnerability of fish in different environments and details of construction of gear and methods of fishing. The distribution of data used (percentage composition of total catch) was not appropriate for multiple correlation analysis, so

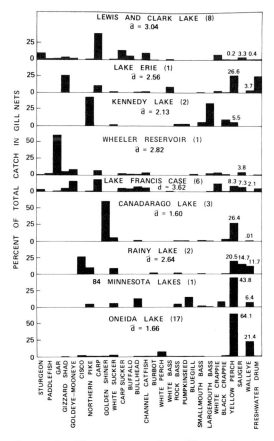

FIGURE 4.—*The species composition (%) and diversity ($\bar{d}$) of catches in gill nets. (See Table 2 for sources of data. Number of years in parentheses.)*

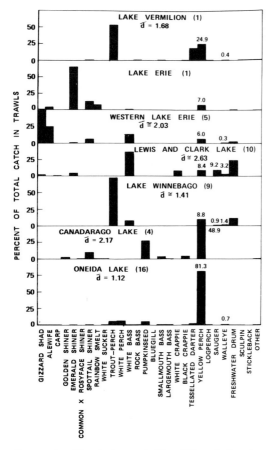

FIGURE 6.—*The species composition (%) and diversity ($\bar{d}$) of catches in trawls. (See Table 2 for sources of data. Number of years in parentheses.)*

that details of interaction among the three species are obscure. Also, space limitations did not allow analysis of species diversity in lakes containing no percids. A comparison of diversity in lakes with and without yellow perch, for example, would be more meaningful than the somewhat redundant correlation of species diversity (including yellow perch) with abundance of yellow perch. This redundancy probably contributes to the strong negative correlation between abundance of the often common yellow perch and diversity but, on the other hand, does not influence the positive correlation between sauger abundance and diversity, since sauger never constituted a significant portion of the fish community.

Despite these weaknesses, several generalizations appear to be warranted. First, yellow perch and walleye populations appear to be most compatible and to thrive in the same environments most often. The presence of sauger usually indicates that the predominance of either the yellow perch or walleye, or both, is diminished.

Second, fish community structure is more diverse in lakes in which saugers are well established and less diverse in lakes where yellow perch are abundant. These relationships could be partially a function of the environment—in the United States yellow perch may thrive in more northerly and mesotrophic natural lakes but the sauger in more southerly reservoirs—but could also be partially due to the ability of the ubiquitous and opportunistic yellow perch (Clady and Hutchinson 1976) to outcompete other fishes. The predominance of yellow perch in all areas of Oneida Lake is undoubtedly a major factor in the low community diversity values (Fig. 8), although this situation may be attributed to walleye predation (Clady and Nielsen 1978, this volume; Forney 1971, 1974) as well as the yellow perch's competitive abilities.

Third, fish communities with different combinations of large percid species (and the type of environment in which they occur) vary in their capacities to support large populations of fishes with certain reproductive requirements (Table 2). In the case of yellow perch-only communities, the absence of other phyto-lithophils may be due to intense competitive interaction with yellow perch in the absence of walleye predation, but lack of pelagophils and lithopelagophils more likely is due to the specialized environment (relatively small, natural lakes) in which the yellow perch is the only large percid present. The occurrence of sauger-only situations largely in southern reservoirs probably also explains the complete lack of representation of 5 of 8 guilds in this community type (Table 2).

The presence of the yellow perch-walleye combination mainly in large, more northerly and perhaps more mesotrophic natural lakes undoubtedly contributes to the abundance of fishes whose reproductive requirements are associated with a firm bottom, and to the absence of pelagophils and

TABLE 1.—Diversity indexes, d̄, of fish communities listed by species of large percid present.

Lake	Trawl	Seine	Gill net	Trap net	All Gear	Reference
Yellow perch						
Canadarago	2.17	2.44	1.60	1.25		Green (1976a, 1976b)
Kennedy			2.13	1.94		Johnson (1963)
St. Marys				2.76		Clark (1960)
Sauger						
Wheeler		2.87	2.82	2.44		Tarzwell (1942)
Yellow perch + walleye						
Oneida	1.12	1.64	1.66	1.30		Clady (1974), unpubl. data
Vermilion	1.68	2.51		2.00		Dobie (1966)
Clear		2.11				Ridenhour (1960)
Yellow perch + walleye + sauger						
Rainy		0.93	2.64			Johnson (1966)
Francis Case		2.64	3.62	2.59		Gasaway (1970)
Oahe		1.25		3.16		June (1976); Gabel (1974)
Lewis and Clark	2.63	1.69	3.04	2.82		Walburg (1976)
Winnebago	1.41					Priegel (1970)
Sandusky Bay				2.57		Chapman (1955)
Erie			2.56			VanVooren and Davies (1974)
Erie (Western)	2.03					VanVooren and Davies (1974)
Yellow perch abundant or not[a]						
34 lakes						
Yellow perch >20%	1.66 (3)	1.90 (6)	2.12 (4)	1.28 (2)[b]	1.89 (16)	
				2.00 (1)[c]		
Other	2.02 (3)	2.22 (3)	2.90 (4)	2.67 (4)[b]	2.51 (18)	
				2.51 (4)[c]		
Sauger present or absent[a]						
34 lakes						
Sauger present	2.02 (2)	2.40 (3)	3.03 (4)	2.99 (2)[b]	2.82 (14)	
				2.52 (3)[c]		
Other	1.75 (4)	1.81 (6)	1.99 (4)	1.81 (4)[b]	1.92 (20)	
				2.22 (2)[c]		

[a] Numbers of lakes are in parentheses.
[b] Spring.
[c] Summer.

phytophils. Johnson et al. (1977) found that, in smaller Ontario lakes, existence of walleye was limited to lakes large enough that wave action was sufficient to expose boulder and gravel substrate for spawning. In larger lakes, however, they found that a combined effect of the morphoedaphic index and lake area probably determined success of walleye, rather than a simple shortage of spawning requirements.

Knowledge of general community types may indirectly aid in management. For example, the potential to manage for a phyto-lithophil like the white bass may be much greater in polytypic large percid situations than in monotypic ones. Introductions of phytophils, such as crappies, into communities containing yellow perch

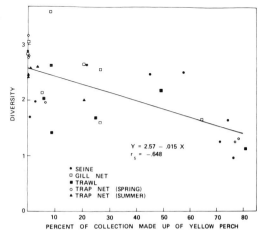

FIGURE 7.—Relationship between community diversity and percentage of catch made up of yellow perch.

TABLE 2.—*Reproductive guilds of species common in various fish communities, listed by species of large percid present. Species are denoted by numerals[a] and methods of capture by letters.[b]*

Lake and (reference)[c]	Reproductive guild							
	Pelago-phil	Litho-pelagophil	Litho-phil	Phyto-lithophil	Phytophil	Psam-mophil	Spele-ophil	Poly-phil
Yellow perch (a phyto-lithophil)								
Canadarago (1)				8cd		10ab		21a
Kennedy (2)			22d		6c,23c,25d			
St. Marys (3)					24d,25d		15d	
Deep (4)			22e					21e
Mill (5)			22d					
Cassidy (6)			22e					21e
Sugarloaf (7)			22d					
Big Bear (8)			11d					21d
Sauger (a lithophil[d])								
Wheeler (9)		3be	22bde		1c,24e,25d			
Douglas (10)		3e	22e					
Watts Bar (11)		3e	22e					
Kentucky (12)		3e	22e					
Cherokee (13)		3e	22e					
Walleye (a lithophil)								
Canton (14)		3e	22e					
Yellow perch + walleye								
Storm (15)			14d,24d	20d				
Oneida (16)		5f,17f	11b,16a	18acf			14d	21a
Vermilion (17)		5d,17d	11bd,16a			10b	27a	
Wilson (18)			11b					
Clear (19)			14bc,22b	19bc				
Winnibigoshish (20)		17f	16a			10b	27a	
Spirit (21)				20c				
Yellow perch + walleye + sauger								
Rainy (22)		5c	11bc		6c			
Francis Case (23)	9b	3b,5c	22d		7cd,24bc,25bd			
Oahe (24)	9b			20b	7d,13d,25d	12d		
Lewis and Clark (25)	9b,26b	3b		20ab	7cd,13d,24d,25d	12cd	15c	
Winnebago (26)	26a		16a	20a				
Sandusky Bay (27)				20d	24d,25d			
Western Erie (28)		3a		2a,20a				
Erie (28)	9af,26c	3cf		20cf	7c	10a		
Shagawa (29)			16a			10a		
Miller (30)		3e	22e					

[a] Species: 1. Gar (*Lepisosteus* spp.)
2. Alewife
3. Gizzard shad
4. Goldeye-mooneye
5. Cisco
6. Northern pike
7. Carp
8. Golden shiner
9. Emerald shiner
10. Spottail shiner
11. White sucker
12. Carpsucker (*Carpiodes* spp.)
13. Buffalo (*Ictiobus* spp.)
14. Bullhead (*Ictalurus* spp.)
15. Channel catfish
16. Trout-perch
17. Burbot
18. White perch
19. Yellow bass
20. White bass
21. Pumpkinseed
22. Bluegill
23. Largemouth bass
24. White crappie
25. Black crappie
26. Freshwater drum
27. Tessellated darter.

[b] Methods of capture: *a*, trawl; *b*, seine; *c*, gill net; *d*, trap net; *e*, rotenone; *f*, fry sampler.

[c] References: (1) Green (1976a, 1976b)
(2) Johnson (1963)
(3) Clark (1960)
(4) Carbine and Applegate (1948)
(5) Schneider (1971)
(6) Schneider (1973)
(7) Cooper (1953)
(8) Crowe (1953)
(9) Tarzwell (1942)
(10) Hayne et al. (1967)
(11) TVA (1965)
(12) TVA (1964b)
(13) TVA (1964a)
(14) Bross (1969)
(15) Rose (1950)
(16) Clady (1974), unpubl. data
(17) Dobie (1966)
(18) Johnson (1977)
(19) Carlander (1954); Ridenhour (1960)
(20) Johnson (1969)
(21) Rose (1955)
(22) Johnson (1966)
(23) Gasaway (1970)
(24) June (1976); Gabel (1974)
(25) Walburg (1976)
(26) Priegel (1970)
(27) Chapman (1955)
(28) VanVooren and Davies (1974); Anonymous (1958)
(29) Swenson (1977)
(30) Christenson and Smith (1965).

[d] Saugers were phytophils in Douglas Reservoir.

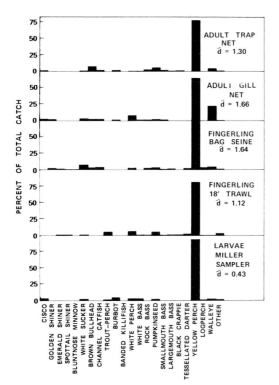

FIGURE 8.—*The percentage species composition and diversity of catches by five gears, Oneida Lake, New York.*

and walleyes would probably not be highly successful. Pelagophils, like the freshwater drum and emerald shiner, and lithopelagophils such as clupeids might be expected to do well in fish communities in which the sauger is well established. Because of the tendency for extreme predominance of yellow perch and walleye in certain lakes (and the resultant low diversity), manipulative opportunities for any species, other than a percid, may be limited by natural factors operating at the community level which are beyond the manager's control.

Acknowledgments

O. E. Maughan reviewed the manuscript, as did J. L. Forney, who also provided unpublished data. This paper presents data collected under Federal Aid in Fish Restoration Project F-17-R, New York.

References

ANONYMOUS. 1958. Preliminary report on larval fishes collected in Lake Erie in 1958. 6 pp. (Mimeo.)
BALON, E. K. 1975. Reproductive guilds of fishes: a proposal and definition. J. Fish. Res. Board Can. 32:821–864.
BROSS, M. G. 1969. Fish samples and year-class strength (1965–1967) from Canton Reservoir, Oklahoma. Proc. Okla. Acad. Sci. 48:194–199.
CARBINE, W. F., and V. C. APPLEGATE. 1948. The fish population of Deep Lake, Michigan. Trans. Am. Fish. Soc. 75:200–227.
CARLANDER, K. D. 1954. Use of gill nets in studying fish populations, Clear Lake, Iowa. Proc. Iowa Acad. Sci. 60:621–625.
CHAPMAN, C. R. 1955. Sandusky Bay report. Ohio Div. Wildl. 84 pp.
CHRISTENSON, L. M., AND L. L. SMITH. 1965. Characteristics of fish populations in Upper Mississippi River backwater areas. U.S. Fish Wildl. Serv. Circ. 212. 53 pp.
CLADY M. D. 1974. Monitoring the ecosystem. N.Y. Fed. Aid Proj. F-17-R-18 (Prog. Rep. Job I-c). N.Y. State Dep. Environ. Conserv. 12 pp. (Multilith.)
———. 1976a. Change in abundance of inshore fishes in Oneida Lake, 1916 to 1970. N.Y. Fish Game J. 23:73–81.
———. 1976b. Analysis of year class formation in a yellow perch population. N.Y. Fed. Aid Proj. F-17-R-20 (Prog. Rep. Job I-b). N.Y. State Dep. Environ. Conserv. 7 pp. (Multilith.)
———. 1976c. Influence of temperature and wind on the survival of early stages of yellow perch, *Perca flavescens*. J. Fish. Res. Board Can. 33:1887–1893.
———. 1976d. Distribution and abundance of larval ciscoes, *Coregonus artedii*, and burbot, *Lota lota*, in Oneida Lake. J. Great Lakes Res. 2:234–247.
———. 1977. Decline in abundance and survival of three benthic fishes in relation to reduced oxygen levels in a eutrophic lake. Am. Midl. Nat. 97:419–432.
———, AND B. HUTCHINSON. 1976. Food of the yellow perch, *Perca flavescens*, following a decline of the burrowing mayfly, *Hexagenia limbata*. Ohio J. Sci. 76:133–138.
———, AND L. NIELSEN. 1978. Diversity of a community of small fishes as related to abundance of the dominant percid fishes. Am. Fish. Soc. Spec. Publ. 11:109–113.
CLARK, C. F. 1960. Lake St. Marys and its management. Ohio Dep. Nat. Resour. Publ. W-324. 107 pp.
COOPER, G. P. 1953. Population estimates of fish in Sugarloaf Lake, Washtenaw County, Michigan, and their exploitation by anglers. Pap Mich. Acad. Sci. Arts. Lett. 38:163–186.
CROWE, W. R. 1953. An analysis of the fish population of Big Bear Lake, Otsego County, Michigan. Pap. Mich. Acad. Sci. Arts. Lett. 38:187–207.
DOBIE, J. 1966. Food and feeding habits of the walleye, *Stizostedion v. vitreum*, and associated game and forage fishes in Lake Vermilion, Minnesota, with special reference to the tullibee, *Coregonus (Leucichthys) artedi*. Minn. Fish. Invest. 4:39–71.
FORNEY, J. L. 1961. Year-class distribution of walleyes

collected by five types of gear. Trans. Am. Fish. Soc. 90:308–311.

———. 1971. Development of dominant year classes in a yellow perch population. Trans. Am. Fish. Soc. 100:739–749.

———. 1974 Interactions between yellow perch abundance, walleye predation, and survival of alternate prey in Oneida Lake, New York. Trans. Am. Fish. Soc. 103:15–24.

———. 1975a. Abundance of larval walleyes (*Stizostedion vitreum*) estimated from the catch in high-speed nets. Pages 581–588 *in* EIFAC Symp. on methodology for the survey, monitoring and appraisal of fishery resources in lakes and large rivers. EIFAC (Eur. Inland Fish. Comm.) Tech. Pap. 23 (Suppl. 1) 2.

———. 1975b. Analysis of year-class size in a walleye population. N.Y. Fed. Aid Proj. F-17-R-19 (Prog. Rep. Job I-a). N.Y. State Dep. Environ. Conserv. 9 pp. (Multilith.)

———. 1976a. Analysis of year-class size in a walleye population. N.Y. Fed. Aid Proj. F-17-R-20 (Prog. Rep. Job I-a). N.Y. State Dep. Environ. Conserv. 13 pp. (Multilith.)

———. 1976b. Year-class formation in the walleye (*Stizostedion vitreum vitreum*) population of Oneida Lake, New York, 1966–73. J. Fish Res. Board Can. 33:783–792.

GABEL, J. A. 1974. Species and age composition of trap net catches in Lake Oahe, South Dakota, 1963–67. U.S. Fish Wildl. Serv. Tech. Pap. 75. 21 pp.

GASAWAY, C. R. 1970. Changes in the fish population in Lake Francis Case in South Dakota in the first 16 years of impoundment. U.S. Fish Wildl. Serv. Tech. Pap. 56. 30 pp.

GREEN, D. M. 1976a. Establishment of fish population baseline. N.Y. Fed. Aid Proj. F-29-R-3 (Prog. Rep. Job I-a). N.Y. State Dep. Environ. Conserv. 9 pp. (Multilith.)

———. 1976b. Initial response of the fish population to nutrient reduction. N.Y. Fed. Aid Proj. F-29-R-3 (Prog. Rep. Job I-c). N.Y. State Dep. Environ. Conserv. 8 pp. (Multilith.)

HAYNE, D. W., G. E. HALL, AND H. M. NICHOLS. 1967. An evaluation of cove sampling of fish populations in Douglas Reservoir, Tennessee. Pages 244–297 *in* Reservoir fishery resources symposium. Am. Fish. Soc., Athens, Ga.

JOHNSON, F. H. 1963. The status of the smelt, *Osmerus mordax* (Mitchill), and native fish populations in Kennedy Lake, Itasca County, four years after the first appearance of smelt. Minn. Dep. Conserv. Invest. Rep. 272. 7 pp.

———. 1966. Status of Rainy Lake walleye fishery, 1965. Minn. Dep. Conserv. Invest. Rep. 292. 22 pp,

———. 1969. Environmental and species associations of the walleye in Lake Winnibigoshish and connected waters, including observations on food habits and predator-prey relationships. Minn. Fish. Invest. 5:5–36.

———. 1977. Responses of walleye (*Stizostedion vitreum*) and yellow perch (*Perca flavescens*) populations to removal of white sucker (*Catostomus commersoni*) from a Minnesota lake. J. Fish. Res. Board Can. 34:1633–1642.

JOHNSON, M. G., J. H. LEACH, C. K. MINNS, AND C. H. OLVER. 1977. Limnological characteristics of Ontario lakes in relation to associations of walleye (*Stizostedion vitreum vitreum*), northern pike (*Esox lucius*), lake trout (*Salvelinus namaycush*), and smallmouth bass (*Micropterus dolomieui*). J. Fish. Res. Board Can. 34:1592–1601.

JUNE, F. C. 1976. Changes in young-of-the-year fish stocks during and after filling of Lake Oahe, an Upper Missouri River storage reservoir, 1966–74. U.S. Fish Wildl. Serv. Tech. Pap. 87. 25 pp.

MALONEY, J. E., AND F. H. JOHNSON. 1957. Life histories and inter-relationships of walleye and yellow perch, especially during their first summer, in two Minnesota lakes. Trans. Am. Fish. Soc. 85:191–202.

MOYLE, J. B., J. H. KUEHN, AND C. R. BURROWS. 1950. Fish-population and catch data from Minnesota lakes. Trans. Am. Fish. Soc. 78:163–175.

NOBLE, R. L. 1968. Mortality rates of pelagic fry of the yellow perch, *Perca flavescens* (Mitchill), in Oneida Lake, New York and an analysis of the sampling problem. Ph.D. Thesis. Cornell Univ., Ithaca, N.Y. 111 pp.

PRIEGEL, G. R. 1970. Reproduction and early life history of the walleye in the Lake Winnebago Region. Wis. Dep. Nat. Resour. Tech. Bull. 45. 105 pp.

RIDENHOUR, R. L. 1960. Abundance, growth and food of young game fish in Clear Lake, Iowa, 1949 to 1957. Iowa State J. Sci. 35:1–23.

ROSE, E. T. 1950. A fish population study of Storm Lake. Proc. Iowa Acad. Sci. 56:385–395.

———. 1955. The fluctuation in abundance of walleyes in Spirit Lake, Iowa. Proc. Iowa Acad. Sci. 62:567–575.

SCHNEIDER, J. C. 1971. Characteristics of a population of warm-water fish in a southern Michigan lake, 1964–1969. Mich. Dep. Nat. Resour. Res. Dev. Rep. 236. 158 pp.

———. 1973. The fish population of Cassidy Lake, Washtenaw County. Mich. Dep. Nat. Resour. Fish. Res. Rep. 1792. 15 pp.

SIEGEL, S. 1956. Nonparametric statistics for the behavioral sciences. McGraw-Hill, New York. 312 pp.

SWENSON, W. A. 1977. Food consumption of walleye (*Stizostedion vitreum vitreum*) and sauger (*S. canadense*) in relation to food availability and physical conditions in Lake of the Woods, Minnesota, Shagawa Lake, and western Lake Superior. J. Fish. Res. Board Can. 34:1643–1654.

TARZWELL, C. M. 1942. Fish populations in the backwaters of Wheeler Reservoir and suggestions for their management. Trans. Am. Fish. Soc. 71: 201–214.

TVA (TENNESSEE VALLEY AUTHORITY). 1964a. Fish inventory data Cherokee Reservoir 1963. 10 pp. (Mimeo.)

———. 1964b. Fish inventory data Kentucky Reservoir 1963. 11 pp. (Mimeo.)

———. 1965. Fish inventory data Watts Bar Reservoir. 13 pp. (Mimeo.)

VANVOOREN, A. R., AND D. H. DAVIES. 1974. Lake Erie fishery research. Ohio Fed. Aid Proj. F-35-R-12 (Prog. Rep. Study II). Ohio Dep. Nat. Resour. 68 pp.

WALBURG, C. H. 1976. Changes in the fish population of Lewis and Clark Lake, 1956–74, and their relation to water management and the environment. Fish Wildl. Serv. (U.S.) Res. Rep. 79. 34 pp.

WILHM, J. 1976. Species diversity of fish populations in Oklahoma reservoirs. Ann. Okla. Acad. Sci. 5:29–46.

Diversity of a Community of Small Fishes as Related to Abundance of the Dominant Percid Fishes

MICHAEL D. CLADY[1] AND LARRY NIELSEN[2]

Department of Natural Resources
Cornell University
Ithaca, New York 14853

Abstract

Number and diversity of species in an inshore community of small fishes in Oneida Lake, New York, were positively correlated with abundance of age 0 yellow perch and walleye during 1959–1975. Although the data are only suggestive because food habits were not studied, buffering or modification of effects of heavy predation by large walleyes appears to be the most likely cause of the correlations.

Inshore areas of lakes are usually different from offshore areas in regard to plant growth, bottom type, wave action, benthic fauna, and chemical and physical properties of the water. This is particularly true in large lakes where cultural eutrophication may produce greater effects on littoral habitats (Beeton and Edmondson 1972). Conditions in Oneida Lake, New York, conform to these general differences between inshore and offshore zones.

In the deeper areas of Oneida Lake, walleyes prey extensively on age 0 yellow perch and other small fishes. Forney (1974) established that high densities of age 0 yellow perch buffered walleye predation on alternate prey species. Consequently, abundant year classes of white perch and walleyes generally coincided with initially abundant year classes of yellow perch. This paper examines the relation between abundance and diversity of the inshore fish fauna and density of the predominant percids to determine if similar interactions are also important inshore. If so, management recommendations based on extensive research in more easily studied offshore waters may be applied to the entire fish community.

Oneida Lake has a surface area of 207 km², a mean depth of 6.8 m and a maximum depth of 16.8 m. Areas deeper than 4 m have muddy bottoms, lack rooted vegetation, and contain a relatively simple community of small fishes dominated by age 0 yellow perch. Bottom trawls fished in these areas in August 1975 captured an average of 4.4 species of small fish per haul. Inshore areas shallower than 4 m constitute about 25% of the lake bottom (Greeson 1971) and are characterized by extensive littoral vegetation and a heterogeneous bottom ranging from silt to sand to boulders. Age 0 yellow perch are also the most abundant species in shallow water, but the fish community is much richer in species. Seine hauls during August 1975 in water shallower than 1.5 m caught an average of 12.6 species of small fish.

Methods

During 1959–1970 and 1975, the same nine sites near Shackelton Point were seined weekly from late June through September (Forney 1971). Total number of hauls in a year varied from 72 to 143 (Table 1). A 22.9-m bag seine with 6.4 mm woven mesh in the bag was positioned parallel to and 15–30 m from shore in water 0.9–1.2 m deep. The seine was pulled directly to shore, where captured fish were identified, counted, and released. Most fish were forage-sized species or age 0 fish of larger-sized species. Age I and older fish of larger-sized species were not included in this study.

Mean annual catch per haul was computed for each species in each year. Species

---
[1] Present address: Oklahoma Cooperative Fishery Research Unit, Oklahoma State University, Stillwater, Oklahoma 74074.
[2] Present address: School of Forestry and Wildlife Resources, Virginia Polytechnic Institute and State University, Blacksburg, Virginia 24061.

diversity ($\bar{d}$) was calculated annually for all species except yellow perch and walleye as

$$\bar{d} = -\sum_{i=1}^{S} p_i \log_2 p_i,$$

where p_i is the proportion of the ith species in the sample and S is the number of species (Shannon and Weaver 1963). Correlations between species number, species diversity, and abundance of percids were based on the Spearman rank correlation, r_s (Siegel 1956). For 13 years of data, $r_s > 0.48$ represents a significant correlation ($P < 0.05$).

Results

Yellow perch were the most abundant fish captured (Table 1). Catches averaged more than 200 per haul in all years, while mean catches of other species seldom exceeded 50 per haul. A total of 41 species was captured during the 13 years, but did not exceed 28 in one year. Sixteen species were captured in at least 11 of the 13 years examined and represent the predominant inshore fauna of the lake (see asterisks in Table 1). Number and abundance of species captured in 1975 were exceptionally high. Highest densities of 14 species occurred in 1975, and densities of most other species were near the highest observed densities.

The total number of species, excluding yellow perch and walleye, captured in a year varied from 16 in 1960 to 26 in 1975 (Table 2). Correlation between number of species collected and the number of hauls was −0.08, indicating that the annual

TABLE 1.—*Mean catch per haul of fish species in a beach seine, Oneida Lake, New York. Asterisk indicates species captured in 11 of 13 years and "T" indicates mean catch lower than 0.1 per haul.*

Species	Year (number of hauls)												
	1959 (143)	1960 (126)	1961 (117)	1962 (96)	1963 (108)	1964 (116)	1965 (126)	1966 (135)	1967 (99)	1968 (90)	1969 (72)	1970 (72)	1975 (90)
Yellow perch*	229.2	328.0	229.7	478.4	408.0	390.1	207.0	200.5	264.0	414.1	564.0	735.6	783.2
Walleye*	15.9	5.3	20.1	12.2	21.0	15.6	6.8	4.6	9.1	34.0	34.6	20.0	6.1
Sea lamprey									T	T			
Bowfin											T		
Alewife													T
Gizzard shad				0.2									0.1
Northern pike		T	T			T				T			0.2
Carp	0.1			T	T			0.3	0.8		0.6	0.1	1.6
Silvery minnow			0.2		T								T
Golden shiner*	1.7	1.0	5.8	0.8	1.0	0.6	7.2	8.2	14.3	3.3	3.2	4.2	40.2
Emerald shiner*	0.8	0.1	0.1	0.1	0.3	0.3	2.2	0.3	7.6	0.3		0.2	
Common shiner		0.2					T		T	0.1	T		
Spottail shiner*	T	T	0.2	0.6	0.1	0.1	T		0.2	0.2	3.1	3.1	7.8
Spotfin shiner	T	0.3	0.1		T							T	T
Sand shiner											0.2		
Bluntnose minnow*	0.8	1.3	1.0	0.4	0.4	0.7	0.5	0.3	0.8	0.3	T	0.1	3.0
Flathead minnow					T					T			
Blacknose dace										T	T		
Creek chub											0.1		
Fallfish		T									0.2		T
White sucker*	19.4	6.1	48.0	10.4	15.8	180.4	45.7	46.4	19.4	20.9	27.2	22.5	35.1
Brown bullhead*	32.3		0.4	17.0	1.7	0.1	0.4	25.4	47.5	4.2	2.0	2.6	64.2
Channel catfish	T		T	0.1	T	T	T	T			T		T
Trout-perch						0.1	T				T	0.1	
Burbot	T			T	T	T	T			T		T	T
Banded killifish*	7.3	2.3	11.2	11.0	9.8	4.2	33.9	39.2	42.5	9.0	12.4	3.4	16.0
Brook silverside	T	T	0.2	T			0.2	0.1	0.1			0.1	0.2
Brook stickleback										T			
White perch*	0.2	0.6	1.0	4.3	0.4	2.2	1.1	0.6	0.8	0.1	5.6	2.7	3.7
White bass			0.1	1.0	0.9		T	0.1			0.2	0.4	0.1
Rock bass*	0.8	T	3.1	3.4	1.9	0.5	0.6	1.5	5.8	0.2	3.5	5.7	14.7
Pumpkinseed*	13.5		10.5	4.0	13.2	64.6	12.7	26.3	3.3	6.3	12.6	10.7	9.7
Bluegill				0.1	T								
Smallmouth bass*	1.7	0.3	1.7	8.3	0.2	T	0.2	2.0	3.4	T	1.0	1.4	7.7
Largemouth bass*	1.3	0.1	1.0	0.6	0.3	T	0.3	0.8	2.4	0.4	0.1	0.4	2.9
Black crappie				0.2				T	T		T	0.1	0.4
Tessellated darter*	3.9	3.4	8.2	3.9	7.8	5.6	3.1	5.5	3.2	3.8	5.5	4.3	5.2
Logperch*	9.3	5.5	21.0	16.2	20.2	15.5	10.0	10.1	4.7	9.0	11.5	10.3	14.5
Mottled and slimy sculpins			T	T	T	0.1	0.1		T		T	T	T

variation in total effort did not affect the number of species captured. More species were captured in years when densities of age 0 yellow perch and walleyes were relatively high. With the exception of 1961, the number of forage-sized species ranged from 16 to 21 when fewer than 400 yellow perch were captured per haul. When mean catch of yellow perch exceeded 400 per haul, 22 to 26 species were collected. Correlation between number of species observed annually and mean catch per haul of yellow perch was significant ($r_2 = 0.77$). Low numbers of species were generally observed in years when mean walleye catch was less than 16 per haul, but high numbers of species were observed during years with mean walleye catches from six (1975) to 35 per haul (1969). Correlation between number of species captured and catch per haul of walleyes was barely significant ($r_s = 0.50$).

Species diversity ranged from 1.46 in 1964 to 3.26 in 1975 (Table 2). The diversity of 1.46 in 1964 was well below values in other years and occurred because of unusually abundant year classes of white suckers and pumpkinseeds (Table 1). Diversity was positively related to abundance of age 0 yellow perch, with values exceeding 3.0 in the four years of highest yellow perch density (Table 2). Correlation between species diversity and mean yellow perch catch per haul was significant ($r_s = 0.78$). Diversity was not obviously related to walleye density, however, and correlation was non-significant ($r_s = 0.22$).

Discussion

Number of Species

Interpretation of species number is complicated by a number of conditions associated with the sampling procedure used in this study. Sampling effort varied somewhat, but number of species was not affected in general by the number of seine hauls which were made. In most years, the majority of species were collected during the first half of the season and additional effort revealed few new species. Sampling in 1960, however, ended during late July; this may account for the fewest number of species occurring in 1960. Seining in August might have collected both pumpkinseeds and white bass, which normally appeared in samples during the late summer. The potential availability of more species in 1960 is consistent with the mean catch of yellow perch, which ranked sixth among 13 years. Species number itself also differs from the usual interpretation because the data concern resident species whose life cycles exceed one year. We interpret variation in species number as variation in the number of species which produced year classes abundant enough to be vulnerable to our sampling.

The positive relations between number of species captured and abundance of percids therefore suggests that in years when yellow perch and walleyes were more abundant, more species were successful in producing relatively large year classes. It also seems probable that rare species would be more easily overlooked in years when yellow perch were very abundant because samples would be less thoroughly examined and chances of escape by individual fish would be greater when large masses of fish were being processed. Thus the relation between yellow perch density and number of observed species is presumably stronger than illustrated by these data.

Several possible explanations exist for this apparent synchrony between abundance of yellow perch and relatively uncommon species in Oneida Lake. Favorable environmental conditions could enhance

TABLE 2.—*Mean catch per haul of yellow perch and walleye, number of species observed and species diversity in Oneida Lake, New York.*

Year	Yellow perch (number/haul)	Walleye (number/haul)	Number of species	Species diversity ($\bar{d}$)
1959	229	16	19	2.73
1960	328	5	16	2.72
1961	230	20	22	2.63
1962	478	12	23	3.21
1963	408	21	23	2.78
1964	390	16	19	1.46
1965	207	7	21	2.42
1966	200	5	18	2.63
1967	264	9	20	2.85
1968	414	34	22	2.75
1969	564	35	24	3.14
1970	736	20	23	3.15
1975	783	6	26	3.26

the reproductive success and survival of yellow perch and other species simultaneously. This is unlikely, however, because larval yellow perch abundance is determined largely by meterological conditions during spawning in May (Clady 1976) and many species of cyprinids, clupeids, and centrarchids spawn at different seasons and in different habitats (see Balon 1975). For example, Forney (1972) showed that seine catches (through 1969) of age 0 smallmouth bass, largemouth bass, and rock bass were correlated with each other and June air temperatures. The catches per haul of all three species in 1975 were the greatest of any year, however, whereas mean air temperature for June in 1975 (19.5 C) was close to the long-term average of 19.1 C. Catches of 11 other species were also highest in 1975 of any year examined and suggest that abundant yellow perch exerted a strong influence on the abundance of other species.

An alternate explanation is that abundant yellow perch buffer walleye predation on other species. Forney (1974) showed that walleyes feeding in offshore waters preyed less heavily on walleye and white perch cohorts when yellow perch were abundant. If walleyes feeding inshore consumed other species in direct proportion to their abundance and predation rate decreased as yellow perch density increased, species would remain relatively more abundant in years of high yellow perch density. Consequently, more species would be present at densities sufficient to permit capture in the seine sampling. Food habits of walleyes feeding inshore have not been studied, however, and this explanation remains speculative.

Species Diversity

According to Regier and Loftus (1972), the diversity index used in this analysis is perhaps the simplest measure of the ecological health of a fish community. It enables the biotic structure of a community to be summarized in a numerical index to which the contribution made by a rare species is very small (Wilhm 1968; Kochsiek and Wilhm 1970). Species diversity in Oneida Lake was directly related to yellow perch abundance. Because diversity is maximum when all species are equally abundant, this relation suggests either that more abundant species were adversely affected by adundant yellow perch or that less abundant species were favored.

Competition which would adversely affect common species seems unlikely. Noble (1975) demonstrated that growth of age 0 yellow perch was density-independent and that density and composition of the zooplankton was largely unaffected by yellow perch predation. Lack of competition among species is consistent with the taxonomic and ecological divergence of the Oneida Lake fish fauna.

Predation by walleyes may also be the mechanism by which species diversity and yellow perch abundance change together. If walleye predation becomes relatively more concentrated on the abundant species when yellow perch are more abundant, species' abundances would become more equal and diversity higher at high yellow perch densities. This interpretation runs counter to the generalization that predation and diversity are directly related. Where strong competition exists among species, as in rocky intertidal habitats, predation has been shown to increase diversity by reducing competition (Paine 1969). In Oneida Lake, however, decreases in predation would presumably be accompanied by no or only slight increases in competition. As with species number, no data regarding walleye feeding can be advanced to support or refute this interpretation.

Inshore-Offshore Relations

Integration of the percid-diversity relations for inshore areas with more extensive studies conducted in offshore waters is difficult. Densities of yellow perch and other species based on inshore seining and offshore trawling do not necessarily correspond (J. L. Forney, personal communication), suggesting either that seining or trawling does not accurately measure abundance or that inshore and offshore fish communities are independent of each other to some degree. This question cannot be resolved from the present data. For species such as yellow perch, walleye, white perch, and white bass, which inhabit the entire lake, relations in shallow water are probably secondary to those occurring in the more extensive offshore habitat.

Many of the species captured inshore, however, are seldom observed in offshore trawl samples, so that inshore sampling is the most appropriate source of data. The relations shown between yellow perch catches and species number and diversity suggest that percids strongly influence the structure of the inshore fish community. This conclusion is consistent with the importance of yellow perch and walleyes as determinants of population dynamics in offshore waters (Forney 1974, 1976), but interactions between these physical areas remain obscure.

If a diverse fauna is desirable for any of a number of reasons, these data suggest that abundant cohorts of age 0 yellow perch are also desirable. Inasmuch as abundant yellow perch also enhance walleye growth and abundance, management designed to increase walleye production is presumably compatible with a diverse inshore fauna. If yellow perch densities in inshore areas fluctuate out of phase with offshore densities, however, results of management would be much less certain.

The mechanisms responsible for these relations remain speculative in the absence of knowledge concerning stock characteristics of various species and food habits of walleyes in inshore waters. The possibility that these relations are merely fortuitous should not be overlooked and it must also be stressed that these results are appropriate only for the range of walleye and yellow perch densities which existed in Oneida Lake during 1959–1975. Recent reduction in biomass of adult walleyes may cause reduced consumption of prey species, competition among inshore fishes, or other unanticipated reactions.

Acknowledgment

Data used in this paper were collected under Federal Aid in Fish Restoration Project F-17-R, New York.

References

BALON, E. K. 1975. Reproductive guilds of fishes: A proposal and definition. J. Fish. Res. Board Can. 32:821–864.

BEETON, A. M., AND W. T. EDMONDSON. 1972. The eutrophication problem. J. Fish. Res. Board Can. 29:673–682.

CLADY, M. D. 1976. Influence of temperature and wind on the survival of early stages of the yellow perch, *Perca flavescens*. J. Fish. Res. Board Can. 33:1887–1893.

FORNEY, J. L. 1971. Development of dominant year classes in a yellow perch population. Trans. Am. Fish. Soc. 100:739–749.

———. 1972. Biology and management of smallmouth bass in Oneida Lake, New York. N.Y. Fish Game J. 19:132–154.

———. 1974. Interactions between yellow perch abundance, walleye predation, and survival of alternate prey in Oneida Lake, New York. Trans. Am. Fish. Soc. 103:15–24.

———. 1976. Year-class formation in the walleye (*Stizostedion vitreum vitreum*) population of Oneida Lake, New York, 1966–73. J. Fish. Res. Board Can. 33:783–792.

GREESON, P. E. 1971. Limnology of Oneida Lake with emphasis on factors contributing to algal blooms. U.S. Geol. Surv., Open-file Rep., N.Y. Dep. Environ. Conserv. Invest. RI-8. 185 pp.

KOCHSIEK, K. A., AND J. L. WILHM. 1970. Use of Shannon's formula in describing spatial and temporal variation in a zooplankton community in Keystone Reservoir, Oklahoma. Proc. Okla. Acad. Sci. 49:35–42.

NOBLE, R. L. 1975. Growth of young yellow perch (*Perca flavescens*) in relation to zooplankton populations. Trans. Am. Fish. Soc. 104:731–741.

PAINE, R. T. 1969. The *Pisastar-Tegula* interaction: prey patches, predator food preference, and intertidal community structure. Ecology 50:950–961.

REGIER, H. A., AND K. H. LOFTUS. 1972. Effects of fisheries exploitation of salmonid communities in oligotrophic lakes. J. Fish. Res. Board Can. 29:959–968.

SHANNON, C. D., AND W. WEAVER. 1963. The mathematical theory of communication. Univ. Ill. Press, Urbana.

SIEGEL, S. 1956. Nonparametric statistics for the behavioral sciences. McGraw-Hill, New York.

WILHM, J. L. 1968. Biomass units versus numbers of individuals in species diversity indices. Ecology 49:153–156.

Effects of Environmental Factors on Growth, Survival, Activity, and Exploitation of Northern Pike[1]

John M. Casselman[2]

Department of Zoology, University of Toronto
Toronto, Ontario M5S 1A1

Abstract

Northern pike grow best at 19 C for weight and 21 C for length. Growth rate is low at temperatures below 4 C, but increases rapidly at temperatures above approximately 10 C. As temperature increases above optimum, growth decreases rapidly and ceases at approximately 28 C. Growth in length and weight is stimulated slightly by long periods of daylight (16 h light–8 h dark) but is reduced and even negative in continuous light. The upper incipient lethal temperature for northern pike is 29.4 C. The lower incipient lethal oxygen concentration is directly related to temperature. Northern pike are extremely tolerant of low oxygen and in shallow lakes can survive minimum winter oxygen concentrations of 0.3 mg/liter (2% air saturation). Spontaneous swimming activity of northern pike fed ad libitum in the laboratory was maximal at 19 to 20 C, close to the optimum for growth. Northern pike were significantly less active below 6 C than at temperatures greater than approximately 9 C. Catch per unit effort of gill nets set in shallow lakes at temperatures of 13 to 24 C was high from 15 to 17 C, and nets most frequently contained northern pike and its primary prey, yellow perch, at 14 to 19 C. During oxygen depletion under lake ice in winter, northern pike can detect and avoid low oxygen concentrations. For both northern pike and yellow perch vertical distribution in, and catch per unit effort of, such stationary gear as experimental gill nets is highly correlated with the dissolved oxygen concentration in the immediate vicinity of the net. In late winter when much of a lake was virtually devoid of oxygen the fish congregated in areas where the oxygen concentration was slightly elevated. Although northern pike were captured alive in stationary gear at oxygen concentrations as low as 0.04 mg/liter (0.3% air saturation), activity in oxygen concentrations below 0.7 mg/liter was extremely low. Northern pike can tolerate a wide range of environmental conditions but are mesothermal or "coolwater" fish and are best adapted to the mesotrophic-eutrophic environment.

The effect of the environment on fishes has been widely studied (e.g., Fry 1947, 1971; Brett 1970; Doudoroff and Shumway 1970). However, the direct and even indirect effects of many environmental variables on the biology of many species remain incompletely described. Observations from the natural environment, or from controlled experiments conducted to simulate natural conditions, are required to elucidate and verify the response of fish to environmental conditions. A description of the response of fish to various environmental factors would define the general environmental needs and the niche occupied by the species. A description of the effect of such environmental variables as temperature, illumination, and dissolved oxygen concentration on activity and catch per unit effort of stationary gear can also be used to refine and eventually more precisely model the relations between catch per unit effort and environmental conditions. This would make catch per unit effort a more useful quantitative measure of relative abundance of fish stocks. The relationship between catch per unit effort and the major environmental parameters could also be applied to increase exploitation efficiency.

This paper relates such major environmental parameters as temperature, photoperiod, and dissolved oxygen concentration to growth, survival, and activity of northern pike and catch of this species as well as of yellow perch per unit effort of stationary gear. Data were collected from studies conducted in the natural environment and under controlled laboratory conditions. The results are also supplemented from the published literature. This paper has two purposes: to specifically docu-

---

[1] Contribution No. 78-14 of the Ontario Ministry of Natural Resources, Fisheries Research Section, Fisheries Branch, Box 50, Maple, Ontario L0J 1E0.

[2] Present address: Ontario Ministry of Natural Resources, Fisheries Research Section, Fisheries Branch, Box 50, Maple, Ontario L0J 1E0.

ment some of the environmental requirements of northern pike, and to define in general the more basic environmental needs of a typical "coolwater" fish.

Methods

Somatic growth of northern pike in relation to temperature was measured in the laboratory through a simulated natural seasonal temperature cycle of decreasing to low, and increasing to high, water temperature. Eight young-of-the-year and yearling northern pike were reared individually in two series of temperature experiments for a total of 30 experimental periods, each lasting 25 days. The specific or instantaneous rates of growth (Weatherley 1972) of total body weight (live) (TW) and fork length (FL) were measured. Northern pike were reared in a 12-h light–12-h dark photoperiod regime and were fed live golden shiners ad libitum. All environmental conditions were held constant in the experimental series except temperature, which was altered between experimental periods. However, temperature was held constant during each period (±0.2 C), and was changed an average of 2.6 C between periods. Test temperatures ranged from 2.9 to 29.4 C. Specific details of the experimental design and facilities used in these temperature studies can be found in Casselman (1978).

Growth results from the controlled laboratory study were compared to growth of northern pike tagged and recaptured in the natural habitat in which environmental conditions were monitored. The tag-recapture studies were conducted in two shallow lakes on Manitoulin Island, Ontario—Wickett Lake (45°54′N, 83°08′W; 71 hectares; mean depth 0.86 m) and Smoky Hollow Lake (45°38′N, 82°04′W; 30 hectares; mean depth 0.89 m). Limnological conditions in these lakes have been described by Harvey and Coombs (1971) and Casselman (1978). Specific details of the tag-recapture study of growth conducted on northern pike in these lakes are available in Casselman (1978).

Two series of photoperiod experiments were conducted in the laboratory to examine the effect of day length on the rate of somatic growth of northern pike. In one series, fish were exposed to constant photoperiods simulating natural day lengths (8 h light–16 h dark, 12 h light–12 h dark, and 16 h light–8 h dark). In this series eight fish were reared for ten experimental periods, each 25 days, at constant temperature ($\bar{x} \pm SE$, 19.9 ± 0.03 C) and fed ad libitum. In the second experimental series eight northern pike were reared in day lengths of extreme duration (4 h light–20 h dark and 24 h light–0 h dark) for four experimental periods, each 29 days, at constant temperature (25.2 ± 0.08 C) and fed ad libitum. In the two series of photoperiod experiments all environmental conditions were held constant except day length, which was changed between experimental periods. The specific experimental sequence of day lengths to which the fish were exposed is presented in the results. Northern pike reared in photoperiods of natural duration were smaller (beginning—FL, 283 ± 14.4 mm; TW, 202.8 ± 37.29 g; end—FL, 411 ± 24.8 mm; TW, 664.1 ± 60.47 g) than were northern pike exposed to photoperiods of extreme duration (beginning—FL, 423 ± 28.6 mm; TW, 681.7 ± 115.71 g; end—FL, 431 ± 29.6 mm; TW, 728.7 ± 107.54 g).

Incandescent light tubes (30-watt) illuminated the surface of the water with approximately 500 lux. An automatic intensity control gradually increased or decreased light intensity over a 30-min period at the beginning and end of the light period. Specific details of the rearing facilities and type and intensity of illumination can be found in Casselman (1978).

Equal numbers of male and female northern pike were used in the laboratory studies, and originated from the populations studied in the natural environment. The study was begun with young-of-the-year and yearling fish, which by the end of these studies were sub-adult, and subsequently matured when water temperature was decreased at the end of the experimental series. The effect of dissolved oxygen and temperature on growth and survival of embryo and larval northern pike has already been described in detail (Siefert et al. 1973; Hokanson, McCormick, and Jones 1973).

The upper lethal temperature was measured during the growth studies on northern pike that were reared in increasingly higher temperatures, eventually elevated to the upper incipient lethal. The fish had been acclimated for a period of at least 25 days to temperatures that were sublethal (e.g., 27.0 C). When the temperature was increased approximately 2.5 C according to the experimental design of the particular series, a temperature was reached that caused at least a 50% mortality in 24 hours.

Survival of northern pike at critically low oxygen concentrations was evaluated from observations made during late winter and early spring in Wickett and Smoky Hollow lakes.

Toleration of critically low oxygen concentrations was examined experimentally under simulated natural conditions. Northern pike were live-trapped in Wickett and Smoky Hollow lakes in late winter 1970. These fish were acclimatized to low oxygen (<2 mg/liter) at winter temperatures (<3 C). After they were removed from the trap nets they were immediately placed in a closed system containing water of a similarly low oxygen concentration and similar temperature to determine the dissolved oxygen level at which they lost equilibrium and could not be revived in well oxygenated water at a slightly elevated temperature (10 C).

Critically low oxygen concentrations were measured at summer temperatures in Minesing Swamp, Ontario (44°24′N, 79°53′W), under overcast skies in drainage ditches that were clogged with aquatic vegetation (*Elodea canadensis*, *Ceratophyllum demersum*, and *Myriophyllum exalbescens*). Regular observations and dissolved oxygen measurements in these drains during July 1974 documented the oxygen level at which mortalities of young-of-the-year northern pike (FL, 125 ± 2.9 mm) first occurred.

Spontaneous swimming activity in relation to temperature was measured in the laboratory in the 120-liter tanks used to study growth (Casselman 1978). Swimming activity was measured by continuously monitoring, by visual observation, the actual distance swum by six northern pike (FL, 458 ± 6.2 mm) reared individually in constant conditions and acclimated to a constant temperature (±0.2 C) for at least 25 days prior to observation. Activity was measured at seven different temperatures (27.3–4.0 C) in a decreasing regime. Forage minnows were constantly available. Swimming activity is expressed as the total distance (fish lengths converted to meters) swum during the entire 12-hour light period of a 12-h light–12-h dark photoperiod regime. Position in the tank was observed frequently during the dark period; the fish appeared to be virtually inactive during that time.

Activity was measured indirectly in the natural environment from the relative catch per unit effort of stationary gear in Wickett and Smoky Hollow lakes. The populations of northern pike and yellow perch were virtually stable when these data were collected. These systems are ideally suited to a study of catch per unit effort because they are uniformly shallow and small, the habitat is homogeneous, and vegetative cover is sparse. Since the results for catch per unit effort were averaged for several nets set at different locations throughout the lake and since this stationary gear captures moving fish, then the catch indicates the activity of the fish.

Catch per unit effort was calculated from the number of fish caught per hour in a dark green multifilament experimental gill net (50 × 1.5 m), containing six panels of equal length, of mesh sizes ranging from 38 to 90 mm. The net was set for a total of 24 consecutive hours and included one complete nocturnal and one complete diurnal period. The net and catch were removed at the end of the first nocturnal or diurnal period. The net was immediately set back if the catch was small; otherwise, another net was substituted for the remaining period. This procedure reduced net avoidance and selection associated with diel activity patterns and 24-h net sets (Casselman, unpublished data). Catch per unit effort is provided not only for northern pike but also for yellow perch, the only major prey species for northern pike in these lakes. These experimental gill nets select for northern pike of fork length greater than ca. 200 mm and yellow perch of fork length greater than ca. 125 mm.

Mean catch per unit effort and mean daily water temperature were averaged by month and compared.

Temperature and oxygen gradients were measured midway through each nocturnal and diurnal set. The mean temperature and dissolved oxygen concentration were determined for the mean depth of capture of the fish to more precisely assess activity and spatial and vertical distribution in relation to these environmental parameters.

To eliminate the influence of visual orientation, only nets set during the nocturnal period were used to determine the vertical distribution of northern pike and yellow perch in relation to oxygen concentration. The nets were set at locations where the depth of the water column was approximately equal to, or slightly less than, the depth of the gill net; hence the float line was at the ice-water interface. Fish captured in these nets were disentangled carefully, and the exact depth at which the fish had first become wedged or entangled was measured.

Results and Discussion

Growth

The optimum temperature for growth of northern pike, as determined by the maximum instantaneous increase in weight, was approximately 19 C (Fig. 1). Growth was low but positive at the lowest temperature examined. However, at these temperatures, which simulated winter conditions (<4 C), growth was only 3.9% of the maximum. A rapid increase in growth occurred in an increasing temperature regime after the water temperature reached approximately 10 C. At temperatures above the optimum the growth rate of northern pike decreased rapidly, and ceased at approximately 27.5 C. Approximately the same rate of growth was attained at 27 C as at 4 C.

The optimum temperature for growth as determined by the maximum instantaneous increase in length was approximately 21 C (Fig. 2), slightly higher than

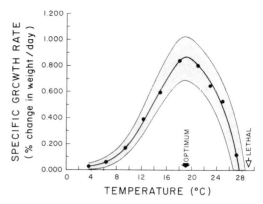

FIGURE 1.—Relation between specific rate of growth of body weight of northern pike and temperature. Fish were fed ad libitum and were reared in a 12-h light–12-h dark photoperiod regime. Laboratory results for 30 experimental periods, each of 25 days duration, each containing eight northern pike, were averaged for ten temperature values. Mean values and the curve of best fit are shown. Curves of best fit for the 95% confidence limits and interval (shaded area) are also illustrated. Arrows mark the optimum temperature for growth in weight and the upper incipient lethal temperature.

for weight. Linear growth continued even at low temperatures, although at a slow rate. This was achieved because the fish were young and had a high potential for

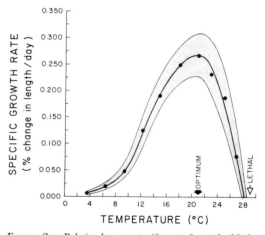

FIGURE 2.—Relation between specific rate of growth of fork length for northern pike and temperature. Fish were fed ad libitum and were reared in a 12-h light–12-h dark photoperiod regime. Laboratory results for 30 experimental periods, each of 25 days duration, each containing eight northern pike, were averaged for ten temperature values. Mean values and the curve of best fit are shown. Curves of best fit for the 95% confidence limits and interval (shaded area) are also illustrated. Arrows mark the optimum temperature for growth in length and the upper incipient lethal temperature.

growth and protein anabolism. Linear growth at temperatures that simulated winter conditions was only 2.4% of the maximum linear growth rate. As expected, specific growth rate in length was very much less than that of weight, but also increased rapidly in an increasing temperature regime at temperatures greater than 10 C. Linear growth ceased at approximately 28.2 C. Approximately the same rate of linear growth was attained at 28 C as at 4 C.

In other controlled growth studies that have been conducted on northern pike, growth rate was maximal near the highest experimental temperatures (Johnson 1966; Frost and Kipling 1968); however, these were lower than the optimum temperature observed here.

The optimum temperature for linear growth of northern pike tagged and recaptured in lakes on Manitoulin Island was very similar to that determined in the laboratory for fish from the same lakes. When northern pike of the same ages (calendar years 2 and 3) from the laboratory and field studies were compared, the optimum temperature for linear growth for the laboratory northern pike was 20.9 ± 0.05 C and under natural conditions in Wickett Lake was 19.8 ± 0.60 C (Casselman 1978). This slight difference may reflect a difference in food supply, since Brett et al. (1969) have shown that maximum growth rate occurs at a lower temperature if the ration is reduced. Direct field observations of the amount of cannibalism in this population, as well as population estimates of both predator and prey and predator-prey ratios, indicate that the northern pike population in Wickett Lake was undernourished (Casselman, unpublished data).

These relations, which describe maximum growth rate and its association with temperature, define the "scope for growth" (Warren and Davis 1967), or "metabolic scope" (Fry 1947) for northern pike. Temperatures ranging from 10 to 23 C are the most favourable for northern pike, although the optimum temperature for growth, or the general physiological optimum, is approximately 19 C. The physiological optimum for larval northern pike is higher (26 C) (Hokanson, McCormick, and Jones 1973), as expected, because the optimum temperature for growth decreases with age.

If fish are grouped by preferred temperature or the temperature of their midsummer habitat (Ferguson 1958), it is apparent that northern pike have a physiological optimum that is intermediate between cold and warmwater fishes, and can be truly classified as coolwater fishes. Coldwater fishes, as exemplified by salmonids and coregonids, have physiological optima for juveniles and adults of approximately 12 to 17 C: e.g., brook trout yearlings, 13 C (Baldwin 1957); brook trout adults, 16.1 C (Hokanson, McCormick et al. 1973); sockeye salmon fingerlings, 15 C (Brett et al. 1969). Warmwater fishes, as exemplified by centrarchids and cyprinids, have physiological optima for juveniles and adults that may exceed 28 C: e.g., bluegill juveniles, 30.1 C (Lemke 1977; review by Hokanson 1977). Very generally, coolwater fishes can be defined as those with physiological optima that range from 18 to 25 C for subadults and adults. The optimum temperature for growth of northern pike (19 C) is slightly lower than that of other fishes included in this coolwater grouping, such as walleye, 22.6 C, and yellow perch, 24.7 C (Hokanson 1977).

Light has a demonstrable effect on growth, although the response it normally elicits is not as great as that observed for temperature. Although the results of the effect of day length on growth of northern pike were not completely conclusive, probably because the experiments were relatively short-term, some consistent responses were apparent. In the experimental series that examined day length of natural duration, the growth rate in both weight and length showed a decreasing trend with time (Fig. 3). Growth decreased from the beginning of the study because the fish were small, young, and fast growing and had a very high but decreasing metabolic rate. When the fish were exposed to longer day length (16 h light–8 h dark) there was a positive shift in the decreasing trend, although the increase in growth rate was not significant at the 5% level. Maximum growth rate of northern pike occurs in the natural environment during long day length. Long day length of natural duration has also been shown to stimulate growth in other fish: e.g., green sunfish (Gross et al. 1965). It undoubtedly functions through the endocrine system (Saunders and Henderson 1970) and possibly through thyroxine and growth hormone. However, it has been shown that photoperiod has no effect on the response of fish to exogenous bovine growth hormone (Adelman 1977).

The results seem sufficiently consistent (Fig. 3) to indicate that the increase in growth rate would probably have been more significant if longer experimental periods had been used. The northern pike used in this experimental series were relatively young, fast growing, and had a high rate of food consumption. Since northern pike feed by sight, a longer light period might be expected to result in a higher rate of food consumption if food were constantly available and particle size were relatively small. The number of food items eaten and rate of food consumption were higher during long day length. However, total swimming activity, although not rate, was greater during this period. This increase in total

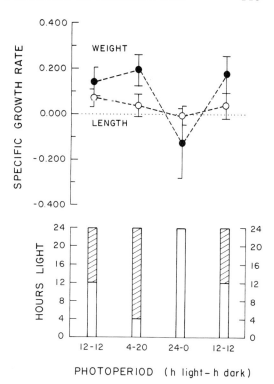

FIGURE 4.—Specific growth rates in weight and length for northern pike in relation to photoperiods of extreme duration. Means and 95% confidence intervals are shown above the photoperiod regime. Each of the four consecutive experimental treatments (29 days duration) used the same eight northern pike, reared at constant temperature (25.2 ± 0.08 C) and fed ad libitum.

activity with increased day length may have stimulated an increase in appetite. Since food was constantly available, this would result in an increased rate of food consumption and growth.

Although long day length of natural duration appeared to stimulate growth (Fig. 3), continuous light caused a significant decrease in weight and a slight decrease in length (Fig. 4). Abnormally short day length produced inconclusive trends. Northern pike reared under continuous light were extremely active, appeared to swim almost continually, and were hyperactive when disturbed. Long day length has been shown to be associated with high maintenance requirements (Johnson 1966). Continuous light appears to be detrimental to growth, and results in a loss in weight that is associated with some linear shrinkage.

It is not known why the extremely short

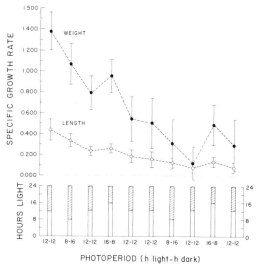

FIGURE 3.—Specific growth rates in weight and length for northern pike in relation to photoperiods of natural duration. Means and 95% confidence intervals are shown above the photoperiod regime. Each of the ten consecutive experimental treatments (25 days duration) used the same eight northern pike, reared at constant temperature (19.9 ± 0.03 C) and fed ad libitum.

day length in this set of photoperiod experiments did not result in a marked decrease in growth rate. However, the northern pike used in this experimental series were larger, slower growing, and had a lower rate of food consumption. It is possible that they had adequate time to capture their daily requirements within this contracted period.

Laboratory experiments were not specifically conducted to examine the effect of light intensity on growth. However, data from angling studies conducted during the open-water period indicate that northern pike feed more actively on cloudy, overcast days than on bright, sunny days (Casselman, unpublished data). On days when light intensity is high they feed more actively during evening, and to a lesser extent morning, twilight, when light intensity is low. When growth rate, food consumption, and appetite are highest, differences in feeding rate in relation to illumination are less apparent.

Although northern pike appear to be more active and feed more intensively at lower light intensities, they are especially efficient at capturing, and select for, fish that swim high in the water column and are silhouetted against a lighter background. Observations from the St. Lawrence River indicate that northern pike prefer such soft-rayed species as alewives that swim in midwater. This preference is especially apparent when alewives move in over the aquatic vegetation of the littoral zone at spawning time.

Results showing the relation between dissolved oxygen concentration and growth rate of subadult northern pike obtained during the present study substantiate those of Adelman and Smith (1970) and others (e.g., Stewart et al. 1967), who have shown that food consumption, conversion efficiency, and growth rate of fishes are maximal at 100% air saturation. Conversion efficiency was greatly reduced at dissolved oxygen concentrations below 4 mg/liter (49% air saturation) when juvenile largemouth bass were reared at 26 C (Stewart et al. 1967). Adelman and Smith (1970) reported a marked decrease in food consumption, conversion efficiency, and growth rate for northern pike reared in water of less than approximately 2.8 mg/liter oxygen (30% air saturation) at 18.6 C. In studies conducted to determine the survival of northern pike at low oxygen concentration Petit (1973) reported that northern pike ceased feeding when reared at an oxygen concentration of 2 mg/liter (22% air saturation) at temperatures of approximately 21.5 C. Any decrease in oxygen concentration effects a decrease in growth rate. However, food consumption and conversion efficiency decrease markedly when oxygen concentrations fall below approximately 2 to 3 mg/liter (20 to 30% air saturation).

Survival

The upper incipient lethal temperature was observed in laboratory studies of growth in which the test temperatures were progressively increased. The upper incipient lethal of subadult northern pike was 29.4 C. Similar values have been reported for northern pike of approximately the same size (Gardner and King 1923—30 C; Scott 1964—33 C; W. B. Horning, USEPA Fish Toxicology Station, Newtown, Ohio, personal communication—29 C). The upper incipient lethal for northern pike (29.4 C) falls within the lower portion of the range of upper incipient lethal temperatures (28 to 34 C) that Hokanson (1977) attributed to temperate mesothermal freshwater fishes.

The lower incipient lethal temperature is more difficult to determine. Northern pike can tolerate water temperatures very close to freezing. They show no apparent stress when subjected to a temperature of 0.1 C for extended periods of time prior to freeze-up in the shallow lakes on Manitoulin Island. A rapid decrease in water temperature over a sufficiently large range can result in mortality. Mortalities resulting from "cold shock" have been reported when northern pike that had congregated in a thermal discharge channel and were acclimatized to a temperature of 21.8 C were exposed to a temperature of 4.9 C when the heated effluent was shut off (Ash et al. 1974).

The effect of critically low oxygen concentration on the survival of northern pike

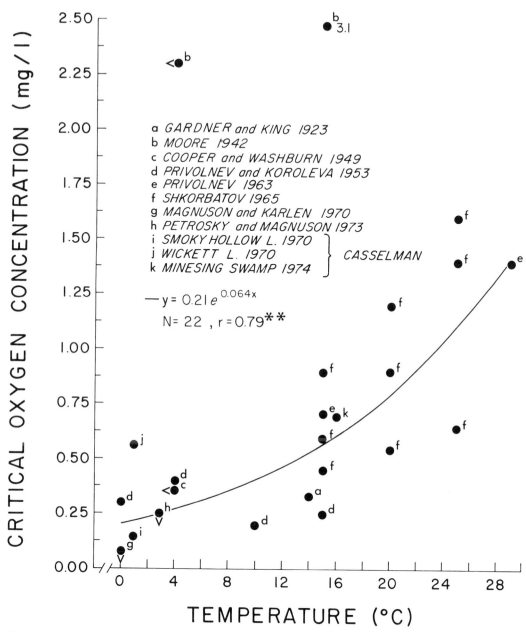

FIGURE 5.—*Relation between lower incipient lethal oxygen concentration for northern pike and temperature. Data are from published literature and the current study. The symbol "<" indicates "less than" and applies to the variable in the direction opposite to which it points. Regression excludes the results of Moore (1942).*

has been widely studied. The relation between critical oxygen concentration and temperature is highly significantly curvilinearly related (Fig. 5). Moore's (1942) results were not used in this analysis because the fish had not been acclimatized to low oxygen concentrations. The report of the study conducted by Petit (1973) was not available when this relation was calculated, so his results were used to test its consistency. For fish tested at approximately 21.4 C Petit reported that the lethal dissolved oxygen concentration was 0.83 mg/liter. The exponential equation presented here estimates the critical oxygen concentration at this temperature to be 0.84 mg/liter. The difference between the observed and the expected is remarkably small, enforcing the reliability of the relation.

The lower incipient lethal oxygen concentration is frequently encountered at low temperature in association with winterkill. This mortality occurs in lakes where oxygen is depleted under the ice cover in late winter to a level that results in fish mortality. Northern pike are extremely tolerant of low winter oxygen, and in shallow lakes where oxygen is greatly reduced during winter they can readily survive minimum oxygen concentrations of 0.3 mg/liter (2% air saturation). In fact, some northern pike were captured alive in Smoky Hollow Lake in stationary gear at oxygen concentrations of 0.04 mg/liter (0.3% air saturation) when the maximum oxygen concentration in the lake was 0.8 mg/liter. No mortality of northern pike was associated with these conditions, although some adult yellow perch succumbed. Many factors affect the ability of northern pike to tolerate low winter oxygen, and mortalities that occur are usually partial and selective (Casselman and Harvey 1975). From data supplied by Johnson and Moyle (1969) it was calculated that complete winterkills occurred in lakes in Minnesota when the oxygen concentration was 0.21 ± 0.06 mg/liter, and partial winterkills when the concentration was 0.85 ± 0.26 mg/liter.

The ability of northern pike to withstand critically low winter oxygen is probably also genetic, since an experimental determination of critically low oxygen concentration conducted under field conditions was higher for northern pike from Wickett Lake (0.55 mg/liter) than for fish from Smoky Hollow Lake (0.15 mg/liter). Minimum winter oxygen concentrations are consistently lower, and partial winterkills occur more frequently, in Smoky Hollow Lake than in Wickett Lake.

Activity and Catch per Unit Effort

Spontaneous swimming activity of northern pike in relation to temperature was measured directly under laboratory conditions (Fig. 6). Maximum swimming activity occurred at temperatures (approximately 19–20 C) that were very close to the optimum temperature for growth for these fish (Fig. 1). Maximum swimming ability of fish as measured by cruising speed and the optimum temperature for growth appear to coincide (see review by Fry 1964). Northern pike were significantly less active at temperatures below 6 C than at temperatures greater than approximately 9 C. If total daily swimming activity occurs during the light period under laboratory conditions, as appears to be the case, then in these laboratory facilities the total daily movement of the fish in relation to temperature was very similar to the gross daily

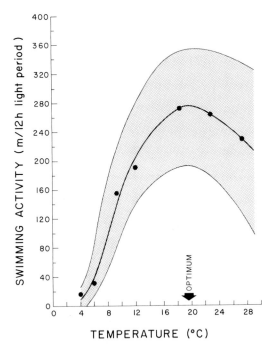

FIGURE 6.—Relation between swimming activity and temperature for six northern pike reared under controlled laboratory conditions. The temperature was constant (±0.2 C) and the fish had been acclimated for at least 20 days, the temperature regime was decreasing, and food was constantly available—ad libitum. Spontaneous swimming activity is expressed in total distance (m) swum during the 12-h light period of a 12-h light–12-h dark photoperiod regime. Curves of best fit about the means and 95% confidence limits and interval (shaded area) are illustrated. Arrow marks the temperature for maximum spontaneous swimming activity.

movement in the natural environment for northern pike tagged with ultrasonic transmittors and monitored during winter and summer (Diana et al. 1977).

Northern pike are least active at high light intensities. Catch per unit effort for gill nets in the natural environment indicates that northern pike are most active during twilight and nocturnal periods during the open-water portion of the year. However, during the ice covered portion of the year they are most active during the diurnal period (Casselman, unpublished data). Northern pike were most active during the light period under laboratory conditions. This diel activity pattern in the laboratory, even at high water temperatures, more closely corresponds to diel activity patterns that occurred during winter under the ice. These results, however, are not paradoxical but are complementary because the intensity of illumination in the laboratory facilities (300–700 lux) was more similar to the illumination just under the ice at midday in January (400–800 lux) than to summer midday intensities.

Swimming activity in the laboratory was spontaneous and independent of foraging activity because forage minnows were constantly available and activity was not caused by seeking out or capturing prey. Although the fish were all fed ad libitum, individuals fed at slightly different rates. A direct correlation exists between the rate of food consumption and swimming activity, which was independent of the activity associated with garnering of food (Casselman, unpublished data). Swimming activity of *Ophiocephalus striatus* reared in cylindrical aquaria was also related to food intake (Vivekanandan 1976).

The effect of various environmental factors on activity patterns of fish has been rather widely examined (e.g., Crossman 1959; Alabaster and Robertson 1961; Ryder 1977; Ryder and Kerr 1978). Activity of fish can be assessed by examining changes in the catches of stationary gear, provided the population remains relatively stable over the period studied, the habitat is relatively uniform, and the fish distribution is reasonably homogeneous. These conditions were fulfilled in the shallow lakes used in this study, justifying a comparison of catch per unit effort in relation to various environmental characteristics and an interpretation of results in relation to activity.

Monthly mean catches of northern pike and their primary prey, yellow perch, per unit effort of experimental gill net in relation to monthly mean temperatures in Wickett Lake were maximal in April, the month when spawning occurred (Fig. 7). Catch per unit effort during the winter months of January and February was also relatively high. Catches were larger because at that time of year the oxygen in the shallows is depleted and the fish were moving to areas of higher dissolved oxygen. Catches at that time were not, however, strictly comparable to those in open-water periods, since the population was constricted by a decrease in the habitable volume of the lake because of both a decrease in oxygen in the shallower areas and an increase in the thickness of the ice cover. During the open-water period, May to October, in Wickett Lake catch per unit effort for northern pike was highest during those months when the mean daily water temperatures averaged from 14 to 18 C (Fig. 7).

The relationship between catch per unit

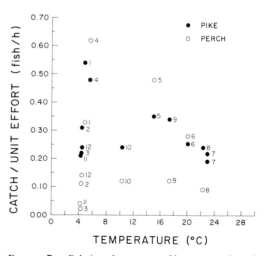

FIGURE 7.—*Relations between monthly mean catches of northern pike and yellow perch per unit effort of experimental gill net and monthly mean water temperature in Wickett Lake, 1969–1971. Results are for 339 24-h experimental gill net sets. The numerical code for the month is located beside the mean.*

effort and temperature was measured more precisely during the open-water period when temperature was monitored at the net. These results exclude such extraneous effects as spawning activity, low winter oxygen, and winter crowding. For nets set at temperatures from 13 to 24 C, catch per unit effort was high from 15 to 17 C (Fig. 8) and was lowest at the maximum temperature. Nets most frequently contained northern pike and yellow perch at temperatures ranging from 14 to 19 C.

There may be several explanations for the difference between maximum activity with temperature as indicated by catch per unit effort in the natural environment (14–19 C) and the temperature at which maximum activity occurred in the laboratory study (19–20 C). Since the food supply for northern pike in these lakes is somewhat limited, then maximum activity as indicated by catch per unit effort would not be expected to be directly related to the temperature for maximum activity (19–20 C) (Fig. 6) or the optimum temperature for growth (19 C) (Fig. 1) as determined on fish fed ad libitum under laboratory conditions. Activity would not be entirely spontaneous when food is limiting, but would in part be directed to its garnering. Also, if food supply is limited, then it would be more efficient if activity were decreased, especially at higher temperatures. In a naturally increasing temperature regime rapid somatic growth commences at approximately 14 C for nothern pike of adult size (Casselman 1978). At this temperature appetite is stimulated and fish actively seek food. The northern pike captured in stationary gear were adults and appreciably older than the subadults used to determine the optimum temperature for growth under laboratory conditions. Since the optimum temperature for growth and activity decreases with age, then the maximum activity of fish captured in the natural environment would occur at a lower temperature than that observed under laboratory conditions.

The vertical distribution of not only northern pike but also yellow perch under the ice in Smoky Hollow Lake was related to the concentration of dissolved oxygen (Fig. 9). As winter advanced and oxygen became progressively more depleted from the substrate to the ice-water interface, fish moved increasingly higher in the water column. This response has been previously observed for this species both in the natural environment (Magnuson and Karlen 1970) and in simulated winter conditions in tanks (Petrosky and Magnuson 1973). At oxygen concentrations of approximately 2 to 3 mg/liter both northern pike and yellow perch were near (10 to 20 cm) the ice-water interface. Privolnev (1954) observed a depression in respiration at these dissolved oxygen concentrations (2 to 3 mg/liter). All data indicate that at oxygen con-

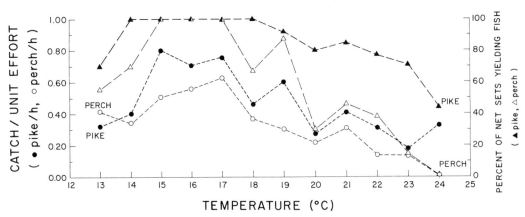

FIGURE 8.—Catch per unit effort and percent of the sets that contained northern pike and yellow perch when experimental gill nets were set in Wickett Lake, 1969–1972, at temperatures of 13 to 24 C. From 8 to 12 24-h sets were made at each temperature, and a total of 120 sets was used.

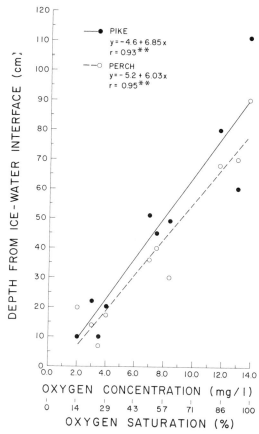

FIGURE 9.—Relations between vertical distribution of northern pike and yellow perch and oxygen concentration at the mean depth of capture. Results are from ten nocturnal sets of experimental gill nets fished during the ice-covered period in Smoky Hollow Lake, 1969–1971. Percent of air saturation was calculated considering 14 mg/liter equal to 100%.

tive of activity only when the nets were not set in areas that contained higher dissolved oxygen concentrations than most other areas of the lake, hence contained congregating fish. In late winter in Smoky Hollow Lake when much of the lake was devoid of oxygen, fish congregated in areas where the oxygen concentration was slightly elevated, ranging from 0.7 mg/liter (5% air saturation) to 4.0 mg/liter (30% saturation). In decreasing oxygen northern pike congregate slightly sooner than yellow perch. As oxygen depletion progresses, congregations of yellow perch remain detectable at a lower oxygen concentration than do northern pike. When the highest oxygen concentrations in the lake are less than approximately 0.7 mg/liter, activity becomes extremely low and congregations can no longer be detected with stationary gear. This value corresponds to the oxygen concentration at which these species were calculated to move up against the ice-water interface. Petit (1973) reported that activity decreased as lethal concentrations were approached. Johnson and Moyle (1969) observed that northern pike were most effectively attracted to an aerated discharge when oxygen concentrations in the lake were less than 1 mg/liter. If an oxygen gradient exists, fish are able to avoid low oxygen. In poorly oxygenated water, threespine sticklebacks became uneasy and moved randomly, but when they moved into well oxygenated water the stimulus to swim vanished (Jones 1952).

centrations below these levels, northern pike, and probably yellow perch, begin to show noticeable stress. Extrapolation of these data indicates that as oxygen depletion progresses yellow perch would be expected to meet the ice-water interface at a slightly higher oxygen concentration (0.86 mg/liter) than would northern pike (0.67 mg/liter).

Catch per unit effort of gill nets set under the ice in relation to the dissolved oxygen concentration in the immediate vicinity of the net showed highly significant linear relations for both northern pike and yellow perch if congregating fish were excluded (Fig. 10). Catch per unit effort was indica-

Exploitation

Angling success, as well as catch per unit effort of stationary fishing gear, is influenced by environmental conditions. In addition to the general associations that have been described here, other specific examples can be cited. Angler success in winter is directly related to oxygen concentration and is adversely affected if the oxygen falls below approximately 2 mg/liter and the northern pike cease feeding. One of the best examples of how environmental factors may be used to facilitate the harvest of young northern pike from shallow winterkill lakes may be seen in a management procedure that

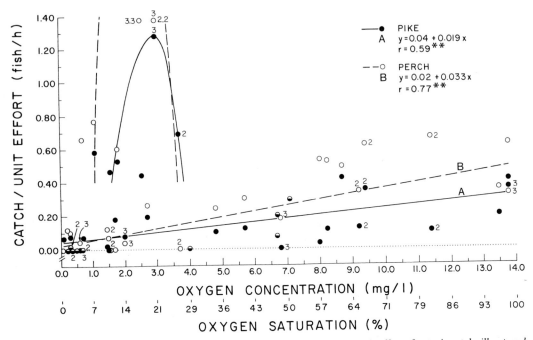

FIGURE 10.—Relations between catches of northern pike and yellow perch per unit effort of experimental gill net and oxygen concentration at the mean depth of capture. Linear regression results are for 46 24-h sets (A—northern pike, B—yellow perch) in Smoky Hollow Lake, 1969–1972. The curvilinear relations for nine 24-h experimental gill net sets that captured congregating fish in low oxygen concentrations (ranging from 0.7 to 4.0 mg/liter) that were appreciably elevated in comparison to other areas of the lake are also illustrated. Numbers indicate the number of times that nets were reset at the same location and same oxygen concentration. The average for the sets is given. Percent of air saturation was calculated considering 14 mg/liter as equal to 100%.

has been practised in Minnesota (Hanson 1958; Johnson and Moyle 1969). Fingerling and yearling northern pike are harvested during late winter, when the oxygen is less than 2 mg/liter, by artificially elevating the oxygen in a pumping channel and along the adjacent shoreline to which the young fish are subsequently attracted and are captured.

Conclusions

Northern pike can tolerate a wide range of environmental conditions. Each environmental parameter has a distinct and demonstrable influence on the physiology of the fish. Although the growth optima for northern pike are 19 to 21 C, this species grows well at temperatures ranging from 10 to 23 C. Northern pike are typical mesothermal or "coolwater" fish that are most productive in temperate mesotrophic-eutrophic environments in which transparency is high, vegetative cover is abundant, and the development of the vegetation-open water interface is high.

Acknowledgments

I wish to thank E. J. Crossman, F. E. J. Fry, and H. H. Harvey of the University of Toronto for their interest and assistance in my Ph.D. program during which most of the data referred to here were collected. This program was funded by the Fisheries Research Board of Canada and the Canadian National Sportsmen's Show. I would like to thank my colleagues of the Ontario Ministry of Natural Resources, Fisheries Branch, especially my supervisor, A. H. Lawrie, for their helpful comments on the data and the manuscript.

References

ADELMAN, I. R. 1977. Effects of bovine growth hormone on growth of carp (*Cyprinus carpio*) and the influences of temperature and photoperiod. J. Fish. Res. Board Can. 34:509–515.

———, AND L. L. SMITH. 1970. Effect of oxygen on growth and food conversion efficiency of northern pike. Prog. Fish-Cult. 32:93–96.

ALABASTER, J. S., AND K. G. ROBERTSON. 1961. The effect of diurnal changes in temperature, dissolved oxygen and illumination on the behaviour of roach (*Rutilus rutilus* (L.)), bream (*Abramis brama* (L.)) and perch (*Perca fluviatilis* L.). Anim. Behav. 9:187–192.

ASH, G. R., N. R. CHYMKO, AND D. N. GALLUP. 1974. Fish kill due to "cold shock" in Lake Wabamun, Alberta. J. Fish. Res. Board Can. 31:1822–1824.

BALDWIN, N. S. 1957. Food consumption and growth of brook trout at different temperatures. Trans. Am. Fish. Soc. 86:323–328.

BRETT, J. R. 1970. Environmental factors, Part 1. Pages 513–560 *in* O. Kinne, ed. Marine ecology, Vol. 1. Wiley-Interscience, London.

———, J. E. SHELBOURN, AND C. T. SHOOP. 1969. Growth rate and body composition of fingerling sockeye salmon, *Oncorhynchus nerka*, in relation to temperature and ration size. J. Fish. Res. Board Can. 26:2363–2394.

CASSELMAN, J. M. 1978. Calcified tissue and body growth of northern pike, *Esox lucius* Linnaeus. Ph.D. Thesis. Univ. Toronto, Toronto, Ontario. 782 pp.

———, AND H. H. HARVEY. 1975. Selective fish mortality resulting from low winter oxygen. Verh. Int. Verein. Limnol. 19:2418–2429.

COOPER, G. P., AND G. N. WASHBURN. 1949. Relation of dissolved oxygen to winter mortality of fish in Michigan lakes. Trans. Am. Fish. Soc. 76:23–33.

CROSSMAN, E. J. 1959. Distribution and movements of a predator, the rainbow trout, and its prey the red shiner in Paul Lake, British Columbia. J. Fish. Res. Board Can. 16:247–267.

DIANA, J. S., W. C. MACKAY, AND M. EHRMAN. 1977. Movements and habitat preference of northern pike (*Esox lucius*) in Lac Ste. Anne, Alberta. Trans. Am. Fish. Soc. 106:560–565.

DOUDOROFF, P., AND D. L. SHUMWAY. 1970. Dissolved oxygen requirements of freshwater fishes. FAO Fish. Tech. Pap. 86. 291 pp.

FERGUSON, R. G. 1958. The preferred temperature of fish and their midsummer distribution in temperate lakes and streams. J. Fish. Res. Board Can. 15:607–624.

FROST, W. E., AND C. KIPLING. 1968. Experiments on the effect of temperature on the growth of young pike, *Esox lucius* L. Salmon Trout Mag. 184:170–178.

FRY, F. E. J. 1947. Effects of the environment on animal activity. Univ. Toronto Stud. Biol. Ser. 55, Publ. Ont. Fish Res. Lab. 68. 62 pp.

———. 1964. Animals in aquatic environments: fishes. Pages 715–728 *in* J. Field, ed. Handbook of physiology. Williams & Wilkins, Baltimore.

———. 1971. The effect of environmental factors on the physiology of fish. Pages 1–98 *in* W. S. Hoar and D. J. Randall, eds. Fish physiology, vol. 6. Academic Press, New York.

GARDNER, J. A., AND G. KING. 1923. Respiratory exchange in freshwater fish. Part VI. On pike (*Esox lucius*). Biochem. J. 17:170–173.

GROSS, W. L., E. W. ROELOFS, AND P. O. FROMM. 1965. Influence of photoperiod on growth of green sunfish, *Lepomis cyanellus*. J. Fish. Res. Board Can. 22:1379–1386.

HANSON, H. 1958. Operation fish rescue. Prog. Fish-Cult. 20:186–188.

HARVEY, H. H., AND J. F. COOMBS. 1971. Physical and chemical limnology of the lakes of Manitoulin Island. J. Fish. Res. Board Can. 28:1883–1897.

HOKANSON, K. E. F. 1977. Temperature requirements of some percids and adaptations to the seasonal temperature cycle. J. Fish. Res. Board Can. 34:1524–1550.

———, J. H. MCCORMICK, AND B. R. JONES. 1973. Temperature requirements for embryos and larvae of the northern pike, *Esox lucius* (Linnaeus). Trans. Am. Fish. Soc. 102:89–100.

———, ———, ———, AND J. H. TUCKER. 1973. Thermal requirements for maturation, spawning, and embryo survival of the brook trout, *Salvelinus fontinalis*. J. Fish. Res. Board Can. 30:975–984.

JOHNSON, F. H., AND J. B. MOYLE. 1969. Management of a large shallow winterkill lake in Minnesota for the production of pike (*Esox lucius*). Trans. Am. Fish. Soc. 98:691–697.

JOHNSON, L. 1966. Experimental determination of food consumption of pike, *Esox lucius*, for growth and maintenance. J. Fish. Res. Board Can. 23:1495–1505.

JONES, J. R. E. 1952. The reactions of fish to water of low oxygen concentration. J. Exp. Biol. 29:403–415.

LEMKE, A. E. 1977. Optimum temperature for growth of juvenile bluegills. Prog. Fish-Cult. 39:55–57.

MAGNUSON, J. J., AND D. J. KARLEN. 1970. Visual observation of fish beneath the ice in a winterkill lake. J. Fish. Res. Board Can. 27:1059–1068.

MOORE, W. G. 1942. Field studies on the oxygen requirements of certain fresh-water fishes. Ecology 23:319–329.

PETIT, G. D. 1973. Effects of dissolved oxygen on survival and behavior of selected fishes of western Lake Erie. Bull. Ohio Biol. Surv. N. S. 4:1–76.

PETROSKY, B. R., AND J. J. MAGNUSON. 1973. Behavioral responses of northern pike, yellow perch and bluegill to oxygen concentrations under simulated winterkill conditions. Copeia (1):124–133.

PRIVOLNEV, T. I. 1954. Physiological adaptations of fishes to new conditions of existence. Tr. Soveshch. Ikhtiol. Kom. Akad. Nauk SSSR (3):40–49. (Transl. from Russian by Fish. Res. Board Can. Transl. Ser. No. 422, 1963.)

———. 1963. Threshold concentration of oxygen in water for fish at different temperatures. Dokl. Akad. Nauk SSSR, Biol. Sci. Sect. 151:439–440. (Transl. from Russian, Nat. Sci. Found.)

———, AND N. V. KOROLEVA. 1953. Critical concentration of oxygen for fish at different temperature in different seasons of the year. Dokl. Akad. Nauk SSSR 89:175–176. (Transl. from Russian by E. Jermolajev).

RYDER, R. A. 1977. Effects of ambient light variations on behavior of yearling, subadult, and adult walleyes (*Stizostedion vitreum vitreum*). J. Fish. Res. Board Can. 34:1481–1491.

———, AND S. R. KERR. 1978. The adult walleye

in the percid community—a niche definition based on feeding behaviour and food specificity. Am. Fish. Soc. Spec. Publ. 11:39–51.

SAUNDERS, R. L., AND E. B. HENDERSON. 1970. Influence of photoperiod on smolt development and growth of Atlantic salmon (*Salmo salar*). J. Fish. Res. Board Can. 27:1295–1311.

SCOTT, D. P. 1964. Thermal resistance of pike (*Esox lucius* L.), muskellunge (*E. masquinongy* Mitchill), and their F_1 hybrid. J. Fish. Res. Board Can. 21:1043–1049.

SHKORBATOV, G. L. 1965. Intraspecific variability in respect to oxyphilia among freshwater fishes. Gidrobiol. Zh. 1:3–8. (Transl. from Russian by Fish. Res. Board Can. Transl. Ser. No. 701, 1966.)

SIEFERT, R. E., W. A. SPOOR, AND R. F. SYRETT. 1973. Effects of reduced oxygen concentrations on northern pike (*Esox lucius*) embryos and larvae. J. Fish. Res. Board Can. 30:849–852.

STEWART, N. E., D. L. SHUMWAY, AND P. DOUDOROFF. 1967. Influence of oxygen concentration on the growth of juvenile largemouth bass. J. Fish. Res. Board Can. 24:475–494.

VIVEKANANDAN, E. 1976. Effects of feeding on the swimming activity and growth of *Ophiocephalus striatus*. J. Fish Biol. 8:321–330.

WARREN, C. E., AND G. E. DAVIS. 1967. Laboratory studies on the feeding, bioenergetics, and growth of fish. Pages 175–214 *in* S. D. GERKING, ed. The biological basis of freshwater fish production. Blackwell Scientific Publications, Oxford. 495 pp.

WEATHERLEY, A. H. 1972. Growth and ecology of fish populations. Academic Press, New York. 293 pp.

Ecologic Separation of Sympatric Muskellunge and Northern Pike[1]

E. J. Harrison and W. F. Hadley

Division of Biology
State University of New York at Buffalo
Buffalo, New York 14214

Abstract

With the exception of a few one-year-old fish, no muskellunge were found in tributaries of the upper Niagara River during 1975–1977. In contrast, nearly all (97%) northern pike were collected in tributaries despite extensive river sampling. The species are spatially segregated throughout much of their life cycles. Differential adaptation to river current may be the most important factor permitting coexistence of the two species. The geologic history of the Niagara River suggests that the muskellunge population may have been established rather recently (since 5,500 years before present). However, the present distribution of northern pike indicates that this population was probably established much earlier (12,300–10,400 BP).

Examination of regional distributions of northern pike and muskellunge show that water bodies containing major natural populations of both species are rare. Rarity of sympatric populations of these species is generally attributed to competitive superiority of northern pike. Among the reasons postulated for this superiority are (1) earlier spawning of northern pike on common spawning grounds and subsequent predation by northern pike fry on newly hatched muskellunge (Threinen and Oehmcke 1950), and (2) less size selectivity for food items and superior food conversion by northern pike (Scott and Crossman 1973).

Because of unsound management practices and water-body modifications, northern pike have gained access to some waters containing native populations of muskellunge (Threinen and Oehmcke 1950). Williamson (1942), Threinen and Oehmcke (1950), and Threinen et al. (1966) stated that these introductions have been detrimental to the muskellunge populations, and Johnson (1975) demonstrated an inverse relationship between the abundance of a native muskellunge population and that of an expanding, introduced northern pike population.

In the upper Niagara River, angler records show that the muskellunge population supports an annual angler harvest of several hundred fish. Newspaper archives suggest that a muskellunge fishery of some importance existed as early as the 1850's, but harvest statistics prior to 1968 are not available. Northern pike are also popular with some local anglers, but little information on annual harvest of that species is available. Neither Niagara River population had been studied extensively.

Between 1960 and 1972, the New York State Department of Environmental Conservation (NYSDEC, unpublished data) examined 150–200 muskellunge and approximately 160 northern pike of various sizes taken by electrofishing, trap nets, seines, and creel surveys. All of those fish were taken in the river, and no collections were made in tributaries.

Additional samples of Niagara River muskellunge were taken during creel surveys by the Ontario Ministry of Natural Resources (OMNR) from 1966–1974 (Craig 1976). That study included scale-age determinations. No samples of northern pike were collected in the OMNR survey.

From the data cited above, it was apparent that sympatric northern pike and muskellunge populations existed in the upper Niagara River. Since both populations were naturally established, we postulated that some ecologic factor(s) segregated the two local populations and permitted coexistence. Since little data on the local populations existed and since other sympatric, natural populations had not been studied, we initiated a study to investigate

---

[1] Research supported in part by New York Sea Grant.

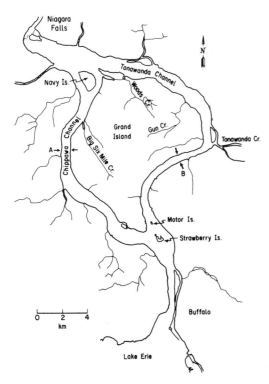

FIGURE 1.—*The Upper Niagara River.*

as many aspects of the life histories and ecologic factors affecting the status of these populations as possible.

The Study Area

The Niagara River arises from Lake Erie and flows north to Lake Ontario. Our research focused on the upper section of the river from its source to a point approximately 4.5 km above Niagara Falls (Fig. 1). The upper Niagara River is 31 km long (measured along the international boundary), has a surface area of 5,000 hectares and a total shoreline (including islands) of about 130 km. Grand Island divides much of the upper river into two channels: the Tonawanda or East Channel and the Chippawa or West Channel. Average flow is 5,550 m³/s (Sibley and Stewart 1969) and apportionment between East and West channels is roughly 42% and 58%, respectively (B. G. DeCooke, U.S. Corps of Engineers 1976, personal communication). Mean (averaged over the river cross section) and maximum current velocities measured in the West Channel (at site A of Fig. 1) were 0.91 m/s and 1.22 m/s, respectively (U.S. Corps of Engineers, 1967, unpublished data). Similar data for the East Channel (at site B of Fig. 1) were 0.71 m/s and 0.88 m/s, respectively.

The shore of the East Channel has extensive urban and industrial development; that of the West Channel remains rural or in parkland. Except for Strawberry, Motor, and Navy Islands, most river islands have been developed for industrial or residential use.

The river area bounded by Strawberry Island, Motor Island, and the southern tip of Grand Island consists primarily of shoals (1–2 m depth) that we sampled extensively by trap netting and electrofishing. Data we collected in 1975 and 1976 suggest that these shoals were a principal muskellunge spawning ground. The shores of northeastern Navy Island and northwestern Grand Island form the second major muskellunge spawning ground discovered during our research. If other major spawning areas once existed, their importance may have been reduced by human alteration of the habitat.

Two major modifications to the river environment have occurred in recent years. These are construction of large hydroelectric generating facilities on both shores of the river and the related annual installation of the Lake Erie ice boom. Lack of historical data prevented quantitative assessment of the impact of those modifications on the *Esox* populations, but some effect seems likely.

The present water-diversion capability of the generating facilities is about one-half of the volumetric flow, but to preserve the esthetic quality of Niagara Falls, diversion is primarily limited to nighttime. Periodic opening and closing of the intakes (located 1–2 km above Niagara Falls) change the water level of the river and of the lower reaches of the creeks. In more northerly tributaries, water level changes of 0.3–0.5 m were observed. Periods of increasing water level are marked by reverse flow in the creeks.

Each year since 1964 an ice-retaining

boom has been placed across the outflow of Lake Erie to prevent large quantities of lake ice from flowing into the river and clogging power-plant water intakes. Retention of ice appears to have altered the spring water-temperature regime of the river (Rumer 1975).

Fourteen major tributaries flow directly into the portion of the river that we studied. All of these are slow, meandering creeks with heavily silted and debris-strewn bottoms. With the exception of Tonawanda Creek, which forms a portion of the Erie-Barge Canal, none of the creeks is more than about 2 m deep. Water quality of the creeks has been severely reduced by human development of the watersheds.

Although most creeks were sampled on at least one occasion, systematic sampling concentrated on three Grand Island creeks: Gun, Woods, and Big Six Mile. These creeks were selected because of accessibility and because they are typical of most Niagara River tributaries.

Sampling Methods

During 1975, 1976, and early 1977 we recorded data on 312 muskellunge and 623 northern pike collected by seines, trap nets, electrofishing, and creel surveys. Data recorded included total length, weight, sex, and comments on injuries and external evidence of parasitism. All fish, except young-of-the-year and creeled fish, were tagged and released. Fish were aged from scales, and curvilinear body-scale relationships were used for back calculations.

We also obtained catch statistics from a local boat livery where most of the muskellunge anglers return with their catch. These data included weight and date of capture for each fish returned to the livery since 1970. Based on tag returns, we estimate that about 70% of the muskellunge captured annually are returned to this livery.

Results and Discussion

Growth of Niagara River muskellunge was comparable to that of rapidly growing populations in other areas (Harrison and Hadley, in preparation). Most muskellunge entered the fishery (legal minimum = 71 cm total length) during the fourth growing season. Males matured by age IV and females by age V. Annual survival was good (65%), and exploitation was moderate (13%). Mark-recapture studies suggest that population density in more heavily exploited portions of the river typically averaged 2–3 legal-size fish/hectare. Therefore, muskellunge seem to be well adapted to the Niagara River habitat.

Annual harvest of muskellunge by anglers averaged about 250 fish from 1970 to 1975, but in 1976 and 1977 harvest increased to 485 and 600 fish, respectively. The increased harvest apparently resulted from an exceptionally successful 1973 year class. The abundance of subsequent year classes (1974 and 1975) appeared to be more similar to that of years prior to 1973.

Northern pike seem poorly adapted to the local environment. For example, growth of local northern pike was similar to that of slower growing populations from more northern latitudes such as Saskatchewan (Koshinsky 1972) and northern Wisconsin (Threinen et al. 1966). Although exploitation was light (7%), survival was poor (40%) compared to 62–65% derived from Ohio data of Clark and Steinbach (1959).

There were no statistically different growth rates or length-weight relationships among northern pike from the three main sampling creeks. However, 87% of tagged fish returned to the same creek in consecutive years, which suggests some degree of isolation between creek populations.

There is little overlap in habitat utilization between the two species. Other than a few age 1+ fish, no muskellunge were found in tributaries (Fig. 2). Muskellunge spawned exclusively in the main river, and a current velocity of 0.2 m/s was typical of spawning sites. Current exceeded 0.1 m/s in habitats where young of the year were most frequently found. Based on angler success, the summer habitat of adults appeared to be restricted to portions of the river where current velocity reached 1 m/s. Recapture of fish tagged on spawning grounds suggests that net movement of most muskellunge was less than 2 km after a mean time at large of 107 days.

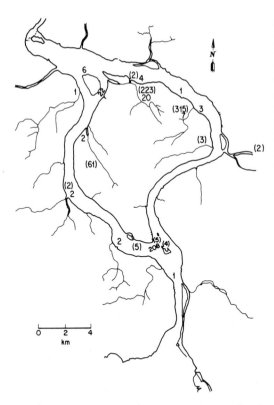

FIGURE 2.—*The number of northern pike (in parentheses) and muskellunge collected in various portions of the Niagara River and its local watershed.*

We found little evidence of river spawning by northern pike. However, the NYSDEC collected northern pike and muskellunge in nearly equal proportions in some areas where we collected seven muskellunge to each northern pike, and many of those northern pike were ripe or gravid. Perhaps river-spawning portions of the populations were eliminated by effects of the Lake Erie ice boom on spring water temperatures. Despite extensive river sampling that extended from early spring into the fall, less than 3% (18 fish) of our northern pike sample came from the river (Fig. 2). Most northern pike left the tributaries after spawning, but their summer habitat could not be determined. All angler tag returns came from creek areas, and 95% of these returns were from fish recaptured in May and June. Only one northern pike was captured by a muskellunge angler. Failure of northern pike to enter the muskellunge fishery suggests that adult northern pike did not inhabit typical muskellunge habitat during the summer and fall, that their feeding habits differed considerably from those of muskellunge or that they were not susceptible to the angling methods used. Young-of-the-year northern pike were rarely captured despite extensive seining and electrofishing of creeks and island shorelines. No young-of-the-year northern pike were found in association with young-of-the-year muskellunge.

Therefore, it appears that the two populations are sympatric only in a broad geographic sense and that they are spatially segregated throughout much of their life cycles by differential adaptation to the river environment. This segregation is most pronounced during the spawning season. Spawning-ground segregation may be the most important factor permitting the successful coexistence of the two species.

To trace the development of the observed segregation, we examined literature (Hough 1963; Calkin 1966; Calkin and Miller 1977) on the geologic history of the river. The relevant period began about 12,300 years before present (BP) with emergence of the Niagara Escarpment from glacial lakes that previously inundated the area (Table 1). At that time, the river received water from the precursors of Lakes Erie and Huron, and expanded eastward between the Niagara and Onondaga escarpments to form Lake Tonawanda. Lake Tonawanda spilled over the Niagara Escarpment through several outlets, but, isostatic uplift gradually shifted all outflow to the outlet at Lewiston, New York. During most of the period prior to 5,500 BP, flow was reduced by diversion of outflow of the upper Great Lakes through alternate routes.

It is unlikely that either species inhabited the local area prior to formation of Early Lake Erie (ca. 12,300 BP), because the glacial lakes that inundated the area were probably biologically inhospitable (Calkin and Miller 1977). However, one or both species may have been in the watershed of these early Great Lakes. This possibility is strongest for northern pike which are more tolerant of near-arctic conditions.

Since both species probably reached the Niagara River from the west (from Missis-

TABLE 1.—*Estimated times for major geological events affecting development of the Niagara River.*[a]

Approximate date (years BP)	Event	Consequence
13,000	Last glacier retreated from Niagara River watershed; local area inundated by Glacial Lake Dana.	Local waters probably physically inhospitable.
12,300	Niagara Escarpment emerged from Glacial Lake Dana.	Immigration and subsequent habitation by *Esox* possible.
11,500	Trent River route opened; upper Great Lakes disconnected from Early Lake Erie.	Immigration into Lake Erie and Niagara River impossible.
11,200	Trent River route closed; upper Great Lakes reconnected to Early Lake Erie.	Immigration possible, some riverine habitat likely.
10,900	Lake Tonawanda outflow entirely through Lewiston outlet.	Extent of riverine habitat increased.
10,400	Ottawa Valley route opened; upper Great Lakes disconnected from Early Lake Erie.	Immigration impossible, river flow substantially reduced.
5,500	Ottawa River route closed; upper Great Lakes reconnected to Early Lake Erie.	Immigration possible, substantial riverine habitat present.

[a] Summarized from Hough (1963), Calkin (1966), and Calkin and Miller (1977).

sippi Valley refugia, Crossman and Harington 1970), development of the modern Great Lakes provided two periods when immigration into Lake Erie and the Niagara River could have occurred: 12,300 BP to 10,400 BP and 5,500 BP to the present. Therefore, the alternatives for sequential establishment of the local populations are: (1) both species entered in the earlier period; (2) both species entered after 5,500 BP; (3) muskellunge entered prior to 10,400 BP and northern pike after 5,500 BP; and (4) northern pike entered prior to 10,400 BP and muskellunge after 5,500 BP.

The present local distribution of northern pike is much more extensive than that of muskellunge and includes areas above sizeable waterfalls (over the Onondaga escarpment) in some tributaries. Unless these creek populations were artificially introduced, northern pike must have been present when these areas were inundated by Lake Tonawanda. Therefore alternatives 2 and 3 are unlikely.

Prior to 5,500 BP, little riverine habitat could have existed in the Niagara River because of reduced flow (10–20% of modern volume, Calkin and Miller 1977). The apparent inability of muskellunge to compete with northern pike in a lentic habitat (Williamson 1942; Threinen and Oehmcke 1950; Johnson 1975) makes alternative 1 unlikely.

Therefore, alternative 4 is the most likely. However, this alternative also has some inherent problems, because it requires that muskellunge were preadapted for spawning in riverine habitats. River and stream populations of muskellunge have been reported (Parsons 1959; Hourston 1952; Crossman 1956), but we know of no data suggesting that those populations spawn in substantial currents. Until additional data become available, early arrival of northern pike followed by a rather recent arrival of muskellunge seems the best explanation for the establishment of the sympatric local population.

Summary

Our study of sympatric populations of northern pike and muskellunge in the Niagara River showed that, although the two species inhabit the same watershed, they are actually spatially segregated throughout much of their life cycles. This segregation is most pronounced during spawning and appears to result from adaptation by muskellunge to nearly exclusive use of the riverine environment. Northern pike appeared to be intolerant of the lotic characteristic of the river, and their distribution was restricted to lentic habitats in the river watershed.

The geologic history of the Niagara River shows that little riverine habitat existed prior to 5,500 BP. This suggests that the muskellunge population may have been established rather recently (within the last

5,000 years). By extrapolation from their present distribution within the river watershed, local northern pike were probably successfully established long before the immigration of muskellunge. If this presumed chronology is correct, muskellunge must have either been preadapted to spawning in substantial currents or developed that ability very quickly. Perhaps closer examination of other river populations of muskellunge will reveal a similar riverine spawning capacity.

References

CALKIN, P. F. 1966. Late Pleistocene history of northwestern New York. Pages 58–68 in E. J. Buehler, ed. Guidebook, Geology of Western New York. N.Y. State Geol. Assoc. 38th Annu. Meeting. Buffalo, N.Y.

———, AND K. E. MILLER. 1977. Late Quaternary environment and man in Western New York. Ann. N.Y. Acad. Sci. 288:297–315.

CLARK, C. F., AND F. STEINBACH. 1959. Observations on the age and growth of the northern pike, *Esox lucius* L., in East Harbor, Ohio. Ohio J. Sci. 59(3): 129–134.

CRAIG, R. E. 1976. The Niagara River maskinonge (*Esox masquinongy*). Ont. Min. Nat. Resour. Rep. 17 pp.

CROSSMAN, E. J. 1956. Growth, mortality and movements of a sanctuary population of maskinonge (*Esox masquinongy* Mitchell). J. Fish. Res. Board Can. 13(5):599–612.

———, AND C. R. HARINGTON. 1970. Pleistocene pike, *Esox lucius* and *Esox* spp., from the Yukon territory and Ontario. Can. J. Earth Sci. 7(4): 1130–1138.

HOUGH, J. L. 1963. The prehistoric Great Lakes of North America. Am. Sci. 51:84–109.

HOURSTON, A. S. 1952. The food and growth of the maskinonge (*Esox masquinongy* Mitchell) in Canadian waters. J. Fish. Res. Board Can. 8(5):347–368.

JOHNSON, L. D. 1975. Statewide fisheries research. Population levels of natural muskellunge populations. Wis. Dep. Nat. Resour. Dingell-Johnson Proj. F-83-R-9 Study 211, 1973. 7 pp.

KOSHINSKY, G. D. 1972. An evaluation of two tags with northern pike (*Esox lucius*). J. Fish. Res. Board Can. 29:469–476.

PARSONS, J. W. 1959. Muskellunge in Tennessee streams. Trans. Am. Fish. Soc. 88(2):136–140.

RUMER, R. R., C. H. ATKINSON, AND S. T. LAVENDER. 1975. Effects of Lake Erie-Niagara River ice boom on the ice regime of Lake Erie. Pages 289–299 in G. E. Frankenstein, ed. Proc. Third Int. Symp. Ice Prob., Int. Assoc. Hydrol. Res. November, 1975. Hanover, N.H.

SCOTT, W. F., AND E. J. CROSSMAN. 1973. Freshwater fishes of Canada. Bull. Fish. Res. Board Can. 184 966 pp.

SIBLEY, T. H., AND K. M. STEWART. 1969. Some variation in the quality of water from the source and mouth of the Niagara River. Proc. 12th Conf. Great Lakes Res. 1969:774–785.

THREINEN, C. W., AND A. OEHMCKE. 1950. The northern invades the musky's domain. Wis. Conserv. Bull. 15(9):10–12.

———, C. WISTROM, B. APELGREN, AND H. SNOW. 1966. The northern pike. Life History, ecology and management. Wis. Dep. Nat. Resour. Publ. 235–268. 15 pp.

WILLIAMSON, L. D. 1942. Spawning habits of muskellunge, northern pike. Wis. Conserv. Bull. 7(5): 10–11.

Survival, Growth, and Vulnerability to Angling of Northern Pike and Walleyes Stocked as Fingerlings in Small Lakes with Bluegills or Minnows

George B. Beyerle

Michigan Department of Natural Resources
Wolf Lake State Fish Hatchery, RR #1
Mattawan, Michigan 49071

Abstract

For northern pike stocked as fingerlings in small lakes with bluegills, survival was high after 3 years for the initial plant but very low for the succeeding two plants. Growth was moderate through age I but slow thereafter, despite an abundance of edible size bluegills. Survival and growth were generally better when northern pike were stocked in lakes with minnows. In one lake supplementary stocking of young salmonids was necessary to stimulate rapid growth of larger northern pike. In a lake with bluegills survival of age II and age III walleyes was high but growth was slow. In a lake with minnows survival of all three age groups of walleyes was consistently higher and growth was considerably faster than in the lake with bluegills. Northern pike were more vulnerable to angling than walleyes. However, both species were relatively easy to catch. Small lakes offer excellent opportunities to optimize survival, growth, and return to the angler of predator fishes stocked as fingerlings.

Northern pike and walleyes are in relatively short supply in most lakes in southern Michigan. Northern pike occur naturally in many lakes but heavy angling pressure and destruction of spawning areas have depressed most populations. Walleyes are present in only a few lakes but anglers have expressed considerable interest in having walleyes as a sport fish. With recent development of intensive rearing techniques for coolwater predatory fishes, the potential now exists for hatchery production of large numbers of fingerling northern pike and walleyes.

The challenge to fisheries managers is to wisely use these valuable fingerlings when augmenting existing fish populations or when establishing new fisheries. Widespread stocking of fingerling predatory fish may give rise to some ecological concerns among biologists but the angler's primary concern usually is the number caught. Thus, information is needed on potential return to the angler of northern pike and walleyes stocked as fingerlings under a variety of ecological conditions.

From 1963 through 1976, six studies were conducted in southwestern Michigan, to measure the survival and growth of northern pike or walleyes stocked as fingerlings in small lakes containing no other predator fish species and only one type of food fish, either bluegills or minnows (golden shiners and fathead minnows). Vulnerability to angling was determined for walleyes in two studies and for northern pike in one study. Additional data on vulnerability of northern pike to angling were available from a study not summarized here. Two lakes were used for all six studies. Daggett Lake has 6.1 surface hectares, a maximum depth of 6.1 m and an average methyl-orange alkalinity of 17 mg/liter. Emerald Lake covers 2.3 surface hectares, and has a maximum depth of 2.4 m and an average methyl-orange alkalinity of 114 mg/liter.

Methods

The basic procedure for each study was as follows. The study lake was treated with rotenone in the fall to remove the existing fish population. The following spring adult food fish of one type, either bluegills or minnows (golden shiners and fathead minnows), were stocked in sufficient quantity to produce a large population of young. Later in the year fingerling northern pike (mean total length 8.9 cm) or walleyes (mean total length 10.2 cm) were stocked at the rate of 111 fish per hectare. Identical stocking of fingerlings were made in the succeeding 2 years. Because of water level fluctuations in Daggett Lake, the actual stocking rate varied from 111 to 136 fingerlings per hectare, but the same total number of fingerlings (675) was stocked each year. Studies involving northern pike lasted 3 years, while studies with walleyes were extended through 4 years. During the studies no angling was permitted on the lakes.

In the study with northern pike and minnows at Emerald Lake, adult golden shiners and fathead minnows produced relatively large numbers of young only during the first year. Thus it was necessary to stock additional prey fish. No minnows were easily available, but a surplus of young coho salmon existed at the Platte River State Fish Hatchery. Consequently, it was arranged to obtain approximately 60,000 young coho salmon in six lots, to be stocked periodically over a period of 14 months. Mean length of the coho salmon ranged from 5.1 to 13.2 cm, depending on the time of year.

Angling vulnerability tests were made in summer of the final year of a study. Several personnel of the Michigan Department of Natural Resources fished for a specified number of hours using only artificial lures. All captured predatory fish were weighed, measured, and aged. Stomach contents were determined.

In September of the final year of each study the lake was treated with rotenone. All visible predatory fish were collected over a period of 1 week following treatment. In all but one study, all prey fishes were also collected to estimate the standing crop.

Results and Discussion

Northern Pike

Survival of age II northern pike was somewhat higher in the two lakes with bluegills (Beyerle 1971) than in the two lakes with minnows (Beyerle 1973a). Conversely, survival of age I and age 0 northern pike was significantly lower in lakes with bluegills (Table 1). This phenomenon can be explained by an understanding of the differential population dynamics of bluegills and minnows. The predicted result of stocking adult bluegills in a lake containing no other fish is the production of a large year class. The many young bluegills provide a surplus of food for the initial stock of northern pike fingerlings and high survival of this predator can be anticipated. However in the following year, or years, survival of young bluegills is often minimal. Thus, if no other acceptable food items are available, survival of northern pike fingerlings stocked in those years will also be minimal. This was true in our studies where young bluegills were the only available food fish.

Minnows have a lower reproductive potential than bluegills and also seem to be more vulnerable than bluegills to predation. Thus minnow populations usually show less variation in the size of successive year classes but are more susceptible to population reduction by heavy predation. Both of these factors influenced the survival of northern pike in the lakes with minnows; e.g., the relatively high survival of initial stocks of northern pike (but less than in the lakes with bluegills) and the cropping of adult minnows by the larger northern pike, resulted in a decline in reproductive potential of minnows and a concurrent dimin-

TABLE 1.—*Survival, growth, and standing crops of northern pike stocked as fingerlings during 3 consecutive years in small lakes containing either bluegills or minnows as prey. Data were obtained at the end of the third summer.*

Measurement and lake	Prey and age group of northern pike					
	Bluegills			Minnows		
	II	I	0	II	I	0
Survival (%)						
Daggett	44.1	2.7	1.5	54.7	35.5	16.1
Emerald	60.4	0.8	9.2	31.6	11.2	4.4
Mean length (cm)						
Daggett	48.8	47.8	35.1	51.3	44.7	31.2
Emerald	45.7	40.1	27.2	57.9	47.8	25.9
Growth index[a] (cm)						
Daggett	−0.5	+8.1	+9.1	+2.0	+5.1	+5.3
Emerald	−3.6	+0.5	+1.3	+8.6	+8.1	0.0
Number per hectare						
Daggett	59.8	3.7	2.0	65.2	42.2	19.3
Emerald	66.7	1.0	10.1	34.8	12.4	4.9
Weight (kg) per hectare						
Daggett	40.7	2.4	0.7	41.2	17.7	2.5
Emerald	31.7	0.3	1.0	47.5	7.4	0.6

[a] An "average growth rate" has been computed for most of the sport fishes in Michigan. For each age group of northern pike the growth index is merely the difference between the actual mean length and the "average mean length" for northern pike in Michigan.

ishing survival of succeeding stocks of northern pike. The combination of a larger population of minnows and considerably more vegetative cover in Daggett Lake very likely contributed to relatively higher survival of all age groups of northern pike compared with survival in Emerald Lake.

In both lakes with bluegills, the mean length of age II northern pike was less than the Michigan state average length (Table 1). However, from samples collected 1 year previous to termination of each study, we know that in both studies these same northern pike at age I averaged 43.4 cm in length, or 3.8 cm above state average. Growth of the few surviving age 0 and age I northern pike was also above state average, especially in Daggett Lake. Lack of competition may explain the rapid growth of age 0 and age I northern pike, while a surplus of food apparently was available through age I for the first stock of northern pike.

In lakes with minnows, growth of all three age groups of northern pike was above state average except for age 0 fish in Emerald Lake. The coho salmon stocked in Emerald Lake apparently enhanced growth of age I and age II northern pike to the extent that the mean length of both groups was more than 8 cm above state average length. However, the larger salmon soon decimated the supply of small minnows, resulting in low survival and slow growth of age 0 northern pike in Emerald Lake. In Daggett Lake growth of age II northern pike was decreasing, and indications were that larger northern pike (over 50 cm) were running out of food items of suitable size. Larger food fish, such as suckers, would probably be necessary to prolong rapid growth of northern pike larger than 50 cm.

In lakes with bluegills, the total standing crop of northern pike (43.8 kg per hectare in Daggett Lake; 33.0 kg per hectare in Emerald Lake) was directly proportional to the standing crop of bluegills in each lake (219.1 and 146.3 kg per hectare, respectively). The number per hectare of age II northern pike was remarkably similar in both lakes (Table 1). The population of bluegills small enough (under 10 cm) to be eaten by northern pike was also substantial: 40,081 per hectare in Daggett Lake and 35,457 per hectare in Emerald Lake. However, the growth of these bluegills was exceptionally slow, the mean lengths being 7.4 and 3.6 cm, respectively, less than state average. Despite the abundance of edible size bluegills, analysis of stomach contents collected in the final year of each study indicated that more than half the diet of larger northern pike consisted of bullfrog tadpoles, crayfish, and insect larvae. The high survival of fingerlings and relatively rapid growth of the initial stock of northern pike are evidence that small bluegills (less than 6 cm in length) may be an acceptable food item for northern pike less than 43 cm in length. However, larger northern pike may select an alternative food item when bluegills are the only available food fish.

In lakes with minnows, the total standing crops of northern pike (by weight) were considerably higher than in the bluegill lakes. However, the total number of northern pike in minnow lakes varied considerably, from 52.1 per hectare in Emerald Lake to 126.7 per hectare in Daggett Lake. Rapid growth of larger northern pike feeding on coho salmon in Emerald Lake compensated for relatively low survival to produce a total standing crop (55.5 kg per hectare) almost equal to the total standing crop of 61.4 kg per hectare in Daggett Lake (Table 1). After 3 years the standing crop of minnows in Daggett Lake was 115.4 kg per hectare, consisting of 48,770 fish per hectare of which 99.3% were less than 10 cm in length. Conversely, in Emerald Lake only 18.7 kg per hectare of minnows remained and all were over 10 cm in length. No coho salmon were found in Emerald Lake but moderate populations of largemouth bass, bluegill, and green sunfish were present. In lakes with minnows, both survival and growth of northern pike were related to the abundance of food fishes of edible size.

Walleyes

Survival of age III and age II walleyes in Emerald Lake with bluegills was much higher than expected (Beyerle 1976). How-

TABLE 2.—*Survival, growth, and standing crops of walleyes stocked as fingerlings during 3 consecutive years in Daggett Lake, with minnows as prey, and Emerald Lake, with bluegills as prey. Data were obtained at the end of the fourth summer.*

Measurement and lake	Age group of walleyes		
	III	II	I
Survival (%)			
Daggett	42.1	29.3	7.6
Emerald	35.1	21.2	0.0
Mean length (cm)			
Daggett	39.4	36.6	31.0
Emerald	35.3	27.7	
Growth index[a] (cm)			
Daggett	+0.8	+2.8	+6.9
Emerald	−3.3	−6.1	
Number per hectare			
Daggett	46.7	32.6	8.4
Emerald	38.8	23.5	0.0
Weight (kg) per hectare			
Daggett	25.1	13.1	2.1
Emerald	12.8	4.1	

[a] The growth index is the difference between the actual mean length and the "average mean length" for walleyes in Michigan (see footnote to Table 1).

ever no age I walleyes were found. Growth of the two surviving age groups was relatively slow (Table 2). Survival of all three age groups of walleyes in Daggett Lake with minnows was moderately but consistently higher than in Emerald Lake (Beyerle 1977). Growth of walleyes also was considerably greater in Daggett Lake (Table 2). Standing crops totaled 62.3 walleyes (16.9 kg) per hectare in Emerald Lake and 87.7 walleyes (40.3 kg) per hectare in Daggett Lake.

As in the studies with northern pike and bluegills, an abundance of small centrarchids was available as food for walleyes in Emerald Lake. The standing crop of bluegills (and green sunfish) was estimated at 163 kg per hectare, of which 95.1% by number (67,666 fish per hectare) were under 7.6 cm in length. Centrarchids were the only food item found in the stomachs of the seven walleyes collected prior to chemical treatment. Of 71 walleyes examined following chemical treatment, 55 contained food and 54 of these contained centrarchids. The relatively high survival but slow growth of walleyes in Emerald Lake was similar to the survival and growth pattern of age II northern pike in lakes with bluegills.

No actual measurements were made of the standing crop of minnows in Daggett Lake, but the population was estimated to be roughly equivalent to the 115 kg per hectare of minnows found in the northern pike-minnow study. Crayfish and minnows were the only important food items found in 75 walleyes collected prior to chemical treatment. Because survival was higher and growth was faster, the resulting total standing crop of walleyes (by weight) was over 2.3 times larger in Daggett Lake than in Emerald Lake.

Vulnerability to Angling

As expected, northern pike were more vulnerable to angling than walleyes. In Emerald Lake 10.6% of 153 "catchable" northern pike were captured in only 11 man hours of angling, while 30 man hours were required to catch only 5.0% of the 141 catchable walleyes (Table 3). In Daggett Lake both predator fishes were subjected to 131.5 man hours of angling, or 21.6 man hours per hectare. Walleyes were somewhat more numerous than northern pike (Table 3). Nevertheless, 28.0% of the catch-

TABLE 3.—*Comparative vulnerability to angling of "catchable size"[a] northern pike and walleyes in Daggett Lake and Emerald Lake.*

Predator species and lake	Man hours of angling		Number catchable size fish per hectare		Fish caught per man hour	Percent catchable fish taken per man hour	Total % caught
	Total	Per hectare	Available	Caught			
Northern pike							
Emerald Lake	11.0	4.8	67.7	7.2	1.5	0.95	10.6
Daggett Lake[b]	131.5	21.6	73.8	20.7	1.0	0.21	28.0
Walleye							
Emerald Lake	30.0	13.0	62.3	3.0	0.2	0.16	5.0
Daggett Lake	131.5	21.6	79.3	14.3	0.7	0.14	18.0

[a] All age I and age II northern pike and age II and age III walleyes.
[b] From Beyerle 1973b.

able northern pike were hooked, compared with a surprisingly high 18.0% of the walleyes.

Conclusions

In typical small lakes in Michigan, dominated by bluegills, survival of stocked northern pike and walleye fingerlings, to a large extent, is dependent on the density of young-of-the-year food fishes. Growth of northern pike and walleyes may be relatively fast through age I but is slow thereafter. Neither predator species has any important beneficial effect on survival or growth of bluegills. High survival and fast growth of northern pike and walleyes occur when suitable soft-rayed food fishes are available at optimum sizes and densities. However, large standing crops of predator fishes will soon decimate most populations of minnows. Therefore a logical procedure for managing small lakes may be to stock only enough adult bluegills to produce several successive year classes of moderate size. Then adult minnows, and possibly suckers, could also be stocked to create a more balanced population of food fishes. Finally, each year the number of fingerling northern pike or walleyes stocked should be matched with the abundance of food fishes of edible size.

Excellent fishing can result from northern pike and walleyes stocked as fingerlings in small lakes. The one obvious advantage of small lakes for angling is the relative ease of locating the fish. Unfortunately, this also makes the predator fish susceptible to rapid harvest. Therefore, maximum benefit to the angler may occur when fingerling northern pike or walleyes are stocked in well managed small lakes where relatively large size limits, or "catch and release" regulations, are enforced.

References

BEYERLE, G. B. 1971. A study of two northern pike-bluegill populations. Trans. Am. Fish. Soc. 100:69–73.

———. 1973a. Growth and survival of northern pike in two small lakes containing soft-rayed fishes as the principal source of food. Mich. Dep. Nat. Resour. Fish. Res. Rep. 1793. 16 pp.

———. 1973b. Comparative growth, survival, and vulnerability to angling of northern pike, muskellunge, and the hybrid tiger muskellunge stocked in a small lake. Mich. Dep. Nat. Resour. Fish. Res. Rep. 1799. 11 pp.

———. 1976. Survival, growth and vulnerability to angling of walleyes stocked as fingerlings in a small lake with bluegills. Mich. Dep. Nat. Resour. Fish. Res. Rep. 1837. 11 pp.

———. 1977. Survival, growth, and vulnerability to angling of walleyes stocked as fingerlings in a small lake with minnows. Mich. Dep. Nat. Resour. Fish. Res. Rep. 1853. 12 pp.

A Life History Study of the Muskellunge in West Virginia[1]

ROBERT L. MILES

West Virginia Division of Wildlife Resources
Route 2, Belleville, West Virginia 26133

Abstract

The native range of the muskellunge in West Virginia is restricted to streams of the Ohio River drainage. Native muskellunge populations are currently present in 41 streams which comprise 1,100 km and 2,935 hectares of muskellunge habitat. A life history study of the muskellunge in Middle Island Creek was conducted from 1966 to 1974. A 28.7-hectare study area had a minimum muskellunge population of 1.5 fish/hectare and 4.3 kg/hectare. An intensively sampled 6.2-hectare pool contained a minimum muskellunge population of 4.4 fish/hectare and 9.1 kg/hectare. Adult muskellunge in Middle Island Creek were heavily exploited by anglers and showed a great deal of upstream and downstream movement. Males matured at age III or IV and at lengths of 61–64 cm. Females matured at age IV or V at lengths of 66–71 cm. Spawning occurred during April when daily water temperatures averaged 10 C or higher for 4–8 days. Spawning sites were located at the lower or upper ends of pools in slack water near riffles. The time period between egg fertilization and fry swim-up ranged from 17 to 30 days.

The muskellunge is the largest and one of the most desirable of West Virginia's major game fish species. The large size to which this species can grow, its excellent fighting ability and the difficulty encountered in catching a legal-sized fish all contribute to its trophy status.

The muskellunge native to West Virginia is referred to as the Ohio muskellunge, *Esox masquinongy ohioensis*, although there is no valid description of this subspecies (Parsons 1959). It inhabits the Ohio-Tennessee river system, including Chautauqua Lake, New York (Hubbs and Lagler 1964).

Native muskellunge populations in West Virginia are currently present in 41 streams which comprise approximately 1,100 km and 2,935 hectares of muskellunge habitat. Most of these streams are in the Middle Island Creek, Little Kanawha River, and Kanawha River drainages.

Very little was known of the basic life history requirements of this important species in West Virginia prior to 1966. Research had not been conducted on stream muskellunge populations in West Virginia and most published studies from other areas had been confined to lake populations. Parsons (1959) studied stream muskellunge populations in Tennessee, but his findings in the rocky, fast-flowing streams of the Cumberland Plateau in east-central Tennessee did not seem applicable to West Virginia's low gradient, sandy-bottomed muskellunge streams.

A life history study of the muskellunge in West Virginia was initiated in 1966 to collect the basic information necessary for the future management and protection of this species.

Study Area

Two major study areas were established on Middle Island Creek near Middlebourne in Tyler County, West Virginia. These were a 13-km stream section upstream from Middlebourne and 3.2-km stream section just above the Tyler-Doddridge County line (Fig. 1). The longer study was composed of 14 major pools that ranged from 0.3 to 2.9 km in length and the 3.2-km study area contained three major pools that ranged from 0.3 to 1.4 km long.

Middle Island Creek flows into the Ohio River at St. Marys in Pleasants County, West Virginia. The stream length is 123 km and average gradient is 0.5 m/km. Throughout most of its length, it is a slow-flowing, meandering stream characterized by short riffles and long, deep pools of from 15 to 45 m in width and up to 3 km in length. It is one of West Virginia's best known muskellunge streams.

United States Geological Survey data (1972) for Middle Island Creek 8 km west of Middlebourne showed a 45 year average flow of 17.4 m³/s. It is a fairly productive stream with a pH of 7.1 to 7.6, total alka-

[1] A contribution from Federal Aid in Fish Restoration Project F-10-R.

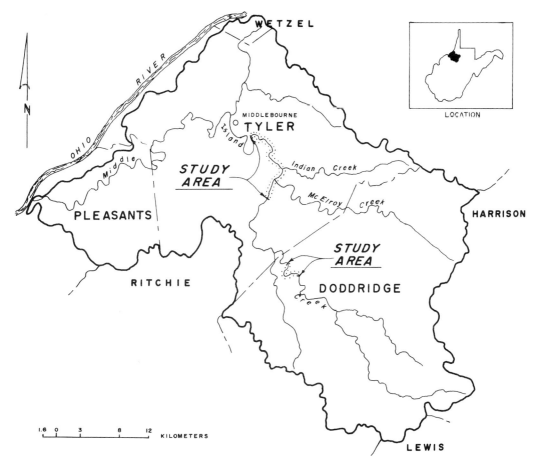

FIGURE 1.—*Middle Island Creek showing the location of study areas.*

linity of 15 to 45 mg/liter and hardness of 35 to 100 mg/liter. Previous fish population surveys have revealed an average standing crop of 133 kg/hectare.

Methods

Data collection began during the spring of 1966 and extended through 1973. Most data were collected during the period of March 1–May 15 in order to collect fish prior to and during the spawning season. Information on muskellunge eggs collected during the spring of 1974 is also included in this report.

Trap nets and electrofishing gear were used to collect muskellunge. The majority of fish were taken with a 230-volt AC boat shocker. Each fish was weighed to the nearest 0.1 pound and measured (total length) to the nearest 0.1 inch. These data were later converted to metric units for reporting purposes. Fish collected for the first time were tagged at the anterior base of the dorsal fin with a numbered monel strap tag identified as belonging to the West Virginia Department of Natural Resources. A number one tag was used for fish less than 56 cm and a number three tag for those over 56 cm.

News releases and personal contacts were used to publicize the tagging program, and fishermen were asked to watch for and report the capture of a tagged fish.

Each fish was examined externally to determine the state of reproductive development. When ripe muskellunge were collected, eggs were stripped, fertilized and transported to Palestine Hatchery for hatching. Techniques used in fertilization and egg preparation were similar to those described by Sorenson et al. (1966). Fertilized eggs were hatched in hatching baskets suspended in hatchery troughs.

A maximum-minimum thermometer was used to collect daily water temperatures from Middle Island Creek during the annual spring study periods.

Scale samples were taken from a key scale area directly above the pelvic fin and midway between the lateral and mid-dorsal lines. Growth rates were determined from a nomograph adjusted to a length of 63.5 mm at the time of scale formation. This corresponds to Erickson's (1967) proposed correction factor of 66 mm for Ohio's muskellunge and Johnson's (1971) length of 76.2 mm at time of scale formation for Wisconsin's muskellunge. Growth rates were not calculated by sex due to uncertainty of sex determinations and the large percentage of immature fish.

Results

Muskellunge Sampling Efforts

During 8 years of shocking operations on Middle Island Creek, 216 muskellunge were collected during 67 survey days totaling 244 hours. Muskellunge were collected at a rate of 0.9 fish per hour of boat shocking. Larger fish were more easily stunned by the electrofishing equipment and 62% of the muskellunge collected were of legal size (66 cm) or longer.

Nine muskellunge were also collected by trap nets in 1966, 1967, 1968, and 1971. This brought the total number of muskellunge collected to 225 and the total number which were tagged and released to 160.

Attempts to collect muskellunge from throughout the two study areas each spring were often hampered by high water conditions. Fishing pressure was generally heavy on Middle Island Creek, and to avoid conflicts with fishermen, concentrated collecting efforts were seldom made in any one pool.

Population Studies

Stream conditions were excellent during the spring of 1971, and from March 14 to April 20 a concentrated sampling effort was made in a 10.8-km section of the lower study area. The stream both above and below this section was very shallow, and with the low stream flows which occurred that spring, movement of fish into and out of the area seemed unlikely. Sport fishermen in this area were also interviewed and all muskellunge catches made during this period were recorded.

Forty-three muskellunge were collected by electrofishing and angling in this 28.7-hectare study area. This stream section contained a minimum of 29 legal and 14 sub-legal muskellunge weighing 123.6 kg. The minimum muskellunge population was 1.5 fish/hectare and 4.3 kg/hectare.

Previous sampling efforts had shown that a 2.6-km pool at the mouth of Indian Creek contained a better muskellunge population than any other section of the two study areas. Twenty-seven muskellunge were collected by electrofishing and by anglers from this 6.2-hectare pool between March 14 and April 15, 1971. This "Indian Creek Pool" contained at least 15 legal and 12 sub-legal muskellunge weighing 56.3 kg. The minimum muskellunge population was 4.4 fish/hectare and 9.1 kg/hectare.

Five days were spent tagging and recapturing muskellunge from this pool during the period of April 7-April 20. The Schumacher-Eschmeyer formula produced a population estimate of 36 muskellunge (5.8/hectare), with a 95% confidence interval of 30-43.

Tag Returns

A total of 160 muskellunge were tagged and released in Middle Island Creek from January 1, 1966 to June 30, 1973. Angler tag returns were received for 56 (35%) of these fish by June 30, 1974. The percentage of returns for any one-year's tagging operations ranged from 15 to 83%.

Discounting tag loss and nonreturn of tags, these returns show heavy angler exploitation of the muskellunge population. This is especially true for those fish which were legal size when tagged. Tag returns were received for 46 (49%) of the 94 legal-sized muskellunge, as compared to 15% of the 66 fish which were sub-legal when released.

Fifty-five percent of the total tag returns were made within 1 year, 75% within 2 years and 86% within 3 years. The shortest time beween tagging and tag return was 7 days and the longest was 5 years and 4 months. Fifty-two percent of the tag returns for muskellunge which were legal size when tagged were made within 4 months and 65% within 1 year. Forty percent of tag returns for fish of sub-legal size were made within 2 years and 70% within 3 years. The longer time interval for tag returns of sub-legal muskellunge reflected the time necessary for the fish to grow to legal size.

Angler catches of tagged fish were made during every month except October and December. Twenty-seven percent of the tag returns were made during April and 18% during May. Seventy-two percnt of the returns were made during March, April, May, and June, while 9% were made in January and February.

Movement

Sufficient data were obtained to calculate the movement patterns of 77 tagged muskellunge. Fifty-one of these were from angler tag returns and 26 were electrofishing recaptures. Seventy-four percent of these fish were legal size when tagged.

To allow for movement possibly resulting from the electrofishing operations, a muskellunge caught or recaptured within 0.3 km of its initial tagging site in the same pool was classified as showing no movement. Muskellunge which had moved out of the original pool in which they were tagged and released were classified as showing movement regardless of the distance. Movement was calculated only for those recaptured fish which had been released at least 14 days.

Forty-seven (61%) of the 77 muskellunge had moved at least 0.3 km from the initial tagging site and 31 (40%) had moved out of the pool in which they had been released.

Upstream and downstream movements were almost equal in frequency (Table 1). Ten fish (21%) had moved less than 0.6 km, 28 (60%) had moved more than 1.6 km and 8 (17%) had moved more than 6.4 km.

The greatest recorded movement was by an 82.3-cm female muskellunge which had moved 33.8 km downstream when it was caught by an angler 71 days after release. A 75.7-cm male muskellunge moved 20.1 km downstream in 43 days and a 50.5-cm immature muskellunge moved 16.7 km upstream in 176 days.

Only three of the muskellunge tagged in Middle Island Creek were recaptured in tributary streams; two in McElroy Creek and one in Indian Creek.

TABLE 1.—Number and percentage of recaptured Middle Island Creek muskellunge taken at various distances from their tagging sites.

Distance between tagging and capture sites	Upstream movement		Downstream movement		Total	
	No. fish	% of total	No. fish	% of total	No.	%
0.3–0.6 km	7	15	3	6	10	21
0.6–0.9 km	1	2	2	4	3	6
0.9–1.2 km	1	2	3	6	4	9
1.2–1.6 km			2	4	2	4
1.6–3.2 km	5	11	8	17	13	28
3.2–4.8 km	2	4	2	4	4	9
4.8–6.4 km	2	4	1	2	3	6
over 6.4 km	3	6	5	11	8	17
Totals	21	45	26	55	47	100

Age and Growth

The average calculated lengths for 77 muskellunge collected from 1967 to 1971 were 29.2, 48.3, 60.7, 70.1, 76.5, 80.8, and 87.9 cm at annuli I–VII, respectively (Table 2). Average calculated lengths compared closely with averages of actual lengths taken when the fish were measured in early spring. Middle Island Creek muskellunge generally reached legal size (increased from 66 cm to 71 cm as of January 1, 1974) at age IV or V.

Maturity and Spawning

Male muskellunge in Middle Island Creek matured at age III or IV and at lengths of 61–64 cm. The smallest ripe male collected was 60.5 cm. Females matured at age IV or V and at lengths of 66–71 cm. The two smallest ripe females collected were 66.0 and 66.3 cm.

Mature males were generally ripe throughout March, April, and May. Mature females became ripe during April when daily water temperatures averaged 10 C or higher for

TABLE 2.—Age and growth data for 77 muskellunge collected from Middle Island Creek, July 1, 1967 to June 30, 1971. Numbers in parentheses are averages of actual length measurements taken just prior to annulus formation.

Age group	No. fish	Total length (cm) at end of year						
		1	2	3	4	5	6	7
I	15	(30.2)						
II	7	28.2	(50.0)					
III	8	28.7	48.0	(60.2)				
IV	16	30.2	48.0	61.0	(69.9)			
V	18	27.9	47.0	60.7	70.6	(77.0)		
VI	11	29.9	50.3	61.2	70.4	76.2	(80.8)	
VII	2	26.7	44.7	58.9	68.3	74.9	81.0	(87.9)
Average		29.2	48.3	60.7	70.1	76.5	80.8	87.9

4–8 consecutive days and it was at this time that spawning began. This occurred as early as April 3 (1974) and as late as April 16 (1970). The exact length of the spawning season could not be delineated due to the small number of ripe females collected, but it appeared to extend for a period of 2–3 weeks.

The difficulty experienced in collecting ripe females was apparently because they spawn very shortly after becoming ripe. A green female released on April 14, 1971 had already spawned when recaptured 6 days later.

Adult muskellunge moved to the upper or lower ends of pools in Middle Island Creek in late March and early April and spawning apparently occurred in slackwater in the vicinity of riffles. This movement and concentration of adult fish was especially evident in the larger pools. There was, however, no evidence of a concentrated spawning movement between pools or up tributary streams. Spawning probably occurred in most of the individual pools of Middle Island Creek.

Egg and Fry Development

Eggs from Middle Island Creek muskellunge were collected, fertilized, and successfully hatched at Palestine Hatchery during 4 years (Table 3). Eggs eyed-up in 7–14 days and hatched in 10–19 days after fertilization, and the fry swam-up 7–11 days later. The time between egg fertilization and fry swim-up ranged from 17 to 30 days, depending on hatchery water temperatures. These temperatures closely approximated those of Middle Island Creek and the resulting data should be applicable to natural muskellunge egg and fry development in this stream.

Discussion

Middle Island Creek is fairly typical of most of West Virginia's major muskellunge streams. Information derived from this study should be applicable to muskellunge management on a statewide basis, although variations undoubtedly occur from stream to stream.

The current muskellunge fishing regulations, which include a 71-cm size limit (increased from 66 cm in 1974) and a two-per-day creel limit, appear adequate. The creel limit probably has little effect on native populations, since few fishermen catch two legal fish per day. The 71-cm size limit increases the trophy status of the species and provides an opportunity for all males and most females to spawn at least once before they can be harvested.

The estimated population of 5.8 muskellunge/hectare and minimum population of 9.1 kg/hectare in the "Indian Creek Pool" study area probably approached the maximum level that can be expected in West Virginia's best muskellunge streams or stream sections. Brewer (1970) reported muskellunge population levels of 2.0–2.1 kg/hectare in eight Kentucky streams. Popu-

TABLE 3.—Hatching and development dates for muskellunge eggs collected from Middle Island Creek and hatched at Palestine Hatchery.

Stage	Year and hatchery water temperatures			
	1970 13–17 C	1971 13–18 C	1973 8–17 C	1974 8–19 C
Eggs fertilized	April 17	April 13	April 3	April 8
Eggs eyed	April 24	April 20	April 17	April 19
Eggs hatched	April 27	April 23	April 22	April 24
Fry swam up	May 4	May 3	May 3	May 1

lation samples, taken with sodium cyanide in six Kentuky streams, yielded about 2.5 muskellunge per hectare of pool habitat.

The high percentage of angler tag returns (35%) revealed that Middle Island Creek muskellunge are easily and effectively harvested. The study area is one of the most accessible, best known, and heavily fished of West Virginia's muskellunge streams and exploitation rates in most other streams or stream sections are probably somewhat lower.

The frequent and often long movement of this species both upstream and downstream was somewhat surprising. Brewer (1970) reported that 11 of 13 recollected muskellunge in Kentucky streams were taken from the same pool where the fish were initially tagged. Most fishermen maintain that West Virginia's stream muskellunge show little movement and generally remain in one "home" pool. While movement within pools and possibly some movement between pools in Middle Island Creek was associated with spawning, this does not explain the movement of fish for several kilometers through many pools that contained suitable spawning habitat. Such movements may have been associated with high stream flows or flood conditions.

The muskellunge spawning season in Middle Island Creek occurred at approximately the same time as that reported for stream muskellunge in Tennessee and Kentucky. Parsons (1959) reported that over a period of several years in Tennessee, the muskellunge spawned during April when water temperatures were 10 C. Brewer (1970) reported that muskellunge in northeastern Kentucky typically spawned during the last half of April or early May after the water temperatures reached 10 C and had averaged at or above 10 C for about 3 weeks. This was not the case in Middle Island Creek, where spawning began 4–8 days after water temperatures averaged 10 C or higher.

Parsons (1959) reported an upstream prespawning movement of adult fish in Tennessee streams during late March and early April. He did not, however, delineate the distances moved. Brewer (1970) also noted a prespawning upstream movement of muskellunge in Kentucky streams. He reported that adult muskellunge in moderately sized streams spawned in the immediate waters of their normal habitat or moved upstream to suitable spawning habitat, with such movements seldom being more than a few km. The lack of concentrated prespawning movement between pools in Middle Island Creek probably resulted from the presence of suitable spawning areas in most individual pools. These spawning areas were similar to spawning sites in Kentucky streams but differed from sites in Tennessee streams which consisted of shallow pools with gradients usually much less than other segments of the stream.

High stream flows are a frequent occurrence in Middle Island Creek during the muskellunge spawning season. The 17 to 30 days between egg fertilization and development to the swim-up fry stage is an extremely critical period and high flows could have a disasterous effect on reproductive success. Additional studies are needed to determine any correlation between stream flows during this period and resultant year-class strength. Regulated stream flows during April and May, perhaps by upstream impoundments, could be beneficial to muskellunge reproduction.

References

Brewer, D. L. 1970. Musky studies. Ky. Dep. Fish Wildl. Resour., Fed. Aid in Fish Restoration. Prog. Rep. Proj. F-31-3, 1969–1970. 83 pp.

Erickson, J. 1967. Age growth procedures. Ohio Dep. Nat. Resour., Div. Wildl. 12 pp.

Hubbs, C. L. and K. F. Lagler. 1964. Fishes of the Great Lakes region. Univ. Mich. Press, Ann Arbor. 213 pp.

Johnson, L. D. 1971. Growth of known-age muskellunge in Wisconsin and validation of age and growth determination methods. Wis. Dep. Nat. Resour. Tech. Bull. 49. 24 pp.

Parsons, J. W. 1959. Muskellunge in Tennessee streams. Trans. Am. Fish Soc. 88(2):136–140.

Sorenson, L., K. Buss, and A. D. Bradford. 1966. The artificial propagation of esocid fishes in Pennsylvania. Prog. Fish-Cult. 28(3):133–141.

United States Geological Survey. 1972. Water resources data for West Virginia. U.S. Dep. Interior. 191 pp.

Home Range and Seasonal Movements of Muskellunge as Determined by Radiotelemetry

JOHN D. MINOR

Zoology Department, University of Toronto
Toronto, Ontario, M5S 1A1

E. J. CROSSMAN

Department of Ichthyology and Herpetology
Royal Ontario Museum, Toronto, Ontario M5S 2C6

Abstract

Sixteen adult muskellunge were studied in a shallow 35-hectare lake (Nogies Creek) and a deeper, 3,725-hectare lake (Stony Lake) within the Kawartha Lakes region of central Ontario. All fish established summer and winter home range areas. Areas of 0.6 to 1.1 hectare, in water less than 2.0 m deep, were used by all Nogies Creek fish and the Stony Lake male. The Stony Lake female utilized a 7.2-hectare area in water of 12.5 m average depth. Some areas overlapped; however, there was never more than one fish in an overlap area at one time. All fish established and used home ranges when water temperatures were less than 5 C. Males again established home ranges when water temperatures exceeded 15 C, however, not all females established home ranges in temperatures of 15–28.5 C. All fish were absent from their home ranges in spring and fall at water temperatures of 8–15 C. Distances travelled outside home ranges were maximum for both male and female fish at temperatures of 10–15 C. Females travelled greater distances than males during summer water temperatures of 20–28.5 C.

The utilization of available habitat by various fish species remains an important issue in planning and technical application of fish management programs. Present day esocoid management evaluations have generally used mark and recapture information. This information does not provide continual monitering of fish movements and is limited by the low frequency of multiple recaptures. Ultrasonic and radiotelemetry provide a method by which fish movements can be quickly and continually monitored. Investigation of the use of habitat by northern pike has provided evidence of transient movements with little or no long term localization (Malinin 1969, 1970; Diana et al. 1977). Crossman (1977) presented evidence that muskellunge may have restricted summer movement patterns. These studies represent the extent of esocoid research using telemetry techniques. Only ultrasonic equipment was used in these studies, even in heavily vegetated waters which severely attenuates the signal transmission. This technical limitation may contribute to the differences noted for northern pike (Diana et al. 1977) and muskellunge (Crossman 1977).

Radiotelemetry signal transmission is not attenuated by vegetation and does not require a direct line of sight in water for signal reception. This greatly facilitates tracking fish in heavily vegetated water and in meandering rivers.

The purpose of this study was to determine by radiotelemetry whether or not muskellunge establish home range areas, and if they do, to investigate what environmental factors effect size and utilization of these areas. Comparisons were made of activity and habitat utilization by fish in a large lake and a small lake in order to evaluate the effect of available area, depth and vegetative cover. Differences between sexes were also studied.

Study Area

This radiotelemetry study was conducted within the Kawartha Lakes Region of south central Ontario. Fourteen fish were studied at Nogies Creek (44°30′N, 78°30′W), a 35-hectare, shallow (depth: $\bar{x} = 1.8$ m; max = 5.0 m), heavily vegetated lake previously described by Muir (1963) and Crossman (1977). This lake is fed by 6.4 km of meandering stream containing 47 hectares of littoral zone. Two fish were studied in Stony Lake (44°40′N, 78°20′W), a eutrophic lake of approximately 3,725 hectares containing many granite shoals and islands, and having a maximum depth of 32 m in certain channels. Mean depth is 12.5 m. Shallows less than 3 m comprise approximately 1,800 hectares, with 50% habitat cover provided by submergent vegetation (*Potamogeton* spp., *Myriophyllum* spp.). Thermal stratification occurs at depths around 15 m in August and September at which time the hypolimnion becomes anaerobic.

Methods

Radiotelemetry began in May 1975 and continued through October 1976. Radiotransmitters were purchased from AVM Instrumentation Company (Champaign, Illinois). Individual fish were identified by transmitters of different frequencies at 5-kHz intervals within the 50-MHz band. Transmitters with a useful lifetime of 4–8 mo weighed 14 g in air, whereas transmitters with a 14–18-mo lifetime weighed 21 g in air. Transmitters were padlock-shaped with approximate dimensions of 60 × 25 × 12 mm.

An AVM 50 MHz LA12 receiver was coupled to a 5-element vertical polarized Yagi antenna. This equipment provided effective ranges of 300 m and 1.5 km at fish depths of 5 m and 0.5 m, respectively. Reception was unhindered by 100% vegetative cover in summer. During winter, tracking range decreased to about 200 m due to increased signal reflection and refraction through multiple air/snow/ice/water interfaces. A bidirectional, diamond shaped, hand-held, loop antenna had a range of 50 and 250 m over open water at fish-depths of 5 and 0.5 m respectively. This antenna had a directional "null width" of approximately ±5 degrees, which allowed location of fish to ±0.5 m.

Fish were captured by trap nets during the post-spawn period of May and June, anaesthetized in tricane methane sulfonate (100 mg/litre), and transported to the laboratory. The anaesthetized fish were held inverted on an operating table equipped with recirculating pumps and tanks that provided continuous irrigation of the gills. Fish were identified with numbered plastic tags. Length, weight, and sex of each fish was recorded. Standard surgical procedure, described by Summerfelt (1972) and Crossman (1977), was used to implant the transmitters into the body cavity. Fish were allowed to regain equilibrium in fresh water and were then immediately returned to their exact site of capture and released.

Fourteen muskellunge, 5 males and 9 females, of 700–920 mm fork length were radio-tagged and tracked at Nogies Creek between May 1975 and October 1976. Two larger fish, one female and one male, 1,005 and 960 mm fork length, were tracked at Stony Lake between May 1976 and October 1976.

Enlarged charts of the study area were prepared from aerial photographs. A grid system was established. Grid lines were laid out with reference to local land marks, plastic ribbons, or blinking light markers at night. Radiolocations of fish were determined by triangulation based on line of sight to shore or floating markers. At each location, water temperature, dissolved oxygen concentration, depth, and available habitat cover was determined. Each fish was located four to six times daily during each two-week monitoring period. Tracking times were distributed equally throughout a 24-h time period. Fish exhibiting greater movements were monitored more frequently.

Radiolocations were converted to values for frequency of occurrence within each grid area. The primary home range (Winter 1977) was calculated by the grid-square method used by Rongstad and Tester (1969). For all locations outside the primary home range, distances were calculated and expressed as daily distance travelled. For each fish, the data for each-two week monitoring period was summarized and expressed as a mean value. Data for all fish during each two-week period was expressed as a mean of means. Differences between sexes were determined by multiple Student's t tests on the mean-of-mean values. The relationships of home range area and daily distance travelled to environ-

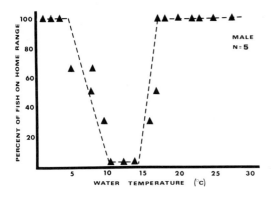

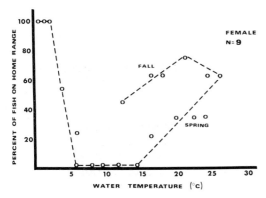

FIGURE 1.—Home range use by muskellunge as a function of water temperature at Nogies Creek.

established in water depths of 2.5–4.5 m. All areas chosen were 30% to 50% covered by submergent vegetation (principally *Potamogeton* spp.) with a few logs or tree stumps. Dissolved oxygen concentration, measured at each fish location and other shallow areas (less than 1.0 m), was never less than 45% saturation. During the last two weeks in March, female fish were significantly more active than males (Table 1). Coincident with this increase in distance travelled, the mean size of home range was significantly reduced.

As spring thaw developed, more fish of both sexes left their home range areas (Fig. 1) resulting in increased travel distances (Table 1). By mid-April, at water temperatures of 8.0–10.5 C, all fish had left their winter home ranges. Most fish movements were directed upstream (4.5 km) to specific areas. The transmitterized individuals, and others, were observed mental factors, fish size, and sex were determined by regression and correlation analysis.

Results

All 14 Nogies Creek fish established a home range during winter months from January to March at water temperatures of 0.5–2.0 C (Fig. 1). Both male and female fish resided in home areas of similar size ($\bar{x}$ = 0.32 hectare; SD = 0.05). Each fish established one individual winter home area. Distances travelled by both sexes outside these areas averaged less than 20 m/day (Table 1). Winter home ranges established in the lake at Nogies Creek did not overlap (Fig. 2). No interaction between fish of adjacent home areas was ever observed, even during short excursions outside their areas. Winter home ranges were

TABLE 1.—*The seasonal variation of distance travelled outside home range areas and the surface area of primary home ranges of muskellunge in Nogies Creek. Data are the mean of means for all fish during each two-week time period. Values that differ significantly (t-test; P ≤ 0.05) between sexes are indicated by an asterisk (*). NH = no home range.*

Date, 1975–1976	Distance travelled (m/day)		Home range area (hectare)	
	Male	Female	Male	Female
Jan. 1–15	8.8	16.1	0.31	0.29
16–31	8.8	12.9	0.31	0.29
Feb. 1–15	10.2	5.3	0.26	0.27
16–28	11.5	7.3	0.24	0.27
Mar. 1–15	20.8	8.1	0.35	0.27
16–31	20.8*	62.3*	0.35*	NH*
Apr. 1–15	158.2	361.4	0.22*	NH*
16–30	407.7	562.3	NH	NH
May 1–15	147.6	401.9	0.51	0.31
16–31	169.2	198.2	0.33	0.39
Jun. 1–15	61.8*	176.3*	0.84	0.46
16–30	26.1*	94.7*	0.82	0.48
Jul. 1–15	31.9*	121.3*	0.77	0.48
16–31	36.7	91.1	0.75	0.50
Aug. 1–15	83.2	115.2	0.87*	0.53*
16–31	81.1	90.3	0.90*	0.57*
Sep. 1–15	227.4	152.3	0.83*	0.58*
16–30	394.7	156.2	0.70	0.65
Oct. 1–15	380.2	268.4	NH*	0.68*
16–31	450.2	337.7	NH*	0.64*
Nov. 1–15	300.2	142.3	0.44*	NH*
16–30	268.9	138.4	0.41*	NH*
Dec. 1–15	88.7	82.1	0.41*	NH*
16–31	21.1	32.3	0.39*	0.09*

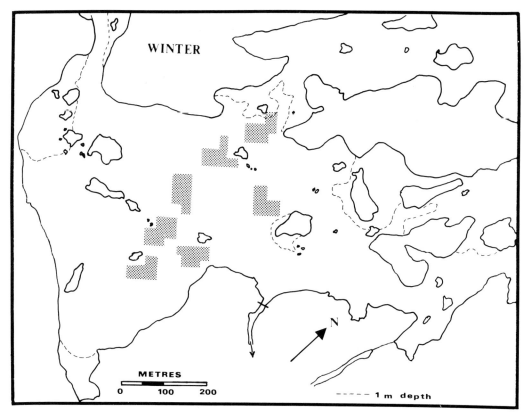

FIGURE 2.—A map of Nogies Creek lake showing primary winter home range of seven muskellunge as shaded areas.

there several days later, and were spawning. Spawning areas were less than 1.0 m deep ($\bar{x} = 0.65$ m) and generally bare of new vegetation. Spawning substrate consisted of matted vegetation, and tree leaves that had fallen the previous fall. Female fish were observed spawning in one or more areas; however most males spawned at only one site. Spawning activity occurred only during daylight hours when water temperatures were 10.5–15.5 C. When water temperatures reached 16.0 C, male fish had abandoned the spawning grounds and were returning downstream to the lake. On reaching the lake, they immediately established home range areas (Fig. 1). This behaviour was duplicated by the Stony Lake male, who moved off his spawning site in mid-May in water temperatures of 15.5 C. This large male moved directly across open water and established a home range area on the leeward side of a small island in water with a depth less than 1.5 m. For a distance of approximately 0.5 km around this home range, the water was less than 3 m deep; the closest area with depths of 5 m or more was more than 0.75 km away. Less than 30% of this home area contained vegetation (*Potamogeton* spp., *Myriophllum* spp.); the remaining area was covered with irregular shaped rock in shelves and outcroppings.

Female muskellunge at Nogies Creek did not immediately return downstream after spawning, but wandered up and down the creek channel in water temperatures of 14.0–23.5 C. At this time, females travelled significantly greater distances ($\bar{x} = 121$ m/day) than males ($\bar{x} = 32$ m/day). As summer progressed, males remained on their home range areas in shallow water ($\bar{x} = 1.2$ m) where water temperatures were as high as 28.5 C. Females gradually wandered downstream, reaching the lake by mid-August, where each fish set up one or two home range areas. These areas were significantly smaller (Table 1) than areas used by males; however, the females were in the upper portion of water of significantly greater depth ($\bar{x} = 2.8$ m). Habitat cover consisted of dense (75%) submergent vegetation (*Potamogeton* spp., *Ceratophyllum* spp.) with many logs and tree stumps. Dissolved oxygen concentrations everywhere in the lake were always greater than 58% saturation.

Summer home range areas of both male and female fish at Nogies Creek were well defined. More than 87.5% of all radio-locations occurred within the calculated home range area. Some areas overlapped; however, there was never more than one fish in an area of overlap at any one time. Movements of a single individual between its two home areas were generally direct, and in August occurred approximately once every 6.5 days (Fig. 3).

The behavior of the large female at Stony Lake differed from that of females at Nogies Creek. Immediately after spawning, this fish travelled great distances (1.2 km/day) around the perimeter of a section of lake and after six days of travel established a summer home range. This home range was over a 0.6-hectare sunken island in a main channel of 20 m depth. The top of this island was 4.5 m below the surface. There was an inverse relationship between frequency of contact and distance from the island.

The mean size of the summer home range for this one female (set at 80% of all locations) was 7.2 hectares. Summer mean water temperature was 18.5 C, and daily distance travelled was 150 m. In late October, at a water temperature of 14.5 C, this fish was no longer on her home range and travelled 2.2 km around the lake shore. Subsequent attempts at radiolocation failed, possibly due to transmitter failure.

As water temperatures decreased in fall,

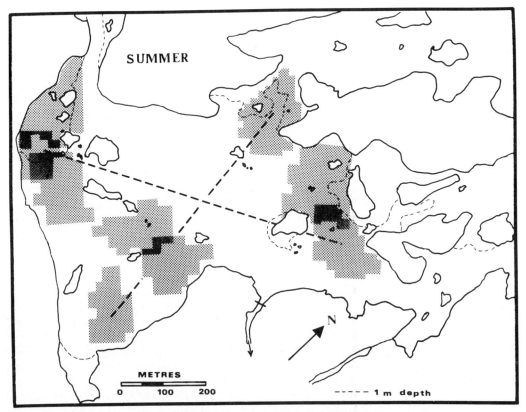

FIGURE 3.—*A map of Nogies Creek lake showing primary summer home range of seven muskellunge as shaded areas. Two fish had double home range areas, which are shown connected by dotted lines. Areas of home range that overlap are shown by darkest shading.*

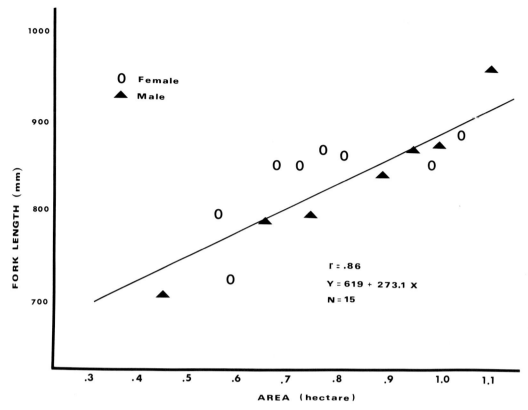

FIGURE 4.—*The relationship of muskellunge size (fork length) to primary home range size during summer months. All fish from Nogies Creek are represented, as well as the male fish from Stony Lake (960 mm FL). The female Stony Lake fish (1,005 mm FL) used an area of 7.2 hectares.*

daily distance travelled increased for both male and female muskellunge in Nogies Creek. Males abandoned their summer home range areas significantly sooner than females (Table 1) and wandered in no apparent pattern until November when they gradually set up winter home range areas. Females also wandered about the lake and did not establish winter home ranges until mid December, when the water temperature was 3.5 C. Winter home ranges were very well-defined, with over 92% of all radiolocations occurring within the calculated home range areas.

Home range size of both sexes (excluding the female from Stony Lake) increased linearly with fish size (Fig. 4). The fish size and size of home range for the female from Stony Lake exceeded that for all other fish. Daily distance travelled was related to seasonal water temperature (Fig. 5), with maximum travel distances occurring between 10.0 and 15.0 C during the fall breakdown of home range.

Discussion

The surgically implanted radiotags did not appear to affect growth or survival; fish recaptured 15 months after implantation exhibited normal length and weight gain. Healing of the incision was complete in seven to twelve weeks; a faint but visible scar could still be detected after eight months. The use of radiotelemetry proved to be highly successful for studying daily and seasonal behavior of the muskellunge in its natural habitat.

The ability to repeatedly locate the same fish throughout 18 months provides evidence that muskellunge establish home ranges in summer and during winter. Spring breakdown of home range occurs during spawning which coincides with increased daily travel distances. A second

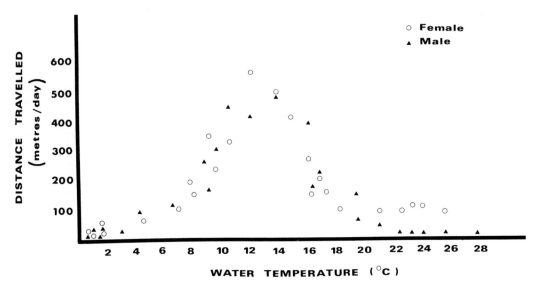

FIGURE 5.—*The variations of distance travelled by muskellunge outside home range areas at different temperatures. Values shown are the mean of means for all fish during each two week time period.*

breakdown in home range occurs in fall at water temperatures similar to those of spring spawning conditions. Distance travelled during these two periods (max = 2.3 km/day; mean of means = 530 m/day) exceeded values reported for northern pike (Malinin 1969) and muskellunge (Crossman 1956). The discrepancy between these Nogies Creek data and previously published data might result from our intensive monitering in 1975–1976 that provided a greater number of radiolocations at shorter time intervals.

Male muskellunge were sedentary in both Nogies Creek and Stony Lake. Various characteristics of home range, including depth, were similar, even though Stony Lake offers greater diversity in depth and opportunity for travel over great distances. Continual residency in water less than 1.0 m deep at temperatures exceeding 25.5 C indicates a high tolerance or possibly a preference for warm water. Movements outside these areas were reduced (Fig. 5) at temperatures above 20.0 C, whereas the size of the primary home range remained unchanged (Table 1). The movements and home range establishment of the single female in Stony Lake followed the male pattern, although her home range was larger, deeper, and more barren of vegetation.

Males and females of Stony Lake, and males at Nogies Creek, follow the same time sequence in establishment and breakdown of home ranges. Females at Nogies Creek follow a different time sequence of these same events.

Males at Nogies Creek and Stony Lake not only follow the same time sequencing, but also utilize similar habitats in the two lakes. Females at Stony Lake use deeper habitats than males at Stony Lake. Females at Nogies Creek, although following a different time sequence, eventually appear on home ranges having characteristics identical to those of the home range of males at Nogies Creek.

These behavioral patterns involve movements similar to those described by Malinin (1970) and Diana et al. (1977) for northern pike, and by Crossman (1956, 1977) for muskellunge. However, in addition, Diana et al. (1977) suggested that their fish were exhibiting "transient" movements. The unique behavior of females at Nogies Creek fits this concept of transient movement. If the behavior of the female at Stony Lake is considered more typical, then the temporary transient movements by females at Nogies Creek could result from their search for greater depths and lower temperatures that are not only available to, but sought out by, the female at Stony Lake.

This study suggests, therefore, that in bodies of water with depths greater than that of Nogies Creek, all male and female muskellunge will follow the same sequence of events at the same time. However, males will probably utilize areas that are shallower and more heavily vegetated than areas used by females.

Acknowledgments

This study was supported financially by research grants made to the junior author from the Canadian National Sportsmen's Show and the National Research Council of Canada (Grant A-1705). Outboard motors were supplied by Outboard Marine Corporation, Peterborough, Ontario. Antenna design and manufacturing was aided by Lindsay Antenna Company, Lindsay, Ontario.

Research quarters and living quarters were supplied by the Ontario Ministry of Natural Resources. Research during the summer of 1976 was also financially supported by a Molson's Brewery Research award.

References

CROSSMAN, E. J. 1956. Growth, mortality and movements of a sanctuary population of maskinoage (*Esox masquinongy* Mitchill). J. Fish. Res. Board. Can. 12:599–612.

———. 1977. Displacement, and home range movements of muskellunge determined by ultrasonic tracking. Environ. Biol. Fish. 1(2):145–158.

DIANA, J. S., W. C. MACKAY, AND M. EHRMAN. 1977. Movements and habitat preference of northern pike (*Esox lucius*) in Lac Ste. Anne, Alberta. Trans. Am. Fish. Soc. 106(6):560–565.

MALININ, L. K. 1969. Home range and homing instinct of fish. Zool. Zh. 48:381–391. (Fish. Res. Board Can. Transl. Ser. 2050.)

———. 1970. Use of ultrasonic transmitters for the marking of bream and pike. Biol. Vnutr. Vod Inf. Byull. 8:75–78 (Fish Res. Board Can. Transl. Ser. 2146.)

MUIR, B. S. 1963. Vital statistics of *Esox masquinongy* in Nogies Creek, Ontario. I. Tag loss, mortality due to tagging, and the estimate of exploitation. J. Fish. Res. Board Can. 20:1213–1230.

RONGSTAD, O. J., AND J. R. TESTER. 1969. Movements and habitat use of whitetail deer in Minnesota. J. Wildl. Manage. 33:366–379.

SUMMERFELT, R. C. 1972. Flathead catfish movements. Completion report, Oklahoma project 4-60-R. Nat. Oceanic Atmos. Adm. (U.S.), Nat. Mar. Fish. Serv. 76 pp.

WINTER, J. D. 1977. Summer home range movements and habitat use by four largemouth bass in Mary Lake, Minnesota. Trans. Am. Fish. Soc. 106(4):323–330.

Implications of Water Management in Lake Oahe for the Spawning Success of Coolwater Fishes

WILLIAM R. NELSON

U.S. Fish and Wildlife Service
North Central Reservoir Investigations
Pierre, South Dakota 57501

Abstract

Lake Oahe, a Missouri River reservoir, was closed in 1958. As it filled, newly inundated prairie grasslands provided ideal spawning habitat for northern pike and yellow perch, and large populations of these species developed. Since the reservoir reached operational level in 1967, fluctuations and wave action have reduced the amount of vegetational spawning substrate, so these species have declined in abundance. Sauger and walleye deposit their eggs over gravel or rubble and are only known to spawn in tributary rivers rather than in the reservoir. Anticipated diversions of water for agricultural and industrial development will reduce both the inflows to and the average level of Lake Oahe. This will affect both riverine and reservoir spawning habitats. To maintain the present fish populations, minimum spring stream flows must be assured and critical reservoir spawning areas must be protected and artificially enhanced.

I review here the spawning requirements of northern pike, yellow perch, walleye, and sauger in Lake Oahe, a main stem Missouri River impoundment. Population trends of these species are related to changes in the quantity and quality of spawning habitats, which will be further altered by demands for water by agricultural and industrial interests. I recommend management programs to maintain and enhance critical habitats that will sustain the reproductive success of these coolwater fishes.

Lake Oahe was formed in 1958 by impoundment of the Missouri River near Pierre, South Dakota. The reservoir was filled to the base of flood control, 490 m above mean sea level, in late 1967; its volume was 23.7×10^9 m^3 and its surface area 126,800 hectares. Annual variations in precipitation and runoff through 1975 caused reservoir fluctuations of about 6 m in elevation and 34,400 hectares in area. The physical and chemical characteristics were described by Selgeby and Jones (1974).

The 4 species under consideration were native to this stretch of the Missouri River before Lake Oahe was formed. Their reproductive success in the reservoir has been followed by means of routine sampling. Larval and juvenile fishes were collected with 0.5-m plankton nets (0.76-mm bar mesh) and 30.5-m bag seines (0.6-cm bar mesh). Larvae were sampled at stations in the upper, middle, and lower reaches of the Cheyenne, Moreau, and Grand embayments, in their respective rivers just upstream from the reservoir, and in the Moreau River 40 and 106 km upstream (Fig. 1). Sampling occurred from April through June, 1972–1975, at the Moreau stations and 1974–1975 at the Cheyenne and Grand stations. Larger young of the year were sampled with seines during July and August at 15 locations (Beckman and Elrod 1971) and with bottom trawls during October at 9 locations (Nelson and Boussu 1974).

Northern Pike

Northern pike require submerged vegetation on which to deposit their adhesive eggs. In Lake Oahe this habitat has been provided by inundated terrestrial vegetation since wave action and fluctuating water levels preclude the establishment of aquatic plants. Re-establishment of terrestrial vegetation on shorelines after water level recession requires at least two growing seasons before a significant ground cover develops (Stanley and Hoffman 1975). The most extensive areas of vegetation which can potentially be inundated occur where the shoreline slopes are most gradual or in the upper reaches of the reservoir and embayments.

During the filling of Lake Oahe, spring water levels generally rose and inundated prairie grassland that was ideal for egg deposition. Particularly strong year classes were established in 1959, 1962, and 1965 that subsequently supported an excellent sport fishery (Hassler 1969; Gabel 1974). After Lake Oahe reached operational pool level and the water level fluctuated within a relatively narrow range, northern pike reproductive success was poor. June (1970) reported a high incidence (71 to 77% from 1966 to 1968) of atretic eggs in northern pike, and field studies by Hassler (1970) revealed that high mortality of northern pike eggs occurred if water temperatures fluctuated greatly or small amounts of silt (1 mm/day) were deposited on the eggs.

From 1972 to 1975 no larval northern pike were collected with plankton nets, and only 16 young of the year were captured in 893 seine hauls (June 1976). The annual catch of young with seines from 1967 to 1975 was significantly correlated ($r = 0.87$; $P < 0.01$) with the area flooded in April and May that had not previously been flooded, or where revegetation had occurred because of exposure for 2 years or longer. These conditions occurred only in 1969, 1971, and 1975 and consequently the northern pike population has declined to a low level (Table 1). The species is sought only occasionally by anglers in early spring when the fish are concentrated in the upper reaches of embayments for spawning.

Yellow Perch

Yellow perch eggs are deposited in gelatinous ribbons or masses in relatively clearwater habitats—usually over grassy or woody vegetation but sometimes on rocks or gravel (Thorpe 1977). Eggs deposited on bare substrates have a reduced chance of hatching because of severe wave action in Lake Oahe and the prevalence of clay and silt substrates. Catches of larvae indicated that yellow perch reproduction has been most successful near the mouths of embayments, where turbidity was low, and decreased toward the more turbid upper reaches of embayments (Fig. 2).

Yellow perch young of the year were the most abundant species seined in 1965–1974 (June 1976). Abundance was highest

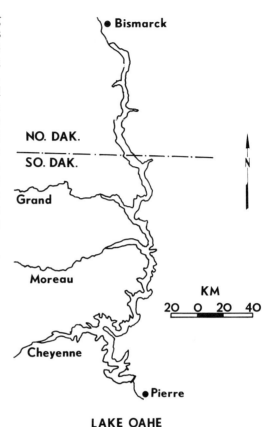

FIGURE 1.—Lake Oahe, showing the Grand, Moreau, and Cheyenne rivers and embayments.

during the years when the reservoir was filling and declined more than 95%, as has the abundance of adults (Table 1), after the reservoir reached full pool. June (1976) believed that this decline occurred because the preferred spawning habitat—brush-covered draws and bottoms inundated

TABLE 1.—Mean number of northern pike, yellow perch, walleye, and sauger older than young of the year captured per 15-min tow with a 10.7-m bottom trawl, Lake Oahe, 1967–1975.

Year	Northern pike	Yellow perch	Walleye	Sauger
1967	0.56	28.34	3.04	1.18
1968	0.60	16.72	6.82	1.86
1969	0.28	5.57	3.75	2.36
1970	0.11	2.72	1.43	0.50
1971	0.14	2.72	4.25	0.50
1972	0	1.72	5.54	1.57
1973	0.14	2.82	14.28	2.39
1974	0	4.46	11.57	1.18
1975	0	3.50	6.32	0.64

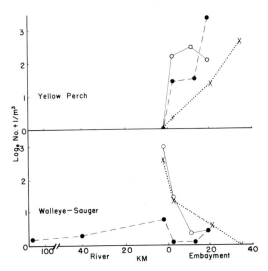

FIGURE 2.—*Mean annual number of larval yellow perch (upper panel) and walleyes and saugers (lower panel) captured in three embayments and rivers: Moreau, 1972–1975 (dashed line); Cheyenne, 1974–1975 (dotted line); and Grand, 1974–1975 (solid line).*

during filling—were destroyed by scouring or sedimentation at full pool. Nelson and Walburg (1977) supported this hypothesis with the finding that the amount of newly inundated terrestrial vegetation and the change in water levels during spawning accounted for 79% of the fluctuation in the year-class strength of yellow perch from 1968 to 1974.

Walleye and Sauger

The kind of substrate on which walleye and sauger randomly scatter their eggs—rocks, rubble, or gravel—was not available along the shorelines of Lake Oahe until the reservoir was filled and wave action had removed the lighter soil particles, leaving the heavier or larger particles along the shoreline. However, this substrate was available in the tributary rivers, and catches of larvae indicated that walleye and sauger spawned primarily, if not exclusively, in these tributaries (Fig. 2). In some Missouri River reservoirs, walleye spawn along wave-washed shorelines, but sauger are only known to spawn in tributary rivers or unaltered reaches of the Missouri River (Nelson and Walburg 1977; Nelson et al. 1978).

The walleye was uncommon in the unimpounded Missouri River, but with increased water clarity, more lake-like conditions, and the development of suitable spawning habitat its abundance increased (Table 1) and it became the primary sport fish in the reservoir. Sauger were abundant in the Missouri River, but the increased water clarity and lack of riverine conditions caused the population to decline to a relatively low level (Table 1). Mean river flows and air temperature during spawning (April) accounted for 55% of the variation in year-class strength of walleyes in 1968–1974 (Nelson and Walburg 1977).

Future Water Depletions

The historical mean annual flow of the Missouri River at Sioux City, Iowa, was estimated to be 26.5 km^3 at the 1970 level of upstream water use (Missouri Basin Inter-Agency Committee 1971). By the year 2000 this flow has been projected to be depleted 7.1 km^3, primarily because of the increased use of water for irrigation and energy development (Northern Great Plains Resources Program 1975). About 90% of this water depletion will occur in the watershed above Oahe Dam.

The United States Army Corps of Engineers (1976) estimates that these projected depletions will lower the level of Lake Oahe 6–18 m below its present mean operational level. This would reduce the quantity and quality of reservoir spawning habitat, especially as the depletions are increasing and before the reservoir has stabilized at a lower level. As depletions increase, the frequency and magnitude of above-average elevations inundating terrestrial vegetation will tend to be reduced. The quality of spawning substrate will also be reduced because the finer soil particles introduced by tributary streams, or eroded from existing shorelines, have settled into the deeper waters which will become the future shorelines. When depletions are no longer being increased, the frequency and magnitude of below and above average water years will be similar to the present situation and wave action will expose rubble shorelines.

A portion of the projected water depletion will be in the Cheyenne, Moreau, and Grand rivers as well as in drainages further upstream. These reduced flows could affect river spawning species by: (1) impeding migration of adults to spawning grounds; (2) retarding the return of eggs and larvae to the reservoir; (3) increasing water temperature; (4) decreasing the dissolved oxygen content of water; and (5) increasing silt deposition in the streams.

The increased use of Missouri River water will require intake structures to remove the water. Although projected construction of industrial plants on Lake Oahe is minor, the number of irrigation intakes is expanding rapidly. Before 1976, few irrigation intakes were present on the reservoir but 27 were installed in 1976, and 18 in 1977. Because of reduced installation expense and ease of maintenance, these intake structures have commonly been placed in shallow embayments which greatly increases the potential for entraining or impinging larval fishes.

Ecological Implications

Reduced water levels and stream flows will affect each of the coolwater species in Lake Oahe differently. The northern pike population is already greatly reduced because of the lack of suitable spawning substrate. Decreased water levels will reduce the frequency and quantity of vegetation flooded and deposits of silt will comprise the new shoreline until wave action exposes gravel and rubble areas. Therefore, the population of northern pike will decline in abundance unless management programs are instituted.

Yellow perch are now relatively abundant and provide an excellent forage base for piscivorous species. Although their reproductive success has been greater when a vegetated substrate was available, they have successfully used barren substrates. This flexibility should reduce the adverse effect of lowered water levels on the population, but their abundance will probably decline.

Walleye and sauger populations are presently abundant and since these species spawn in tributary rivers, increased water use should affect them less than those that spawn in the reservoir. However, adequate stream flows during April and May are required for these species to continue at their present level of abundance.

To minimize the deleterious effect of reduced water levels on the more susceptible northern pike, managers should consider constructing subimpoundments for spawning ponds, the seeding of newly exposed shorelines, and the exclusion of cattle from the upper reaches of embayments to protect vegetation. Establishment of minimum instream flows for the tributary rivers will be necessary to protect the spawning habitat of walleyes and saugers. Permanent water diversion intakes should be 6 m below the water surface projected for low-flow years in order to minimize entrainment and impingement of larvae. Temporary intakes, such as those for irrigation, should be excluded from the littoral zone for the same reason. Without such measures to protect and enhance critical habitats, water depletions will result in reduced populations of these coolwater species.

References

BECKMAN, L. G., AND J. H. ELROD. 1971. Apparent abundance and distribution of young-of-year fishes in Lake Oahe, 1965–69. Am. Fish. Soc. Spec. Pub. 8:333–347.

GABEL, J. A. 1974. Species and age composition of trap net catches in Lake Oahe, South Dakota, 1963–67. U.S. Fish Wildl. Serv. Tech. Pap. 75. 21 pp.

HASSLER, T. J. 1969. Biology of the northern pike in Oahe Reservoir, 1959 through 1965. U.S. Bur. Sport Fish. Wildl. Tech. Pap. 29. 13 pp.

———. 1970. Environmental influences on early development and year-class strength of northern pike in Lakes Oahe and Sharpe, South Dakota. Trans. Am. Fish. Soc. 99:369–375.

JUNE, F. C. 1970. Atresia and year-class abundance of northern pike, *Esox lucius*, in two Missouri River impoundments. J. Fish. Res. Board Can. 27:587–591.

———. 1976. Changes in young-of-the-year fish stocks during and after filling of Lake Oahe, an upper Missouri River storage reservoir, 1966–74. U.S. Fish Wildl. Serv. Tech. Pap. 87. 25 pp.

MISSOURI BASIN INTER-AGENCY COMMITTEE. 1971. The Missouri River Basin comprehensive framework study. Vol. 1. Comprehensive framework study. Washington, D.C. 274 pp.

NELSON, W. R., AND M. F. BOUSSU. 1974. Evaluation of trawls for monitoring and harvesting fish populations in Lake Oahe, South Dakota. U.S. Fish Wildl. Serv. Tech. Pap. 76. 15 pp.

———, D. B. MARTIN, L. G. BECKMAN, D. W. ZIMMER, AND D. J. HIGHLAND. 1978. Prediction of the effects of energy development on the aquatic resources of two upper Missouri River reservoir ecosystems. U.S. Fish Wildl. Serv. Rep. FWS/OBS 0-33-76 and EPA-1AG-D6-F079. Denver, Colo. 89 pp.

———, AND C. H. WALBURG. 1977. Population dynamics of yellow perch (*Perca flavescens*), sauger (*Stizostedion canadense*), and walleye (*S. vitreum vitreum*) in four main stem Missouri River reservoirs. J. Fish. Res. Board Can. 34:1748–1763.

NORTHERN GREAT PLAINS RESOURCES PROGRAM. 1975. Effects of coal development in the Northern Great Plains: a review of major issues and consequences at different rates of development. Prepared in cooperation with federal, state, regional, local, and private organizations. Denver, Colo. 165 pp.

SELGEBY, J. H., AND W. E. JONES. 1974. Physical and chemical characteristics of Lake Oahe, 1968–69. U.S. Fish Wildl. Serv. Tech. Pap. 72. 18 pp.

STANLEY, L. D., AND G. R. HOFFMAN. 1975. Further studies on the natural and experimental establishment of vegetation along the shorelines of Lake Oahe and Lake Sakakawea, lakes of the mainstem Missouri River. Dep. Biol., Univ. S.D., Vermillion. 116 pp. (Mimeo.)

THORPE, J. E. 1977. Morphology, physiology, behavior, and ecology of *Perca fluviatilis* L. and *Perca flavescens* Mitchill. J. Fish. Res. Board Can. 34:1504–1514.

UNITED STATES ARMY CORPS OF ENGINEERS. 1976. Main stem reservoir regulation studies. Series 12-75. Omaha, Neb. 25 pp.

Management of Endangered Coolwater Fishes

JAMES M. ENGEL

U.S. Fish and Wildlife Service
Twin Cities, Minnesota 55111

Abstract

Management of 11 coolwater fishes presently on the United States Endangered Species List is discussed.

Of the thirty fishes on the United States Endangered Species List as of November 1, 1977, eleven may be categorized as coolwater fishes (Table 1). Present management of these species includes artificial propagation, restocking programs, transplanting, removal of competitors, and protection of watershed. Similar management is likely to continue in the future with protection of watersheds being the primary factor in assuring survival of most species. By 1983, it is projected that an additional 100 fish may be listed as endangered or threatened, and it is likely that at least 30% of these will fall within the coolwater classification.

Specific management of the listed species follows.

Blue Pike

Primary threats to the blue pike are pollution, overharvesting, and possibly introduction of the sea lamprey. A recovery plan, completed in June 1976, focuses on artificial propagation, stocking efforts, and protection of natural and planted stocks. To date, attempts to secure broodstock have failed.

Longjaw Cisco

There is a question whether or not the longjaw cisco should be classified as a species. Few specimens with morphological characteristics similar to those originally described have been collected in the past decade. Management of the species, if found, will likely include regulation of take within a given geographic boundary. Regulation would be administered on a local basis, and a recovery plan for the species is probably not feasible.

Scioto Madtom

The scioto madtom had been found only in a single riffle in Big Darby Creek, Ohio. Only 19 specimens have been taken, the last being in 1957. Attempts to collect the species have been made periodically since 1943, and most recently a concerted effort to collect specimens was conducted between October 1976 and February 1977. This effort proved unsuccessful. Work with other species indicates artificial propagation would probably be possible. Protection of the watershed is of primary importance until such time as the species is found or declared extinct.

Shortnose Sturgeon

Primary threats to the shortnose sturgeon include habitat alteration, pollution, and overfishing. Protection by harvest regulation and pollution control is the only ongoing management effort for this species.

Snail Darter

Considerable attention has been given to the snail darter (*Percina tanasi*) because its only known habitat coincides with the Tellico Dam, a flood control and hydroelectric project which is close to completion. Efforts to translocate the species have been made, and there is some indication the species will reproduce in captivity.

Humpback Chub and Colorado Squawfish

Mainstream dams have altered the watershed inhabited by humpback chub and Colorado squawfish from turbulent fast-flowing rivers to deep cold-water impound-

TABLE 1.—*Endangered fishes of the United States and their known distributions.*

Taxon	Known distribution
Blue pike	Lakes Erie & Ontario
Longjaw cisco	Lakes Michigan, Huron, & Erie
Scioto madtom	Upper Darby Creek, Ohio
Shortnose sturgeon	Atlantic Coast (USA & Canada)
Snail darter	Little Tennessee River, Tennessee
Humpback chub	Arizona, Utah, Wyoming
Colorado squawfish	Arizona, California, Colorado, New Mexico, Utah, Wyoming
Cui-ui	Nevada
Fountain darter	Texas
Maryland darter	Maryland
Threespine unarmored stickleback	California

ments. A recovery plan for fishes of the Colorado River focuses on habitat protection, regulated releases of cold water from upstream dams, and control of competing nonnative fishes. Artificial propagation of the Colorado squawfish was begun at the Willow Beach National Fish Hatchery, and future plans call for the production of 150,000 Colorado squawfish per year for restocking. Propagation of the humpback chub is also being considered.

Cui-ui

The primary threat to the cui-ui is a diversion dam which blocks the historic spawning runs. A small hatchery was constructed and wild fish were used for broodstock. In addition, a fish ladder was constructed for re-establishing the spawning run, and regulations closely control the harvest. The fish ladder has not been entirely successful, however, and investigations into the problem have been initiated.

Fountain Darter

Management programs for the fountain darter include maintenance of a captive broodstock population and reintroduction of the species into its former range.

Maryland Darter

The primary threats to the Maryland darter are habitat alteration and pollution. A draft recovery plan has been developed, periodic surveys have been conducted, and a bank-stabilization program has been initiated.

Threespine Unarmored Stickleback

Efforts to recover the threespine unarmored stickleback, *Gasterosteus aculeatus williamsoni*, focus on watershed protection, transplanting the species above unnecessary barriers, and control of competitors (c.g. the African clawed frog).

CULTURE

Historical Perspective of Propagation and Management of Coolwater Fishes in the United States

JOSEPH WEBSTER

U.S. Fish and Wildlife Service
Twin Cities, Minnesota 55111

ARDEN TRANDAHL

U.S. Fish and Wildlife Service
Spearfish, South Dakota 57783

JOHN LEONARD

U.S. Fish and Wildlife Service
Washington, D.C. 20240

Abstract

Propagation for expanding the ranges of coolwater fish began in the late 1800's in the western states. Since 1940, the construction of reservoirs throughout the United States has resulted in intensive propagation and management of coolwater species outside their natural ranges. Many states report an inability to establish naturally reproducing populations, but have established continuing supplemental stocking programs because the public demand has been high. Most coolwater species are top predators and are useful in managing overabundant populations of forage fish. More than half the states are involved in coolwater fish production, and others are planning coolwater fish production programs. Nationwide utilization of coolwater fishes seems imminent.

To meet the demands of a growing population and increasing fishing pressure in the United States, over the last one hundred years species of fish have been introduced into areas where they did not occur due to geographical barriers or lack of suitable habitat.

Action by the reservoir construction agencies has resulted in tremendous habitat changes throughout the country. Development of the water resources has resulted in replacement of existing stream habitats with new lake and reservoir habitats and the creation of new aquatic habitats where little, or none, existed previously. Habitat alteration has affected various fish species throughout the country, creating favorable conditions for some and unfavorable conditions for others. It has enabled certain species to be successfully introduced in areas where suitable habitat was not previously available and the species did not exist.

A group of fish which have always received considerable attention within their native range include walleye, sauger, yellow perch, northern pike, muskellunge, and northern pike/muskellunge hybrid. In the last few years this group has become collectively known as the "coolwater" species. These fish were originally confined to a native range in the midwest and eastern United States, generally extending from the Dakotas to the Appalachians and from the Tennessee River drainage and Arkansas northward into Canada. In fact, man has expanded their range by introducing them throughout most of the continental United States and they have become important aspects of fishery programs in places where they previously were nonexistent.

In their native range, artificial propagation has played an important role in the management of these species and has provided for the expansion of these species throughout most of the United States. Traditionally, propagation of these fish has involved "extensive" propagation methods, that is, pond-type culture generally requiring extensive areas of water. In recent years, as interest in the species has developed and pressures of increasing human population have demanded more intensive resource management, the need for better and more efficient "intensive" cultural methods has become apparent. In the last several years, initial thrusts have been made to develop intensive culture methods for coolwater fish similar to those utilized in the propagation of salmonids. Although some of the initial results have been impressive, a concentrated effort will be required to continue rapid development of methods for intensive culture.

There are currently 16 federally operated national fish hatcheries in eleven states that are involved in coolwater fish production. Although the hatcheries with the large production programs are generally confined to the north central states, four of these facilities are located in southern states. Federal production of coolwater fishes began near the start of the 20th century and has steadily increased. In 1946, federal hatcheries propagated three species of coolwater fish, totaling slightly over 19 million fish (Table 1). In 1976, production involved five species and totaled more than 71 million. These coolwater fish are used in management programs on nearly a nationwide basis.

As conditions and needs for management of our fishery resources change, resource administrators charged with managing the fishery resource must make program decisions as to where effort will be expanded. If the intensive culture of coolwater species is to develop, we must show that coolwater fish play an important role in management of the fishery resources of the nation and focus attention on the need to develop and refine cultural and management techniques. This study was initiated to review the spread of coolwater species by natural and artificial introduction and determine their present status and extent of artificial culture in their expanded ranges.

Methods

The information presented in this report was obtained by a survey. A questionnaire was sent to the 48 state fisheries agencies in the contiguous United States. The questions were formulated for easy answering and the information obtained was intended to provide general information on where coolwater species are native or introduced, the present and future status of the species, and the general extent of propagation programs.

The questionnaire contained 13 questions, all pertaining to walleye, sauger, yellow perch, northern pike, and muskellunge. Briefly, the questions included:

(1) Were the following species native to your state?

(2) If not native, what year were they first introduced and what was the stock origin?

(3) Were there other later introductions, and if so, what were the stock origins?

(4) Have introductions been successful in the state? If not, why not?

(5) What is the present status in the state?

(6) Do the fish presently support a significant sport and/or commercial fishery? Man-days of fishing (annual)? Monetary value—sport or commercial? Public acceptance and demand? Other comments?

(7) Within the state, what is the future sport or commercial potential for each species?

(8) Does the agency have a production program for these species?

(9) If the agency has a production program, what has been the approximate production for the past five years?

TABLE 1.—*Production of coolwater fish at national fish hatcheries (thousands of fish).*

Species	Year			
	1946	1956	1966	1976
Walleye	17,303	2,475	26,557	54,735
Sauger				124
Northern Pike	200	14,395	15,555	16,892
Muskellunge				125
Yellow Perch	1,957			103

(10) What do you project as future production?

(11) What type of role do you foresee for these fish in your agency's fishery management plans?

(12) From a socioeconomic aspect, what do you see as the present and future impact of coolwater fish?

(13) Other comments or recommendations regarding coolwater fish?

Results and Discussion

Completed questionnaires were received from 40 of the 48 states contacted. No response was received from Georgia, Kansas, Kentucky, Mississippi, Montana, North Dakota, Tennessee, and Virginia. The following is a brief discussion of the information received on the returned questionnaires.

Eleven of the responding states had only one of the fish species native to their area. Nineteen states had two or more species native to their area. Generally, the information corresponded to the species ranges described in the recognized ichthyological texts. As a group, these fish were generally restricted to the northeastern and north-central states, with several species extending as far south as Arkansas, Mississippi, Alabama, and South Carolina. The sauger extended as far west as Wyoming and Montana and as far south as Louisiana. Although the recorded ranges of these fish may be quite extensive, natural populations that historically supported significant sport or commercial fisheries were generally restricted to northeastern, central, and north-central United States.

Coolwater fish stockings in the United States outside their native ranges were first recorded over 100 years ago. A flurry of stocking occurred during the late 1800's and early 1900's when stocking was often looked upon as the ultimate answer to fishery management problems. Most of the records of these early introductions were not available. Some were lost, and no doubt many stockings made by individuals, groups, and agencies never were recorded.

The walleye was one of the first coolwater species to gain the interest of "fish managers" outside its native range. A number of the eastern states reported extensive stockings of walleye fry in the late 1800's and early 1900's. Several eastern states also indicated introduction of northern pike were made during this period. It is doubtful these early stockings involved much scientific consideration and most were not very successful. However, they did extend the range of the coolwater species in this region of the United States.

A number of the western states also reported introduction of several coolwater species in the late 1800's and early 1900's. California reported the first introduction of walleye (ultimately unsuccessful) of Vermont origin as long ago as 1874. Introductions of yellow perch, northern pike, and muskellunge were also made in California in the 1880's. Wyoming also reported walleye stockings in the 1880's. It was surprising that a number of western states reported introductions of yellow perch in the late 1800's and early 1900's. Apparently, during this period the yellow perch carried a fair reputation as a desirable game and food fish. In a number of western states, the yellow perch is the only coolwater species still surviving from these early stockings.

In reviewing the information obtained, we found that a second period of interest in the coolwater species outside of their native range developed in the late 1940's and 1950's and is still continuing. During this time, these species were introduced and intensively managed in many parts of the United States where they previously were nonexistent or rare. This interest corresponds to the intense reservoir construction and water development activity which occurred in the 1940's, 1950's and 1960's. The construction of large mainstream reservoirs on the Tennessee River drainage, the Missouri River drainage, and other large river systems created excellent habitat for walleye, northern pike, and other coolwater fish where little or none existed previously. Also, the pressures of increasing human populations, and the increasingly intensive management of our resources forced fishery managers to utilize a wide variety of species to meet the demand for fishing. Coolwater fishes normally survive at higher water temperatures than salmonids. Thus some waters which had supported salmonids and were subjected to natural or human degradation now will not support salmonids, but will support coolwater species.

Generally, the coolwater species receiving the greatest emphasis in the last 30 years, at least in terms of sport fishery management, are walleye, northern pike, and muskellunge. Sauger appear to have received much less management consideration, although in certain instances very successful sauger fisheries have been established or re-established through introductions. The yellow perch, due perhaps to its widespread classification as a panfish and sometimes as a nuisance, has received less specific management effort and has been relatively unimportant in terms of artificial propagation. However, in the last several years, it has received attention as a species possibly well suited to commercial propagation as a food fish and is now being investigated for artificial propagation and intensive culture.

In the last 30 years, coolwater fish have received varying attention in the eastern and northeastern United States. In states such as New York, Pennsylvania, Ohio, Michigan, Wisconsin, and Minnesota, however, the coolwater species always have and probably always will have a top role in fishery programs. Many of the other states where these species were less abundant or absent have become involved in coolwater fisheries to a greater or lesser degree. Several New England states have established new walleye fisheries with good future potential and others report the successful stocking of northern pike for a trophy fishery. However, a reported common problem is the inability to establish a reproducing population, even though survival is good. West Virginia reported very good success in managing native walleye and muskellunge fisheries and in establishing new fisheries in various parts of the state. West Virginia also had good initial success with northern pike introductions and sees an increasingly important role for this species. One New England state where the northern pike was not native reported a put-take and grow trophy fishery as the most significant new fishery program in the last 50 years. In recent years, the southern states have been involved in development of some of the greatest coolwater fisheries in the United States. The development of the walleye fishery in Tennessee Valley Authority reservoirs on the Tennessee River drainage has been one of the most successful. Texas and Arkansas have also been very successful in establishing walleye populations in large multi-habitat reservoirs. Although difficulty has often been encountered in establishing naturally reproducing populations, successful fisheries are being maintained by artificial propagation and stocking.

Several states have also initiated management activities involving the stocking of northern pike. The use of northern pike as an additional predator to reduce or keep in check stunted forage populations has received attention in many states. Other states are evaluating the stocking of fingerling northern pike/muskellunge hybrids to create a trophy fishery in suitable waters. Interestingly, Alabama, a southern state removed from the natural muskellunge range, indicated definite interest in the potential use of the hybrid.

Interest in the coolwater species has developed in some western states in recent years, primarily resulting from the construction of large reservoirs on many major river systems. Reservoirs at lower elevations have provided habitat for good walleye fisheries established through stocking programs. Natural habitat suitable for coolwater fishes is limited in many of the western states, and consequently, these fish will probably never assume a dominant role in most of the western state fishery programs. However, this is not to say that they are not useful tools of the fishery manager. They will, and rightly so, be used in special situations involving man-made habitats. Good coolwater fisheries have been established, or are developing, on the Platte River, the upper Missouri River, a portion of the

Columbia River, and other drainages in the west.

The midwest has, of course, always been the center of the coolwater fisheries. The coolwater species are the backbone of the fishery programs of most of the states in this area. Most of the propagation techniques for the coolwater species have developed here and this is where the largest propagation and stocking programs now exist.

The survey form requested information concerning stock origin from states where coolwater fish were not native. However, after reviewing the information obtained, definitive determinations on the specific origin of different introductions could not be made, at least within the scope of this project. Exact information on the origin of many early introductions may not exist or would be very difficult to find. This is also true of many of the introductions of the later decades. Introduced fish in many cases were obtained from stock introduced previously from other areas. In some cases, three or four or more introductions are involved before the stock can be traced back to its native origins. It is known, however, that Lake Erie stocks were transferred on a large scale to many United States locations. Also, documentation exists on many transfers from several different Canadian stocks into the United States.

Determining a social or economic value of the coolwater fish on a national basis is complex and difficult. Many agencies have not addressed the social impact of coolwater fisheries. Texas, however, indicated that acceptance and interest in coolwater fishes stemmed partially from retired people from the north who were interested in and familiar with these species. Another state also referenced the outsiders to the state program that stimulated demand for coolwater fish. Most of the states replying to the questionnaire did not have a useful estimate of angler days. A few states involved in coolwater species had made rough estimates of angler days and economic impact. For example, Wisconsin estimated coolwater fisheries supported 15,000,000 man-days of fishing with a monetary value in excess of $100,000,000. Michigan estimated 11,000,000 angler-days were directed toward the esocids and percids in 1975, but did not provide any estimate of economic impact. At any rate, in states where a large tourist and resort industry is heavily dependent on fishing, the economic impact of the coolwater fishery is highly significant.

Summary

The extent of propagation of the coolwater species in the United States is considerable. Of the 40 responding state agencies, 23 had active walleye production programs, 14 had northern pike production programs, and 12 had muskellunge production programs. Two states had active yellow perch and sauger production programs. Also, we know at least four non-responding states are actively engaged in coolwater fish production. In total, more than half the contiguous 48 states are involved in production of coolwater fish. Several states that do not have a production program indicated they are involved in limited stocking programs with coolwater fish obtained from other sources. A number of states also indicated they are considering or planning coolwater production programs for the future. In addition to these state programs, there are currently 16 federally operated national fish hatcheries in 11 States that are involved in coolwater fish production. Although the large production programs are generally confined to the north-central states, hatchery-reared coolwater fish are used in management programs on nearly a nationwide basis.

Coolwater species are a major part of the fishery resources of the United States, and must receive prime consideration by administrators when allocating money and personnel for fish management and development. All aspects of management of coolwater species must be considered. Proposals such as the use of northern pike and muskellunge in nonnative areas as a predator to control overabundant forage fish are worthy of more investigation. However, when introducing these nonnative species, we must exercise due caution and evaluate both the positive and negative effects the action may have on native populations. As mentioned by several states in the questionnaire, the potential for the northern pike/muskellunge hybrid in a put-grow-and-take trophy fishery in suitable situations is exciting and worthy of developing with additional investigation. Directly related to this would be a more intensive effort in developing intensive culture methods for coolwater species. Advancement of coolwater culture to the level of salmonid culture would open many management alternatives for our coolwater resources. To make the necessary advancements will require the cooperation of all individuals and agencies involved in the development of coolwater fisheries. Each must be willing to contribute where they can and to accept and build on the contribution of others.

This paper demonstrates that coolwater fish are becoming a significant and necessary part of the national fishery resource, and we believe the intensive propagation, culture, and management of these fish has the potential to rival salmonids.

The Status of Coolwater Fish Diets

LEO E. ORME

Diet Development Center
U.S. Fish and Wildlife Service
Route 1, Box 205
Spearfish, South Dakota 57783

Abstract

There is an increasing demand for more and larger coolwater fish. Intensive culture with a formulated feed is the only feasible method to meet the demand. Trout feeds are not satisfactory for coolwater fish. Seven diets were formulated specifically for coolwater fish and tested. The W-7 feed has been the most satisfactory. Cultural techniques and disease control have advanced with the development of the feeds. Success has been achieved with fingerlings but not fry. Results vary with the different fish. Formulated feeds now are fed to tiger muskellunge on a production scale. Economics are comparable to those for trout. Many improvements still are needed and basic nutritional information is sorely lacking.

Most people who go fishing can remember a river or lake which held a much better stock years ago. There are certainly fewer wild fish about of catchable size and practically no fisheries, except in remote places, which maintain a good natural stock. What is happening is too many fishermen are chasing too few fish and nature is incapable of producing fish fast enough to satisfy the demands of anglers and provide the same quality of fishing as in the past.

A lake or river can only grow a certain weight of fish in a given time. The production of fish depends on the food animals available for the fish.

Restocking used to mean planting small fish, in order that they would grow big enough to give sport later on. Since overfishing is not easily remedied by stocking fry or fingerlings, this suggests the need for stocking larger fish on a "put, grow, and take" basis. This type of stocking is defined by fishery management as the stocking of harvestable-size fish to provide an immediate fishery where conditions are suitable for growth, but angling mortality is high and natural reproduction or stocking of small size fish will not produce an acceptable fishery.

We must, therefore, look for the best size of fish to stock in order to yield the maximum returns. Greater emphasis is being placed on hatchery production of larger sized coolwater fish to supply the increasing demands for this fishery.

There also is increasing interest in commercial production of market sized walleye and yellow perch. The demand for these fish is being created by the dwindling supply of fish from the commercial fishing of wild stocks.

Culturalists have primarily utilized extensive cultural techniques to rear fingerlings of the coolwater species. The size and number of fingerlings that could be raised was limited by the type and amount of natural food available. Small fingerlings can be reared on high densities of zooplankters (Davis 1953). To rear larger fish, a supplemental food source is required. Live forage fish are the natural food for the larger fingerlings of coolwater species, but their cost is prohibitive and the supply of natural feeds uncertain.

The most dependable and economical technique for the production of large fish, as exemplified by the trout and catfish industries, is to use a formulated feed. Many advantages are gained by using formulated feeds rather than natural feeds. Formulated feeds are available throughout the year in a nearly unlimited supply and a variety of feed sizes for growing fish; their application is easily regulated and can be automated; they are convenient to store, handle, and augment with medications; they do not introduce parasites and diseases to fish like natural feeds do. They greatly reduce the cost of the feeding program. Also, ponds and facilities used to raise natural feeds can be used for additional production of coolwater sport fish.

TABLE 1.—*Formulation of trout feeds (% by weight) used in Coolwater Testing Program.*[a]

Ingredient	SD3A	PR4	PR9
Herring meal	42	35	35
Soybean oil meal	5	10	20
Corn fermentation solubles	5	8	8
Dried whey		4	10
Dried skim milk	5		
Brewer's yeast	4	4	5
Dehydrated alfalfa meal		3	3
Trace mineral salt		2	2
Vitamin premix[b]	4	4	4
(Premix no.)	(21)	(21)	(28)
Fish oil	8	4	5
Wheat middlings	8	20	8
Blood flour	2		
Condensed fish solubles	5		
Corn gluten meal (60%)	5	6	
Kelp meal	2		
Animal liver meal	5		

[a] Proximate analyses are given in Table 5.
[b] The composition of vitamin premixes is given in Table 2.

Culturists had always believed that coolwater fish would accept only live food and could not be reared on a formulated feed. This attitude had to be dealt with before any progress could be made toward developing a feeding program for coolwater fish based on formulated feeds.

When we started our program there was no information on feeding formulated feeds to coolwater fish. The only literature available on feeding artificial feeds to fish other than trout, salmon, and catfish were reports by Snow (1968) and Lewis et al. (1969) on largemouth bass.

In the spring of 1970, a series of feeding trials were set up to test natural and red-colored SD3 and PR4 trout feeds with walleye and northern pike (Tables 1 and 2). The fish did not take either feed, and the response of the fish indicated that neither the texture nor the flavors of these feeds were acceptable.

It was evident that a diet had to be developed specifically for the coolwater fishes. There was no information on the nutritional requirements of these species, but it was logical to compare them to trout and try to adjust the diet to meet their needs. The coolwater fishes are raised in warmer water, and have higher metabolic rates and faster growth, than trout. Therefore, their protein and energy requirements are probably greater. They do not feed as aggressively as trout and prefer a softer feed. It is assumed that the nutrients in a formulated feed would be utilized as efficiently by the coolwater fishes as by trout. The coolwater fish do accept a larger feed particle than do comparable sized trout. We also had to consider feed manufacturing characteristics and availability of ingredients when formulating a diet for the coolwater fish.

Based on the above factors, the W-1, W-2, and W-3 test diets were formulated by the Diet Development Center for the 1971 testing program (Table 3). The W-1 feed contained gelatin so it could be reconstituted to a semi-moist feed with the addition of water. The 25% blood flour in the W-2 diet produced a dark red color and a meaty odor and flavor. The W-3 diet contained the highest level of protein and energy. The vitamin fortification of all three diets was the same as for trout feeds.

The results of the feeding trials varied greatly due to a lack of refinement of intenive cultural techniques and disease control for these coolwater species. However, the W-3 diet showed some promise as a successful feed.

The W-3 diet and Oregon Moist Pellet (OMP) diet (Table 4) were selected for testing in the 1972 program. The nutrient content of the OMP diet is similar to that of the W-3 diet, but the OMP is a soft semi-moist feed and contains a fresh fish product. It was felt that these factors would make the OMP diet more acceptable to the cool-

TABLE 2.—*Vitamin premix formulations (units per kg premix).*

Vitamin	No. 21	No. 25	No. 28
D-calcium pantothenate (g)	1.32	1.32	1.98
Pyridoxine (g)	1.10	1.10	0.88
Riboflavin (g)	4.40	1.32	1.32
Niacin (g)	13.75	13.75	8.80
Folic acid (mg)	220.0	220.0	220.0
Thiamin (g)	1.65	1.10	1.10
Biotin (mg)	11.00	11.00	11.00
B_{12} (mg)	0.55	0.55	0.55
Menadione sodium bisulfite (mg)	275.00	275.00	275.00
BHT (mg)	550.00	550.00	
Vitamin E (IU)	4,400.00	4,400.00	8,800.00
Vitamin D_3 (IU)	4,400.00	33,000.00	11,000.00
Vitamin A (IU)	176,000.00	165,000.00	165,000.00
Choline chloride (g)	27.50	8.80	22.00
Ascorbic acid (g)	6.60	11.00	16.50
Inositol (g)			9.90

water fishes. The overall results of the feeding trials indicated that neither the W-3 nor OMP feeds were the answer.

A new set of feeds, W-4, W-5 and W-6 (Table 3) were formulated for the 1973 test program. The OMP feed was again evaluated by several hatcheries. The ingredients used and nutrient content of the test diets in this series varied from previous diets and each other.

Disease control and improved intensive-culture rearing techniques for the coolwater species improved the success and validity of the feeding trials. The best results were obtained with the W-5 diet, but the texture of the feed appeared to be too hard. There still was something lacking in the test feeds because they did not perform as well as they should. A comparison of the metabolic rate between trout and the coolwater fishes indicated that there may be a greater need for vitamins by the coolwater fishes.

Based on the information from the previous tests, the W-7 diet (Table 3) was formulated to be a high protein-high energy feed (Table 5). Compared with the W-5 diet, W-7 has 5% more fish meal, 5% less whey, no wheat middlings, addition of 5% blood flour, and a 50% increase in vitamin fortification. The protein and energy levels remained the same. What is the function of each ingredient in the W-7 feed?

The fish meal (50%) is a readily available source of high quality protein. Soybean oil meal (10%), a protein source, provides

TABLE 4.—*Oregon Moist Pellet feed.*

Ingredient	Pellets	Starter
Meal mix		
Herring meal	28	46
Cottonseed meal	15	
Dried whey	5	10
Shrimp meal	4	
Wheat germ meal	4	10
Corn distillers solubles	4	4
Vitamin premix	1.5	1.5
Wet mix		
Fish meal (wet)	30	16
Kelp meal	2	2
Soybean oil	6	10
Choline chloride	0.5	0.5

amino acids to balance the diet. Brewer's yeast (5%) is a source of protein, vitamins, and unidentified factors. Delactosed whey (5%), a source of animal sugar, serves as a pellet conditioner. Blood flour (5%) is a source of essential amino acids, as well as a natural binding agent and a flavoring agent. Condensed fish solubles (10%) are a source of protein and a flavoring agent. Fish oil (9%) is a source of readily available energy and essential fatty acids. The vitamin premix (6%) supplies essential vitamins.

The W-7 feed has a softer texture than trout feeds when properly manufactured.

In the 1974 feeding trials, greater success was achieved in getting the coolwater fishes to accept the dry feed and grow well by using the W-7 feed and improved cultural methods. The accomplishments in 1974 generated new enthusiasm in the testing program and provided the basis for expanding the testing to a production scale.

The growth of the testing program and success of the formulated feeds is exemplified by the increase in the quantity of feed used each year. From 270 kg in 1970, use of the feeds has steadily increased to 45,750 kg in 1977.

The preliminary success of the W-7 feed did not diminish the effort to develop a better and more efficient diet for the coolwater fishes.

The greatest success with dry feeds has been achieved only when the fish are started as fingerlings. The tiger muskellunge (northern pike × muskellunge hybrid) is the most adaptable to a formulated feed. There has been moderate success with northern pike

TABLE 3.—*Formulation of experimental coolwater diets (% by weight).*[a]

Ingredient	W-1	W-2	W-3	W-4	W-5	W-6	W-7
Herring meal	30	30	50	45	45	30	50
Wheat middlings	16	8			2	14	
Soybean oil meal	10	5	5		10	10	10
Brewer's yeast	5	5	5	5	5	5	5
Dried whey	10	5	7	7	10	10	5
Blood flour			25	10	20		5
Condensed fish solubles	10	10	10	10	10	15	10
Corn gluten meal					5	5	
Fish oil	3	4		9	9	7	9
Soybean oil	3	4	9				
Vitamin premix[b]	4	4	4	4	4	4	4–6
(Premix no.)	(25)	(25)	(25)	(25)	(25)	(25)	(28)
Salt	1.5						
Water	5						
Gelatin	3						

[a] Proximate analyses are given in Table 5.
[b] The composition of vitamin premixes is given in Table 2.

TABLE 5.—*Proximate analysis (% by weight) of coolwater test feeds.*

Diet	Protein	Fat	Moisture	Ash	Fiber	NFE[a]
W-1	43	11.5	9	10	2	24.5
W-2	54	13	7	8	3	15
W-3	57	14	6	11	2	10
W-4	62	12	7	8	2	9
W-5	51	15	7	9	2	15.5
W-6	43	13	7	8	4	24
W-7	51	15	7.5	9	3	14
SD3	49	13	6	11	2	19
PR4	43	8	7	11	4	27
PR9	41	11	6.5	10.5	4	27
OMP[b]	42	14	20	9	3	12

[a] Nitrogen free extract—calculated.
[b] Oregon Moist Pellet.

and little success with walleye and muskellunge (Nagel 1974, 1976; Graff and Sorenson 1970).

Most of the attempts to start fry on dry feeds have resulted in failure. In an effort to improve the acceptance of the W-7 feed by fry, the feed has been coated with a dry egg product, colored with an orange-red dye, and flavored with shrimp meal. Tiger muskellunge fry readily accepted all feeds with good survival. The northern pike, walleye, or muskellunge fry did not start on the dry feeds (Beyerle 1975; Cuff 1977; Cheshire and Steele 1972; Pfister 1976).

The Oregon Moist Pellet feed has also been tested as a starter feed for fry with the results being similar to that for the W-7 feeds.

In trout feeding programs, the fish are converted from a high protein-high energy feed to a lower quality feed as advanced fingerlings. Presently the coolwater fishes are fed the same high protein-high energy feed throughout the entire feeding program. Instituting a program similar to that for trout would result in substantial savings in feed costs. A series of feeding trials were conducted to compare feeding the W-7 for the entire feeding program as opposed to switching to the lower-protein PR9 (Table 1) trout feed for advanced fingerlings. Northern pike had a conversion factor (kg dry feed/kg wet-weight gain) of 1.9 for the W-7 feed and 2.7 for the PR9 trout feed. Tiger muskellunge converted the W-7 at a rate of 1.7 and PR9 at 2.6. The growth rate of the fish fed the PR9 feed was substantially less than for those receiving the W-7 feeds.

No economical benefits were gained from feeding the trout feed to the advanced fingerlings.

The rearing of tiger muskellunge in intense culture on the W-7 feed has advanced out of the experimental stage and is routinely conducted on a production basis.

The feed costs for the coolwater fish feeding programs using the W-7 feed can be calculated from the feed conversions and feed prices. Representative feed conversions have been obtained from production scale feeding trials. Reported conversions are 1.6–3.3 for tiger muskellunge and 1.9 for northern pike.

The average price of the W-7 feed was US $0.73/kg. Therefore, the feed cost for rearing tiger muskellunge and northern pike was $1.16 to $1.45 per kg of weight gain. This compares favorably with the cost of US $0.62 per kg of gain for trout.

Significant advancements have been made in developing a formulated feed for the coolwater fishes. But, we do not have a satisfactory feed to start first-feeding fry. Very limited success has been obtained in feeding walleye and muskellunge on dry feeds. The cost of the feeding programs could be reduced by changing the diet and using the lower priced vegetable proteins to replace part of the fish meal and other animal proteins.

Disease control, cultural methods, and genetics must also advance with nutrition for the intensive production of coolwater fish to be successful.

Plans are to formulate and test new practical diets in a continuing effort to improve the feeding programs. Different feeding procedures will also be developed and tested.

Presently the greatest need is to establish the basic nutritional requirements for the coolwater fishes. Even preliminary information on the protein and energy requirements for the various sized fish would significantly aid in formulating the practical feeds.

We have made some great accomplishments in developing a formulated feed for the coolwater fishes and have a lot yet to do. The easy tasks are behind and the more difficult ahead so progress will be slower and require more effort. But, the challenge is still as great as it was in the beginning.

References

BEYERLE, G. G. 1975. Summary of attempts to raise walleye fry and fingerlings on artificial diets, with suggestions on needed research and procedures to be used in future tests. Prog. Fish-Cult. 37(2).103-105.

CHESHIRE, W. F., AND K. L. STEELE. 1972. Hatchery rearing of walleyes using artificial food. Prog. Fish-Cult. 34(2):96-99.

CUFF, W. R. 1977. Initiation and control of cannibalism in larval walleye. Prog. Fish-Cult. 39(1):29-32.

DAVIS, H. S. 1953. Culture and diseases of game fishes. Univ. Calif. Press, Los Angeles. 322 pp.

GRAFF, D. R., AND L. SORENSON. 1970. The successful feeding of a dry diet to esocids. Prog. Fish-Cult. 32(1):31-35.

LEWIS, W. M., R. HEIDINGER, AND M. KONIKOFF. 1969. Artificial feeding of yearling and adult largemouth bass. Prog. Fish-Cult. 31(1):44-46.

NAGEL, T. 1974. Rearing of walleye fingerlings in an intensive culture using Oregon Moist Pellets as an artificial diet. Prog. Fish-Cult. 36(1):59-61.

———. 1976. Intensive culture of fingerling walleye on formulated feeds. Prog. Fish-Cult. 38(2):90-91.

PFISTER, P. J. 1976. Production of tiger muskies from swim-up fry on an artificial diet. Report presented at Coolwater Diet Testing Meeting, Toledo, Ohio, Feb. 1977.

SNOW, J. R. 1968. The Oregon Moist Pellet as a diet for largemouth bass. Prog. Fish-Cult. 30(4):235.

Culture of Yellow Perch with Emphasis on Development of Eggs and Fry

GRADEN WEST

Senecaville National Fish Hatchery
U.S. Fish and Wildlife Service
Senecaville, Ohio 43780

JOHN LEONARD

Division of National Fish Hatcheries
U.S. Fish and Wildlife Service
Washington, D.C. 20240

Abstract

Yellow perch were cultured intensively from spawn through incubation of the eggs to harvest of the fingerlings. During the spring of 1975, female and male yellow perch weighing an average of 38.6 and 13.6 g, respectively were collected and held in a metal tank for spawning. Eggs were incubated on trout egg hatching trays in troughs and in a commercial incubator. A return of 35% was realized from stocked eggs to fingerling harvest. Fingerlings were trained to accept dry feed and from initial feeding on July 13, until they were distributed in early September, grew from an average of 0.38 to 1.36 g with a survival of 37%.

Although yellow perch are not normally cultured at hatcheries, they are raised at Lake Mills National Fish Hatchery, Wisconsin, for use by various agencies. Our attempts at intensive culture of yellow perch began in 1973 when the fish we were holding in a tank spawned and we successfully incubated the eggs. The study was repeated in 1975 to gather additional data.

Raising perch by intensive methods will increase the production options available to the hatchery manager. As methods are developed which permit perch to be cultured in facilities other than ponds, ponds will become available for raising other game fish or forage fish.

A feasibility study on the commercial production of yellow perch from the University of Wisconsin Sea Grant Program estimates that 23,700 kg of yellow perch could be produced annually at a cost of $2.31/kg. Although 1976 retail prices for yellow perch fillets were as high as $8.58/kg, the market price for yellow perch in the round ranged from $1.65 to $2.76/kg. Refinements in the intensive culture of yellow perch are necessary if successful commercial production is to be realized.

Parental Stock

Adult yellow perch were acquired from Rock Lake, the source of pond water for the Lake Mills National Fish Hatchery, by angling through the ice. The fish were held in an indoor concrete rearing tank about 2 months before being transferred to a metal spawning tank on February 26. Volume of the spawning tank was 912 liters with dimensions of 0.89 m × 2.29 m and a water depth of 0.46 m. Water drawn from Rock Lake flowed into the tank at a rate of 12.5 liter/min. The only food offered the fish from time of capture to April 6 was No. 4 trout granules (1.68–2.83 mm). The fish were skittish and feeding was not observed although it might have occurred. No food was provided to the broodfish after April 6.

On March 6 the fish were sorted according to sex, weighed, measured, and given a prophylactic 3% salt dip. No other treatments were necessary as there was no mortality until after spawning when several spent females succumbed to a heavy fungus infection.

On April 4, 50 females averaging 158 mm in total length (range 140–191 mm) and 38.6 g in weight, and 50 males averaging 108 mm total length (range 98–165 mm) and 13.6 g were selected as broodfish. Prior to sorting, two females were studied to determine fecundity, the ovaries were removed from each female and weighed. Portions of each ovary were then weighed and the number of eggs in each portion was counted. This was expanded to estimate the egg count per ovary pair. A female weighing 45 g contained approximately 5,120 eggs and an 80-g female 8,960 eggs, or 114.8 and 112.0 eggs per g of body weight, respectively. We assumed the size of the ovary is proportional to the body weight, and estimated an average spawn of 4,377 eggs per female by multiplying the average fecundity (113.4 eggs/g) times the average body weight of the females (38.6 g). After broodfish selection, but prior to any spawning activity, two black synthetic plastic mats with a total of 1.5 square meters were placed on the bottom as a spawning substrate. A small section of fencing with a square mesh of 150 mm was also placed in the tank for a mid-water spawning structure.

Spawning

Periodic inspections of the spawning tank were made from 0730 to 1600 h daily. Spawning rarely took place in midday as only 3 of 46 spawns were collected other than first thing in the morning. Although human activity during the day may have affected fish behavior, Terry Kayes (Aquaculture Laboratory, University of Wisconsin, Madison, personal communication) reported that early morning was the preferred spawning time for yellow perch.

The first two egg envelopes were collected April 19 at a temperature of 3.6 C (Table 1). Four days later at 6.7 C spawning again occurred and continued each day, with two exceptions, through May 5. The largest number of spawns was collected on May 2 when the water temperature was 10 C, after having risen from 7.8 C the day before. Temperatures prevailing during the majority of spawning were similar to the range (7.2 C–11.1 C) recorded during the reproductive activity of wild yellow perch in Wisconsin (Herman et al. 1959).

The egg envelope is tube-shaped resembling an elongated pickle with transverse accordian-like folds. Often dead eggs were noted inside the folds of the envelopes. Eggs within the folds either received poor water circulation or were poorly fertilized. Egg fertility was checked April 28 on the spawns of April 19 and 23. The sections of the egg envelopes were chosen randomly and of the 54 eggs selected 51 developed, for a 94% fertility.

TABLE 1.—*Relationship of water temperature to the frequency of collection of yellow perch egg envelopes.*

Date	Noon temperature C	Number of egg envelopes collected
April 19	3.6	2
23	6.7	1
24	6.1	2
25	6.1	2
26	7.2	0
27	7.5	3
28	6.7	2
29	7.2	5
30	7.8	3
May 1	7.8	5
2	10.0	12
3	10.3	6
4	10.6	0
5	10.9	3

The synthetic mats were removed April 24 as the egg envelopes became entangled in the matting and were difficult to collect. However, the mid-water spawning structure caused no problem and was not removed.

Egg Incubation

Upon discovery, half of the egg envelopes were transferred to trays in trout hatching troughs and half to a Heath Incubator. The trough contained 184 liters of water and the Heath with 7 shelves contained about 40 liters of water. City water was used at 4.5–9.5 liters/minute through the Heath and 6 liters/minute through the trough. A constant flow of formalin at 166–200 mg/liter for one-hour intervals was used occasionally to combat fungus infections. Development was measured in time and temperature units (t.u.). One temperature unit is 1 C above 0 C for 24 hours. As the temperature increased, fewer days and temperature units were required for hatching (Table 2). Our first spawns required 228 t.u. over an average daily temperature of 9.9 C.

Table 2 also presents information describing features of the developing larvae. During early development of the larvae, melanin was easily seen with magnification or by close observation aided by an artificial light, but eye-up was not recorded until pigmentation could be easily seen with the naked eye. Prior to hatching, egg envelopes lost rigidity and became flaccid, gold iris

TABLE 2.—*Temperature units and observations during the development of yellow perch eggs from date of spawning to date of hatching.*[a]

| Date of observation | Temp C | Date of spawning ||||||||||||
| | | April |||||||| May ||||
		19	23	24	25	27	28	29	30	1	2	3	5
April 19	8.8												
20	8.7	8.7											
21	8.7	17.4											
22	8.9	26.3											
23	9.2	35.5											
24	9.4	44.9	9.4										
25	9.4	54.3	18.8	9.4									
26	9.5	63.8	28.3	18.9	9.5								
27	9.4	73.2	37.7	28.3	18.9								
28	9.4	82.6	47.1	37.7	28.3	9.4							
29	9.2	91.8	56.3	46.9	37.5	18.6	9.2						
30	9.5	101.3	65.8	56.4	47.0	28.1	18.7	9.5					
May 1	9.7	111.0[b]	75.5	66.1	56.7	37.8	28.4	19.2	9.7				
2	10.0	121.0	85.5	76.1	66.7	47.8	38.4	29.2	19.7	9.7			
3	10.4	131.4[d]	95.9	86.5	77.1	58.2	48.8	39.6	30.1	20.1	10.4		
4	10.6	142.0	106.5	97.1	87.7	68.8	59.4	50.2	40.7	30.7	21.0	10.6	
5	10.6	152.6	117.1[bc]	107.7[b]	98.3	79.4	70.0	60.8	51.3	41.3	31.6	21.2	
6	10.6	163.2	127.7[d]	118.3	108.9	90.0	80.6	71.4	61.9	51.9	42.2	31.8	10.6
7	10.4	173.6	138.1	128.7	119.3	100.4	91.0	81.8	72.3	62.3	52.6	42.2	21.0
8	10.5	184.1	148.6	139.2	129.8	110.9	101.5	92.3	83.8	72.8	63.1	52.7	31.5
9	10.6	194.7	159.2	149.8	140.4	121.5[c]	112.1[b]	102.9	93.4	83.4	73.7	63.3	42.1
10	11.1	205.8[ef]	170.3	160.9	151.5	132.6[d]	123.2	114.0[c]	104.5	94.5	84.8	74.4	53.2
11	11.3	217.1[g]	181.6[e]	172.2	162.8	143.9	134.5	125.3	115.8	105.8[c]	96.1[b]	85.7[b]	64.5
12	11.3	228.4	192.9	183.5[e]	174.1	155.2	145.8	136.6	127.1	117.1[d]	107.4[c]	97.0	75.8
13	11.3		204.2	194.8	185.4	166.5	157.1	147.9	138.4	128.4	118.7	108.3	87.1
14	11.3		215.5	206.1	196.7[g]	177.8	168.4	159.2	149.7	139.7	130.0	119.6[d]	98.4
15	11.5				208.2	189.3[e]	179.9[e]	170.7[e]	161.2	151.2	141.5	131.1	109.9
16	11.7					201.0[g]	191.6[g]	182.4[g]	172.9[g]	162.9[g]	153.2	142.8	121.6
17	12.2					213.2[f]	203.8[f]	194.6	185.1	175.1	165.4	155.0	133.8
18	12.2							206.8	197.3	187.3	177.6	167.2	146.0
19	12.6							219.4	209.9	199.9	190.2	179.8	158.6

[a] Daily temperature equals the temperature units (t.u.) for that day. Daily entries are cumulative t.u.'s. The last entry for each column represents the total t.u.'s required to hatch eggs in a given day's spawn.
[b] Larval movement.
[c] Weekly eyed.
[d] Well eyed.
[e] Gold iris pigmentation.
[f] Soft and flaccid egg envelope.
[g] Loss of vigor.

pigmentation became visible surrounding the melanin in the eyes, larval movement decreased, and finally, bubbles accumulated in the egg envelope giving the mass a tendency to rise. Although not observed during this study, Hokanson (1977) reported that mouth and opercular movements of the larva became syncronized just prior to hatching. On the hatching date, which was determined when several fry hatched, the eggs were transferred to the rearing pond.

Pond Rearing

The 0.2-hectare rearing pond had an average depth of 0.3 meters prior to stocking of eggs, which began on May 12 and continued through May 19. Eggs were placed on anchored, floating screens (30 cm × 30 cm × 3.8 cm) of 6 mm mesh, large enough to allow newly hatched fry to fall through. Eleven of the 12 spawns produced through April 28 provided an estimated 48,147 eggs. Half the eggs in spawns produced between April 28 and May 5 appeared opaque in color and were judged dead; nevertheless, 28 of 34 spawns from this period were stocked and provided an estimated 61,278 additional live eggs. Based on a 94% fertilization rate, a total of 102,860 viable eggs and fry were placed in the pond. However, the fertility of yellow perch eggs reportedly lowers as the spawning season progresses (Terry Kayes, personal communication). Therefore, fertility may have been lower than 94% in the eggs from late spawns, especially in eggs stocked on May 19.

Fry were first sighted May 25 and were approximately 16 mm in length. They were next observed June 6 when 25 mm long. Because there appeared to be little zooplankton available as feed for the fry, 250 kg of fertilizer, in the form of hay, was added to the pond from June 6 through the 20th.

Harvest and Fingerling Production

On July 11, 35,789 fingerlings weighing a total of 13.6 kg or 0.38 g per fish were harvested from the pond. This represents a return of 35% from the number planted. The rearing pond contained extensive beds of aquatic vegetation including *Potamogeton pectinatus*, *Najas* sp., and *Elodea* sp. This made harvesting the fish difficult and in combination with the generally poor zooplankton blooms, which occur with heavy rooted vegetation, probably decreased the yield of fingerlings from the pond. Fingerlings appeared emaciated and exhibited loss of equilibrium and slow recovery when handled.

From this harvest, 20,700 fingerlings were retained for a feeding study which commenced on July 13. They were held in concrete tanks at a temperature of about 25 C. Feed was an open formula production diet (PR9-27) of No. 1 granules designed for trout. A switch was made to No. 2 granules after a few days when preference was noted for the larger size particles. The fish were feeding well after 12 days, and on July 25 were transferred to a 1.2-m (800-liter) circular tank containing water at a temperature of 16 C and held until September. A total of 7,900 fingerlings were distributed September 3 and 9 for a survival of 38% over the 2-mo feeding period.

Project Evaluation

We were unable to monitor percent fertility and egg survival as closely as desired; however, they did appear to vary considerably. This may have been due to inadequate water circulation into the folds of the egg envelope or more likely poor fertilization of the eggs in the area of the fold.

Midwater structures were used to entangle and catch the egg envelope as it was expressed and to stretch it allowing the accordian-like folds to open. Spraying the structure with an abrasive substance like non-slip paint may increase its effectiveness. Better, or more consistent, fertilization might be achieved with different sex ratios. In this experiment, a one-to-one sex ratio was used. To assure a supply of viable males and increase fertility, a ratio of two or more males per female is recommended. If this is attempted, two tanks should be used to decrease crowding and minimize the possibility of battering the egg envelopes prior to collection.

Physical measurement of the envelopes was inconsistent due to the accordian effect. A volumetric method of determining envelope size and subsequent number of eggs would be more accurate.

The occasional 1-h constant-flow formalin treatments at 166–200 mg/liter were not adequate to control fungus. Future tests should evaluate the effectiveness of daily treatments with more concentrated formalin for fungus control.

The Heath Incubator was more effective than troughs, as eggs were totally contained and did not float away. Since we had small females which produced a maximum envelope length of 27 cm, we had sufficient incubation space with either facility. However, spawns from large adults may reach 213 cm (Herman et al. 1959). A spawn this size would probably not fit into trays of either incubation unit of the size used in this study.

A criterion of 50% hatch may seem more appropriate for setting a hatching date than that based on initial hatch, but too many fry would be lost if transfer to a pond were delayed to this stage. To minimize loss from handling, fry should hatch in the pond.

Early stocking may also overcome some of the problems associated with loss of prehatching vigor by larvae, since wave action would agitate the eggs, helping them to break out of the shell. However, sometimes if eggs are stocked at an earlier stage, fungus may occur, and it is difficult to treat fungus in a pond.

New hatched fry are pelagic (Ney and

Smith 1975) and upon stocking in the pond were noted to be photo-positive. This would aid in holding fry in tanks if jar culture methods used for walleye are atempted for yellow perch.

When the first eggs are produced, the pond should be filled with water to the greatest practical depth and fertilized as soon as possible. Ideally, plankton blooms should be monitored prior to stocking fry and during rearing. Various fertilizers may be applied to assist in triggering the bloom. Chopped hay applied at a rate of 1,100 kg/hectare would be suitable initially. This should be followed in two weeks with an additional application of 550 kg/hectare. Fading blooms may be restimulated with lesser amounts of hay or small quantities of brewers yeast, alfalfa meal, or other organic fertilizers. Herbicide might be considered for pre-flood or post-flood application to control rooted vegetation, which hinders the harvest of fingerlings.

Post-harvest feeding by fingerlings was noted after one or two days. Initially, it seems best to drop a few particles into small areas rather than to disperse food over the entire surface. This attracts fish to specific areas and when feeding begins it helps to teach others. This has been successful with largemouth bass fingerlings and seems to apply to yellow perch. Largemouth bass prefer live, moving food and when a pellet or granule splashes to the surface they rush to inspect it. If the food does not move, but only sinks, it may not be taken. Recently thawed Oregon Moist Pellets rolled into small "worms" spin as they sink. This entices the largemouth bass to sample them, which eventually leads to acceptance of regular dry feed.

In our project, time was not available to do the best job of feeding, which requires that small amounts of feed be provided the fiingerlings at closely spaced intervals. Automatic feeders should work well provided the feed is not spread over too large an area initially.

Chronic mortality of fingerlings occurred throughout the feeding period. Most mortality appeared to involve emaciated fish which never went on feed. If the fish had been in better shape at harvest, they would have taken production feed more readily, and the survival (38%) to distribution would probably have been much higher.

Bacterial infections often occur when warmwater fish are held at normal summer temperatures. Therefore, we purposely held the yellow perch fingerlings at a lower temperature. However, since some fish will not readily accept dry feed at cooler temperatures, the water could not be cooled enough to eliminate bacterial infections. A feed containing teramycin may be useful in controlling bacteria.

Recommendations

In order to achieve intensive culture of yellow perch, the following steps should be followed:

(1) Adults should be acclimated to tank culture. This means that fingerlings which accept feed readily should be selected for broodstock.

(2) If broodstock are not domesticated, forage should be provided during pre-spawning and post-spawning periods.

(3) A male/female sex ratio of 2 or 3:1 should be evaluated.

(4) Adult fish should number 75 or less per tank.

(5) Fertility of eggs in the folds of the envelopes should be checked.

(6) Individual spawn sizes should be determined volumetrically.

(7) The rearing pond should be fertilized and herbicide added to prevent excessive rooted vegetation.

(8) Techniques for spawning yellow perch by hand should be developed and refined.

References

HERMAN, E., W. WISBY, I. WIEGERT, AND M. BURDICK. 1959. The yellow perch. Its life history, ecology and management. Wis. Dep. Nat. Resour. Publ. 228. 14 pp.

HOKANSON, K. 1977. Optimum culture requirements of early life phases of yellow perch. Paper presented at the symposium on Perch Fingerling Production for Aquaculture, Dec. 12, 1977, Univ. Wisconsin, Madison.

NEY, J. J., AND L. L. SMITH, JR. 1975. First year growth of yellow perch (*Perca flavescens*) in the Red Lakes, Minnesota. Trans. Am. Fish Soc. 104(4):718–725.

Preliminary Observations on the Sperm of Yellow Perch[1]

Steven D. Koenig, T. B. Kayes, and H. E. Calbert

Aquaculture Research Laboratory
Department of Food Science
University of Wisconsin, Madison, Wisconsin 53706

Abstract

Observations were made on the morphology and physical-chemical characteristics of yellow perch spermatozoa. Spermatozoa resembled the primitive type described for other teleosts. Counts for 10 fish ranged from 1.14×10^{10} to 3.02×10^{10} sperm/ml. The seminal plasma of 10 fish had an average pH of 8.50, osmotic pressure of 316.7 milliosmols, and contained 2.64, 0.46, 0.128, and 0.145 mg of Na^+, K^+, Ca^{++}, and Mg^{++}/g, respectively. Corresponding ion concentrations for spermatozoa were 1.81, 1.65, 0.073, and 0.228 mg/g.

Methods of artificially fertilizing yellow perch eggs are currently being evaluated at the University of Wisconsin Aquaculture Research Laboratory for possible use in commercial aquaculture systems. This paper describes preliminary observations on the morphology and physical-chemical properties of perch spermatozoa and seminal plasma. It is hoped that such information will aid in perfecting fertilization techniques and will eventually help to develop methods of preserving yellow perch spermatozoa.

Methods

Sexually mature male perch (242–278 mm total length) were collected in fyke nets from Lake Mendota, Dane County, Wisconsin from mid-March to mid-April, 1977. Fish were brought to the laboratory alive. Milt was stripped from fish by gently massaging the flanks directly over the testes from anterior to posterior. Milt was collected in calibrated 15-ml centrifuge tubes. Contaminated samples were discarded. Testes used in histochemical examinations were removed and frozen in liquid nitrogen.

Scanning electron micrographs were taken of sperm fixed 2 h in a test tube set in an ice bath with a solution containing 0.5 ml milt, 1 ml 5% glutaraldehyde, and 2 ml of 0.2 M phosphate buffer at pH 7.4 (Stanley 1967). Following fixation, a drop of solution was placed on a precleaned coverslip and dried at room temperature in a graded series of ethanol in water: 30% to 100% in 5% intervals for 2 min each. The coverslip was air-dried. Following drying, a 2Å layer of a gold-palladium (60:40) alloy was evaporated onto the coverslip. The spermatozoa were examined with a JEOL model JSM-U3 scanning electron microscope.

Light microscope observations of spermatozoa were made with frozen sections of testicular tissue. Sections 16 μm thick were cut from pieces of tissue previously frozen in liquid nitrogen. Sections were stained with Harris hematoxylin and eosin by standard methods (Luna 1968). Observations were made at 1,250× magnification with a compound microscope.

Sperm counts, and determinations of pH and osmolarity of seminal plasma and major cation concentrations in spermatozoa and plasma were made on milt from 10 fish. The concentration of spermatozoa in milt was determined by standard methods for counting red blood cells (Anonymous 1975). A 0.025-ml aliquot of milt was diluted to 10 ml with a solution of 110 mM NaCl, 5 mM KCl and 0.1 M MOPS[2] buffer at pH 7.4. Following mixing, a drop was placed on an AO Bright-Line hemacytometer. Five replicate counts were made on milt from each fish and the average was calculated.

Two ml of whole sperm were removed for cation determinations. The remainder

---

[1] Research supported by the University of Wisconsin Sea Grant College Program and by the College of Agriculture and Life Sciences, University of Wisconsin-Madison.

[2] Sigma Chemical Co.

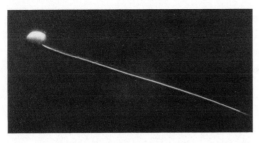

FIGURE 1.—*Scanning electron micrograph of the yellow perch spermatozoan. The length of the head is 2.5 μm.*

was centrifuged at 3,000 rpm for 45 min at room temperature. Two ml of plasma were removed for pH and osmolarity determinations. The remainder of the plasma was removed for cation determinations. The pellet was taken for cation determinations. The pH and osmolarity were measured immediately. Samples for cation determinations were placed in 10-ml fliptop polyethylene vials and frozen (0 to −4 C) for later analysis.

The pH of seminal plasma was determined with a Beckman Zeromatic II pH meter.

Osmolarity of plasma was measured by determining its freezing point depression with an Advanced Instruments osmometer.

The Na^+, K^+, Mg^{++}, and Ca^{++} concentrations in whole sperm, seminal plasma, and spermatozoa were determined with a Perkin-Elmer model 603 flame spectrophotometer. Samples were prepared by wet digestion with nitric and perchloric acid (Anderson 1972; Anonymous 1976).

Results

Morphology of Spermatozoan

Electron micrographs of the yellow perch spermatozoan constitute Figures 1 and 2. The spermatozoan had a rounded head 2.5 μm long with a maximum width of 1.8 μm. There was no morphological distinction between the head and midpiece. No acrosome was present. A flagellum 29 μm long was attached near an indentation in the base of the head. The diameter of the principle piece remained constant. The flagellum terminated with a short end piece.

When stained with hematoxylin, most of the head was dark purple (Fig. 3). An area near the base of the tail was not stained. The tail was not stained.

Characteristics of Semen

Sperm counts, pH, and osmolarity of seminal plasma, and distribution of major cations between spermatozoa and plasma are listed in Table 1. Concentration of spermatozoa in milt of the ten fish ranged widely. There was little variation in pH of the plasma. Osmotic pressure of seminal plasma ranged from 308.3 to 333.7 milliosmols. Concentrations of Na^+ and K^+ were an order of magnitude greater than those of Ca^{++} and Mg^{++}. Amounts of Na^+ and Ca^{++} were greater in plasma than in spermatozoa, whereas amounts of K^+ and Mg^{++} were greater in spermatozoa than in plasma.

Discussion

The morphology of yellow perch spermatozoa is characteristic of the primitive sperm type first described by Retzius (Baccetti and Afzelius 1976). A large portion of the head is composed of the nucleus. The recess at the base of the head has been seen in other teleost spermatozoa (Ginzburg 1972).

The absence of an acrosome is typical of most teleost spermatozoa. Biochemical data have suggested that where present, the acrosome serves a role in recognition and fusion of sperm and egg membranes and consequent penetration of the egg membrane (Anderson and Personne 1973). The acrosome of bovine sperm is sensitive

FIGURE 2.—*Scanning electron micrograph of the head of the yellow perch spermatozoan. The length of the head is 2.5 μm.*

to osmotic shock (Baccetti and Afzelius 1976). Since teleost spermatozoa come into direct contact with the micropylar canal, the acrosome is not needed to penetrate the egg (Ginzburg 1972). Presumably, because of osmotic shock, the acrosome of teleost sperm has been sacrificed during evolution.

Organization of the midpiece varies among teleosts. In most, the midpiece exists as a small globule which is attached at the base of the head and may extend into a concavity in the head (Ginzburg 1972). Electron micrographs of spermatozoa from channel catfish have revealed mitochondria and vesicles irregularly distributed in a collar-like midpiece of the sperm (Jaspers et al. 1976). Cytological investigations revealed that one large granule, identified as a modified mitochondrion in *Fundulus*, existed in *Perca fluviatilis* (Ginzburg 1972). Presumably the midpiece (mitochondrion) in yellow perch sperm was located in the unstained region near the recess in the head. In a similar recess two centrioles, components of the midpiece, have been found in spermatozoa of *Salmo trutta fario* (Ginzburg 1972).

The tail of many teleost spermatozoa, e.g., of salmon and trout, has an end piece similar to that of yellow perch spermatozoa. In others, including pike and carp, the tail gradually thins toward the end (Ginzburg 1972).

The physical-chemical properties of sperm have not been reported for many fish. The concentration of spermatozoa in milt varies considerably: from 2.03×10^7 sperm/ml in rainbow trout injected with pituitary extracts (Clemens and Grant 1965) to 3.0×10^{10} sperm/ml in Atlantic salmon (Truscott and Idler 1969). The

TABLE 1.—*Spermatozoan concentration, pH, and osmotic pressure of seminal plasma, and the distribution of the major cations in whole semen, seminal plasma, and spermatozoa of yellow perch.*

Measurement	Material	Mean[a] ± SD	Range
Sperm count (sperm × 10^{10}/ml)		2.27 ± 0.37	1.14–3.02
pH (25 C)	Plasma	8.50 ± 0.002	8.43–8.57
Osmolarity (mosmols)	Plasma	316.7 ± 38.2	308.3–333.7
Na^+ (mg/g)[b]	Semen	2.31 ± 0.32	1.16–3.55
	Plasma	2.64 ± 0.30	1.96–3.72
	Sperm	1.81 ± 0.06	1.19–2.12
K^+ (mg/g)[b]	Semen	1.43 ± 0.07	1.08–1.92
	Plasma	0.46 ± 0.02	0.24–0.64
	Sperm	1.65 ± 0.11	0.96–2.03
Ca^{++} (mg/g)[b]	Semen	0.050 ± 0.004	0.000–0.104
	Plasma	0.128 ± 0.009	0.000–0.337
	Sperm	0.073 ± 0.004	0.000–0.171
Mg^{++} (mg/g)[b]	Semen	0.135 ± 0.001	0.077–0.171
	Plasma	0.145 ± 0.011	0.057–0.339
	Sperm	0.228 ± 0.007	0.116–0.385

[a] Average value of 10 samples.
[b] mg of cation per g wet weight of sample.

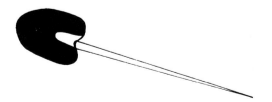

FIGURE 3.—*Schematic diagram depicting the area of the head of the yellow perch spermatozoan stained by Harris hematoxylin.*

range observed in yellow perch (1.14×10^{10}–3.02×10^{10} sperm/ml) is typical. Concentrations for several other fishes have been reported (Guest et al. 1976). The pH of seminal plasma appears to be a parameter closely maintained within a species. Hwang and Idler (1969) reported a mean and standard error of 8.25 ± 0.03 for Atlantic salmon. The pH of yellow perch seminal plasma was 8.50 ± 0.002 (SD). The osmolarity of yellow perch seminal plasma was higher than that reported for goldfish (289.6 ± 2.7 mosmol) (Grant et al. 1969) and Atlantic salmon (232 ± 13 mosmol) (Hwang and Idler 1969). An increase in the sodium ion concentration in yellow perch semen over that of Atlantic salmon accompanied the difference in osmolarities. Gregory (1970) reported a freezing point depression of −0.489 C for walleye seminal plasma.

The relatively high sodium ion and low potassium ion concentrations in seminal plasma accompanied by the reverse in spermatozoa indicate the presence of a transport mechanism. Similar results have been found in mammals as well as Atlantic salmon (Hwang and Idler 1969). Distribution of Ca^{++} and Mg^{++} was also similar to that in Atlantic salmon.

Yellow perch sperm are adapted to the

environment into which they are thrust. To be effective, efforts to artificially propagate perch must incorporate natural adaptations. Billard (1977) stated that pH and osmolarity were the most important parameters for a NaCl diluent used in a wet method of artificial insemination. The basic medium used for cryopreservation of Atlantic salmon spermatozoa was based on the concentrations of major cations in seminal plasma (Truscott and Idler 1969). Morphological and physical-chemical characteristics of yellow perch spermatozoa have been observed and measured. These measurements, coupled with observations on the physiology of perch spermatozoa, will provide the framework for future developments of techniques of artificial insemination of yellow perch eggs and preservation of sperm.

Acknowledgments

We thank M. B. Garment, Department of Entomology, M. P. Meyer, Department of Zoology, and A. W. Andren and N. T. Loux, Department of Civil Engineering, University of Wisconsin—Madison, for their help in supplying equipment and technical assistance.

References

ANDERSON. J. 1972. Wet digestion versus dry ashing for the analysis of fish tissue for trace metals. At. Absorp. Newsl. 11(4):88–91.

ANDERSON, W. A., AND P. PERSONNE. 1973. The form and function of spermatozoa: a comparative view. Pages 3–14 *in* B. A. Afzelius, ed. The functional anatomy of the spermatozoan. Pegamon Press, Oxford.

ANONYMOUS. 1975. AO bright-line hemacytometer counting chamber. American Optical, Buffalo. 19 pp.

ANONYMOUS. 1976. Analytical methods for atomic absorption spectrophotometry. Perkin-Elmer Corp., Main Ave., Norwalk, Conn.

BACCETTI, B., AND B. A. AFZELIUS. 1976. The biology of the sperm cell. Monogr. Dev. Biol. 10. 181 pp.

BILLARD, R. 1977. A new technique of artificial insemination for salmonids using a sperm diluent. Fisheries 2(1):24–25.

CLEMENS, H. P., AND F. B. GRANT. 1965. The seminal thinning response of carp (*Cyprinus carpio*) and rainbow trout (*Salmo gairdnerii*) after injections of pituitary extracts. Copeia 1965:174–177.

GINZBURG, A. S. 972. Fertilization in fishes and the problem of polyspermy. Israel Program of Scientific Translations, Jerusalem. 366 pp. (Academy of Sciences of the USSR, Institute of the Development Biology.)

GRANT, F. B., P. K. T. PANG, AND R. W. GRIFFITH. 1969. The twenty-four-hour seminal hydration response in goldfish (*Carassius auratus*)—I. Sodium, potassium, calcium, magnesium, chloride and osmolarity of serum and seminal fluid. Comp. Biochem. Physiol. 30:273–280.

GREGORY, R. W. 1970. Physical and chemical properties of walleye sperm and seminal plasma. Trans. Am. Fish. Soc. 99(3):518–525.

GUEST, W. C., I. W. AVAULT, JR., AND I. D. ROUSSEL. 1976. A spermatology study of channel catfish, *Ictalurus punctatus*. Trans. Am. Fish. Soc. 105(3): 463–468.

HWANG, P. C., AND D. R. IDLER. 1969. A study of major cations, osmotic pressure, and pH in seminal components of Atlantic salmon. J. Fish. Res. Board Can. 26(2):413–419.

JASPERS, E. J., J. W. AVAULT, JR., AND J. P. ROUSSEL. 1976. Spermatozoal morphology and ultrastructure of channel catfish, *Ictalurus punctatus*. Trans. Am. Fish. Soc. 105(3):475–480.

LUNA, L. G. 1968. Manual of histologic staining methods of the Armed Forces Institute of Pathology, third edition. McGraw-Hill, New York. pp. 32–41.

STANLEY, H. P. 1967. The fine structure of spermatozoa in the lamprey *Lamptera planeri*. J. Ultrastruct. Res. 19:84–99.

TRUSCOTT, B. R., AND D. R. IDLER. 1969. An improved extender for freezing Atlantic salmon spermatozoa. J. Fish. Res. Board Can. 26(12):3254–3258.

Comparative Growth of Male Versus Female Yellow Perch Fingerlings Under Controlled Environmental Conditions[1]

E. F. SCHOTT, T. B. KAYES, AND H. E. CALBERT

*Aquaculture Research Laboratory
Department of Food Science
University of Wisconsin, Madison, Wisconsin 53711*

Abstract

Female yellow perch fingerlings outgrew males when raised under controlled environmental conditions. Six groups of 102 fish each were grown in 110-liter aquaria on two different formulated feeds, under conditions of a constant temperature of 21 to 22 C and a 16-h light:8 h dark photoperiod. On both feeds, females became significantly larger than males at about 110 mm total length and about 15 g total weight. It is postulated that body size, rather than environmental cues or chronological age, may control the onset of sex-related size dimorphism in perch and that reproductive hormones are involved in this process.

Factors influencing growth of yellow perch are being investigated at the University of Wisconsin Aquaculture Research Laboratory. The objective of this study was to assess the effect of sex on growth of young-of-the-year fish. Numerous field observations on yellow perch and other North American coolwater species indicate that adult females are generally larger than adult males (Scott and Crossman 1973). The age or size at which sex-related differences in growth rate appear is an important consideration in developing systems for commercial aquaculture. Such variations may alter product uniformity and indicate divergent patterns of food consumption and nutrient utilization.

Methods

Capture and Holding Procedures

Approximately 4,000 young-of-the-year yellow perch, seined from Willow Flowage, Oneida County, Wisconsin by the Wisconsin Department of Natural Resources, were obtained on September 9, 1976. They were held in the laboratory for 3 months in two 700-liter, fiberglass tanks. Water temperature was maintained at 19 to 21 C for an initial 3-week period while the fish were habituated to a mixed diet of ground beef heart, 1.6-mm W-3 (Huh et al. 1976) and 1.6-mm Oregon Moist[2] pellets. After this, the fish were held at 14 to 15 C and fed 2% of wet body weight per day. The photoperiod was 16 h light:8 h dark.

Acclimatization and Experimental Procedures

Three to six weeks before the experiment, 918 perch of 64.6 to 86.5 mm total length were selected from the holding tanks, and 102 fish within a 3- to 4-mm size range were assigned to each of nine 110-liter aquaria. Individual aquaria were shielded from external light by black plastic screens and were supplied with 21 ± 1 C water at 1.5 liters/min. The water in each aquarium was aerated; the photoperiod was 16 h light:8 h dark. Dissolved oxygen levels were monitored with a Yellow Springs Instrument Model 54A meter and probe. Oxygen concentrations were generally maintained at 6.8 to 7.0 mg/liter. Throughout the study, solid wastes were siphoned twice daily from each aquarium; sides of aquaria were brushed clean twice a week; and the water was clarified by operating a Cosmic Multi-Flow filter on each aquarium for 1 h, as needed, every 2 to 4 days.

The experiment ran for 91 days. Before it, fish were fed various amounts of W-3 or Oregon Moist (OM) pellets (Orme 1978, this volume) at 3 to 4% of wet body weight

---

[1] Research supported by the University of Wisconsin Sea Grant College Program and by the College of Agriculture and Life Sciences, University of Wisconsin, Madison.

[2] Bioproducts Incorporated, Box 429, Warrentown, Oregon 97146.

per day. After day 1, the fish in six aquaria were given the OM diet exclusively; fish in the remaining three aquaria were supplied only with W-3 feed. In both instances, the daily ration was 4% between days 1 and 68, and 3% between days 68 and 91. Equal portions of the daily ration were given at 0830, 1230, and 1630 h, 6 days a week. Ration sizes were adjusted monthly to correct for weight gain. Dry weights of feeds were used in calculating ration sizes.

Measurement and Autopsy Procedures

Determinations of total length (TL), standard length (SL), and total body weight (TWt) were made at intervals both before and after fish were assigned to the experimental aquaria. Erosion of caudal fins was not observed. Standard length was defined as the distance from the tip of the snout to the last pore of the lateral line. Lengths were measured to the nearest 0.5 mm. Fish were weighed to within 0.05 g after being carefully blotted in a standard fashion on absorbent paper. Before measurement or autopsy, fish were anaesthetized in 1.2 mg/liter tricaine methanesulfonate.[3] Measurements were made on several hundred, randomly selected fingerlings taken from the holding tanks before the fish were trained to feed, and 3 months later, before fish were assigned to the experimental aquaria. Fish used in the experiment were measured when first assigned to aquaria, five days before day 1, and at monthly intervals thereafter. Each time, length determinations were made on all the experimental fish, but only 20 to 50% of the fish in each aquarium were weighed.

The gonads of male and female yellow perch fingerlings are morphologically dissimilar and can be readily identified on autopsy. The initial size distribution of the sexes was estimated by autopsying representative small, medium, and large size fish from the holding tanks at the time of the first measurement (see Fig. 1). During the experiment, about 30 fish from each aquarium were randomly selected and autopsied on day 57; the remainder, except in two aquaria (see Fig. 2), were autopsied on day 91. The significance of differences in relative numbers of males versus females, as well as the variation of both sexes from 50% of a sample population were analyzed by z tests for differences between proportions (for details see Bruning and Kintz 1968). Student's t test and the Mann-Whitney U test served to compare the average sizes of males versus females.

Results

Fish grew during both the holding and experimental periods. Throughout the study, variations in size were essentially the same, regardless of whether size was expressed as TL, SL, or TWt. Accordingly, only TL or TWt data are presented in this paper.

Fish of both sexes were similar in size before the experiment. At the initial autopsy (Fig. 1, September), males outnumbered females in the three size ranges examined, but the difference in the relative number of males versus females within each size range was not statistically significant. During the holding period (Fig. 1, September to December), the average TL of the overall population increased ($P < 0.001$), as did the spread or variance of the TL's recorded ($P < 0.001$; F-test for difference in variances). There was no evidence that this increase in size variance was caused by sexually dimorphic growth. Males were eventually found to outnumber females in all nine experimental groups. In three of these, the differences in proportion of males versus females were significant ($P < 0.05$). However, such differences were not related to size of fish at the beginning of the experiment.

Growth of fish during the experiment was influenced both by sex and diet. The perch fed OM pellets invariably grew better than those on the W-3 feed (Fig. 2). By day 57 (Figs. 2 and 3), females were larger than males in four of the nine experimental groups. The same was true in six of the seven groups examined on day 91. In most instances, the significance of differences between the sexes increased between days 57 and 91. A careful analysis of the data in Fig. 2 revealed that sexually dimorphic growth, regardless of diet, first became

---

[3] Crescent Research Chemicals Inc., 7050 Fifth Ave., Scottsdale, Arizona 85251.

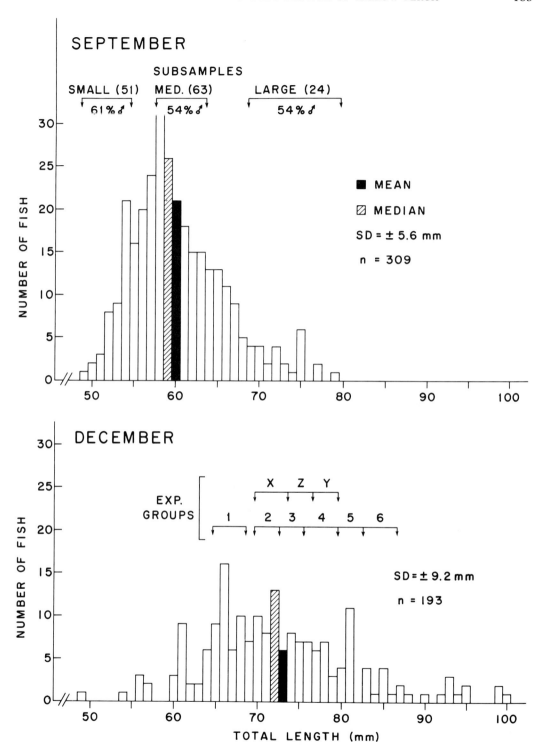

FIGURE 1.—Size distribution of yellow perch at the beginning (September) and end (December) of the holding period. Sample sizes (n) and standard deviations (SD) are indicated. Arrows bracket the size ranges of groups that were initially autopsied (September) or later assigned to the experiment (December). Numbers of small, medium-size, and large fish initially autopsied are enclosed by parentheses. During the experiment, Groups 1 through 6 were fed Oregon Moist pellets; Groups X, Y, and Z were given W-3 feed.

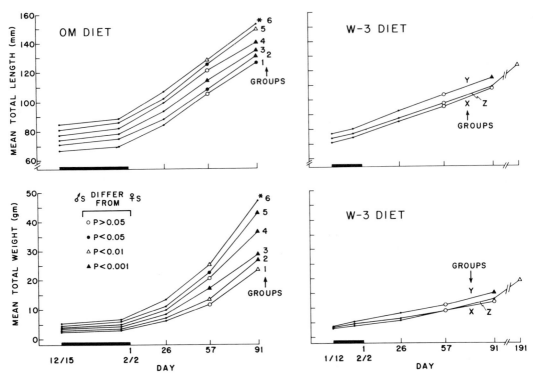

FIGURE 2.—Growth of yellow perch on Oregon Moist (OM) and W-3 diets. Black bars on the abscissas symbolize the lengths of time fish were acclimatized to experimental conditions. Groups were assigned to experimental aquaria on December 15 and January 12. The experiment began on February 2. Mean lengths and weights are designated by points and geometric figures. About 30 fish from each aquarium were autopsied on day 57; the remainder, except in (*) Group 6, were autopsied on days 91 and 191. Females were larger than males as indicated in the figure.

apparent at about 110 mm TL and 15 g TWt. The size spread of fish within each group increased significantly ($P < 0.001$) during the course of the experiment (Fig. 3).

Discussion

Females of many North American coolwater species of fish grow faster than males and reach a greater ultimate size. This has been observed for northern pike, pickerel, and muskellunge (Scott and Crossman 1973), walleye (Rawson 1956; Priegel 1969; Niemuth et al. 1972), sauger (Carlander 1950), and yellow perch (Herman et al. 1964; Scott and Crossman 1973). Our study demonstrated that such sex-related size dimorphism can develop in perch under artificial conditions within the first year of life. Size variation between the sexes was first detected at about 110 mm TL and 15 g TWt, and occurred among fish on two different feeds. Temperature, light intensity, and photoperiod were carefully controlled, since all these factors can greatly influence yellow perch growth (Huh et at. 1976).

Body size in yellow perch, rather than environmental cues or chronological age, may be the major factor controlling the onset of differential growth between the sexes. Two lines of evidence support this hypothesis. First, the size variations noted developed under artificial conditions among fish that had never experienced a full annual cycle of temperature and photoperiod fluctuation. Second, these variations were detected at a size rather than age threshold. Thus, size dimorphism was generally observed at an earlier age in larger, rapidly growing fish (those in Groups 2, 3, 5, and 6, see Fig. 2) than in smaller or slower growing ones (those in Groups 1, X, Y, and Z). As further support

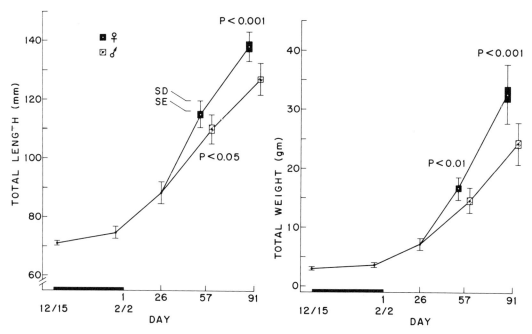

FIGURE 3.—*Growth and size divergence during the experiment of the yellow perch in Group 2. Data are given contrasting males and females. Standard deviations (SD) and standard errors (SE) about the means are indicated, as are the P values for statistical comparisons of males versus females. Patterns of growth and size divergence were similar in all the experimental groups.*

for the concept of a size threshold, we observed that there was no difference in the size distribution of males versus females when they were first brought to the laboratory (Fig. 1), nor was there any meaningful variation in the relative numbers of males versus females assigned to the different experimental groups. In all the latter, males outnumbered females by 12 to 30%, regardless of the initial size of the fish.

Field studies of yellow perch indicate that the size threshold for sexually dimorphic growth may vary in different populations. For example, Carlander (1950) found that young-of-the-year females in a Minnesota lake were significantly larger than males by August 17, at a mean SL of 43.1 mm. In our study, males and females from the Willow Flowage, Wisconsin did not differ until they reached approximately 110 mm TL and 15 g TWt. In contrast, LeCren (1958) concluded that size dimorphism in Lake Windermere perch (*Perca fluviatilis*) did not develop until their third year, when they were about 115 to 144 mm TL. Males and females in Lake Mendota, Wisconsin also remained similar in size until their third year, when they exceeded 200 mm TL and 120 g TWt (Hasler 1945). Considering this uniformity, yellow perch stocks originating from Lake Mendota may be particularly desirable for commercial aquaculture. This prospect is currently being examined.

LeCren (1958) suggested that the onset of sexually dimorphic growth in perch is closely associated with the onset of sexual maturity. The fish used in our study were clearly not mature, but, by days 57 and 91, the gonads of many of the larger females did contain yolky eggs (no such eggs were seen during earlier autopsies). Vitellogenesis in fish is triggered and maintained by reproductive hormones secreted from the pituitary gland and its target tissues (de Vlaming 1974; Emmersen and Emmersen 1976). If this is true in yellow perch, gonadotropic and estrogenic hormones must have been circulating in some of the fish in our study before sexually dimorphic growth was detected. Various androgens and estrogens can influence fish growth—

either positively or negatively, depending on species (Atz 1957; McBride and Fargerlund 1976; Yamazaki 1976). Perhaps androgens retard and estrogens accelerate the growth of coolwater fishes.

References

ATZ, J. W. 1957. The relation of the pituitary to reproduction in fish. Pages 178–270 *in* G. E. Pickford and J. W. Atz. The physiology of the pituitary gland of fishes. N.Y. Zool. Soc., New York.

BRUNING, J. L., AND B. L. KINTZ. 1968. Computational handbook of statistics. Scott, Foresman and Co., Glenview, Ill.

CARLANDER, K. D. 1950. Growth rate studies of saugers *Stizostedion canadense canadense* (Smith) and yellow perch *Perca flavescens* (Mitchell) From Lake-of-the-Woods, Minnesota. Trans. Am. Fish. Soc. 79:30–42.

DE VLAMING, V. L. 1974. Environmental and endocrine control of teleost reproduction. Pages 13–83 *in* C. B. Schreck, ed. Control of sex in fishes. Va. Polytech. Inst. State Univ., Blacksburg.

EMMERSON, B. K., AND J. EMMERSON. 1976. Protein, RNA and DNA metabolism in relation to ovarian vitellogenic growth in the flounder *Platichthys flesus* (1). Comp. Biochem. Physiol. 55B:315–321.

HASLER, A. D. 1945. Observations on the winter perch populations of Lake Mendota. Ecology 26:90–94.

HERMAN, E., W. WISBY, AND M. BURDICK. 1964. The yellow perch: its life history, ecology and management. Wis. Conserv. Dep. Publ. 228.

HUH, H. T., H. E. CALBERT, AND D. A. STUIBER. 1976. Effects of temperature and light on growth of yellow perch and walleye using formulated feed. Trans. Am. Fish. Soc. 105:254–258.

LECREN, E. D. 1958. Observations on the growth of perch (*Perca fluviatilis*) over twenty two years with special reference to the effects of temperature and changes in population density. J. Anim. Ecol. 27:287–334.

MCBRIDE, J. R., AND U. M. H. FAGERLUND. 1976. Sex steroids as growth promoters in the cultivation of juvenile coho. Pages 145–162 *in* J. W. Avault Jr., ed. Proc. Seventh Annu. Meeting, World Maricult. Soc. La. State Univ. Div. Continuing Educ., Baton Rouge.

NIEMUTH, W., W. CHURCHILL, AND T. WIRTH. 1972. Walleye, its life history, ecology and management. Wis. Dep. Nat. Resour. Publ. 227–72 D-1847.

ORME, L. E. 1978. The status of coolwater fish diets. Am. Fish. Soc. Spec. Publ. 11:167–171.

PRIEGEL, G. R. 1969. Age and growth of walleye in Lake Winnebago. Wis. Acad. Sci. Arts Lett. 57:121–133.

RAWSON, D. S. 1956. The life history and ecology of the yellow walleye *Stizostedion vitreum* in Lac la Ronge, Saskatchewan. Trans. Am. Fish. Soc. 86:15–37.

SCOTT, W. B., AND E. J. CROSSMAN. 1973. Freshwater fishes of Canada. Bull. Fish. Res. Board Can. 184. 966 pp.

YAMAZAKI, F. 1976. Application of hormones in fish culture. J. Fish. Res. Board Can. 33:948–958.

Intensive Culture of Walleyes: The State of the Art

JOHN G. NICKUM

*New York Cooperative Fishery Research Unit
Cornell University
Ithaca, New York 14853*

Abstract

Although intensive culture of walleyes has developed considerably during the past 10 years, it is still an incomplete art and an even more incomplete science. Walleye fingerlings have been reared in ponds and then converted to intensive culture; they have attained lengths of 100–125 mm over a 10–15 week growing period with survival rates in excess of 60%. Several forms of pelleted dry diets have been used; the W-7 of the U.S. Fish and Wildlife Service is the diet of choice at this time. Temperatures of 20–22 C and water exchange rates of at least two per hour have produced the most rapid growth and have limited disease problems. A variety of tanks, troughs and raceways have been used successfully for intensive walleye culture. Walleye fry have accepted dry feeds; however few have survived to fingerling size on such diets. Better diets, improved control of diseases, and different designs for rearing units hold promise for bringing intensive culture of walleye fry to satisfactory levels. Post-fingerling walleyes have been reared successfully under intensive conditions, thus indicating that culture of walleyes to lengths of 175 mm and larger is technically feasible. Full development of intensive walleye culture to a level comparable with that of the culture of channel catfish or rainbow trout may occur during the next 10 years, if the problems of fry culture are solved and economical sources of 20–22 C water are obtained.

The intensive culture of walleyes is an incomplete art, and an even more incomplete science. The National Task Force for Public Fish Hatchery Policy (1974) identified "The inability to rear the tiny delicate larvae of species like striped bass and walleye on artificial diets" as "the most critical bottleneck in the national fish-culture program." No one has yet reported success in rearing walleyes from eggs to adult size entirely under intensive hatchery conditions. Very few controlled, replicated studies have been completed with either fry or fingerlings. Several hatcheries and laboratories have reared fingerlings on dry, prepared diets under hatchery conditions (McCauley 1970; Cheshire and Steele 1972; Graves 1974; Nagel 1974, 1976; Beyerle 1975; Huh et al. 1976); however, all attempts to rear walleye fry on dry diets have failed to produce more than a few hundred fish. In spite of such failures, walleye culturists are optimistic. The gap between success and failure has diminished with each new set of experiments. Walleye culturists now speak in terms of "when we succeed" rather than "if we succeed."

A discussion of the state of the art of intensive walleye culture may be conveniently divided into three parts: fry culture; fingerling culture; and post-fingerling culture. In all cases I must rely heavily upon unpublished results obtained through informal mimeographed reports and personal communications, plus unpublished results from studies in progress by the New York Cooperative Fishery Research Unit (NYCFRU). The "Coolwater Dry Diet Steering Committee," coordinated by Leo Orme, has been an invaluable mechanism for communication of unpublished records of successes and failures. The conclusions and insights of experienced, practical fish culturists are often highly accurate and represent an excellent point from which to start the formal experimentation needed to supply the statistically sound data demanded by the scientific community.

Walleye Fry Culture

Beyerle (1975) stated that "attempts to raise walleye fry on various 'artificial' diets in 1973 were uniformly unsuccessful." Although new information has been obtained, final results of feeding trials in 1974–1977 were not appreciably different. Fry have ingested dry diets, but no one has reared substantial numbers to fingerling size. An estimated 50–75% of walleye fry

TABLE 1.—*Composition and proximate analysis of the W-7 diet. Total available energy = 3,400 kilocalories/kg. (Source: Diet Development Center, U.S. Fish and Wildlife Service, Spearfish, South Dakota.)*

Ingredients	Mg/kg diet	Percent
Herring meal		50
Blood flour		5
Soybean flour		10
Dried whey		5
Brewer's yeast		5
Fish solubles		10
Fish oil		9
Supplement, vitamin premix 28		6
Ascorbic acid	660	
Choline Cl	530	
d-Ca pantothenate	120	
Pyridoxine	66	
Riboflavin	79	
Niacin	826	
Folic acid	13	
Thiamine	66	
Biotin	0.7	
Vitamin B_{12}	0.03	
Vitamin D_3 (IU/kg diet)	1,980	
Vitamin A (IU/kg diet)	9,910	
Vitamin E (IU/kg diet)	396	
Menadione sodium bisulfite	17	
Antioxidant	33	
Proximate analysis		
Protein		51.3
Fat		14.4
Moisture		8.6
Ash		8.5
Total carbohydrates		17.2

accepted dry diets in 1977 tests at the Wolf Lake (Michigan) State Fish Hatchery; however, essentially all fish died at the age of about 3 weeks and less than 1% survived to fingerling size. No pathogen or other lethal agent could be identified. In similar tests at Cornell University and the Oneida (New York) State Hatchery no more than 21% of the fry accepted dry diets. Ten days' feeding on brine shrimp and zooplankton followed by gradual conversion to a dry diet over a 10-day period produced acceptances as high as 63%; however, few fish survived longer than 3 weeks after their first feeding. Results in other hatcheries and in other years have been roughly similar.

Facilities and General Conditions

Most attempts to rear walleye fry have been conducted in troughs of various sizes, ranging from 5-liter "mini-troughs" to "standard" hatchery troughs. There is no reliable evidence to indicate the superiority of one type of unit over any other. In tests by NYCFRU, results were very poor in circular tanks where water depth was as great as 50 cm, since fry became too weak to swim to the surface where most of the fine feed particles remained.

Stocking density, water exchange rates, and feeding intervals have also been thought to influence the feeding of walleye under intensive rearing conditions. No density guidelines can be offered as yet, but a typical density is 200,000 fry per "standard" trough. "Standard" trough sizes vary; however, dimensions of about 3 × 0.4 × 0.15 m (water depth) are reasonably typical. Some workers believe that crowding aids initial feeding; however, no systematically gathered evidence supports this belief. Water exchange rates of less than two exchanges per hour apparently contributed to disease problems in tests by NYCFRU but, again, the evidence is limited. It seems obvious that low water exchange rates might lead to poorer water quality and additional stress on fish. Regular feeding at brief intervals (i.e., 5 min or less), with automatic feeders of various types, apparently has contributed to better acceptance of dry diets by walleye fry. Feeding small quantities of feed at brief intervals seems preferable to feeding larger amounts at longer intervals in terms of both water quality and nutritional benefits. Since the feed particles used with walleye fry must be very small, leaching of water-soluble nutrients can be substantial if the particles are in the water for relatively long periods before they are consumed.

Feeds and Feeding

A variety of diets and environmental conditions have been used when feeding walleye fry under intensive culture conditions. A series of diets (W-1 to W-7) developed by Leo Orme at the Spearfish (South Dakota) Diet Development Center, U.S. Fish and Wildlife Service, have probably been used most widely (Table 1). Oregon Moist Pellets, chicken egg yolk, Atlantic salmon diet (pellets), various trout pellets, liver slurry, egg yolk and Farina slurry, and a variety of live diets (Beyerle 1975) have not been fully successful. Failures with both dry and live diets have tended to take one of two forms. Immediate failures to accept the feeds have been followed by cannibalism, "tail-biting," and early starvation. When the feed has been accepted by some of the fry, a phenomenon labeled the "dwindles" by hatchery workers has been observed repeatedly. Fish 2 to 3 weeks old die in substantial numbers for no discernible reason and the population dwindles to near zero. Malnutrition is the most probable explantion for the "dwindles."

There has been considerable speculation over the importance of the size, texture, color, movement, density, and continual availability of feed particles offered to walleye fry. The size of natural food of walleye fry provides clues to the acceptable size of formulated feed granules. Houde (1967) found that copepods were the primary item in the diet of walleye fry (7–24 mm long) in Oneida Lake. Smith and Moyle (1945) found rotifers to be significant in the diet of walleye fry in rearing ponds, and Hohn (1966) and Paulus (1969) found diatoms in the stomachs of walleye fry in Lake Erie. Nauplii of brine shrimp have been used commonly to feed walleye fry in hatchery and laboratory tests. It may be concluded that walleye fry can ingest particles of at least 0.2 mm in diameter (the typical smallest axis of brine shrimp nauplii), although they will accept smaller items. Tests in our laboratory indicated that walleye fry can ingest particles of W-7 diet that averaged 0.25 mm in diameter.

No differences in acceptance of red, brown, or yellow particles occurred in studies at our laboratory, thus indicating that color was not important, at least with the W-7 diet. Hatchery workers generally have believed that walleye fry accept soft textured feeds more readily than hard textured ones, but no controlled tests have been completed.

Continual availability of feed appears to be important in walleye fry feeding. Acceptance of dry diets has been highest when the fish were fed at 2–5-min intervals with automatic feeders. Kostomarova (1961) suggested that "pike" fry require feeding during the final stages of yolk sac absorption to prevent mass starvation. Although it is improbable that he was working with walleye fry, it may be useful to note that a continual supply of required nutrients was extremely critical when the fish were very young. Cuff (1977) concluded that greater food availability significantly reduced the amount of cannibalism in walleye less than 20 days old, primarily because feeding fish more successfully avoided cannibalistic attacks. Walleye fry may respond favorably to systems similar to those mentioned by Lovell (1977), who achieved up to 95% survival of striped bass fry by suspending feed particles in convection currents. Tim Nagel (London [Ohio] State Hatchery, personal communication) improved acceptance of dry feeds by a system which provided both internal lighting and water currents. These suggestions have not been tested systematically. However, I speculate that survival of walleye fry can be enhanced by a system that maintains feed particles in the water column close to the walleye fry.

The quality of the diet is certainly of critical importance. The W-7 diet (Table 1) is superior to earlier versions (W-1 through W-6), but there is at least circumstantial evidence to indicate that it is not an optimal diet for walleye fry. Improvements may be expected. The NYCFRU and the Tunison Laboratory of Fish Nutrition (U.S. Fish and Wildlife Service, Cortland, New York) are cooperating in the development of a walleye fry diet that will approximate the nutrient composition of walleye eggs. Previous studies of Atlantic salmon indicated that a diet based on the nutrient composition of its egg produced superior growth by this species. Since the egg has nourished the fry up to the point of yolk sac absorption, it seems only logical to take this approach. Initial testing of an egg-derived diet for walleye fry will be conducted in 1978.

Physical Factors: Light and Temperature

The importance of various environmental factors has been stressed in a number of unpublished reports although there have been few controlled, replicated experiments. Tests by NYCFRU have demonstrated that walleye fry respond very positively to light and to light-colored backgrounds. Over 95% of walleye fry chose the white halves of troughs that were divided equally into black and white sections. Walleye fry in lighted white or pale-colored troughs were so attracted to the sides of their troughs that they ignored all forms of live and formulated feeds. Fry in dark troughs (black, gray) were distributed uniformly throughout the troughs and fed readily on live food. Fry in non-lighted troughs fed poorly, if at all. Our observations, under laboratory conditions, support the statement of Houde (1969) that walleye larvae behave photopositively during the first few weeks following hatching. Our trials suggest that walleye fry are most likely to accept feed when held in containers that are uniformly lighted. An internal lighting system, in which a light is suspended inside a plastic cylinder but below the water surface, has also been credited by Tim Nagel (personal communication) with improving the acceptance of dry feeds by walleye fry. The intensity of the light seems to be less important than the uniformity, so long as the level of light is sufficient for the fry to see the individual food items. Photoperiods at least equal to those experienced by walleye fry in nature and perhaps considerably longer, but short of continuous light, appear to be warranted.

No controlled, replicated studies have tested the effects of temperature on feeding of walleye fry. In natural environments they may by assumed to begin to feed at temperatures between 10 C and 15 C, if a normal warming trend follows spawning. It seems reasonable to suggest similar temperatures for cultured walleye fry. Temperatures of about 20 C were used in 1977 tests by NYCFRU with generally poor results. The higher temperatures and resultant higher metabolism apparently caused the fry to exhaust all nutrient reserves before they were able to adapt to dry feeds, thus causing massive starvation and cannibalism. Although temperatures of about 20 C have been desirable for rearing walleye fingerlings, a lower initial temperature for fry seems desirable.

Pathology

Pathological problems in walleye fry culture have not been investigated in detail. "Myxobacterial" and "fungal" infections have been noted, but not confirmed, at several hatcheries. A fungal infection involving the oral cavity and the gills was prevalent in one set of 1977 trials by NYCFRU. A similar problem was noted at Linesville, Pennsylvania. It is not known whether the infection was primary or secondary. The flush rate in all troughs was doubled in later trials in our laboratory and no further problems were noted. Preliminary tests of the toxicity of standard therapeutic agents conducted in our laboratory were inconclusive; however, walleye fry seemed to be unusually sensitive to most "standard" chemicals. If this observation is sustained by further investigation, the necessity for maintaining high standards of water quality will be further emphasized, since even the agents on the limited list of approved chemicals may be toxic to walleye fry at levels necessary to achieve control.

Suggested Methods for Walleye Fry Culture

It is extremely speculative to suggest a culture method for walleye fry at this time; however, standard procedures may be needed for comparisons in the future and the following suggestions are offered.

(1) Any hatchery trough or tank which can be fitted with appropriate feeders and lights should be an acceptable rearing unit. Troughs should be of a neutral (i.e. neither light not dark) color and uniformly lighted.

(2) At least two exchanges of 10–15 C water per hour should be provided. Once feeding is established the temperature may be raised to 20 C, the temperature which appears optimal for fingerlings.

(3) Feed should be given at regular, brief intervals by an automatic feeder. Feed particles should be maintained in suspension in the water.

Walleye Fingerling Culture

Walleye fingerlings have been reared on formulated diets for at least 10 years at several hatcheries (McCauley 1970; Cheshire and Steele 1972). Feeds which have been accepted by walleye fingerlings have included Oregon Moist Pellets (Nagel 1974, 1976), trout granules (Cheshire and Steele

1972), W-3 (Huh et al. 1976), W-4, W-5, and W-6 (Beyerle 1975), and W-7 (unpublished data, ours and others). Results have varied also, but survival rates of 60% (based on numbers of fingerlings) and growth to lengths of 100–125 mm during a 10–15 week feeding period have been obtained (Cheshire and Steele 1972; Nagel 1976; our unpublished data).

Facilities and General Conditions

Walleye fingerlings have been reared in aquaria, troughs, tanks, and raceways of various materials, designs, and capacities. No particular unit has been shown to be superior to the others.

Water exchange rates of less than two complete exchanges per hour have proven unsatisfactory in our laboratory; however, Huh et al. (1976) provided only one exchange per 7.3 hours with no discernible adverse effects. Apparently the flow required varies with water quality and the specific culture system. Cheshire and Steele (1972) reported that survival of walleye fingerlings was best at stocking densities of 12 kg/m^3; however, they employed stocking rates as high as 24.2 kg/m^3. T. Nagel (personal communication) found no adverse effects with stocking densities up to 22.4 kg/m^3. Temperatures, flow rates, and feeding rates influence the optimal stocking density. At this time it is reasonable to suggest that densities of at least 24.2 kg/m^3 can be used; however, further research is needed to determine desirable stocking rates for each system. Studies by NYCFRU indicate that no particular density of fish is needed; however, other workers have believed that "crowding" is useful when the fingerlings are first offered dry feeds.

Pond-reared walleye fingerlings have been converted to intensive culture at total lengths of 35 to 75 mm. Best results have usually been obtained with fish about 50 mm long. Fingerlings normally go through a period of starvation (up to 2 weeks) when offered dry feeds, even when the conversion is gradual. Larger fish (<50 mm) seem better able than smaller fish to withstand the stress of this starvation period; however, if the fingerlings have begun to feed on fish, the conversion becomes difficult. The size at which walleye fingerlings become piscivorous varies with food supply, but can be expected to occur by a length of 60 mm, even in productive water (Walker and Applegate 1976).

Feeds and Feeding

A conversion period during which fingerlings were first fed brine shrimp nauplii and mixed zooplankton and then gradually offered increasing ratios of dry feed to live feed over 5- to 10-day periods generally produced better survival than direct conversion, in tests by NYCFRU. However, other workers have achieved similar results with direct conversion when the fingerlings were in good condition and thus able to endure an initial period of starvation. Even though active feeding may not begin for several days, it seems to be necessary to offer the fish feed during the conversion period, rather than starving them and offering feed after they are sufficiently hungry. It has been suggested that saugers or walleye-sauger hybrids may be more adaptable to dry feeds and might be useful in training walleye fingerlings to accept dry diets; however, this suggestion has not been tested. Automatic feeders appear to be nearly essential to the intensive culture of walleye fingerlings, since adequate hand feeding is not practically possible; they should supply small quantities of feed either continuously or at brief intervals (<5 minutes).

Few systematic comparisons of survival and growth of fingerlings on various diets have been completed. The W-7 diet is adequate to sustain growth similar to that observed for non-cannibalistic pond-reared fish (i.e., fingerling length of 100–125 mm by mid-September in New York). If the growth obtained by cannibalistic fingerlings (up to 250 mm) is considered to be a practical maximum, it is obvious that the W-7 diet or the rearing systems now in use, or both, can be improved considerably. No difference was detectable between W-7 and shrimp-flavored W-7 in 1977 feeding tests in New York. Growth, survival, and conversion rates obtained with W-7 appear to be at least comparable to results obtained with other diets in independent tests conducted elsewhere.

Walleye fingerlings have been fed at rates

varying from 3% of body weight (Huh et al. 1976) to 50% (Cheshire and Steele 1972), to "feeding to excess"; however, no systematic comparisons of various rates have been made. Huh et al. (1976) obtained a gross conversion rate of 1.62 for walleye fingerlings reared at 22 C on W-3 diet. Conversion rates comparable to those obtained with channel catfish or rainbow trout (e.g. 1.2–1.5) probably can be expected as diets and rearing systems are improved. Efficiency may also be increased as the appropriate daily rations and sizes of pellets for various sizes of walleyes are systematically determined.

Variations in color and texture, as well as flavor enhancers have been used to increase the acceptance of dry feeds by walleye fingerlings; however, most such tests have lacked adequate replication and controls. It is the opinion of most workers that softer-textured feeds are the more acceptable. No consistent effects have yet been attributed to flavor enhancers or to various colors.

Physical Factors: Light and Temperature

Walleye fingerlings were not attracted to light in the same fashion as walleye fry in rearing units operated by NYCFRU. Fingerlings normally sought out shaded areas of troughs, but entered lighted areas when feed was presented. T. Nagel (personal communication) found that internal lighting of troughs seemed to enhance feeding and also made walleye fingerlings less excitable. Huh et al. (1976) found no significant differences in growth of walleyes subjected to 16-h and 8-h photoperiods. A dim light of an intensity approximating dusk and dawn has been assumed to be optimal for walleye fingerlings, but definite proof is lacking. In the absence of better information, "dim" light and a photoperiod of 12 to 16 hours for rearing walleye fingerlings seems warranted. Some workers, however, prefer dark troughs, lighted only at feeding stations.

Huh et al. (1976) obtained more rapid growth and better feed conversion by walleye fingerlings at 22 C than at 16 C. Unpublished results from various hatcheries support the conclusion that walleye growth increases as temperatures increase to 22 C and perhaps as high as 27 C (Cheshire and Steele 1972). Although systematic studies have not been conducted, it is generally accepted that temperatures higher than 20 C lead to greater incidence of disease, particularly bacterial infections. Careful sanitation should reduce such problems, since Huh et al. (1976) and Cheshire and Steele (1972) apparently did not encounter substantial disease problems. Results at various hatcheries and laboratories have indicated clearly, however, that 20 C is adequate for "acceptable" growth.

Pathology

Myxobacterial infections, particularly *Flexibacter columnaris*, have been the primary pathological problems in walleye fingerling culture. *Ichthyophthirius multifilis*, *Trichodina* sp., *Scyphidia* sp., bacterial gill disease, furunculosis, fin rot, and fungal infections of walleyes have also been reported (Hnath 1975). At least 68 additional parasites and diseases of walleyes in natural populations have been reported (E. M. Davis 1973, Cornell University, unpublished manuscript); however, the majority do not appear to be serious threats in intensive culture situations.

Various treatments have been employed to combat pathological problems in walleye culture. Nagel (1976) controlled outbreaks of *Flexibacter columnaris* with 10-s dip treatments in 500 mg/liter copper sulfate and 1-min baths of copper sulfate at 30 mg/liter. Hyamine 3500 at 2 mg/liter for 45 minutes and Diquat at 16 mg/liter have also been reported as successful against *F. columnaris* infections (Hnath 1975). Malachite green, formalin, Acriflavine, potassium permanganate, and Furanace have been used with variable success in various unpublished studies. Hnath (1975) reported that Roccal at 2 mg/liter or Hyamine 3500 at 2 mg/liter for 1 hour effectively controlled bacterial gill disease. He also reported that Acriflavine at 5 mg/liter for 1 hour or Hyamine 3500 at 2 mg/liter for 1 hour controlled fin rot, that Terramycin at 4% in feed controlled the symptoms of furunculosis, but that formalin treatments for *Ichthyophthirius* were not successful. Nagel (1974) controlled bacterial gill diseases with Roccal at 2 mg/liter and fungus disease with formalin at 1:6,000.

Disease problems with fingerling walleyes have apparently been reduced by minimizing the handling of the fish when water temperatures exceed 20 C, by strict sanitation, and by maintaining high water quality and flush rates. Dietary insufficiencies probably have contributed to the disease problems encountered to date; however, systematic studies of dietary effects and of pathological problems in walleye culture probably will not be conducted until intensive culture is developed on a larger scale.

Suggested Methods for Walleye Fingerling Culture

The intensive methods summarized below undoubtedly will be improved as practical experience and research advance; however, they can produce acceptable survival and growth of walleye fingerlings and may serve as both an interim production technique and as a base from which to develop better methods.

(1) Harvest pond reared walleyes at 50–55 mm length or when zooplankton populations sharply decline.

(2) Do not use light colored troughs, tanks, or raceways. Maintain dim lighting for no more than 16 hours per day, stock less than 24.2 kg/m^3, and have at least two water exchanges/hour of 20–22 C water.

(3) Feed W-7 or similar feeds with automatic feeders, at least during daylight hours. Feed a quantity of at least 3–6% of body weight per day at regular, brief intervals. A training period to convert the fish from live to dry feeds is desirable, but not essential.

Post-Fingerling Culture

Studies of intensive culture of walleyes generally have been terminated when the fish reached a length of 100–125 mm. It is obvious that commercial farming of walleyes for human food will require rearing them to much larger sizes. It also seems desirable to rear larger fingerlings for management, and to develop domesticated brood stocks which can be held under intensive rearing conditions.

T. Nagel (unpublished manuscript) has reared walleye fingerlings intensively, overwintered them in ponds, and returned them to raceway culture the following spring. The fish quickly readapted to intensive culture conditions and to dry diets when returned to the raceway. One lot of fish has been reared through three growing seasons and has learned to feed on dry feeds in the pond environment as well. Growth rates have been slower than might be expected for "wild" walleyes, however.

Practical culture of post-fingerling walleyes will require facilities for year-round growth. Maintenance of adequate temperatures apparently is the most immediate requirement for post-fingerling culture; practical production diets and effective disease control techniques are other major needs. The cost of heating large volumes of water to maintain temperatures of 20–22 C seems to require either the development of recirculating systems or the location of walleye rearing facilities near sources of large volumes of warm water (e.g., steam electric stations). There seems to be no fundamental biological or technical problem with either approach; however, the economics of such development have not been explored systematically.

Summary and Prognosis

As recently as 10 years ago, the nearly unanimous opinion of fishery managers and fish culturists concerning intensive walleye culture was "It can't be done." Progress with fingerling and post-fingerling culture indicates that "It can be done." Although only a few thousand fry have been reared intensively to a length of 100 mm, it has been clearly established that some walleye fry will accept dry diets under intensive rearing conditions. New diets and rearing systems probably will bring fry survival to acceptable levels under intensive rearing conditions.

Development of acceptable fry-rearing

systems seems to be the most immediate problem in the development of intensive walleye culture. Although a survival rate of at least 60% can be expected for intensive rearing of fingerlings, the pond-rearing phase of an extensive-intensive system is too unpredictable for large-scale development. Acceptable techniques for post-fingerling rearing appear to require only slight modification of fingerling techniques and an economical source of 20 C water to become established.

Although walleye fingerlings have been reared under intensive culture conditions, many basic questions remain to be answered and there are a number of major immediate needs: improved diet formulations; determination of rates of oxygen consumption and metabolite production under various conditions; establishment of tolerance levels for various metabolites; determination of maximum densities; and establishment of effective prophylaxis and disease control techniques.

The prognosis for walleye culture is, in spite of present obstacles and serious gaps in knowledge, very favorable. The commercial market for the fish already exists. It is a leading sport fish throughout the Great Lakes region and has attracted considerable angler interest when introduced elsewhere. Research is needed to determine the most effective use of large walleye fingerlings (e.g., 200 mm long and longer); it is widely felt that such fish could be important in controlling some populations of forage fishes in selected waters. Given the current state of the art, the management need, and the potential commercial market, I believe that intensive walleye culture will reach maturity during the next 10 years.

References

BEYERLE, G. B. 1975. Summary of attempts to raise walleye fry and fingerlings on artificial diets, with suggestions on needed research and procedures to be used in future tests. Prog. Fish-Cult. 37(2):103–105.

CHESHIRE, W. F., AND K. L. STEELE. 1972. Hatchery rearing of walleyes using artificial food. Prog. Fish-Cult. 34(2):96–99.

CUFF, W. R. 1977. Initiation and control of cannibalism in larval walleyes. Prog. Fish-Cult. 39(1):29–32.

GRAVES, G. 1974. They said it couldn't be done. Farm Pond Harvest 8(1):6–8, 17, 22.

HNATH, J. G. 1975. A summary of fish diseases and treatments administered in a cool water diet testing program. Prog. Fish-Cult. 37(2):106–107.

HOHN, M. H. 1966. Analysis of plankton ingested by *Stizostedion vitreum vitreum* (Mitchill) fry and concurrent vertical plankton tows from southwestern Lake Erie, May 1961 and May 1962. Ohio J. Sci. 66(2):193–197.

HOUDE, E. D. 1967. Food of pelagic young of the walleye, *Stizostedion vitreum vitreum*, in Oneida Lake, New York. Trans. Am. Fish. Soc. 96(1):17–24.

———. 1969. Sustained swimming ability of larvae of walleye, *Stizostedion vitreum vitreum*, and yellow perch, *Perca flavescens*. J. Fish. Res. Board Can. 26:1647–1659.

HUH, H. T., H. E. CALBERT, AND D. A. STUIBER. 1976. Effects of temperature and light on growth of yellow perch and walleye using formulated feed. Trans. Am. Fish. Soc. 105(2):254–258.

KOSTOMAROVA, A. A. 1961. Significance of the phase of mixed feeding for the survival of pike larvae. Pages 344–347 *in* Proceedings of the Conference on the Population Dynamics of Fishes. Ministry of Agriculture, Fisheries and Food. (English Translation prepared on behalf of Fisheries Laboratory, Lowestoft, Suffolk, England.)

LOVELL, R. T. 1977. Physical aspects of food important in feeding fish. Commer. Fish Farmer Aquacult. News 3(6):32.

MACAULEY, R. W. 1970. Automatic food pellet dispenser for walleyes. Prog. Fish-Cult. 32(1):42.

NAGEL, T. O. 1974. Rearing of walleye fingerlings in an intensive culture using Oregon Moist Pellets as an artificial diet. Prog. Fish-Cult. 36(1):59.

———. 1976. Intensive culture of fingerling walleyes on formulated feeds. Prog. Fish-Cult. 38(2):90–91.

NATIONAL TASK FORCE FOR PUBLIC FISH HATCHERY POLICY. 1974. Report of the National Task Force for Public Fish Hatchery Policy. U.S. Fish Wildl. Serv., Washington, D.C. 295 pp.

PAULUS, R. D. 1969. Walleye fry food habits in Lake Erie. Ohio Fish Monogr. 2, Ohio Dep. Nat. Resour. 45 pp.

SMITH, L. L. JR., AND J. B. MOYLE. 1945. Factors influencing production of yellow pike-perch, *Stizostedion vitreum vitreum*, in Minnesota rearing ponds. Trans. Am. Fish. Soc. 73:243–261.

WALKER, R. E., AND R. L. APPLEGATE. 1976. Growth, food and possible ecological effects of young-of-the-year walleyes in a South Dakota prairie pothole. Prog. Fish-Cult. 38(4):217–220.

Intensive Culture of Esocids: The Current State of the Art

Delano R. Graff

Division of Fisheries
Pennsylvania Fish Commmission
Bellefonte, Pennsylvania 16823

Abstract

The intensive culture of esocids has been profoundly influenced by the use of artificial diets. Adequate diets have been developed and automatic mechanical feeders have come into wide use. Most culture programs are emphasizing the production of tiger muskellunge (northern pike × muskellunge). Hatchery design has changed to reflect increased knowledge of requirements of esocid fishes and to accomodate techniques developed to feed artificial diets. Trends in Michigan and Pennsylvania are to incorporate esocid culture and salmonid culture into an integrated or combination hatchery design. Fiberglass tanks, concrete raceways, and silos are being used to rear esocids. Techniques and hatchery design have changed dramatically over the past decade; fisheries pathology as related to intensive culture of esocids has not kept pace with this change and basic information is still being developed. Achievements in the art of intensive culture of esocids should serve as an inspiration to culturists working with other difficult-to-rear fishes.

Northern pike and muskellunge are highly regarded sports fish and are often utilized by fishery managers to provide recreational fishing and as a management tool for control of stunted or undesirable fishes. Muskellunge tend to be more highly regarded by both anglers and managers; muskellunge are the big game trophy of freshwater fishing. In general, the natural range of muskellunge is restricted and population densities are relatively low. To satisfy the needs of anglers and fishery managers, a number of conservation agencies have developed esocid culture programs.

The hatchery propagation of esocid fishes, particularly muskellunge, can be traced back to 1890 in New York and 1899 in Wisconsin (Buss 1961). Programs in other states followed these pioneering efforts. Early efforts to culture esocids concentrated on muskellunge because anglers preferred muskellunge and there seemed a greater need for maintenance or increase of muskellunge populations through stocking. Also, due to the similarity of the species, most techniques developed for muskellunge culture were applicable for culture of northern pike.

Rather successful techniques were developed for capturing wild brood stock, spawning, incubating eggs, and stocking fry into nursery ponds where optimum conditions for survival had been established. When fry, feeding on zooplankton in the nursery ponds, achieved 3.8 to 4.5 cm in length, they were transferred to growing ponds where minnows were abundant. This was extensive culture, or more simply, "pond culture." There were variations in procedure, but the principle was the same: put the esocids in a quasi-natural environment and be sure food was always available. The best description of extensive culture—in fact, the standard reference on extensive culture—was published in Wisconsin (Johnson 1958).

Pond culture provided the beginnings of large-scale artificial propagation of esocids. There were, however, some major problems inherent in pond culture. As the name implies, pond culture requires a significant investment in ponds, not only to rear esocids but also to provide forage. The somewhat uncontrolled environment involved in rearing esocids in ponds produces an element of unpredictability.

The uncertainty and lack of control involved in extensive culture were unacceptable to fish culturists who had experience in salmonid propagation. A trout hatchery symbolized what fish culture should be—mechanized, efficient, and reduced to a science. Trout, provided an adequate water supply, thrive under intensive culture methods. Trout feed readily on pelleted food, are tolerant of crowding, and are generally adaptable to what literally constitutes mass production. The challenge to the esocid culturist was to achieve similar success. The response to that challenge has given rise to the art of intensive culture of esocids.

The foundations for intensive culture of esocids were laid in 1966 when Sorenson, Buss, and Bradford published *The Artificial Propagation of Esocid Fishes in Pennsylvania*. The art was profoundly influenced by the discovery in 1968 that esocids could be reared on artificial or "dry" diets (Graff and Sorenson 1970). Since those not-so-long-ago days of the late 1960's, a revolution has taken place in the culture of esocids. The art has gone from dirt ponds to concrete raceways, from daphnia and minnows to crumbles and pellets.

In the last decade, the state of the art of intensive culture of esocids has been dynamic and exciting. Assessing the state of that art is challenging because intensive culture of esocids is changing rapidly. Much of what has been accomplished has not been published formally; some has been presented in mimeographed job reports, some has been offered at task force or steering committee meetings, and some has only been reported verbally on that jungle telegraph that seems to exist among the coterie of esocid enthusiasts who are actively involved or interested in intensive culture. I can make no claim that this assessment of the state of the art is comprehensive. In this rapidly evolving field of fisheries science, the risk of overlooking an innovation in culture techniques or facility design is high.

Spawning and Hatching

Historically, spawning of esocids and incubation of eggs was done by people who had learned it through firsthand experience. Techniques were handed down from one fish culturist to another. Some of the accepted methods placed as much emphasis on strength as on skill. One can imagine that little precision or scientific method was possible while wrestling with a large, vigorous muskellunge in an effort to strip eggs. The simple expedient of utilizing an anesthetic or immobilizing agent made a major contribution to an orderly spawning procedure. The relaxed fish were easier to spawn, injury to brood fish was minimized, and quality of eggs was improved (Sorenson et al. 1966). The days when a big female muskellunge could give a mighty flop and send an entire pan of eggs flying across the room were gone forever.

Spawning techniques involving a blood pressure cuff as used by physicians have been developed. Anesthetized females are placed on an inclined rack, and the blood pressure apparatus is inflated to gently apply pressure and expel eggs. This technique has been generally unchanged from that first described by Sorenson et al. (1966).

Despite their relatively large size, male esocids produce small quantities of milt. Sorenson and his associates developed an innovative approach for collecting milt. Briefly, the technique involves the insertion of a catheter into the urogenital opening to drain off urine. This permits manual expression of urine-free milt which can be collected by vacuum. Other methods of obtaining sufficient sperm include sacrifice of the male, removal of the testes, and mincing of the testes into eggs or the squeezing of testes in cheesecloth over the eggs. The technique originally described in 1966 remains the choice of preference (Sorenson et al. 1966).

Lack of sufficient sperm, particularly in the production of interspecific hybrids, remains somewhat a problem. There appears to be a legitimate need for the preservation of esocid sperm. A limited amount of work has been done with storage of esocid sperm utilizing 5% dimethyl sulphoxide and 5% ethylene glycol as preservatives. Ethylene glycol appears to show promise (Gustafson, unpublished). Should successful sperm preservation techniques be developed, they could have practical application in assuring availability of sperm for production of tiger muskellunge (northern pike male × muskellunge female) or for producing crosses of geographically isolated strains.

Virtually all esocid culture programs utilize wild brood stock. Some states capture brood stock from lakes open to public fishing; others maintain brood stock lakes or "sanctuaries" where angling is prohibited. In both instances, the brood fish are returned to the wild after spawning. Some efforts have been made by the United States Fish and Wildlife Service, Valley City National Fish Hatchery, and at least one state conservation agency (Ohio) to maintain a hatchery or "domestic" population of muskellunge as brood stock. The results have been favorable but not totally successful. Injection of carp pituitary has been attempted to increase success but with mixed results. Most agencies engaged in intensive culture of esocids apparently do not view establishment of domestic brood stock as necessary, or perhaps even desirable.

Esocid eggs are incubated in jars. Fungicides are routinely used; formalin and malachite green are both used. With the exception of fungicides, little has changed in the mechanics of jar incubation since the early days of esocid culture. A great deal has changed in terms of water supplies utilized for incubation of eggs. Not so terribly long ago a river or lake was considered an adequate water source for incubation of esocid eggs. The contribution clean water of the correct temperature makes to a high percentage hatch has been fully recognized. In today's esocid hatchery, one finds well water or spring water sources, controlled temperatures achieved via heat exchange units and blending valves, and, if needed, the best available filters and occasionally even ultraviolet light systems. Disinfection of green eggs of muskellunge and tiger muskellunge can be satisfactorily accomplished with Betadine (Hnath and Copeland, unpublished), thus reducing the risk of disease transmission by transfers of eggs from one hatchery to another.

Feed and Feeding

There can be no appreciation of the state of the art of intensive culture without some knowledge of the evolution of satisfactory diets. In the early stages of intensive culture, the main concern was how to optimize production of appropriate size natural food, i.e., better ways to rear brine shrimp and daphnia, to get better and more reliable production from minnow ponds, to control the spawning time of carp and goldfish, to utilize sucker fry and yellow perch fry to start muskellunge, etc. And so it went. When it was announced that esocids, particularly northern pike and tiger muskellunge, could be reared on artificial diets without elaborate schemes to produce adequate forage, the art of intensive culture of esocids was irrevocably changed.

Artificial diets developed for other fishes (catfish and trout) were not satisfactory for esocids. Early efforts at intensive culture of esocids utilizing dry diets were plagued with problems, some due to nutritional deficiencies and poor acceptance of the diet, others due to lack of knowledge about how to feed. After a few years of independent, uncoordinated efforts, it became apparent to some of the most actively involved participants in the Interstate Muskellunge Workshop that intensive culture of esocids utilizing artificial diets was clearly feasible, but techniques had to be refined and diets adequate for esocids had to be developed.

In 1970 a concerted and cooperative effort of agencies interested in coolwater fish culture was directed toward development of an effective diet and better techniques for intensive culture. Initially, this group consisted of personnel from Michigan, New York, Pennsylvania, and the United States Fish and Wildlife Service. Later other states, including Missouri, Ohio, and Wisconsin, took an active role. Interest and participation grew and the Coolwater Diet Testing Steering Committee was formed. Valuable assistance in the form of expertise in diet formulation and program coordination was provided by experts (most notably Leo E. Orme) from the United States Fish and Wildlife Service's Diet Testing Development Center at Spearfish, South Dakota. Participating states provided the fish, facilities, and testing for evaluation of diets and development of rearing techniques.

The results of the coolwater diet testing program have been gratifying. An adequate diet has been developed. This diet, known only by its formula code, W-7 (Orme 1978, this volume), has provided the esocid culturists with a reliable diet which has a satisfactory level of acceptance by esocids. Other diets from commercial sources, particularly Oregon Moist Pellet and a stabilized form of moist pellet which does not require refrigeration, show promise, at least in terms of being "palatable" and readily accepted by esocids.

Rearing techniques had to be modified to accommodate the feeding habits of esocids. The old maxim of the live forage days, "keep feed in front of them at all times," was difficult with dry diet. Esocids would not pick pellets from the bottom of rearing units, nor would they survive if not fed very frequently. It became evident that the labor involved in having hatchery personnel dribble feed into tanks on an almost constant basis defeated much of the labor-saving aspects of intensive culture. Automatic mechanical feeders are essential for good results in feeding artificial diets to esocids. The feeders are generally set to feed small amounts at very frequent intervals during daylight hours.

Several designs of automatic feeders are being used successfully, including hanging basins with adjustable openings and timed traveling brush, trough-type feeders, and pneumatic feeders which can feed an extensive (linear) area almost simultaneously. Trough-type feeders, incorporating a heat tape to minimize caking, are the most satisfactory for moist diets. For dry diets no single design seems superior for all applications. Feed conversions are quite comparable to those achieved in salmonid culture (Sanderson and Bean, unpublished; Humphrey and Houghton, unpublished).

The major problem attendant to the use of automatic feeders set to feed on a very frequent basis is waste of feed and resultant accumulation of food waste in rearing units. The early solution to this problem was to drain and brush on a daily basis. There is no doubt this contributed to mortality of small esocids. More refined cleaning techniques involve the use of siphoning devices in small units and swimming pool vacuums in larger units. Workers in Michigan (Harry Westers, personal communication) have devised a structural solution to the problem of waste accumulation. By using baffles which cause flow to be accelerated and directed along the bottom of the rearing unit, they have tanks and raceways that are, to a very great extent, self cleaning.

Species Reared

Early in the evaluation of the feasibility of intensive culture of coolwater fish utilizing artificial diets the tiger muskellunge was recognized as having the greatest potential. Northern pike accept dry diet readily and generally adapt well to intensive culture. However, northern pike do not have the same attraction to fishery managers and anglers as muskellunge. Unfortunately, muskellunge have proven difficult to rear on artificial diets. The hybrid tiger muskellunge combines the best features of both species. Tiger muskellunge have the general appearance of muskellunge, are prized by anglers, are regarded by fishery managers as equivalent to muskellunge for value as a predator and, most important, accept dry diet and adjust to intensive culture very well. If ever a fish were "made to order" for successful intensive culture of esocids, it is the tiger muskellunge. Despite increasing success in the intensive culture of purebred muskellunge, the tiger muskellunge is overwhelmingly the predominant esocid being reared on artificial diet, and it appears that will be the pattern for intensive culture of esocids for some time to come. There seems to be limited interest, at this time, in intensive culture of chain pickerel and virtually no interest in working with the redfin or grass pickerels. Pennsylvania has successfully reared Amur pike (*Esox reicherti*) on artificial diet, but this species has no major management role.

Hatchery Design

The design of esocid hatcheries has been greatly influenced by the shift in emphasis to intensive culture with artificial diets. Only a short time ago the ideal esocid hatchery had a lake water supply, daphnia units, forage ponds, nursery ponds, and rearing ponds. If the hatchery was designed for intensive culture, the hatch building would include steel tanks and perhaps, if it were a "modern" facility, some larger concrete tanks. A few of the more advanced hatcheries included a temperature control system, primarily for use during incubation of eggs. The esocid hatchery of a decade ago was considered a "warmwater" design. The influence of changing techniques based on utilization of artificial diets in intensive culture of esocids has been to make esocid hatcheries very similar in appearance to salmonid hatcheries.

The trend in Michigan (Jack Hammond and Harry Westers, personal communication) and Pennsylvania is to incorporate esocid culture and salmonid culture into an integrated or combination hatchery designed to rear both coolwater and coldwater fishes. Development of dual capacity for coolwater and coldwater fish production at the same hatchery seems to offer advantages. The use of warming ponds (or warmwater sources) and coldwater sources through a blending system provides temperature control flexibility not normally available at warmwater hatcheries. The state of the art of intensive culture of esocids utilizing artificial diets and automatic feeders is similar enough in theory and principle to salmonid culture to permit fish culturist personnel to transfer expertise in one field to the other with a minimum of training in transition. The feeding of esocids has become mechanized and, consequently, much less labor intensive than in former years. This enhances the fish production per man-year that can be realized in an integrated or combination approach to intensive culture.

Hatch houses designed for intensive culture of esocids tend to be well ventilated with either drive-through capacity or one wall consisting primarily of overhead doors so easy access for vehicles is provided to the rearing tanks. Electrical systems are designed to accommodate automatic feeders at each rearing unit. Lighting is adequate to work, but the tendency is to avoid bright or harsh lights and direct overhead lighting in the immediate vicinity of rearing units. Some fish culturists feel it is desirable to provide covers over a significant portion of individual rearing units. This is believed to minimize nervous stress in young esocids and to promote better acceptance of dry diet. Opinions vary as to the need for cover, but the very clear consensus is that bright or direct overhead light is undesirable.

Indoor rearing units include fiberglass tanks and larger concrete tanks. Dimensions vary, but approximately 4.6 dm × 6 dm × 3 m seems average for fiberglass "starting units" and approximately 7.6 dm × 7.6 dm × 6 m is representative of concrete rearing units (Shyrl Hood, personal communication).

The water supply for a modern esocid hatch house will be of good quality, generally a spring or well source. If necessary, filtration or even ultraviolet treatment will be included. A temperature control unit, generally based on heat exchange systems and thermally controlled blending valves, will be included. Temperature control will be built into the entire system, not just the incubation units. Rearing temperatures will be maintained at 20 C to 21 C as optimum (Shyrl Hood, personal communication). This is somewhat lower than had been advocated previously when fish culturists regarded esocids as warmwater fishes and assumed better performance was achieved at higher temperatures.

Intensive culture of esocids can no longer be considered tank culture. Intensive culture extends beyond tanks in the hatch house; it has gone outside to large concrete rearing units and raceways. The dirt ponds are gone. The esocid hatchery of today is concrete, asphalt, and automatic feeders. Individual clean-out and underdrain systems are utilized in Pennsylvania to handle accumulated feed waste, in Michigan a system of baffles is used to channel flow along the bottom and move waste out of rearing units. In both instances, effluent treatment (primarily solids removal) is incorporated into hatchery design. Additional features which have been incorporated in modern designs of esocid hatcheries include aeration towers and other water reconditioning features to permit recycling or reuse of water. There is some preliminary evidence that esocids produce less ammonia as a by-product of metabolism than salmonids (Pecor 1977). This may influence the role of recycling and reuse of water in esocid culture. Experimentation is underway to determine the effect of increased flows and exchange rates. Future designs for outside rearing units may provide for isolation from general public visitation or, as an alternative, covers over portions of units. There is a general feeling among esocid culturists that esocids tend to be stressed and do not feed well when there is much human activity in the immediate vicinity of rearing units.

Vertical rearing units, silos, appear to be well suited for the intensive culture of esocids. Tiger muskellunge have been reared successfully in silos at the Linesville Fish Cultural Station in Pennsylvania. Three separate experiments (1975, 1976, 1977) have confirmed that, at the Linesville Station, the growth rate of tiger muskellunge in silos is exceptional when compared to tiger muskellunge growth in tanks at the same hatchery. There was less labor involved in silo culture and the silos were (at a flow of 378 liters/min through a 20.8 kl silo) self-cleaning with very little accumulation of feed or excrement (Sanderson, unpublished).

Diseases

Intensive culture and use of artificial feeds in warm water provides an optimum environment for the growth and rapid multiplication of most of the microorganisms causing disease in fish (Hnath, unpublished). The disease, problems encountered in the intensive culture and artificial feeding of esocids can be serious; however, many problems can be eliminated or greatly reduced through appropriate prophylactic measures (Hnath, unpublished). Major disease problems in intensive culture of esocids have been apparent dietary deficiencies resulting in physical deformities such as separation at the isthmus (Sanderson, unpublished), columnaris, gill disease, *Aeromonas* sp., and unidentified bacterial infections. Appropriate treatment has been developed for columnaris, and it can be dealt with even though it is still considered a problem. Better diet, particularly W-7, has eliminated most of the problems attributed to dietary deficiencies. Gill disease remains a serious problem. Most instances of serious gill disease occur quite early.

Immediately after feeding begins gills quickly become infected, and are secondarily invaded by fungus; mortalities can be very high. Losses due to *Aeromonas* sp. and unidentified bacteria tend to be a sporadic problem. Some striking results have been achieved through the use of Terramycin incorporated in the diet, water soluble Terramycin in a bath treatment, or Terramycin used in both a feed and bath treatment. In general, it can be said that fisheries pathology as related to intensive culture of esocids is a young and developing science. Much basic information is needed before this aspect of the pathologist's art will be equivalent to that in salmonid culture.

Summary

Twenty years ago it was the fish culturist's dream to rear esocids in the same manner as trout. Looking at the state of the art of the intensive culture of esocids today, it is safe to say that dream is, in substantial portion, realized. Culturists can now start tiger muskellunge, immediately upon initiation of feeding, on an artificial diet and maintain those fish on an artificial diet until they are of appropriate size for use by fishery managers. These magnificent game fish never see natural food until they are stocked. The same is true of northern pike. Significant advances have been made with pure strain muskellunge, and it appears only a matter of time until the correct combination of diet and techniques are developed to allow large-scale intensive culture of muskellunge on artificial diets.

The current state of the art of intensive culture is exciting. There seems no doubt that progress will continue although perhaps not at the same rate as during the past five or six years. Intensive culture of esocids stands as an example to the fisheries technical community of what can be accomplished through coordinated, directed, and persistent effort. Breakthroughs, innovations, determination, and the daring to defy "conventional wisdom" of fish culture have brought us to the current state of the art in the intensive culture of esocids. These remarkable achievements of the past decade should serve to provide inspiration to culturists working with other difficult-to-rear fishes.

Acknowledgments

Much of the material covered in this paper was brought to my attention by Shyrl Hood, Chief, Warmwater Production Section, Pennsylvania Fish Commission. A great deal of advice on hatchery operations was also provided by Jack Hammond and Harry Westers of the Michigan Department of Natural Resources. I am indebted to these gentlemen for their guidance and assistance.

References

Buss, K. W. 1961. The northern pike. Pa. Fish Comm., Benner Spring Fish Res. Stn., Spec. Purpose Rep. 58 pp.

Graff, D. R., and L. Sorenson. 1970. The successful feeding of a dry diet to esocids. Prog. Fish-Cult. 32(1).31 35.

Johnson, L. D. 1958. Pond culture of muskellunge in Wisconsin. Wis. Conserv. Dep. Tech. Bull. 17. 54 pp.

Orme, L. E. 1978. The status of coolwater fish diets. Am. Fish. Soc. Spec. Publ. 11:167–171.

Pecor, C. H. 1977. Preliminary trials with a coolwater fish in a three pass water reuse system. Mich. Dep. Nat. Resour. Tech. Rep. 77. 13 pp.

Sorenson, L., K. Buss, and A. D. Bradford. 1966. The artificial propagation of esocid fishes in Pennsylvania. Prog. Fish-Cult. 28(3):133–141.

Intensive Culture of Tiger Muskellunge in Michigan During 1976 and 1977

CHARLES H. PECOR

Wolf Lake State Fish Hatchery
Department of Natural Resources
Mattawan, Michigan 49071

Abstract

The hybrid tiger muskellunge (male northern pike × female muskellunge) was reared intensively on artificial diets during 1976 and 1977. The total numbers of 15- to 23-cm fingerlings produced during 1976 and 1977 were 88,000 and 109,000 hybrids, respectively. Survival from eggs ranged from 11 to 15% for 1976, and 22% for 1977. The improved survival during 1977 was attributed to better egg quality and modifications of feeding techniques which resulted in a lower incidence of cannibalism. Growth rates averaged between 0.140 and 0.175 cm/day during 1976, and 0.180 cm/day during 1977. Improved feeding techniques again were responsible for the better growth rates. Columnaris and bacterial gill disease caused mortalities, but these mortalities were low in comparison to the total loss due to cannibalism.

The hybrid tiger muskellunge (male northern pike × muskellunge) has long been recognized as an important management tool. This particular hybrid exhibits "hybrid vigor." These fish are hardy, grow very fast to a size equaling that of purebred muskellunge, are very aggressive, and have limited or no reproductive potential. However, the availability of this hybrid to fisheries managers in the past has been limited due to the space, effort, and money required to rear it extensively on natural foods. It has not been until the past 5 to 7 years that hatcheries have been working experimentally to develop cultural techniques to rear the hybrid intensively on artificial diets.

During 1975, after a number of years of developmental work, the first production-level rearing of tiger muskellunge in Michigan was undertaken at Wolf Lake State Fish Hatchery. Results established the feasibility of rearing tiger muskellunge under intensive rearing conditions although only about 10,000, 17-cm hybrids were produced. During 1976, Wolf Lake was requested to produce from eggs 150,000 tiger muskellunge fingerlings 4.3 cm long. Of these fingerlings, 120,000 were to be transferred to Platte River State Fish Hatchery for further rearing, and the remainder were to be used for further developmental work at Wolf Lake. During 1977, Wolf Lake again was requested to produce 150,000 tiger muskellunge fingerlings 4.3 cm long, but all fingerlings were to be transferred to Platte River for further rearing. This paper presents the results on growth and survival of tiger muskellunge for the two years of intensive rearing on a production basis.

1976 Production

A total of 728,700 tiger muskellunge eggs were collected from wild broodstock trapnetted in Bankson and Murphy Lakes between March 30 and April 6, 1976. The eggs were hand-stripped, fertilized, and water-hardened in the field and then transported to the hatchery. Eggs were incubated at 10.5 C in 6-liter jars. The eggs were gently "rolled" during incubation allowing the dead eggs to be siphoned off after eye-up. Fourteen days after fertilization the eggs were force-hatched with warm water, and produced about 284,000 sac-fry. The precent hatch for the six lots of eggs ranged from 17.6 to 80.5 and averaged 39.0. The egg mortality during this period may have been higher than normal because the eggs were used in disinfection trials (Hnath and Copeland 1976). The sac-fry at hatch were placed in starting troughs at densities generally not exceeding 25,000 per trough.

The fiberglass starting troughs were 2.74 × 0.5 × 0.5 m in size with a water depth of 22 cm and a water volume of about 300 liters. Boiler-heated well water was supplied to these troughs at rates between 7.5 and 17 liters/min, which resulted in water exchange rates of 2.0 to over 4 per hour. Water temperature was maintained near 20.5 C.

The troughs and concrete tanks used later were cleaned twice daily by lowering the water level and brushing the feces and sediments out the end of the tank. Baffles were used in the concrete tanks to aid in the cleaning process. A flow of 140 liters/min was needed in conjunction with the baffles to move sediments through the tank. Baffles were not used in the fiberglass troughs until near the end of the rearing period. These baffles proved to be very efficient in moving sediments through the trough and did not adversely affect the fish.

Formalin was used daily as a prophylactic treatment to control gill fungus. Constant flow treatments were administered at a maximum formalin concentration of 1:4,000 for 1 h. Furanace at 1 mg/liter in a 1-h static bath was used experimentally in an attempt to control other disease problems.

Eight days after hatch, the fry had absorbed most of their yolk sacs and were "swimming up." On the ninth day, food was first presented to the fry. Feed was distributed to the troughs (and later to the tanks) with 2.1-m (troughs) on 3.7-m (tanks) Loudon automatic feeders. Both sizes of feeders were set to feed from dawn to dusk (14–17 h) at 5-min intervals. Feed was placed in the feeders once daily except in some cases when the entire daily ration was too large to be placed in the feeder at one time. Overhead lighting was used between 0800 and 1630 h.

The federal W-7 dry diet (Orme 1978, this volume) manufactured by Glenco Mills was the primary food used to rear the tiger muskellunge. However, the federal PR-9 diet was fed to fish in one tank for comparison of growth rates and survival with W-7-fed fish. All sizes of food larger than starter were sifted to remove fine particles prior to being placed in the feeders. Daily feed quantities were calculated by the Hatchery Constant method, where hatchery constant = 3 × food conversion × daily length increase (cm) × 100; and % of body weight to feed = hatchery constant/length of fish (cm).

A hatchery constant of 114 was used initially to calculate daily amounts of food to be fed to the fish. Later in the rearing cycle, a hatchery constant of 101 was used.

Mortalities for the first 23 days amounted to approximately 463,700 eggs and sac-fry or 63.6% of the original number of eggs taken. Survival through the initial training period up to May 12, 18 days after initial feeding, was considered excellent. Most fry were actively feeding within the first 2 to 5 days. The mortality during this 18-day period amounted to 41.5%.

Cannibalism was a problem only during the early rearing stages when the fish were less than 6.5 cm in length. Cannibalism was most evident during the first 10 to 12 days after initial feeding as size differences between feeding and nonfeeding fry became more pronounced. The removal of weak and dying nonfeeding fry greatly reduced the cannibalism. Cannibalistic behavior decreased to almost zero by the time the fish attained a length of 7 cm.

On May 17, 127,000 fry were transferred to the Platte River State Hatchery for further rearing (see below). On June 2, 18,490 of the 23,300 fry remaining in the fiberglass troughs at Wolf Lake were moved to inside concrete tanks.

The concrete rearing tanks were 5.4 × 1 × 1 m in size with a water depth of 56 cm and a water volume of 2,800 liters. These tanks were supplied with well water via a 0.8-hectare solar heating pond. Flows through these tanks were maintained between 113 and 190 liters/min with corresponding water exchange rates ranging between 2.4 and 4.1 per hour. Water temperatures ranged between 16 and 23 C.

Growth rates for the tiger muskellunge were very good up to June 18, when the fish were 11 cm long. The fry grew 0.20 cm/day between May 12 and June 18. However, after June 18 the growth rates of fingerlings reared in both the troughs and tanks gradually decreased (Table 1). Growth rates during the final inventory period in both rearing systems were less than 0.10

TABLE 1.—A summary of the growth and survival of intensively reared tiger muskellunge at Wolf Lake and Platte River during 1976.

Date	Number	Mortality	Number of fish planted	Total weight kg	Weight per fish g	Food conversion Period[a]	Food conversion Cumulative[b]	Length cm	Daily length increase cm	Density index kg/m³	Flow index kg/liter/min
				Wolf Lake—fiberglass troughs							
4/02	728,700[c]										
4/06	284,100[d]	444,600									
4/24	265,000[e]	19,100		2.65	0.016			1.27			
5/12	155,000	110,000		34.4	0.223	1.63	1.63	3.55	0.12	9.56	0.252
5/25	150,500	4,500	127,000[f]	81.1	1.32	1.56	1.59	6.42	0.22	14.9	0.338
6/02	23,300	300	18,490[g]	61.0	2.62	1.65	1.61	8.05	0.20	16.9	0.336
6/15	4,760	50		28.2	5.92	2.08	1.66	10.6	0.20	23.5	0.466
7/02	4,750	9		58.3	12.3	1.75	1.68	13.5	0.17	32.4	0.642
7/19	3,890	862		80.3	20.7	3.85	1.95	16.0	0.15	38.2	0.690
7/26	3,890	3		92.9	23.9	2.67	1.99	16.8	0.11	38.7	0.691
8/12	3,870	16	668	116	36.3	2.29	2.04	19.4	0.15	48.4	0.867
8/31	3,180	15	3,180	152	47.7	2.85	2.15	21.2	0.09	63.2	1.13
				Wolf Lake—concrete tanks							
6/02	18,490[g]		126	47.3	2.56		1.61	8.00		16.8	0.260
6/18	18,000	350		135	7.49	1.64	1.62	11.4	0.20	16.3	0.285
7/09	17,850	150		278	15.6	1.85	1.72	14.6	0.15	33.4	0.588
7/27	17,800	54	5,525	287	23.4	2.82	1.97	16.7	0.12	25.9	0.505
8/03	12,300	23		324	26.4	3.38	2.09	17.4	0.10	29.3	0.571
8/26	12,200	100	3,350	435	35.9	4.37	2.52	19.3	0.08	38.1	0.836
10/04	8,830	8	8,830	372	52.7	3.91	2.64	21.9	0.07	39.0	0.840
				Platte River—inside tanks							
5/17	126,900[f]			49.9	0.393		1.60	4.3	0.21	9.1	0.175
6/17	98,900	28,000	98,900[h]	565	5.72	1.98	1.95	10.5	0.21	20.5	0.396
				Platte River—outside tanks							
6/17	98,900[h]			565	5.72			10.5		5.7	0.166
7/15	96,900	1,961	66,700	1,913	19.7	2.55	2.34	15.7	0.19	19.1	0.341
8/10	66,700	29,340		2,735	41.0	4.46	3.10	20.2	0.18	27.3	0.481

[a] Weight of food fed during inventory period/weight gain of fish during inventory period.
[b] Weight of food fed since initial feeding/weight gain of fish from initial feeding.
[c] Eggs, date fertilized.
[d] Sac-fry, date hatched.
[e] Swim-up fry, date initially fed.
[f] Fish transferred to Platte River Hatchery.
[g] Fish transferred from fiberglass troughs to concrete tanks.
[h] Fish transferred to outside raceways.

cm/day. Food conversions (weight of food fed/weight gain of fish) followed a similar pattern (Table 1). The simultaneous decrease in growth rates and conversions in both rearing systems on approximately the same date indicate that underfeeding was not the cause of the slower growth rates. The slower growth rates also coincided with higher density indices (total fish weight in kg/total rearing volume in m³) and flow indices (total fish weight in kg/total flow in liters per min). In the fiberglass troughs, density and flow indices were greater than 32 and 0.60, respectively, after growth rates started decreasing. These indices in the concrete tanks were more variable due to the distributing of fish into additional tanks. However, on July 9 when slower growth rates were first observed in the tanks, the density index was 33.4 and the flow index was 0.59. This information suggests that density and/or flow has an effect on growth rates.

The average growth rates for tiger muskellunge reared in the fiberglass troughs (130 days) and the concrete tanks (164 days) were 0.15 and 0.13 cm/day, respectively. The fish in the troughs at final inventory (August 31) were 21.2 cm in length and weighed 47.7 g/fish or 21 fish/kg. The corresponding lengths and weights for fish in the concrete tanks on October 4 were 21.9 cm and 52.7 g/fish or 19 fish/kg.

Disease problems were minimal throughout the rearing cycle. Gill fungus was present in most troughs during early rearing but its incidence was very low, probably due to daily formalin treatments. Furanace did not effectively control gill fungus.

A suspected disease related mortality did occur in 5 of 7 of the fiberglass troughs between June 30 and July 19 when the fish were approximately 14.7 cm long. The affected fish were very lethargic, swimming at the surface of the water with gill covers flared. They failed to react to any external stimuli, such as light and noise, and one could actually touch the fish without eliciting a response. Laboratory examination of the fish showed varying degrees of hyperplasia of the gill epithelium and fusion of the gill lamellae. Columnaris was not cultured from gills or kidneys. The exact cause of this mortality was not identified, but Furanace was effective in controlling the mortality. Furanace was used as a therapeutic treatment for two days after the appearance of symptoms. Mortalities were back to normal by the end of the second day. This problem did not recur in affected troughs with the exception of one trough which received only one Furanace treatment. In this trough, a subsequent mortality occurred four days after the original mortality. The second mortality was more severe and resulted in about 80% loss in this trough. The other troughs incurred losses between 5 and 20%.

Platte River Hatchery was shipped 127,000 fingerling tiger muskellunge weighing 50 kg on May 17. No mortality was observed as a result of the 4.5 hour trip from Wolf Lake Hatchery. The fingerlings were placed in one indoor rearing tank and were actively feeding within 10 minutes after being placed in the tank.

The inside rearing tanks at Platte River were 9.1 × 1.2 × 1 m in size with a water depth of 51 cm and a water volume of 5,500 liters. Boiler-heated spring water was supplied to these tanks at a rate of 285 liters/min which resulted in a water exchange rate of 3.1 per hour. Water temperature was maintained at about 18.5 C.

These tanks were cleaned twice daily by lowering the water level and brushing the sides and bottom until the tiger muskellunge showed an increased incidence of gill problems. Thereafter the tanks were cleaned with siphon hoses. Baffles were installed in the upper third of the tanks to help with the cleaning process. No overhead lighting was used; only natural light from windows lighted the room. Two suspended Nielson automatic feeders were operated over each tank from 0600 to 2100 h.

Three different treatments were used to control disease problems. Formalin was used daily as a 1-h constant-flow treatment at a concentration of 1:4,000 to control gill fungus. Hyamine 3500 was used at 2 mg/liter active ingredient as a 1-h constant flow treatment on 4 consecutive days to control gill disease as it occurred. Commercially prepared medicated food containing 4% Terramycin was fed for a 10-day period at a 2% body weight level to control columnaris outbreaks.

During the 31-day indoor rearing period, the muskellunge were divided into four additional rearing tanks. Mortalities during this period were high (Table 1). A total of about 28,000 fish were lost as indicated by inventory on the last day of the period but a mortality of only 2,880 fish was indicated from daily work sheets. The additional loss of over 25,000 fish was assumed to have been due to cannibalism. However, a week after transfer to Platte River, a substantial number of dead tiger muskellunge was found underneath the screens in the fish planting units used in the transfer. It was estimated that these fish made up 50% of the unaccountable loss during the indoor rearing.

Growth rates were considered very good during this period. The tiger muskellunge grew an average of 0.21 cm/day and attained an average length of 10.5 cm. Food conversions were also good (Table 1). Maximum density and flow indices were 20.5 and 0.396, respectively.

On June 18, 98,950 tiger muskellunge weighing 565 kg were transferred to the two outside raceways. The outside rearing units are recirculation ponds modified into flow-through raceways. These raceways are 27.5 × 3 × 1.5 m with a water depth of 61 cm and a water volume of 50,000 liters. Platte River water was supplied to the race-

TABLE 2.—*Summary of growth, survival, and food conversions for tiger muskellunge reared on W-7 and PR-9 artificial diets during 1976.*

Date	Number	Mortality	Total weight kg	Weight gain cumulative kg	Food fed cumulative kg	Conversion	Conversion cumulative	Length cm	Daily length increase cm
				W-7 diet					
6/02	5,751		15.4					8.1	
6/11	5,751	0	31.6	16.2	22.2	1.37	1.37	10.3	0.22
6/18	5,751	0	43.4	28.0	47.2	2.12	1.69	11.5	0.17
7/14	2,725[a]	6	102	86.6	152	1.79	1.75	15.3	0.15
8/29	2,917	21	105	137	362	4.15	2.64	19.3	0.09
				PR-9 diet					
6/02	5,726	138	14.6					8.0	
6/11	5,588	2	24.4	9.8	21.7	2.21	2.21	9.5	0.15
6/18	5,586	20	41.3	26.7	45.0	1.38	1.69	11.4	0.27
7/09	2,782[a]	32	87.8	73.2	133	1.90	1.82	14.7	0.16
8/29	2,828		105	134	356	3.69	2.66	19.5	0.09

[a] Fish planted.

ways at rates between 2,570 and 2,840 liters/min with corresponding exchange rates between 3.1 and 3.4 per hour. Water temperatures averaged 16.5 C, but ranged from 13 to 19 C during the outside rearing period. Baffles were placed every 3 m for the entire length of the raceways. Additional cleaning with the baffles installed was not necessary. A single Garon automatic feeder (hopper capacity 68 kg) served two raceways with the federal W-7 dry diet from 0400 to 2400 h. The hatchery constant method was again used to calculate daily feed quantities.

The tiger muskellunge were reared for 54 days in these raceways. The mortalities during this period were again higher than expected. Daily work sheets recorded a loss of 3,687 fish but at plant-out the final inventory showed an additional loss of about 27,600 fish. This additional unaccountable loss was attributed to cannibalism and a hole in one of the outlet screens which allowed fish to escape.

Growth rates during the outside rearing were good considering the colder water temperatures. The average growth rate was 0.185 cm/day. Food conversions were poorer during this period (Table 1), which was attributed to lower water temperatures, overfeeding, and inability of the automatic feeders to distribute food in the area of the highest concentration of fish. Maximum density and flow indices were 27.3 and 0.481, respectively.

Bacterial gill disease and columnaris were continual problems throughout the rearing cycle at Platte River but were kept in check with therapeutic treatments.

Growth on Artificial Diets

On June 2, approximately 11,500 tiger muskellunge fingerlings 8.0 cm long were divided equally between two concrete tanks at Wolf Lake. These fish were reared under identical rearing conditions with the exception of the type of artificial diet fed. One tank was fed W-7, an experimental warmwater diet, and the other was fed PR-9, a standard trout diet. Available energy for W-7 and PR-9 are 350 and 285 kcal/100 g, respectively (Orme 1976). A hatchery constant of 101 was used to calculate feeding levels of both diets.

Table 2 summarizes the data collected during this feeding trial. The average total lengths and food conversions were almost identical throughout the 90-day rearing period. The greatest differences were found after the first inventory when both length increase and conversion were poorer for the fish fed PR-9. These differences may have been the result of the fish switching to a less palatable diet or to sampling error. In either case, average total length and conversions were similar for both diets during the rest of the food trials. The higher protein and fat levels in W-7 did not increase the growth rates over the PR-9 fed

TABLE 3.—*Summary of growth data for tiger muskellunge reared at three feeding levels during 1976.*

Date	Number	Mortality	Total weight kg	Weight gain kg	Food fed kg	Conversion	Total length cm	Daily length increase cm
			Hatchery constant = 77					
6/15	1,600		9.5				10.6	
7/02	1,558	42	17.8	8.2	14.2	1.73	13.0	0.14
			Hatchery constant = 93					
6/15	1,561		9.0				10.5	
7/02	1,558	3	19.0	10.0	17.2	1.72	13.5	0.18
			Hatchery constant = 109					
6/15	1,600		9.7				10.7	
7/02	1,596	4	21.6	11.9	22.7	1.91	14.0	0.19

fish. Based on these results, PR-9 could be used to rear tiger muskellunge over 8 cm in length.

Growth at Different Feeding Rates

In an attempt to determine the proper feeding level for maximum growth, tiger muskellunge in three pairs of troughs were fed at three different feeding rates for 17 days. Feed quantities for each pair of troughs were calculated from hatchery constants of 77, 93, and 109.

As shown in Table 3, the highest daily growth rate was attained by the fish in the trough fed the largest quantity of food. However, the fish in the troughs receiving intermediate amounts of food had a growth rate of 0.18 cm/day, only 0.01 cm/day less than the fish receiving the highest level of feed. In addition, the food conversions for the highest level of feeding were poorer than the other two levels of feeding. These data suggest that the fish receiving the largest amount of food were being overfed and that the proper amount of food for maximum efficient growth is intermediate between the two highest levels of feeding. The intermediate hatchery constant was 101 and was used to feed the tiger muskellunge at Wolf Lake during the later rearing stages.

1977 Production

A total of 684,100 tiger muskellunge eggs were collected from wild broodstock in Bankson Lake on March 31, 1977. The eggs were force-hatched and placed in troughs at densities generally not exceeding 25,000 per trough. Feed was placed in the feeders twice daily at 0800 and 1600 h to distribute the feed more evenly through the feeding period. The overhead lights were wired through a timer switch and were set to be on during the entire feeding period. Table 4 summarizes the data on growth and survival of intensively reared tiger muskellunge during 1977.

A total of 401,700 sac-fry were hatched. The percent hatch for this lot of eggs was 59, substantially better than that experienced during 1976. Mortalities during the sac-fry stage were higher than expected and were estimated at about 50,000 sac-fry. The mortalities were caused by what appeared to be fungus growth. The mortality was partially controlled by constant-flow formalin treatment for 1 h at a concentration of 1:6,000.

Survival through the initial training period was very good. Mortalities between April 22 and May 10 amounted to 93,550 (27%) fry with over 75% of this mortality attributed to cannibalism. During the next 7-day period mortalities decreased to about 1% and cannibalism did not contribute substantially to the mortality.

Growth rates for the early rearing periods during 1977 were identical to those recorded during 1976. However, food conversions were better during 1977. Maximum density and flow indices were 26.7 and 0.604, respectively.

On May 17, 164,700 tiger muskellunge fingerlings 5.0 cm long and weighing 104

TABLE 4.—A summary of the growth and survival of intensively reared tiger muskellunge at Wolf Lake and Platte River during 1977.

Date	Number	Mortality	Fish planted	Total weight kg	Weight per fish g	Food conversion		Length cm	Daily length increase cm	Density index kg/m³	Flow index kg/liter/min
						Period	Cumulative				
				Wolf Lake—troughs							
3/31	684,100[a]										
4/14	401,700[b]	282,400									
4/22	351,700[c]	50,000		5.6	0.016			1.27		1.43	0.057
5/10	258,150	93,550	91,600	53.8	0.208	1.11		3.45	0.12	13.8	0.437
5/17	164,700	1,800	164,700[d]	104	0.631	1.17	1.14	5.02	0.22	26.7	0.604
				Platte River—inside tanks							
5/18	164,700[d]			104	0.631		1.14	5.02		9.45	0.183
6/14	126,880	37,821	126,880[e]	807	6.36	3.77	3.45	10.8	0.22	24.5	0.474
				Platte River—outside raceways							
6/15	126,880[e]			807	6.36		3.45	10.8		8.1	0.220
7/07	125,580	1,298		1,753	14.0	5.31	3.78	14.1	0.14	17.5	0.370
7/27	108,950	16,630	108,950	3,004	27.6	2.97	3.48	18.5	0.22	30.0	0.529

[a] Eggs, date fertilized.
[b] Sac-fry, date hatched.
[c] Swim-up fry, date initially fed.
[d] Fish transferred to Platte River Hatchery.
[e] Fish transferred to outside raceways.

kg were transferred to Platte River in two 2.5-ton Manchester planting units. The fingerlings were transferred in fine mesh crates placed inside the units to prevent the fish from getting under the screen in the bottom of the tanks. No hauling losses were observed on arrival at Platte River.

Mortalities at Platte River again resulted from a high incidence of cannibalism. During the inside rearing a total loss of 37,800 fingerlings was indicated by final inventory but only 30.3% of this loss was accounted for on the daily work sheets. Similarly, during the outside rearing only 47.5% of the total loss of 17,900 fingerlings was accounted for on the daily work sheets. The only other significant mortality resulted from an outbreak of columnaris during the inside rearing. This loss amounted to almost 10,000 fish over a 7-day period.

Growth rates were very good with the exception of the first 26 days after the fingerlings were moved to the outside raceways. The growth rate during this period averaged 0.14 cm/day and poor food conversions were attributed to the low water temperatures that averaged 15.4 C. Growth rates and food conversions improved during the next period when water temperatures increased to an average of 17.9 C.

Assessment

A comparison of the survival and growth data for the two production years shows that improvements were made. During 1977, survival from eggs to plant-out fingerlings was 22% and the tiger muskellunge grew an average of 0.18 cm/day. During 1976, survival ranged from 11% at Platte River to 15% at Wolf Lake with corresponding growth rates of 0.175 cm/day and 0.14 cm/day.

Two major factors were considered responsible for the improved survival and growth rates experienced during 1977. The single most important factor was better egg quality. Survival through the egg incubation phase was 59% for 1977, but only 39% for 1976, although egg-take and incubation methods were identical. Egg quality is an environmental variable and probably will not be controlled by hatchery managers.

The second factor was the decreased incidence of cannibalism during the 1977 early rearing stage. This was attributed to a change in the feeding technique. During 1977, feed was placed in the feeders twice daily and lights were set to be on during the entire feeding period. The behavior of tiger muskellunge makes this feeding regime very important. First, tiger muskellunge will not travel laterally a long distance to get food as coldwater fish do, so food has to be dropped near to them. Secondly, tiger muskellunge appear to feed by sight so ample light is required for them to see the pellets. The behavioral requirements of tiger muskellunge were better met during 1977.

The tiger muskellunge production program encountered very few problems with the exception of the high incidence of cannibalism at Platte River and the slower growth rates during the latter rearing phases at Wolf Lake during 1976. The high cannibalism can again be related to the feeding techniques at Platte River. The tiger muskellunge were reared in large tanks and raceways, and in most cases the feed was distributed over too small an area. Techniques need to be developed to evenly distribute food over the entire surface of the tanks in the areas of highest concentrations of fish.

The slower growth rates at Wolf Lake during 1976 are still an unresolved problem. As discussed previously, growth rates started decreasing after the fish attained a length of about 13 cm. However, tiger muskellunge shipped to Platte River for rearing did not show the slower growth rate, even though water temperatures were lower than at Wolf Lake. One major factor which was different between Platte River and Wolf Lake was rearing density. Maximum density and flow indices at Platte River were 27 kg/m^3 and 0.53 kg/liters per min, respectively. Both of these indices are lower than the respective values of 32 and 0.60 encountered at Wolf Lake at the time when growth rates started to decline. Higher rearing densities may have affected the growth rates, although this needs to be confirmed.

References

HNATH, J. G., AND J. A. COPELAND. 1976. The use of iodophors for the disinfection of green eggs of muskellunge and tiger muskellunge. Mich. Dep. Nat. Resour., Wolf Lake Hatchery, Mattawan, Mich. 4 pp.

ORME, L. 1976. Proximate analysis of the W-7 diets used in the 1976 coolwater diet testing program. U.S. Fish Wildl. Serv. Diet Testing Dev. Cent., Spearfish, S.D. 2 pp.

———. 1978. The status of coolwater fish diets. Am. Fish. Soc. Spec. Publ. 11:167–171.

Reviewing the Esocid Hybrids

KEEN BUSS

Box 554, Boalsburg, Pennsylvania 16827

JAMES MEADE III AND DELANO R. GRAFF

Pennsylvania Fish Commission
Bellefonte, Pennsylvania 16823

Abstract

A review of the hybrids of the Esocidae is presented. Artificial crosses of the smaller pickerels, chain pickerel, grass pickerel and redfin pickerel and their reciprocals produced fertile progeny. However, the crosses of the larger pikes, muskellunge, northern pike, and Amur pike (*Esox reicherti*), although successful, produced only one with fertile progeny, the Amur pike × northern pike. Crosses of the larger pike with smaller pickerel were generally unsuccessful or produced sterile young. Natural hybridization has been prevented by difference in distribution, habitat, size of mature fish, behavior, spawning time, spawning sites, and immunological barriers. Crosses of northern pike with muskellunge and Amur pike produced hybrids which appear to have great potential as a sport fish. These hybrids are relatively easy to rear, utilize artificial foods, grow fast, and eventually produce a sport fish which is prized by anglers.

The esocid fishes are probably the most exciting family of freshwater sport fishes in the world. Unlike the much publicized trouts which do little or nothing to control or change associated fish populations for better or worse, the esocids are competitive predators which, in most cases, influence their environment; and when present in sufficient numbers, they insure a healthy population of panfishes and other prey species in their adopted ecosystem.

This family has only six species in the world, five of which are native to North America. The sixth is indigenous to Asia but has been introduced in limited areas of Pennsylvania. Because of this unusual situation, a complete evaluation is possible of all the members of the Esocidae and of F_1 progeny of crosses and reciprocal crosses. This is, indeed, a rare situation in a study of fish phylogeny.

One need only examine the species of the Esocidae to appreciate the gene pool available for the studies and management possibilities of F_1 hybrids. Muskellunge are usually considered the largest of the North American game fishes. Even though fishing returns of muskellunge are low in terms of effort expended, they are highly desired by anglers and considered "braggin stock" by all who catch them. Another large member of the Esocidae is the northern pike. Although their ultimate growth is not as great as that of muskellunge, northern pike can produce prize catches of 9 to 18 kg. They overshadow the muskellunge in sports fishing because they can be found and caught in large numbers on ideal occasions. It has been only in recent years that northern pike has been recognized as a great sport fish in many areas of its natural range. The third and last of the larger esocids is the Amur pike (*Esox reicherti* Dybowski) from the Amur River Basin in China and the U.S.S.R. It was first introduced into Pennsylvania in 1968 and stocked in a limited isolated area (Meade 1976). Although not too much is known about this species, it may have the potential to fill a niche in the sport fisheries. Limited returns reveal it is a good sport fish and reaches a size of 112.5 cm and 61.6 kg.

The chain pickerel is probably one of the most important coolwater fishes in its entire range. It is also the most neglected piscivorous sport fish in all of fishery science. Once a mainstay of lake anglers along the Atlantic drainage, it has, in many cases, been relegated to a second-class position because of bass introductions and the assumed popularity of trout. This is one species that needs revival and study. The little pickerels, the redfin pickerel and the grass pickerel, are generally not considered as game fish but undoubtedly influence their environment more than might be realized for such small species (Buss 1962). For the purpose of review, the grass pickerel will be considered a separate species based on Sexton's (1963) assumptions.

Involved in this family are species which have the potential to produce predators which range in size from unusually small to exceptionally large. Some species are difficult to capture; some are easily caught; some are stream dwellers; and some prefer weedy lakes. Parent species can be crossed to produce sterile hybrids or fertile hybrids. Some species easily adapt to artificial culture and others do not; some are prized by anglers, others are "snakes." Somewhere in this biological complexity of traits may be a combination which is as good or better than the parent stocks for specific management purposes and sport fishing. With purpose and foresight, an evaluation of the hybrids is worthy of the effort involved.

Although the intrinsic value of the esocids has long been recognized by some biologists, little study has been done on them when compared to that on the salmonids. Less has been done with pike hybrids even though the successes of plant breeders and animal husbandmen with hybrid strains and species are almost legend. Since coolwater fisheries form a major portion of the sport fisheries in North America, not excluding trout, it would seem to be reasonable to exploit the innate qualities of this family for fish management purposes and sport fishing.

Natural Barriers to Hybridization

Distribution

The northern pike is the only esocid which is circumpolar in distribution, and it is also the only esocid which is found in Europe or above 60° north latitude in Asia. In North America, it would be possible for the northern pike to hybridize with the muskellunge, the chain pickerel, and the grass pickerel in the southern part of its range and with the redfin pickerel in the St. Lawrence River Basin. However, these areas of overlap are limited.

The Amur pike is geographically isolated in the Amur River Basin south of 60° north latitude in the U.S.S.R. and China. It does not overlap the northern pike, and there is no opportunity to produce natural hybrids.

The muskellunge appears to be supersensitive to sharing habitat with other esocids. Even though the muskellunge's range overlaps those of the northern pike, grass pickerel, and, to a lesser extent, the chain pickerel and redfin pickerel in the St. Lawrence River Basin, habitats appear to be separate.

The range of the chain pickerel overlaps those of the two little pickerels, the grass and redfin; where habitat overlaps, some natural hybridization does occur. The range of the redfin pickerel is east of the Appalachian Mountains and that of the grass pickerel is west of the Appalachians. Overlap and intergrades of the two species occur around the southern extremity of the Appalachians (Buss 1962).

Immunological Barriers and Similarities

Genetic studies in the laboratory of James E. Wright, The Pennsylvania State University, indicated the relative positions of the six species of esocids on the phylogenetic scale. In this series of studies, Sexton (1963) determined antigenic relationship of five species of esocids, excluding Amur pike. He hypothesized that the grass pickerel is immunologically in the center of the Esocidae. Also, he determined that close immunological relationships exist between chain pickerel and northern pike and between the redfin pickerel and muskellunge.

Additional preliminary investigations by Wright (personal communication) of electrophoretic patterns of certain enzymes and other proteins in esocids show that northern pike and Amur pike are practically identical for these traits. These results, together with those of Davisson (1972), have led Wright to postulate that the two fishes are the same species or, at most, subspecies and that Amur pike might be an isolate of northern pike which has developed differences in appearance through gene mutations. The muskellunge was very similar, if not practically identical, to the other two large pikes in these respects.

The smaller pickerels (chain, grass, and redfin) were quite different from the three larger esocids, but were very similar to each other for transferrin; hemoglobin; tetrazolium oxidase; and glucose-6-phosphate, 6-phosphoglutamic, alcohol, lactate, and malate dehydrogenases. This differentiation was substantiated by Eckroat (1974) in a study of interspecific comparisons of eye lens proteins of the esocids.

Davisson (1972) concluded that the karyotypic similarity among esocids suggests a cytological basis for the ease of hybridization among most members of this family. Therefore, since most crosses produce interspecific hybrids in almost all cases, the species whose ranges overlap must be restrained from hybridizing by physical and behaviorial barriers such as size incompatibilities and difference in spawning time and place.

Spawning Time

Muskellunge are later spawners than the other members of the pike family; the only possibility of obtaining natural hybrids of this species would be the intermingling of a late-spawning strain of northern pike and an early-spawning individual or strain of muskellunge. Spawning time appears to be no barrier to hybridization among other North American esocids. Ripe grass pickerel and northern pike have been taken in the same trap nets.

Spawning Sites

Selection of spawning areas can limit hybridization. In Presque Isle Bay, Lake Erie, northern pike have a spawning area distinct from that of muskellunge. Northern pike spawn in the sloughs and muskellunge spawn in areas of the open bay (Buss and Larsen 1961). Grass pickerel, chain pickerel, and northern pike have similar spawning habitats, and often utilize flooded vegetated areas. Redfin pickerel have been found to spawn in shallow spring areas where other esocids probably seldom enter (Buss and Miller 1967).

Spawning Behavior

Crossman and Buss (1965) suggest that other deterrents to natural hybrids are method of egg laying, preferred spawning depth, and territory and general solitary nature of spawning pairs. The size of the parents must have some effect on the spawning stimulus. A small grass pickerel of 10 to 15 cm probably would have little sexual attraction to a northern pike female over 50 cm in length. When this cross does take place, it probably occurs between a small male northern pike and a large female grass pickerel (Crossman and Buss 1965). The size of the egg can be an effective barrier to hybridization (Buss and Wright 1957). Eggs of smaller pike may be too small to contain a large hybrid embryo and physical damage can result. If the egg of one parent species has a shorter or longer incubation period than the other parent, the embryo

TABLE 1.—Results of crossing esocid species[a] and the fertility of F_1 generation.

	Northern pike	Muskellunge	Amur pike	Chain pickerel	Redfin pickerel	Grass pickerel
Northern pike	XX F	XX S	XX F	–	–	X ?
Muskellunge	XX S	XX F	XX ?	X S	X S	XX *
Amur pike	XX ?	XX ?	XX F	–	–	–
Chain pickerel	X S	X S	X ?	XX F	XX F	XX F
Redfin pickerel	X S	X *	0 ?	XX F	XX F	XX F
Grass pickerel	XX S	XX ?	X ?	XX F	XX F	XX F

[a] Key:

Hybridizing

XX = Good hatch and fry survival.
X = Some survivors but poor egg fertility or fry survival.
– = Unsuccessful cross.
0 = Cross not made.
* = Fry hatched and died.

F_1 fertility

S = Hybrid sterile.
F = Hybrid fertile.
? = Hybrid fertility not determined.

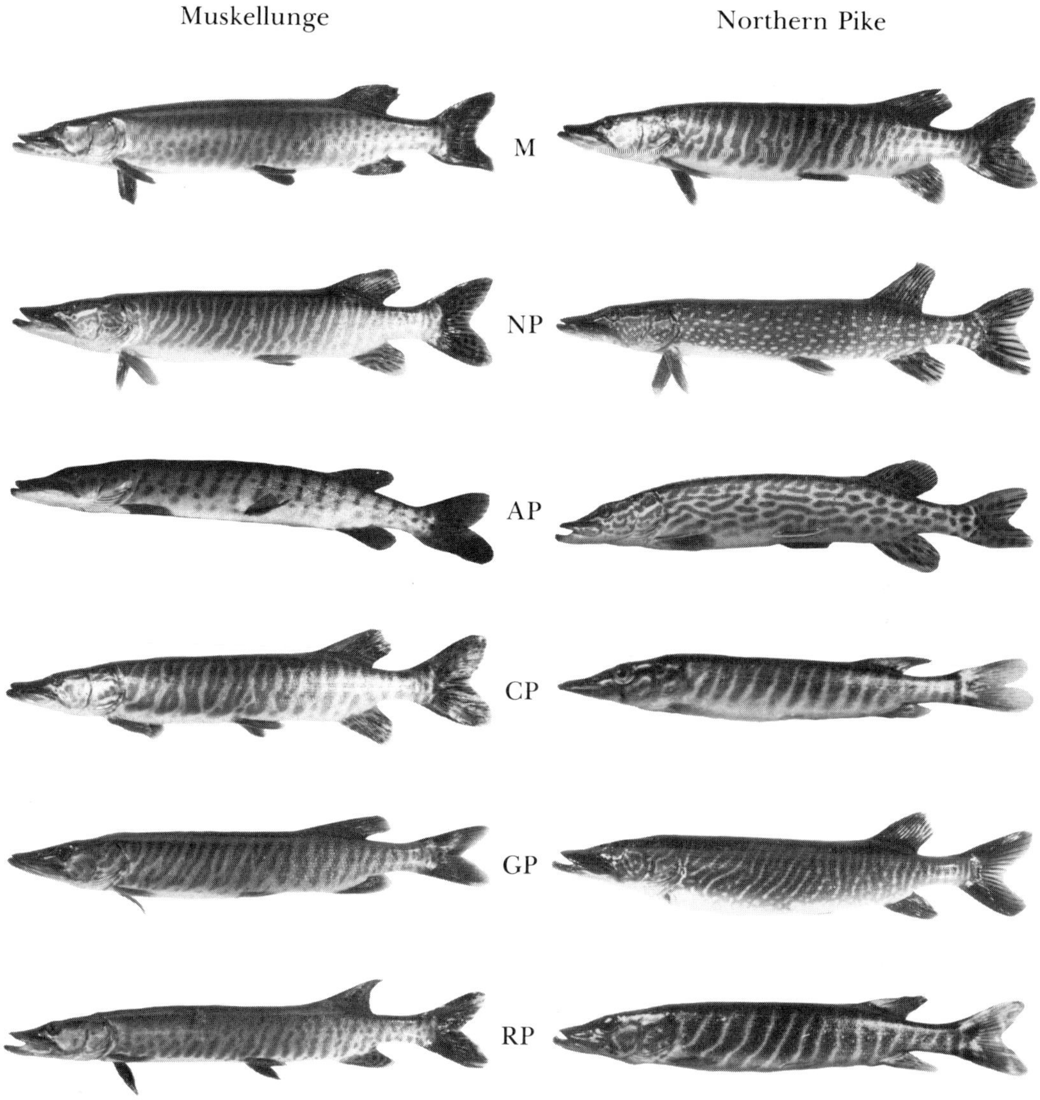

FIGURE 1.—*Hybrids of the larger pike. Symbols:* M = *muskellunge;* NP = *northern pike;* AP = *Amur pike;* CP = *chain pickerel;* GP = *grass pickerel;* RP = *redfin pickerel. Note the influence of parental species on the lateral markings of these hybrids. The sex of the parents or their reciprocal crosses did not influence the markings. Total lengths of fish shown (vertical sequence, left column first):* M × M, *75 cm;* M♀ × NP♂, *66 cm;* AP × M, *44 cm;* M♀ × CP♂, *62 cm;* GP♀ × M♂, *36 cm;* M♀ × RP♂, *54 cm;* NP♀ × M♂, *67 cm;* NP × NP, *50 cm;* NP♀ × AP♂, *53 cm;* CP♀ × NP♂, *16 cm;* GP♀ × NP♂, *43 cm;* RP♀ × NP♂, *21 cm.*

will hatch prematurely or it will die in the egg. Incompatibility of some genomes may produce nonpathogenic diseases, such as white sac or blue sac, which will destroy the embryo or fry (Buss and Wright 1957).

All of these natural barriers explain the small numbers of natural hybrids found.

Natural Hybrids

Natural esocid hybrids have not been abundant. From various reports, the muskellunge × northern pike cross (tiger muskellunge) have occasionally appeared in nature (Threinen and Oehmcke 1950).

Chain Pickerel (CP) Grass Pickerel (GP)

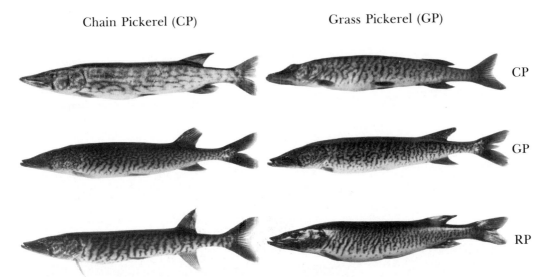

FIGURE 2.—*Hybrids among pickerel. The closely related smaller pickerels (chain and redfin) produce hybrids with similar markings. Total lengths of fish shown (vertical sequence, left column first): CP × CP, 27 cm; GP♀ × CP♂, 28 cm; RP♀ × CP♂, 28 cm; CP♀ × CP♂, 24 cm; GP × GP, 30 cm; RP♀ × GP♂, 24 cm; CP♀ × RP♂, 28 cm; GP♀ × RP♂, 23 cm; RP × RP, 23 cm.*

Because of their unusual markings, their occurrence usually causes a stir among muskellunge fishermen.

Northern pike and chain pickerel hybrids were identified in New York State (Crossman and Buss 1965), but their regular occurrence has not been reported. Natural hybrids of northern pike and grass pickerel have been reported from Nebraska, Michigan, and Wisconsin.

Chain pickerel-redfin crosses are among the best known in nature (Crossman and Buss 1965). Natural hybrids have been reported through much of their range east of the Appalachians. Hybrid crosses of chain pickerel have also been reported from Ohio; but again, minor range overlap diminishes the chances of hybridization.

Redfin and grass pickerel, being closely related, or perhaps subspecific, are in transition form or intergrades wherever their ranges overlap.

Artificial Hybridization

Crossman and Buss (1965) established that total sterility did not exist among esocid hybrids since one or the other reciprocal cross produced progeny (Table 1).

Buss and Miller (1967) established that some F_1 progeny of the esocids are fertile. They noted successful reproduction by hybrids of chain pickerel × redfin pickerel, chain pickerel × grass pickerel, and grass pickerel × redfin pickerel, but no reproduction by F_1 progeny of northern pike × muskellunge cross or of large pike × small pickerels (Table 1). In addition, the F_1 progeny of northern pike × Amur pike is fertile (Crossman and Meade 1977).

All of the larger and older specimens of muskellunge hybrids have a dark vertical bar on a light background. The hybrids of northern pike with species other than muskellunge have light markings on a dark background (Fig. 1). The hybrids of the pickerels usually have broken patterns very similar in appearance (Fig. 2).

Evaluation of Esocid Hybrids

Muskellunge × Northern Pike (Tiger Muskellunge)

Hybrids of muskellunge × northern pike are more easily cultured, have fewer disease problems, and eat dry feed better than muskellunge. The F_1 has an excellent

Redfin Pickeral (RP)

FIGURE 2.—Continued.

growth rate in the hatchery; consequently, larger fish may be stocked with a better chance of survival in most cases. Tiger muskellunge have a higher survival rate than northern pike or muskellunge (Weithman 1975). The faster growth is generally related to increased feeding activity which, in turn, means more predation. Increased feeding activity also provides more easily caught sport fish. The hybrids are sterile and, therefore, can be introduced into watersheds where a spread of large piscivorous fishes is not desired. These hybrids are less vulnerable to angling than northern pike (Beyerle 1973; Weithman 1975). As "tiger muskies," they are a welcomed catch by anglers.

Muskellunge × Amur Pike

The muskellunge × Amur pike cross is a fast growing, hardy fish in hatcheries. It has possibilities for adaption to a stream environment where a good predator and game fish is desirable. Maximum growth obtained has been to 117.5 cm and 10.9 kg (Meade 1976). This hybrid is striking in appearance and beautifully marked.

Northern Pike × Amur Pike

The northern pike male × Amur pike female produces viable, hardy progeny. It is established that the progeny are fertile as Davisson (1972) hypothesized from the 25 bivalent chromosomes found in meiosis. It is potentially a stream inhabitant and thus of value where a coolwater predator may be required.

Northern Pike × Grass Pickerel

Progeny of grass pickerel female × northern pike male hatch and survive well. It is an apparently sterile hybrid with potential for stream utilization.

Chain Pickerel × Larger Pikes

Hybrids of chain pickerel with the larger pikes have no apparent value over parental species.

Chain Pickerel × Redfin or Grass Pickerel

Crosses of chain pickerel with the smaller pickerels are very viable. The F_1 progeny are fertile and are potentially excellent intermediate-size predators for farm ponds (Buss and Miller 1967).

References

BEYERLE, G. B. 1973. Comparative growth, survival, and vulnerability to angling of northern pike, muskellunge, and their hybrid tiger muskellunge stocked in a small lake. Mich. Dep. Nat. Resour. Fish. Res. Rep. 1799. 13 pp.

BUSS, K. 1962. The redfin and grass pickerels. Spec. Purpose Rep. Pa. Fish Comm., Benner Spring Fish Res. Stn. 12 pp. (Mimeo.)

———, AND ALFRED LARSEN. 1961. The northern pike of Presque Isle Bay, Lake Erie. Pa. Angler 30(9):4–6.

———, AND J. MILLER. 1967. Interspecific hybridization of esocids: hatching success, pattern development, and fertility of some F_1 hybrids. U.S. Bur. Sport Fish. Wildl. Tech. Pap. 14. 30 pp.

———, AND J. E. WRIGHT. 1957. Appearance and fertility of trout hybrids. Trans. Am. Fish. Soc. 87:172–181.

CROSSMAN, E. J., AND K. BUSS. 1965. Hybridization in the family Esocidae. J. Fish. Res. Board Can. 22(5):1261–1291.

———, AND J. W. MEADE. 1977. Artificial hybrids of the Amur pike *Esox reicherti* and North American esocids. J. Fish. Res. Board Can. 34(11): 2338–2343.

DAVISSON, M. T. 1972. Karyotypes of the teleost family Esocidae. J. Fish. Res. Board Can. 29:579–582.

ECKROAT, L. R. 1974. Interspecific comparisons of lens proteins of Esocidae. Copeia 1974(4):977–978.

MEADE, J. W. 1976. The propagation, hybridization, and experimental introduction of Amur pike, *Esox reicherti*, in the United States. Pa. Fish Comm., Benner Spring Fish Res. Stn. 19 pp. (Mimeo.)

SEXTON, A. D. 1963. Investigations of the antigenic relationships of the Esocidae and two species of Percidae. M.E. Thesis. Pa. State Univ., University Park. 55 pp.

THREINEN, C. W., AND A. OEHMCKE. 1950. The northern pike invades the musky's domain. Wis. Conserv. Bull. 15(9):10–12.

WEITHMAN, S. 1975. Survival, growth, efficiency, preference, and vulnerability to angling of Esocidae. M.S. Thesis. Univ. Mo., Columbia. 71 pp.

Control of Reproduction in Northern Pike

G. de Montalembert, C. Bry, and R. Billard

Laboratoire de Physiologie des Poissons
Institut National de la Recherche Agronomique
78350 Jouy en Josas, France

Abstract

Several methods to control gamete availability in northern pike were investigated. Precocious induction of ovulation was achieved with a single injection of partially purified salmon gonadotropin, but the treatment should be administered soon after capture in order to avoid ovarian atresia. Ovulated oocytes should be inseminated within 24 hours after ovulation. Loss of fertility due to aging occurred 1–3 days after ovulation. High doses of progesterone induced a threefold increase of sperm release. Cryopreservation of diluted sperm at −196 C led to variable fertilizing capacities after thawing, depending on the donor male. A diluent originally tested for trout was used successfully for artificial insemination.

In France, the main purpose of northern pike breeding is the stocking of natural waters. However, the amount of professional fishing is far from being sufficient to gather ripe breeders in adequate numbers and obtain fry or fingerlings in quantities that would meet the national needs.

The knowledge of spawning migration habits improved the catch of ripe females (Chimits 1947, 1956; Cointat and Darley 1953, Preudhomme 1968). However, the results were very irregular from year to year, depending notably on weather conditions. Therefore, some attempts were made to control reproduction, mostly the phases of maturation and ovulation.

Ovulation does not occur spontaneously under confined conditions. Moreover, most confined females quickly undergo ovarian atresia and may die after a few weeks (Portal 1947; Chimits 1947, 1956; de Montalembert et al. unpublished observations).

In small ponds, ripening can be attained, but a certain range of temperature must be maintained despite the atmospheric conditions (Cointat and Darley 1953; Huet 1972). Natural spawning and fertilization may then also take place (Schaperclaus 1961; Kausch 1976); yet the real efficiency (i.e. the number of fry collected compared to the potential number of eggs) of such an extensive method has not been specified.

Hypophysation with carp pituitaries, performed in one or several injections, can induce ovulation (Sorenson et al. 1966) but precise data concerning the state of maturity of recipient females, as well as the ovulation and hatching rates, are lacking.

Another difficulty commonly experienced lies in the low volume of sperm usually delivered by the males, especially at the end of the spawning season (Portal 1947; Chimits 1956; Sorenson et al. 1966; Preudomme 1975; Huet 1976).

Salmon pituitary extracts, HCG (human chorionic gonadotropin), and various steroids induce sperm release in goldfish (Yamazaki and Donaldson 1969; Billard 1976). Progesterone is more potent than androgens (Billard 1976). In northern pike, sperm release was stimulated within 24 hours following hypophysation (Anwand 1963), but no steroid—a much cheaper treatment—was tested.

Cryopreservation of viable spermatozoa would allow the ova of belated females to be fertilized with semen collected from ripe males early in the spawning season. With the sperm of the northern pike, the only trials reported to date resulted in an average fertilizing capacity of 25% (Stein 1977).

Finally, with a given quantity of semen, the best way to increase the efficiency of artificial insemination is to lengthen the life of spermatozoa and thus greatly reduce the number of spermatozoa needed per ovulated oocyte. Since a buffered and osmotically balanced diluent has been used with success with trout to obtain high fertilization rates with minute quantities of sperm (Petit et al. 1973; Billard 1977), it seemed particularly worthy to examine its possible effectiveness with northern pike.

The present investigation was carried out in order to develop methodologies for (1) hormonal induction of maturation and ovulation in submature females, (2) hormonal stimulation of sperm release, (3) sperm cryopreservation, and (4) artificial insemination with the aid of a diluent.

Methods

The experiments were performed in March-April 1976 and 1977. Northern pike breeders (males: 1–2 yr old, weighing 0.2–0.4 kg; females: 2–3 yr old, weighing 0.8–2 kg) were collected from natural ponds and kept in full darkness at a temperature of about 12 C. Before each handling (injection, gamete sampling), the fishes were anesthetized in a solution of 2-phenoxyethanol (Merck) (0.5 ml/liter).

Induction of Ovulation

The nature of the injected material as well as the determination of oocyte stages, ovulation, and fertilization rates have been described previously (de Montalembert et al. 1978) In brief: the efficiency of various hormonal treatments (gonadotropic preparation and/or steroid) on in vivo maturation and ovulation was investigated in submature females; the germinal vesicle in non-peripheral position was the criterion for initial oocyte stage. Two gonadotropic preparations (partially purified salmon gonadotropin and HCG) and two steroids (17α-hydroxy-20β-dihydroprogesterone and 17α-hydroxyprogesterone) were employed.

Retention of Ova

In the first of two experiments, 5 groups of 2 females were used. Ovulation was induced by salmon gonodotropic extracts (0.1 mg/kg). One thousand ova were sampled from each female of group 1 on the day of ovulation and inseminated with sperm pooled from 10 males, by means of the dilution technique described below (dilution rate: 1/1,000). The eggs were incubated in a fine mesh trough at about 12 C. The same procedure was applied to group 2 one day after ovulation, to group 3 two days after ovulation, and so on until day 4. Fertilization rates were expressed as the percentage of embryos on the third day following insemination (i.e. at about 40 degree-days of development).

The second experiment was carried out with 7 females kept at a constant temperature of 13 C. Ovulation was obtained with a low priming dose of trout gonadotropic extracts followed by 3 mg/kg of 17α-hydroxy-20β-dihydroprogesterone (17α-20β P) 24 hours later. Four hundred ova were sampled from each of the 7 females on the day of ovulation and on the five following days, and inseminated with sperm pooled from 10 males. The fertility of the ova was assessed at the time of hatching.

Stimulation of Sperm Release

HCG was dissolved in saline (0.8% NaCl) Progesterone and testosterone were dissolved in pure ethanol, then precipitated in saline just before injection (20% ethanol in the injected product). The injected volume always was 1 ml/kg.

Every two days, sperm was expelled by gentle handstripping and the volume of sperm was measured with a precision of 10 μl. In a preliminary experiment the efficiencies of HCG (1,000 IU/kg), progesterone (10 mg/kg), and testosterone (10 mg/kg) (2 injections 24 hours apart) were compared in groups of 7 fish by measuring the volume of sperm release until exhaustion. In a second experiment, 3 doses of progesterone (100, 10, 1 mg/kg, a single injection each) were tested on 3 groups of 10 fish. A control group was treated with saline. The volume of sperm and the spermatocrit (volume of spermatozoa in %) were measured. The fertilizing capacity of the sperm was checked 2 days after treatment (peak of the response) by insemination of a 300-egg sample with 10 μl of sperm (dilution rate: 1/1,000).

Cryopreservation of the Sperm

A drop of handstripped sperm was mixed with the insemination diluent (DIA_{TG}; see below) and the motility classified as one of the 5 stages described by Emmens (1947) and Jaspers (1972). The activity of the spermatozoa was defined as the duration (in minutes) of the most active stage following the dilution. The fertilizing capacity of the fresh semen was tested on 300 ova using the dilution technique (dilution rate: 1/100). The fertilization rate was the percentage of embryos at 40 degree days.

For cryopreservation, 50-μl aliquots of sperm were mixed with the V_2 extender described by Stein (1977): 750 mg NaCl; 200 mg $NaHCO_3$; 53 mg $Na_2HPO_4 \cdot 2H_2O$; 23 mg $MgSO_4 \cdot 7H_2O$; 38 mg KCl; 46 mg $CaCl_2 \cdot 2H_2O$; 100 mg glucose; 500 mg glycine; 20% egg yolk; 10% dimethyl sulfoxide; 100 ml H_2O.

In a preliminary experiment, four dilution ratios (volume of sperm: volume of extender: 1:2; 1:3; 1:5; 1:10) were tested with sperm pooled from 10 males. The diluted sperm was equilibrated in an ice water bath during 5 min and the absence of motility of the spermatozoa was checked. The samples were then dropped onto dry ice (-79 C) and stored in liquid nitrogen (-196 C). After a 7-day storage, the motility and activity were assessed in the DIA_{TG} diluent. In order to estimate the fertilizing capacity, the sperm was thawed directly in DIA_{TG} containing the ova.

In a second experiment, the semens of 10 males were individually cryopreserved (dilution ratio 1:2) and checked for fertility according to the same experimental procedure.

Artificial Insemination by a Dilution Technique

Ovulation was induced by a single injection of partially purified salmon gonadotropin (de Montalembert et al. 1978). The ova and the sperm were collected carefully by handstripping, in dry condition. Ova from several females were mixed and divided into batches of 300 eggs. Sperm was sampled from 10 males, examined to assess motility, and pooled. The control lots were inseminated according to the classical dry method.

Each experimental lot (300 ova) was placed in 10 ml of DIA_{TG} diluent and immediatly inseminated. The composition of DIA_{TG} (Billard 1977) was the following: solution of NaCl buffered with 0.02 M Tris and 0.05 M glycine to an osmotic pressure of 250 mosmol and pH 9. Various amounts of sperm (100 μl, 10μl, 1 μl, 0.1 μl; dilution rates: 10^{-2}, 10^{-3}, 10^{-4}, 10^{-5}) were used in a first experiment. The eggs were allowed to stay for 15 min following insemination, then poured into a fine meshed trough where incubation was carried out at about 12 C until day 4 following insemination. The eggs were then cleared in Stockard's solution and the fertilization rates were determined as the percentage of embryos at 40 degree days.

In a second experiment, sperm was prediluted in DIA_{TG} (dilution ratio 1:100 or 1:1,000) or in fresh water (dilution ratio 1:1,000) during various periods of time (0, 1, 2, 4, 8, 16 min) and then mixed with ova.

Statistical Analysis

The mean values of the ovulation, fertilization, and hatching rates and their confidence limits at the 95% level of probability were computed from the angular (arc-sine) transformation of the percentages. The levels of significance were determined by the F test in the induction-of-ovulation experiment, and by the t-test on paired data in the retention-of-ova experiment.

Unless otherwise indicated, differences noted as significant in the text correspond to $P < 0.05$.

Results

Induction of Ovulation

The detailed results concerning induction of ovulation have been described in a previous paper (de Montalembert et al. 1978). Therefore we will only mention here the most important experimental facts.

A single injection of partially purified salmon gonadotropin (0.1 mg/kg) induced maturation and ovulation in submature females (initial stage of the oocytes: nucleus in nonperipheral position). Ovulation occurred 4 days after the treatment (water temperature: 12 C).

In 6 females out of 14 the ovulation was total. In the remaining 8 females the ovulation was only partial but was always over 50% (general mean: 89%). In all cases, the

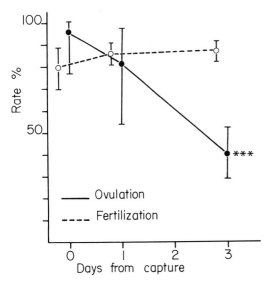

FIGURE 1.—*Effects of time between capture and treatment (partially purified salmon gonadotropin: 0.1 mg/kg) on ovulation and fertilization rates. Vertical bars show confidence limits at the 95% level of probability. *** indicates statistical difference from initial rate ($P < 0.001$).*

fertilization rates were high (mean 83%). Figure 1 shows the effect of various time intervals between capture and treatment. When the treatment was performed within 24 hours after capture no significant changes could be noted in ovulation and fertilization rates but when the females were held under confined conditions of captivity during 3 days before treatment, ovulation rate dropped to 40% ($P < 0.001$). However, the fertility of ovulated oocytes was not affected.

Effect of the Retention of Ova on Fertility

In the first experiment where each female was sampled once only, no significant alteration of fertility occurred during 48 hours after ovulation (fertilization rates at 40 degree-days: 90 and 85%). On the third day following ovulation, the fertilization rate sharply dropped to 30% (Fig. 2).

In the second experiment, where each female was sampled every day during 6 days, a significant alteration occurred one day earlier (day 2; Fig. 2). The medium fertility (50%) obtained on day 0 results from the nature of the hormonal treatment to induce ovulation and from the moment fertility was checked (at hatching).

Stimulation of Sperm Release

The greatest sperm release was obtained with progesterone (10 mg/kg) and occurred three days after the second injection; the average volume delivered was twice as much as in the controls and all 7 animals responded (Fig. 3). With testosterone (10 mg/kg) no change could be noted in 3 males and the response was heterogeneous in the remaining 4 individuals. HCG (1,000 IU/kg) seemed to have a delayed effect after 10 days.

In the second experiment very high doses of progesterone (100 mg/kg) were the most efficient; all the fish were stimulated and the maximum response (a threefold increase $P < 0.001$) was recorded 48 hours after treatment. The spermatocrit was not affected by the hormonal treatment, and insemination performed at the 1:1,000 dilution ratio led to high fertilization rates (mean: 70%); there was no significant difference from the fertilizing

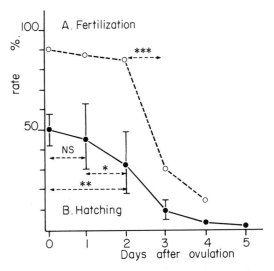

FIGURE 2.—*Effect of the in vivo retention of ova on fertilization and hatching. Curve A: each female was sampled only once. Each point is the mean fertility of ova from 2 females. Fertility was checked at 40 degree-days. Curve B: each of the 7 females was sampled once a day and fertility was checked at hatching time. Vertical bars show confidence limits at the 95% level of probability. Levels of significance: NS: $P > 0.05$; *: $P < 0.05$; **: $P < 0.01$; ***: $P < 0.001$.*

capacity of the control sperm. With the intermediate dose of progesterone (10 mg/kg) only 5 males out of 10 showed a significant response (2 to 3 times increase). With the lowest dose (1 mg/kg) there was a 2 to 3 times increase in 4 animals (Fig. 4).

Cryopreservation of the Sperm

Among the dilution ratios tested during cryopreservation tests, 1:2 gave the best results (Fig. 5).

The motility, activity, and fertilizing capacity of the ten sperm samples cryopreserved individually appear in Table 1. Eight of the sperm samples showed variable fertilizing capacities after cryostorage (1% to 68%), while two had lost all viability (no motility and no fertilizing ability).

Artificial Insemination by a Dilution Technique

The fertilization rate (Fig. 6) stayed at its maximum value (94–98%) with the 10^{-1} to 10^{-3} dilution ratios and dropped at 10^{-4} (41%) and 10^{-5} (10%). The control (dry insemination) was fertilized at 75%.

When fresh sperm was diluted in DIA_{TG} prior to insemination, fertilization was still 80% after 2 min and 15–25% after 8 min (Fig. 7). In fresh water, fertilization dropped to 25% after 1 min and was no longer possible after 2.5 min.

Discussion

Induction of Ovulation

Our data on induced ovulation clearly demonstrate the efficiency of an adequate hormonal treatment (a single injection of partially purified salmon gonadotropin) and bring out the drastic effects of captivity which have been discussed previously (de Montalembert et al. 1978).

Hormonal control of ovulation offers the obvious advantage of independence with regard to such restraints as temperature and space. Moreover, the females become synchronized, since most ovulations occur on day 4 after treatment. However, before a widespread use of the technique can be considered, progress should be made towards economic production of standardized fish gonadotropic extracts.

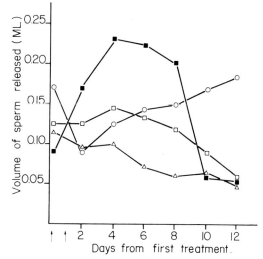

FIGURE 3.—Effect of 2 injections of hormones administered 24 hours apart (arrows) on the volume of sperm released. ○: human chorionic gonadotropin (1,000 IU/kg); ■: progesterone (10 mg/kg); □: testosterone (10 mg/kg); △: controls.

In Vivo Retention of the Ova

The drop in fertility occurred on the third day after ovulation in our first experiment and on day 2 in the second experiment. A slightly higher water temperature in the latter case, as well as the repeated handling of the females (once a day during 6 days), and individual variability might account for such a difference.

The in vivo aging of ova appears to be a rapid phenomenon in northern pike when compared to that in rainbow trout, in which ova fertility is preserved for at least 8 days after ovulation (Sakai et al. 1975; Escaffre et al. 1976; Bry, unpublished data).

Stimulation of Sperm Release

The highest and most efficient dose of progesterone tested was 100 mg/kg. In the goldfish, doses of 10 mg/kg (progesterone) and 100 to 1,000 mg/kg (methyltestosterone) were necessary for a maximum response (Yamazaki and Donaldson 1969; Billard 1976). The requirement for such considerable amounts might result from a gradual absorption of the steroid as Yamazaki and Donaldson (1969) pointed out, and from biological inactivation at the

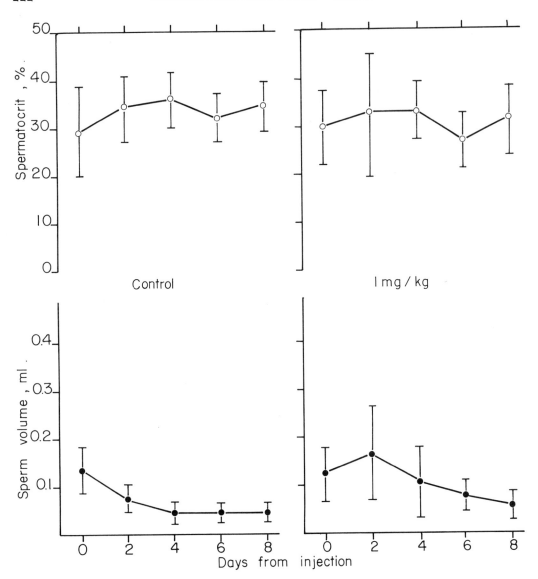

FIGURE 4.—*Effect of a single injection of progesterone (1, 10, 100 mg/kg) on sperm release and spermatocrit. Vertical bars show confidence limits at the 95% level of probability.*

plasma level by protein binding as suggested by the existence of two steroid binding systems with high capacities in female trout plasma (Fostier and Breton 1975).

Furthermore, a lack of responsiveness might occur at the testicular level, because progesterone must probably be metabolized to activate sperm release.

Our stimulation experiments were carried out on males that delivered 100–150 μl of sperm at the first sampling, prior to treatment. The question of the possible reinitiation of sperm release at the end of the reproductive season, when most males cease to give milt, would certainly be worthy of investigation.

Cryopreservation of the Sperm

We obtained considerable variability (0–68%) in the fertilizing capacity of the sperm

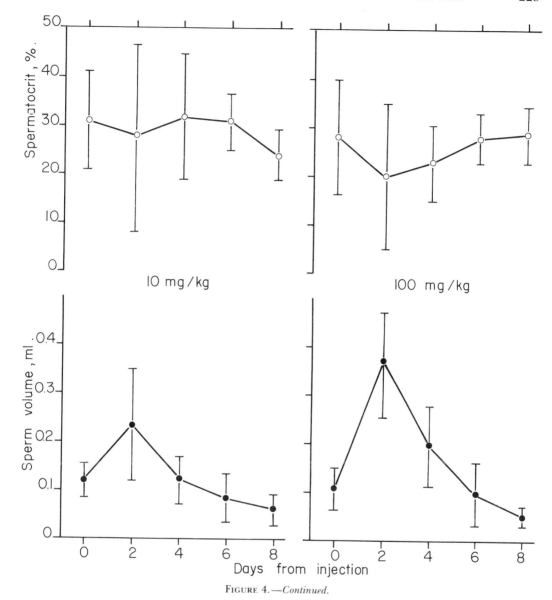

FIGURE 4.—Continued.

after cryopreservation, depending on the donor male, as Mounib et al. (1968) did with Atlantic cod sperm. Some steps of the freezing, freezing-storage, and thawing procedures cause irreversible modifications in some spermatozoa which lose their fertilizing ability, while others retain it. A careful adjustment of each step should allow progress toward the achievement of the ideal conditions. It might also be of interest to examine whether or not the degree of fertility of semen from a given male following cryopreservation is a reproducible feature.

Regarding the estimation of fertility after cryostorage, no rigorous protocol has ever been defined to the authors' knowledge. In order to determine what is the actual rate of fertility preserved after cryostorage one should work in slightly saturating conditions, i.e., inseminate with the minimum amount of fresh sperm required to obtain

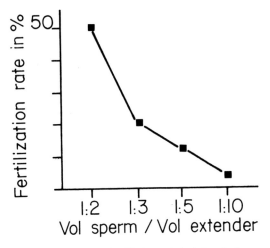

FIGURE 5.—Effect of the dilution ratio (vol sperm/vol cryo-extender) on the fertilizing capacity of sperm after thawing. Sperm was pooled from 10 males. Each dilution was tested with 2 samples.

the highest fertilization rate. The fertilizing ability after storage could then be expressed as a valid percentage of the initial level.

Such an approach could not be realized in our reported experiments, since the influence of high dilutions (10^{-3}, 10^{-4}) on fertilization had not been tested at the time of our cryopreservation attempts.

Artificial Insemination with a Diluent

According to our results the sperm dilution should not exceed 1:1,000. This is a safety measure to reduce the influence of individual variations in the sperm concen-

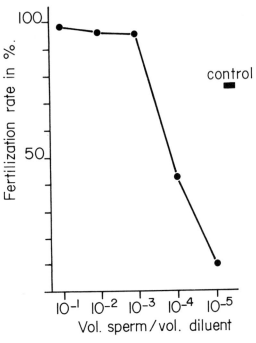

FIGURE 6.—Effect of the dilution ratio (vol fresh sperm/vol diluent) on fertilization.

trations and quality as mentioned by Billard (1975) for trout. In the northern pike, sperm concentration has been found to vary in the range of $18-30 \times 10^9$ spermatozoa/ml (Lindroth 1946; de Montalembert unpublished data). Such concentrations correspond to 800,000 spermatozoa per fertilized egg when the dilution ratio is 1:1,000. An insemination diluent elaborated for northern pike with specific values of osmotic pressure and pH might extend

TABLE 1.—Effect of cryostorage on the motility, activity, and fertilizing capacity of northern pike sperm.

	Fresh sperm			Thawed sperm		
Male no.	Motility in DIA_{TG}[a]	Activity min	Fertilizing capability[b] %	Motility in DIA_{TG}[a]	Activity min	Fertilizing capability[b] %
1	4	1.50	76	3	0.50	50
2	4	1.75	50	2	0.75	48
3	4	2.25	75	0	0	0
4	3	2.25	72	1	0.15	1
5	4	1.75	80	2	0.80	66
6	3	2.50	70	2	0.35	14
7	4	2.75	84	0	0	0
8	5	2.50	82	3	0.75	68
9	4	1.75	72	1	0.25	5
10	5	2.25	74	2	0.15	11

[a] Motility scale: 0 = non motile; 5 = fully motile.
[b] Volume sperm/volume diluent (DIA_{TG}) = 10^{-2}.

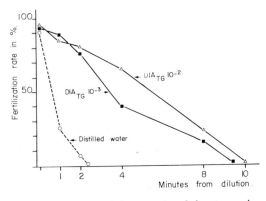

FIGURE 7.—The fertilizing capacity of the sperm when diluted in DIA_{TG} or water before insemination.

the viability of the sperm and consequently allow greater dilution rates. The ratio between the number of ova and the volume of diluent is also an important parameter in artificial insemination (Billard 1975). In the present study, we have only worked with one ratio (300 ova/10 ml diluent) and the optimum value of this parameter remains to be determined.

Conclusions

Due to the timing and synchronization of induced ovulations, artificial insemination can be planned in advance and hormonal activation of sperm release performed in the males, if necessary, 2 days before insemination. Insemination should be carried out within 24 hours of ovulation. Handling of the breeders can be reduced to a minimum: once for hormonal treatment and once for gamete sampling. Fish need to be held in captivity only for a few days. The use of DIA_{TG} as an insemination diluent allows reproducible results with minimum amounts of sperm.

Acknowledgements

This work was supported by a grant from Le Conseil Supérieur de la Pêche. The authors wish to thank B. Breton for providing the partially purified salmon gonadotropin and J. Marcel for secretarial assistance.

References

ANWAND, K. 1963. Die Wirkung von Hypophysen und Gonabionijektionen auf Hechtmilchner. Dtsch. Fisch. Ztg. 10:202–207.

BILLARD, R. 1975. L'insémination artificielle de la truite *Salmo gairdneri* Richardson. V-Effets de la dilution et définition du rapport optimum gamètes/dilueur. Bull. Fr. Piscic. 257:121–135.

———. 1976. Induction of sperm release in the goldfish by some steroids. IRCS Med. Sci. 4:42.

———. 1977. Utilisation d'un système tris-glycocolle pour tamponner le dilueur d'insémination pour truite. Bull. Fr. Piscic. 264:102–112.

CHIMITS, P. 1947. Note sur le repeuplement artificiel du brochet. Bull. Fr. Piscic. 146:16–24.

———. 1956. Le brochet. Bull. Fr. Piscic. 180:81–96.

COINTAT, M., AND R. DARLEY. 1953. Esociculture du Der: observations sur les campagnes 1951 et 1952. Bull. Fr. Piscic. 169:153–163.

DE MONTALEMBERT, G., B. JALABERT, AND C. BRY 1978. Precocious induction of maturation and ovulation in northern pike (*Esox lucius*). Ann. Biol. Anim. Biochim. Biophys. (Submitted for publication.)

EMMENS, C. W. 1947. The motility and viability of rabbit spermatozoa at different hydrogen ion concentrations. J. Physiol. 106:471–481.

ESCAFFRE, A. M., J. PETIT, AND R. BILLARD. 1976. Evolution de la fécondabilité des ovules de truite arc-en-ciel laissés dans la cavité coelomique après ovulation. IIè Congrès Europ. Ichthyol. Paris, Septembre 1976.

FOSTIER, A., AND B. BRETON 1975. Binding of steroids by plasma of a teleost: the rainbow trout, *Salmo gairdneri*. J. Steroid Biochem. 6:345–351.

HUET, M. 1972. Elevage et maturation de géniteurs brochets en petits étangs. Int. Ver. Theor. Angew. Limnol. Verh. 18:1128–1134.

———. 1976. Reproduction, incubation et alevinage du brochet (*Esox lucius*). EIFAC (Eur. Inland Fish. Advis. Comm.) Tech. Pap. 25:147-163.

JASPERS, E. 1972. Some spermatological aspects of channel catfish. Ph.D. Thesis. La. State Univ. 98 pp.

KAUSCH, H. 1976. Breeding habits of the major cultivated fishes of EIFAC region and problems of sexual maturation in captivity. EIFAC (Eur. Inland Fish. Advis. Comm.) Tech. Pap. 25:43–52.

LINDROTH, A. 1946. Zur Biologie der Befruchtung und Entwicklung beim Hecht. Mitt. Anst. Binnenfischerei, Drottningholm 24:1–173.

MOUNIB, M. S., P. C. HWANG, AND D. R. IDLER 1968. Cryogenic preservation of atlantic cod (*Gadus morhua*) sperm. J. Fish. Res. Board Can. 25:2623–2632.

PETIT, J., B. JALABERT, B. CHEVASSUS, AND R. BILLARD 1973. L'insémination artificielle de la truite arc-en-ciel *Salmo gairdneri* Richardson. I-Effets du taux de dilution, du pH et de la pression osmotique du dilueur sur la fécondation. Ann. Hydrobiol. 4:201–210.

PORTAL, J. 1947. Observations sur la pisciculture artificielle du brochet. Bull. Fr. Piscic. 147:61–70.

PREUDHOMME, J. G. 1968. Rapport annuel 1967. Sur la pêche et la pisciculture dans les eaux continentales. Rev. Annu. du Fishing Club du Moyen Atlas.

———. 1975. Troubles sexuels résultant de l'acclimatation chez les poissons, notamment brochet et truite arc-en-ciel. La Pisciculture Française 43:38–46.

SAKAI, K., M. NOMURA, F. TAKASHIMA, AND H. OTO 1975. The over-ripening phenomenon of rainbow trout. II-Changes in the percentage of eyed eggs, hatching rate and incidence of abnormal alevins during the process of over-ripening. Bull. Jap. Soc. Sci. Fish. 41:855–860.

SCHAPERCLAUS, W. 1961. Lehrbuch der Teichwirtschaft. 2te Auf. Paul Parey, Berlin et Hambourg. 582 pp.

SORENSON, L., K. BUSS, AND A. D. BRADFORD. 1966. The artificial propagation of esocid fishes in Pennsylvania. Prog. Fish-Cult. 28:133–141.

STEIN, H. 1977. Cryopreservation of the sperm of some freshwater teleosts. Ann. Biol. Anim. Biochim. Biophys. (Submitted for publication.)

YAMAZAKI, F., AND E. M. DONALDSON. 1969. Involvement of gonadotropin and steroid hormones in the spermiation of the goldfish (*Carassius auratus*). Gen. Comp. Endocrinol. 12:491–497.

A Muscular Dystrophy-Like Anomaly of Walleye

Philip P. Economon

Fish Pathology Laboratory
Minnesota Department of Natural Resources
St. Paul, Minnesota 55155

Abstract

A skeletal muscle abnormality found in a series of 32 walleyes obtained from widely distributed lakes in Minnesota between 1959 and 1974 is described. These discrete to diffuse, yellowish-brown, coarsely fibrous and fatty muscle tissue changes occur more frequently in adult male walleyes caught from comparatively small, fertile lakes that are managed almost exclusively by stocking. The most common and distinguishing histologic features of muscular dystrophy in man and the dystrophy-like myopathies in animals were found in this myopathy of walleyes. A unique feature of the walleye lesion is the formation of concentric laminations of hyalinized muscle fiber substance that appears to develop from the consolidation of aberrant myofibril encirclements known as "ring-binden." This myopathy, known in Minnesota as myofibrogranuloma (MFG), has some basic similarities with genetically acquired dystrophic muscle processes that occur in several domestic species of birds and mammals. MFG remains in a group of dystrophic myopathies in which the pathogenesis requires further elucidation, so it may be considered more accurate to refer to this myopathy as a muscular dystrophy-*like* anomaly.

Myofibrogranuloma (MFG) is a term I introduced in 1961 to identify a unique form of skeletal muscle degeneration that has occasionally been found in adult walleyes from Minnesota waters. This myopathy is characterized by profound alteration of the trunk musculature produced by extensive hypertrophy of its muscle fibers (Figs. 1, 2). The resultant process is a swollen, heavy, coarsely fibrous, lipogranuliform muscle lesion of an opaque to semi-translucent yellowish-brown color (Economon 1970, 1974, 1975).

Included in this pattern of striated muscle deformation is a consolidation and partial fusion of contiguous muscle fibers forming prominent aggregates of rough, cord-like strands of varied lengths interspersed within the muscle, which eventually is reduced to a friable, coagulated, and mineralized state. Neither hyperplasia nor hemorrhage is evident in these non-suppurative muscle fiber processes, and there is no visible evidence that lesion resorption or muscle regeneration takes place.

The lesions are often more severe in the general musculature surrounding the spinal column and in the deeper strata of the large dorsal muscles. The segments of muscle adjacent the skin are frequently affected and such lesions often appear isolated and detached from the more deeply seated myopathic processes (Fig. 3). On the basis of diseased muscle specimens we have collected and examined over several years, it appears that virtually all skeletal muscles of the walleye are susceptible to this muscular dystrophy-like anomaly.

Thirty-two MFG-afflicted walleye specimens, principally males, were investigated over the period 1959 to 1974 from 25 different lakes and ponds in Minnesota (Economon 1974). Their ages ranged from 3 to 10 years, and weights were from 1.5 to 3.6 kg, with an average age of 4.1, and weight of approximately 2.7 kg. Intact walleyes that were available for examination appeared as normally developed examples, having no demonstrable external pathological processes nor physical deformities, excepting occasional posteroventral, and anal petechiasis. Gross examinations of the kidney, liver, spleen, and circulatory and central nervous systems showed these structures to be essentially within normal limits. The gastrointestinal tract and its contents indicated active foraging fishes, typical for their respective ages. The perivertebral and dorsal muscles of the younger, less severely affected walleyes, showed occasional focal to diffuse capillary injection. The gonads exhibited uniform

FIGURE 1.—*Gross appearance of skeletal muscle structure in a typical case of myofibrogranuloma of walleye. From a ten-year-old male of 66 cm length and 3.2 kg body weight. Reeds Lake, Waseca County, Minnesota, 1965. (MFG-13/Economon.)*

development, with superficial erythema as an intermittent symptom. The remaining anatomical structures and tissues were free of any noticeable aberrancies.

Only two of the affected walleyes showed symptoms of paralysis when captured. According to the catch data obtained from the other specimens in this group, no outward manifestations of cachexia, unusual motor function, or partial immobilization were apparent. On the basis of information available at present, it appears that a subclinical

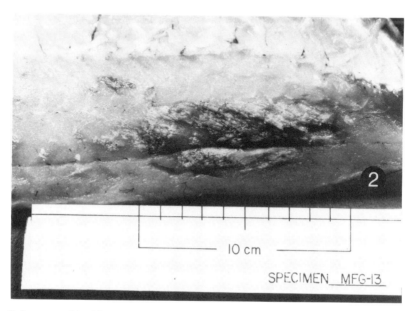

FIGURE 2.—*Enlargement of focal lesion in muscle of Reeds Lake walleye, demonstrating the coarsely fibrous, granular, and swollen condition characteristics of the myopathy. (MFG-13/Economon.)*

state may exist in some specimens with patently gross lesions of MFG. However, the walleye is recognized as a passive swimmer, and somewhat submissive when hooked and removed from the water, so that these traits make it difficult to detect or assess with precision any abnormal behavior or muscular incoordination that is present. Carefully controlled captive conditions are required to study any disability phenomena more fully.

The absence of any notable outward symptoms of the anomaly in live specimens has limited the collection and rapid identification of diseased individuals from larger samples of walleye in the general population. An effective screening method allowing efficient utilization of trap net sampling and tagging procedures has not up to this time been feasible. A method of detecting the myopathy in live walleyes by combining ultrasound and biopsy techniques is presently under investigation at this laboratory. Detailed information on the incidence, distribution, and behavior of affected walleyes is necessary in order to examine what influence heredity, nutrition, and environmental stress might have on the development of the myopathy.

MFG has been found exclusively in walleyes, and it has been recovered from quite widely distributed locations in Minnesota. The walleyes from mesotrophic waters, such as Lake Winnibigoshish in northern Minnesota and other similarly productive walleye lakes, appear to have a significantly lower incidence of the myopathy than those from eutrophic lakes of comparable size such as Lake Minnetonka, Hennepin County, which is situated in a suburban area just west of Minneapolis. Minnetonka has provided six MFG-affected walleyes for this study, the largest number of specimens recorded from the known occurrences of the myopathy in Minnesota lakes. The apparent difference in incidence of the myopathy between eutrophic and mesotrophic waters is about two to one, respectively. However, the actual number of grossly affected walleyes from eutrophic lakes is estimated to be of an order of magnitude probably no greater than one out of 10,000 in the age bracket of 4.5 years and older.

A variation has been observed recently indicating that a higher frequency of this anomaly occurs in walleyes from comparatively small, fertile lakes and ponds in which this species is maintained exclusively by periodic stocking of hatchery-raised walleyes. This trend is evident even when walleye populations in such lakes become relatively low. Even though the small eutrophic lakes that support no natural reproduction of walleyes appear to have a higher incidence of MFG, there is no indication that it causes serious morbidity or measurable mortality in walleyes at any of the lakes in which it has been found.

Microscopy

Representative samples of typically affected skeletal muscle tissues from eight different walleye specimens preserved in a 10% buffered formalin solution were embedded in paraffin; sectioned at six microns; and stained with hematoxylin-eosin, Mallory's phosphotungstic acid hematoxylin, Brown and Brenn (Gram's) stain for bacteria, acid-fast stain (Ziehl-Neelsen), Bauer-Feulgen and Grocott's silver technique for fungi, periodic acid-Schiff method, and Masson's (modified) trichrome stain.

Histologic sections of affected skeletal muscle show extensive focal to diffuse lesions characterized by muscle fibers in various stages of dystrophy.

The incipient lesion exhibits a marked enlargement and rounding of the muscle fibers, with decreased eosinophilia (Fig. 4). The normal uniform size and polygonal outline of muscle fibers disappears, and some of the fibers also become extremely atrophic. Only a few fibers in a fasciculus are affected in this stage of the process. A distinguishing feature of markedly hypertrophied muscle fibers in transverse section is a disarrangement of the peripherally located myofibrils (Figs. 5a and 5b). These peripheral myofibrils form distinct bands or "ringbinden" which encircle the central core of longitudinally arranged myofibrils

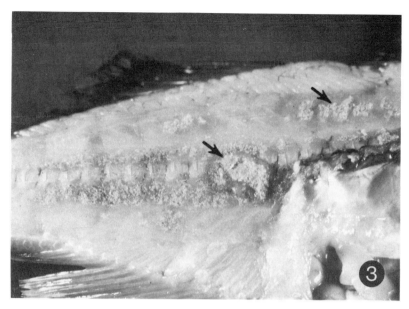

Figure 3.—*View of sagittal plane of Reeds Lake walleye, exhibiting massive focal to diffuse muscle fiber lesions in area of spinal column and dorsal fin. (MFG-13/Economon.)*

of the muscle fiber. Distinct transverse striations are still recognizable at this stage of lesion development with hematoxylin-eosin, and phosphotungstic acid-stained preparations.

The more advanced lesions may involve most or all of the muscle fibers in a fasciculus (Fig. 6). The swollen muscle fibers stain intensely with eosin and the sarcoplasm has coagulated and retracted from the endo-

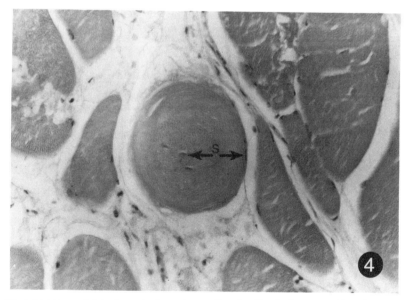

Figure 4.—*The initial phase of skeletal muscle change in an affected walleye is characterized by only a few faintly staining hypertrophic and atrophic fibers in a fasciculus. Note rounded hypertrophic fiber with apparent internalization of sarcolemmal nuclei (s), and circular arrangement of peripheral myofibrils. (Hematoxylin-eosin; arrow length represents 25 microns.)*

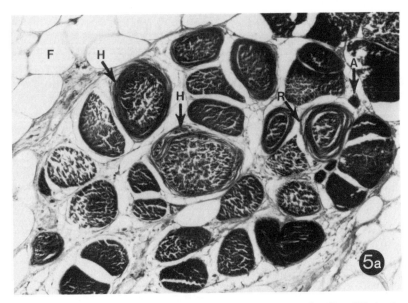

FIGURE 5a.—*Transverse section of a more advanced stage of hypertrophy with swollen fibers (H) showing characteristic peripheral myofibril encirclements. Three markedly atrophic fibers (A) appear along side a swollen pear-shaped fiber exhibiting "ringbinden" (R). (Phosphotungstic acid-hematoxylin; arrow length represents 35 microns.)*

mysium. The fibers are extremely brittle and shatter transversely. The transverse striations become attenuated and finally are no longer apparent with either routine or special staining procedures. In cross section, the more peripheral myofibrils, recognized in less severely affected fibers as "ringbinden," appear as hyalinized concentric laminations (Fig. 7). Mineralization or calcification is evident in the central portion of some fibers in this more granular stage of degeneration. A multi-layered

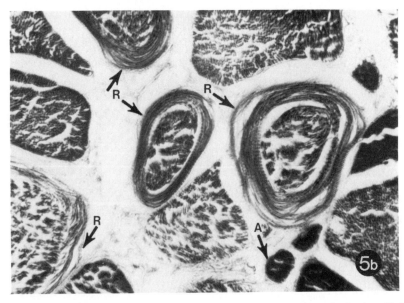

FIGURE 5b.—*Enlargement of Figure 5a, to show structural details of hypertrophic fibers, and striated myofibril annulets (R). (Arrow length represents 15 microns.)*

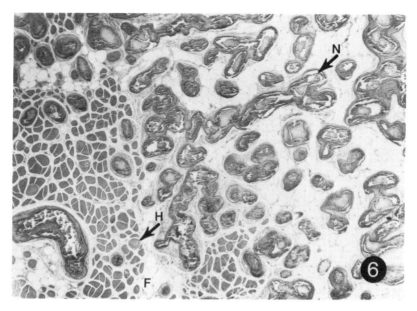

FIGURE 6.—*Transverse section of myopathy showing transition of essentially normal appearing polygonal muscle fibers, on the left side of field, to advanced hyalinized-necrotic fibers (N) on the right. Numerous fibers in intermediate stages of hypertrophy (H) and atrophy occur in a stroma of fatty endomysial connective tissue (F). (Hematoxylin-eosin; arrow length represents 175 microns.)*

zone of epithelioid-like cells with vesicular nuclei, located peripheral to some necrotic fibers, is the only demonstrable evidence of an inflammatory response. In some fasciculi, an extensive coalescence of adjacent necrotic fibers occurs, giving rise to expanded

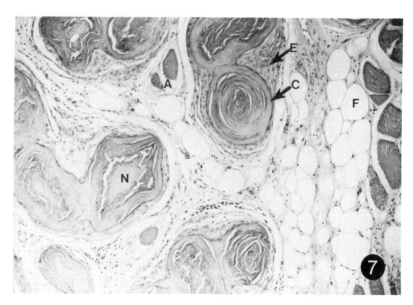

FIGURE 7.—*Early phase of coalescence of markedly hypertrophic muscle fibers. The hyalinized concentric laminations (C) manifested in this stage appear to arise from myofibril encirclements recognized in less severely affected fibers as "ringbinden." A scattering of grossly unaffected muscle fibers may occur in close proximity to atrophic (A) and hyalinized-necrotic fibers (N). (Hematoxylin-eosin; arrow length represents 50 microns.)*

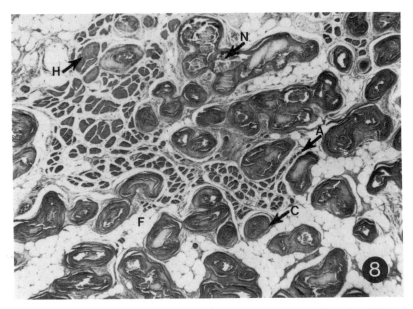

FIGURE 8.—*In the final phase, large areas of whole fascicles have been reduced to markedly swollen (H), laminated (C), and coalesced necrotic fiber masses (N), surrounded by a dense stroma of adipose connective tissue (F). Small clusters of atrophic (A), and less seriously affected muscle fibers are found in endomysial spaces not occupied by fat. (Hematoxylin-eosin; arrow length represents 175 microns.)*

hyalinized masses. A minimal degree of fibrosis is evident between dystrophic muscle fibers. However, marked infiltration of adipose connective tissue is seen in these advanced lesions (Figs. 8 and 9). Abnormalities of the intramuscular nerves and vasculature were not demonstrated by the histologic methods employed.

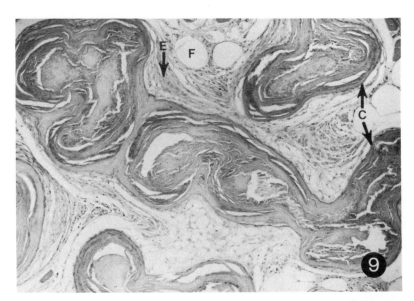

FIGURE 9.—*Advanced state of coalesced muscle fibers showing concentric laminations (C) of hyalinized and necrotic tissue, and fatty composition (F) of endomysium which contains a multi-layered zone of epithelioid-like cells (E) with vesicular nuclei. (Hematoxylin-eosin, arrow length represents 50 microns.)*

Dystrophic Myopathies

Hereditary dystrophy-like myopathy, inherited as an autosomal recessive character and closely resembling forms of human muscular dystrophy, has been reported to occur in the mouse (Michelson et al. 1955; Harman et al. 1963; West et al. 1966), Syrian hamster (Homburger et al. 1962; Homburger et al. 1966), chicken (Asmundson and Julian 1956; Julian and Asmundson 1963; Asmundson et al. 1966), and turkey (Harper and Parker 1964, 1967).

The principal histologic characteristics of these myopathies in both animals and man are the great variation in diameter of muscle fibers, the rounding of muscle fibers in transverse section, and the eventual progression to coagulation necrosis (Bethlem 1970; Walton 1973; Hadlow 1973). Adipose cell and connective tissue proliferation and replacement, increased numbers of internally placed sarcolemmal nuclei, regenerative attempts manifested by myoblast proliferation, vacuolization of dystrophic muscle fibers, and a peripheral disorientation of myofibrils to form so-called ringbinden or striated annulets, are somewhat less consistent histologic features of muscular dystrophy in man (Walton 1973), and dystrophy-like myopathies in animals (Hadlow 1973). The significance of ringbinden, exhibited in transverse sections of muscle fibers as a spiral band of aberrant myofibrils encircling the central core of myofibrils, has been the subject of considerable controversy (Adams et al. 1962; Bethlem 1970; Walton 1973). Investigators have variously considered these structures to be artifacts of either fixation or embedding, or to be a manifestation of in situ functional or anatomic separation of muscle from its tendinous attachment (Walton 1973).

Discussion

The most common and distinguishing histologic features of muscular dystrophy in man and the dystrophy-like myopathies in animals are embodied in the myopathy of walleyes reported in this study. The abnormal rounding of muscle fibers was associated with both markedly hypertrophied and atrophied muscle fibers. Distinctive ringbinden was frequently associated with enlarged fibers that exhibited decreased eosinophilia. The more advanced lesions were characterized by coagulation necrosis of muscle fibers, mineralization, and pronounced infiltration of the endomysium by fatty tissue. Extraordinarily large masses of necrotic sarcoplasm, often with a dense capsular zone of epithelioid-like cells, appears to be the result of coalescence of individual necrotic muscle fibers. Only a slight amount of actual fibrosis, or condensation of endomysial connective tissue, was observed. Muscle regeneration appeared to be entirely absent, and internalization of sarcolemmal nuclei was not a consistent feature.

The histologic similarities of the myopathy in walleye specimens to those of muscular dystrophy in man and hereditary dystrophy-like myopathies in animals, suggests that this abnormality might have a genetic basis. Further investigations beyond the scope of this report are required to illucidate the pathogenesis of this myopathy. Perhaps this walleye anomaly will be useful as a new experimental model for research in allied biomedical investigations.

Acknowledgments

The author gratefully acknowledges the assistance of Kenneth H. Johnson of the University of Minnesota Department of Veterinary Biology and his staff for helpful advice in the pursuit of this investigation, and to Ms. Joan Walz, formerly of the Experimental Surgery Laboratory at University Hospital for assistance in tissue sectioning. I also wish to thank the Fisheries personnel of the Department of Natural Resources who have helped me collect pathological fish specimens and field data, and for John B. Moyle's many valuable suggestions and constructive review of the manuscript which culminated in this presentation.

References

ADAMS, R. D., D. DENNY-BROWN, AND C. M. PEARSON. 1962. Diseases of muscle: a study in pathology, 2d ed. Harper and Brothers, New York. 735 pp.

ASMUNDSON, V. S., AND L. M. JULIAN. 1956. Inherited muscle abnormality in the domestic fowl. J. Hered. 47:248–252.

———, F. H. KRATZER, AND L. M. JULIAN. 1966. Inherited myopathy in the chicken. Ann. N.Y. Acad. Sci. 138:49–58.

BETHLEM, J. 1970. Muscle pathology, introduction and atlas. North-Holland, Amsterdam. 132 pp.

ECONOMON, P. P. 1970. Characteristics of some common fish diseases and parasites. Pages A 13–14 *in* Manual of instructions for lake survey. Spec. Publ. 1 (Rev. 1970). Minn. Dep. Conserv., Sect. Fish., St. Paul.

———. 1974. A survey of myofibrogranuloma in walleyes in Minnesota from thirty-two case reports 1959–1974. Minn. Dep. Nat. Resour. Fish Wildl. Pathol. Arch. 19:25–47.

———. 1975. Myofibrogranuloma, a muscular dystrophy-like anomaly of walleye (*Stizostedion vitreum vitreum*). Minn. Dep. Nat. Resour. Fish Wildl. Spec. Publ. 113. 13 pp.

HADLOW, W. J. 1973. Myopathies of animals. Pages 364–409 *in* C. M. Pearson and F. K. Mostofi, eds. The striated muscle. Williams and Wilkins, Baltimore.

HARMAN, P. J., J. P. TASSONI, R. L. CURTIS, AND M. B. HOLLINSHEAD. 1963. Muscular dystrophy in the mouse. Pages 407–456 *in* G. H. Bourne and M. N. Golarz, eds. Muscular dystrophy in man and animals. Hafner, New York.

HARPER, J. A., AND J. E. PARKER. 1964. Hereditary muscular dystrophy in the domestic turkey, *Meleagris gallopavo*. Poult. Sci. 43:1326–1327.

———, and ———. 1967. Hereditary muscular dystrophy in the domestic turkey. J. Hered. 58:189–193.

HOMBURGER, F., J. R. BAKER, C. W. NIXON, AND G. WILGRAM. 1962. New hereditary disease of Syrian hamsters. Primary, generalized polymyopathy and cardiac necrosis. Arch. Intern. Med. 110:660–662.

———, C. W. NIXON, M. EPPENBERGER, AND J. R. BAKER. 1966. Hereditary myopathy in the Syrian hamster. Studies on pathogenesis. Ann. N.Y. Acad. Sci. 138:14–27.

JULIAN, L. M., AND V. S. ASMUNDSON. 1963. Muscular dystrophy of the chicken. Pages 457–498 *in* G. H. Bourne and M. N. Golarz, eds. Muscular dystrophy in man and animals. Hafner, New York.

MICHELSON, A. N., E. S. RUSSEL, AND P. J. HARMAN. 1955. Dystrophia muscularis: a hereditary primary myopathy in the house mouse. Proc. Nat. Acad. Sci. U.S.A. 41:1079–1084.

WALTON, J. N. 1973. Progressive muscular dystrophy: structural alterations in various stages and in carriers of Duchenne dystrophy. Pages 263–291 *in* C. M. Pearson and F. K. Mostofi, eds. The striated muscle. Williams and Wilkins, Baltimore.

WEST, W. T., H. MEIER, AND W. G. HOAG. 1966. Hereditary mouse muscular dystrophy with particular emphasis on pathogenesis and attempts at therapy. Ann. N.Y. Acad. Sci. 138:4–13.

Lymphosarcoma in Muskellunge and Northern Pike: Guidelines for Disease Control[1]

R. A. SONSTEGARD

Department of Microbiology
University of Guelph
Guelph, Ontario N1G 2W1

JOHN G. HNATH

Michigan Department of Natural Resources
Wolf Lake Hatchery
Mattawan, Michigan 49071

Abstract

Epizootics of a malignant blood cancer (lymphosarcoma) affect feral populations of northern pike and muskellunge. Overall frequencies of occurrence of the disease in northern pike and muskellunge as high as 20.9% and 16%, respectively, were found. The disease in feral muskellunge causes high mortalities while in northern pike spontaneous regressions are common. The disease is transmitted percutaneously during the act of spawning. The disease does not seem to be transmitted to progeny via the egg. These species should be stocked as eggs or fry, not as adults, if the spread of lymphosarcoma is to be restricted.

Today, aquaculture has assumed an increasingly important role in fishery management in view of man's encroachment on the environment and increasing sport fishing pressure. Historically, epizootics have caused devastating losses in a number of hatchery reared species.

In this paper I summarize the biology of one of the most important diseases of coolwater fishes. The disease, lymphosarcoma, is a malignant blood cancer which infects northern pike and muskellunge. It is contagious and may infect as high as 21% of feral populations of northern pike and 16% of muskellunge (Sonstegard 1970, 1975, 1976a), and has obvious importance to aquaculture and fisheries management. I offer guidelines for control of the disease based on field epizootiological studies and laboratory investigations.

Occurrence

Probably the most striking feature of the lymphosarcoma in *Esox* is that it exhibits the highest frequency of occurrence of a malignant cancer in any known free-living vertebrate. The disease has widespread geographical distribution in North America. It has been diagnosed histopathologically in northern pike from Alaska, Northwest Territories, Alberta, Saskatchewan, Manitoba, Ontario, Quebec, New York, and Michigan (Sonstegard 1976a). To date, the muskellunge lymphosarcoma has only been diagnosed in Ontario. In the Old World, the disease has been diagnosed only in Ireland (Malcahy 1963) and Sweden (Ljunberg 1976). However, responses to my questionnaires circulated to fisheries workers throughout Europe and Asia leave little doubt that the disease is widespread in Old World northern pike. The disease is not a modern phenomena in *Esox*, based on the occurrence of the disease in both the Old and New World. The disease has apparently persisted in northern pike since the last glaciation and probably has played an important role in the evolution of fishes of the genus *Esox* (Sonstegard 1976a).

Clinical Characteristics

The following are some generalizations of the clinical course of the *Esox* lymphosarcoma. No fish was found to have the disease without involvement of the skin. The skin lesions may be found on a variety

---

[1] This work was supported by grants from the National Cancer Institute of Canada. R. A. S. is a Research Scholar of the National Cancer Institute of Canada.

FIGURE 1.—*Northern pike with large lesion of lymphosarcoma on the caudal peduncle. The tumor is typically pink in color, elevated, soft to the touch, fragile, and easily ruptured during the process of netting.*

of anatomical sites; however, most are on the posterior one half of the fish (Figs. 1 and 2). The tumor is unicentric, arising in the cutaneous tissues, forming a nodular tumor. The tumor spreads from the cutaneous focus via intracellular spaces to involve the underlying musculature. In the late stage of the disease, the kidney and spleen (major sites of hematopoiesis in *Esox*) become infiltrated and may be accompanied by leukemia. Occasionally, in the muskellunge the superficial cutaneous lesions spontaneously regress, however, in most cases the tumor progresses to the death of the animal (less than 1% survive) (Sonstegard 1970, 1975). The clinical course in the northern pike, although unresolved, appears similar to that of the muskellunge. However, in the northern pike, spontaneous regressions frequently occur (Sonstegard 1976a). It is not known whether the spontaneous regressions described here are a common occurrence or unique to the watershed studied.

Transmission

Both the muskellunge and northern pike lymphosarcomas are highly infectious. Normal northern pike and muskellunge held in aquaria together with tumor-bearing fish of the respective species developed the tumor. Similarly, the disease is readily transmittable by transplantation (Sonstegard 1976a; Ljunberg 1976). Electron microscopy studies of post-mitochondrial preparations made from frozen northern pike tumors revealed virus-like particles banding at 1.15–1.17 g/ml in density gradients (Papas et al. 1976). The viral enzyme (reverse transcriptase) was detected in association with the banded preparations, suggesting a viral etiology of the tumor (Sonstegard and Papas 1977; Papas et al. 1977). The viral enzyme activity dropped rapidly at temperatures above 20 C (Papas et al. 1976; Papas et al. 1977). It is hypothesized that the elevated water temperatures of the summer months are non-permissive temperatures for virus expression and may be the mechanism of spontaneous regression. Although virus has been found associated with the tumor, a viral etiology has not been confirmed.

The stocking of muskellunge fingerlings raised under hatchery conditions from spawn collected from endemic waters failed to introduce the disease into an isolated watershed. This constitutes strong circumstantial evidence that the disease is not transmitted to progeny via gametes from the parental generation. The clinical picture of a neoplasm spreading from a localized cutaneous tumor strongly suggests that the disease is introduced percutaneously. Only sexually mature fish

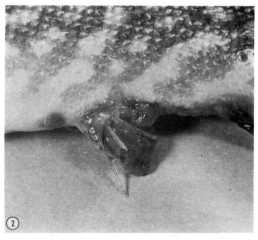

FIGURE 2.—*Northern pike with lymphosarcoma lesion just posterior to the vent.*

develop the neoplasm, and the observation that the tumors were most frequently found on the posterior one half of the body suggests that the disease is transmitted horizontally by contact during the act of spawning.

Recommendations for Control

Under no conditions should adult *Esox* (including normal-appearing specimens) be transported for stocking purposes. Only eggs, fry, and fingerlings, raised under hatchery conditions (isolated from adults), should be used for stocking purposes.

The cooperation of the public should be solicited. Specifically, people should be made aware of the dangers they pose to the fishery by transporting *Esox* to "their" lake, a practice widely followed by fishermen in some regions.

Research should progress toward a vaccine against the tumor, so that stocked *Esox* may be protected against endemic epizootics.

To further document the occurrence and epizootiology of the disease in various geographical realms, fishery biologists should monitor the disease in their regions and preserve suspect animals in 10% formalin for histopathological confirmation.

As the public becomes aware that the "red sores" in *Esox* are cancerous growths, the question invariably asked is whether the disease constitutes a health hazard. To date, laboratory studies have failed to suggest that the disease constitutes a human health hazard. The temperature studies of the viral enzyme associated with the tumor gave no activity at 37 C (human body temperature), and suggest that the likelihood of the virus in fish "jumping" across phylogenetic barriers and infecting man is a very remote one. The public should be advised of such findings as they occur.

References

LJUNBERG, O. 1976. Epizootiological and experimental studies of skin tumors in northern pike (*Esox lucius* L.) in the Baltic Sea. Prog. Exp. Tumor Res. 20:156–165.

MULCAHY, M. F. 1963. Lymphosarcoma in the Pike, *Esox lucius* L. (Pisces: Esocidae). Proc. R. Ir. Acad. Sect. B. 63:103–129.

PAPAS, T. S., J. E. DAHLBERG, AND R. A. SONSTEGARD. 1976. The presence of Type C virus in lymphosarcoma in fish (northern pike, *Esox lucius*). Nature (London) 261:506–508.

———, M. SHAFFER, T. PRY, AND R. A. SONSTEGARD. 1977. DNA polymerase in lymphosarcoma in northern pike (*Esox lucius*). Cancer Res. 37: 3214–3217.

SONSTEGARD, R. A. 1970. Descriptive and epizootiological studies of Infectious Pancreatic Necrosis Virus of salmonids and lymphosarcoma of *Esox masquinongy*. Ph.D. Thesis. Univ. Guelph, Guelph, Ontario.

———. 1975. Lymphosarcoma in muskellunge (*Esox masquinongy*). Pages 902–924 *in* W. E. Ribelin and G. Migaki, eds. Pathology of fishes. Univ. Wis. Press, Madison.

———. 1976a. Studies of the etiology and epizootiology of lymphosarcoma in *Esox* (*Esox lucius* L. and *Esox masquinongy*). Prog. Exp. Tumor Res. 20: 141–155.

———. 1976b. Studies of the etiology and epizootiology of lymphosarcoma in northern pike (*Esox lucius*) and muskellunge (*Esox* masquinongy). Bibl. Haematol. Basel 43:242–244.

———, AND T. S. PAPAS. 1977. Descriptive and comparative studies of C type virus DNA polymerase of fish (northern pike, *Esox lucius*). Proc. 1977 Meet. Assoc. Cancer Res. p. 806.

The Hatchery Development Process

CECIL L. FOX

Kramer, Chin & Mayo Inc.
1917 First Avenue, Seattle, Washington 98101

Abstract

Hatchery development involves five major phases: (1) exploratory studies to establish needs and feasibilities; (2) preliminary studies of siting, funding, and technical criteria; (3) project design; (4) construction; and (5) operation. This process, from determined need to full production, requires 4–8 years. There should be extensive interaction among biologists, managers, and architects through the study and early design phases; later design modifications can be costly of time and money. Construction should be overseen by owner representatives to ensure fidelity to the project concept. A Hatchery Management Manual is important for the project's success; it should be started well before operations begin, and continually updated thereafter.

Although this symposium is directed to coolwater fishes, my presentation is for any temperature of water, in that, when it comes to general procedures, the hatchery development process remains the same.

In order to avoid confusion when talking about this process, I have attempted to generalize identities or roles as follows.

Agency implies any private or governmental body with ownership, funding, and authorization responsibilities.

Management implies that individual or department directly responsible for the administration of the hatchery development process, including architecture and engineering contracts, cost controls, etc.

Using Branch is that subdivision of the Agency that establishes operating and design criteria and has ultimate responsibility for the operation of the facility.

A/E (architect/engineer) and Consultant identify those people or organizations that will do the technical development and documentation of the process. They can be private or a part of the Agency.

I use *Biologist* to include those involved with fisheries management and policy, and those who rear the fish. All questions and functions related to fish propagation are expected to be directed to these individuals. Such people normally are part of the using branch.

The development and building of a fish hatchery is an involved process encompassing many activities and many participants.

The agency, location, and the available resources for each project establish a unique set of conditions for a hatchery development; however, most follow a series of common events. *Exploratory studies* establish identifiable needs, the probability of support for a more detailed study of the proposed project, and the level of agency commitment to the project. Next, *preliminary studies*, which can become quite involved depending on the project and the development work required to verify actual conditions, culminate in a feasibility report, a very important document that sets the direction and character for the remainder of the project. Once the feasibility report is accepted by the agency, the *project design* phase carries through the completion of drawings, specifications, and cost estimates. The *construction phase* includes bidding, building, and the first two phases of the Hatchery Management Manual; the Manual is not a part of the construction contract, but the time schedule requires it be started at this time by the A/E and using branch. Finally, the agency assumes the benefits and responsibilities for a unique prototype hatchery. After the ribbon cutting, it is back to work for the *operational phase*.

This is almost all there is to developing a hatchery project. "Almost," for a few details have been left out, such as how to fund it, how to staff it, how to optimize the operation, etc. I would like to consider a few of those details that seem relevant to me.

Time and its relationship to the state of the art, the continuity of staff, enthusiasm, policy, and cost is an important element to consider (Table 1). The time required for planning and construction probably is a minimum of five years. It takes another two to five years before the facility is fine-tuned and achieving the design production. The decisions and commitments made at the beginning to establish the "basis of design" at the feasibility-report stage are at least five years old before they are realized. Are they still valid? Are they adaptable to today's technology? If not, how do we assure ourselves that the hatchery is the best obtainable? Do we still have funding escalated to scheduled building? These are questions that all those responsible and involved in the design of a hatchery are continually asking. We need an awareness of how this time factor effects each of the participants in the process so that more satisfactory techniques can be developed to allow input and still complete a project.

Participant Input and Design Update

In order to achieve the best facility possible, the relationship between the using branch personnel and the designers should be close and harmonious. When this happens, formality often is lost in the common desire to provide the best. It is very easy for a process to develop that has the biologists continually asking the designers for answers to possible conditions of utility, and the designers trying to respond to these questions by investigating the feasibility of new developments in the ever-changing state of the art. This approach is virtually endless and very difficult if one is ever to complete a project within a budget and without bankrupting a private designer. It is a process that must take place, however, and how this technology transfer occurs and is applied to a defined, fixed-fee contract is one of the most difficult problems to resolve.

The A/E's manager has to establish a list of tasks and define them in such a manner that all participants know what is included in the task and when the task is to be completed. He has to schedule manpower and

TABLE 1.—*General estimates for the duration of hatchery development phases.*

Exploratory studies—varies
Preliminary studies—over 1 year
Project design phase—over 1 year
Project construction period—over 2 years
Initial production phase—2 to 5 years
Usual duration of projects—5 to 7 years
Time from need to production—4 to 5 years
Time from need to production quota—4 to 8 years

budget costs to these tasks. From the manager's point of view, in order to do all of these tasks and budget determinations it is almost imperative that the scope be frozen at the time the fee and manpower commitment is made. Changes beyond that point are "extra work" so far as the fee and contract are concerned, unless other provisions are made for ongoing changes. No matter how he sympathizes with the benefits of the change and its overall merits, he cannot afford to make them or conduct investigations into worthwhile causes unless he receives extra compensation. It is a tough position to be in because the manager ends up saying "No" to everyone. If he weakens, he has lost control of his project, budget, and completion commitment.

I propose that everyone become involved in the process as early as possible so that by the time the feasibility report is complete, the agency and all its subdivisions have had their inputs and have accepted the project to that point. Secondly, I propose that during the project design phase, updates are reviewed, negotiated for cost, and scheduled at the end of each submittal phase. At the start of design development the concept should be firm and only minor changes should be made before the start of the contract documents. At this point, the design should be frozen except for a major technological breakthrough, because any change from this point forward will have a great impact on the designers and the cost of the services provided. An outline for early participation is shown in Figure 1.

The preliminary studies phase (Fig. 2) is comprised of a number of individual studies, investigations, problem solvings, and budgeting activities that are probably the most important in establishing the character of the hatchery. They can be

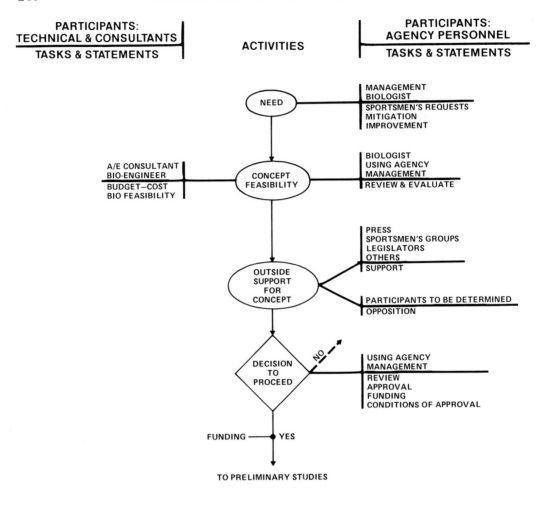

PARTICIPANTS ROLES AND DECISION RESPONSIBILITY

FIGURE 1.—*Participant roles and decision responsibilities during exploratory studies of hatchery development.*

funded in a manner to allow sufficient time and input by all the participants. The agency should consider these preliminary studies as the major effort to establish the state of the art in the culture methods selected, and how these methods are to be accomplished. The recommendations which result from these studies become the "policy statement" and define the commitment to the degree of technological advancement that the Agency is willing to assume in developing the project. The preliminary planning report is the "project definition" that will be used in evaluating the success of all following work. It should be considered very carefully to ensure that it defines an acceptable final product: a functional hatchery.

The fish culturist's input is most required during the preliminary studies and the preparation of the resulting report. In our work, we believe we have to have the biological criteria and the related bio-engineering defined during this phase. All of the known criteria are brought together and evaluated so judgements can be made

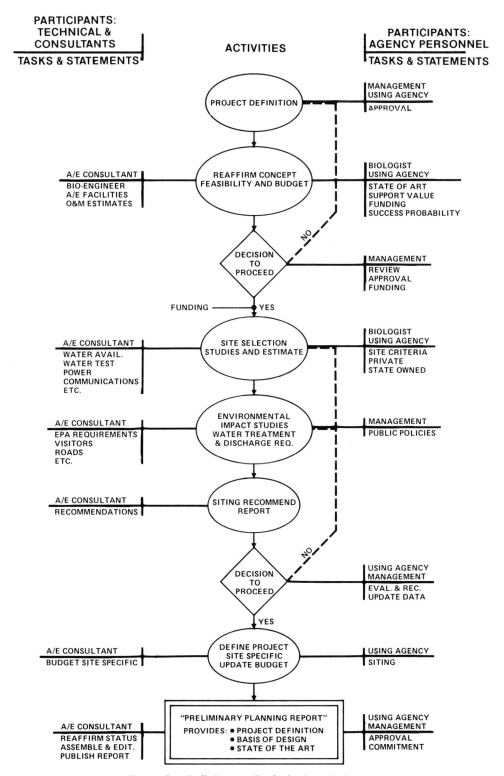

FIGURE 2.—Preliminary studies for hatchery development.

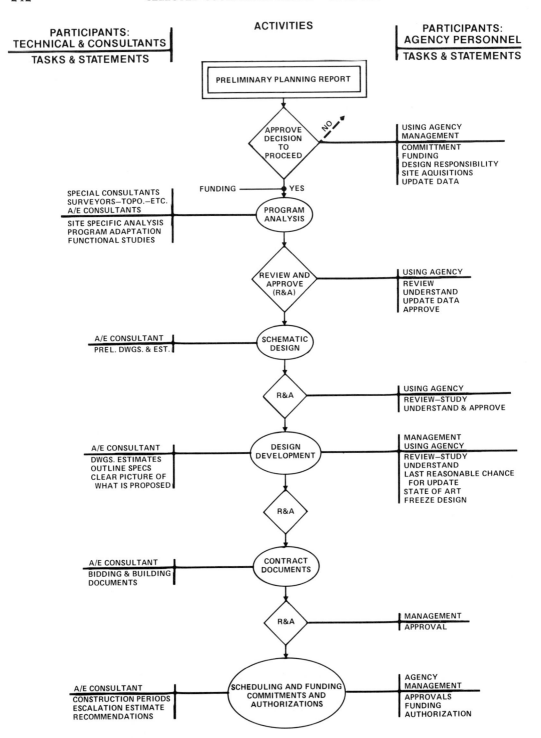

FIGURE 3.—*Project design phase of hatchery development.*

and a basis of design can be established. From that point forward, the basis of design will only be updated to reflect better data or program changes. Those responsible for approving the biological criteria for the basis should take this work very seriously, for it determines what the hatchery will be.

While I have stressed the importance and commitment established by the preliminary studies, I do not want to imply that we have all the answers upon their completion. It is a starting point which provides us with a concept and policy. We then must translate our ideas into definable objects that contractors can buy and build. This is the purpose of the project design phase (Fig 3).

A fish hatchery is generally a one-of-a-kind project for both the biologists and the building contractor. There are very few contractors that have built more than one and in most cases, the contractor interested or available have never done one before. For this reason, design and construction documents for building hatcheries require a great deal more detail and definition than the common types of construction. The A/E's task, therefore, is to translate biological fish requirements into components a good building contractor can recognize in a well-defined manner for building and bidding. It is very important that the contractor understand what he is expected to do if a good job and a reasonable price are to be received by the agency. Things left to the imagination will cost in both dollars and function. Let the studies define the problems early, so they can be resolved in the early design phases by the A/E and presented to the contractor properly.

There are many and varied types of cost involved in the development of a hatchery project. These include agency costs, consultant costs, construction costs, operating costs, and maintenance costs. I believe one of the most misunderstood or controversial costs involves the type of construction. Should it be a hole in the ground or a totally enclosed 50-year facility? Should it be labor intensive or automated? Concrete or plastic? All of these questions and many more must be resolved in establishing the basis of design, the contract documents, and the character of the facility. We have designed for both ends of the spectrum as seemed appropriate to the situation and the client's needs. In most cases., however, the savings achieved in using less expensive facilities are in the initial capital investment. The cost for a biologist, architect, or engineer per hour is the same regardless of what type of construction is being used. The operating and maintenance costs for a facility generally reflect the quality of the facility.

In many states and for the federal government, the life expectancy of buildings and projects is established by policy. Whether the facility is considered permanent or temporary determines the expected life. Facilities funded by bond money are expected to last the life of the bonds. Private companies, on the other hand, often have different funding polices and construction techniques; their operations are tailored to suit these polices. Many hatchery programs, including their promotion, funding, building, and operation, have become major achievements for their agencies and have been used for many years. Each one is a policy statement of the agency. Judgements as to their adequacy, costs, and impact should be made according to the project criteria and basis of design developed during the project design. When it comes time to discuss the costs for the project, the options, goals, and methods should be considered and then, in concert with the participating personnel, construction criteria can be established.

The construction phase or the "bid-build" phase (Fig. 4) is another topic of major importance. No matter how good the basis of design, the construction documents, and the site are, unless the project is built to the intent of these concepts, the agency has suffered a great loss. Therefore, we strongly recommend that all projects be built under the observance and with the suggestions and contract compliance requirements of a full-time agency representative. This representative should be highly qualified with special expertise in hatcheries and particularly in the project being developed. The construction observer should have easy and frequent communication

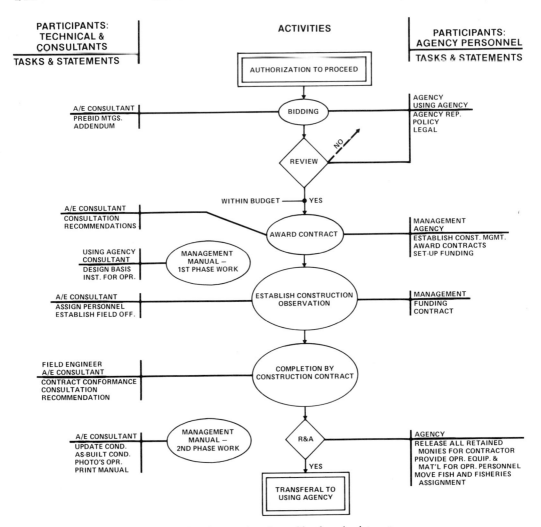

FIGURE 4.—*Construction phase of hatchery development.*

with the designers and the ability to give guidance in advance to the contractors. The cost of such service is a small price to pay for the assurance of the quality and feature intended. The person filling this role should be carefully selected. Our experience has been that good inspectors in general construction do not necessarily make good fish hatchery observers because they are not sensitive to the special requirements of fish-related details, and that the observer should come from the design agency. But, even more important, we believe there should be continuous guidance in the field at all times.

When the contractor "turns over the key" to the operating branch, personnel should be ready to take over with a sense of confidence and purpose. A Hatchery Management Manual should be ready for them prior to this event. The preparation and updating of the Manual is a continuing process. The first phase identifies the instructions and the basis of design complete with bio-criteria and bio-engineering data. It is intended for use by operating personnel after the contractor has gone. The second phase incorporates the actual "as-built" conditions into the Manual for operational purposes so that operating personnel can relate to actual conditions.

The Management Manual should be up-

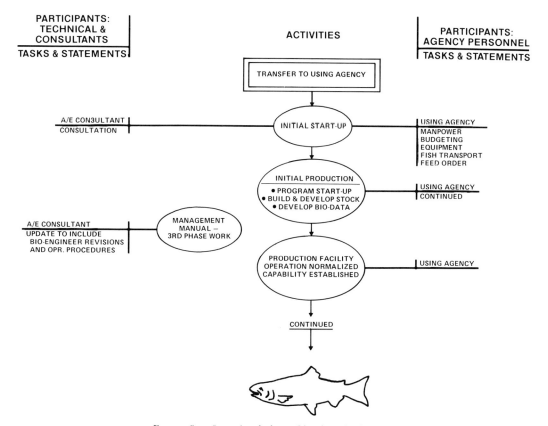

FIGURE 5.—*Operational phase of hatchery development.*

dated to reflect actual operating conditions and advanced biological data following full production loading conditions in the hatchery. Knowledge gained from conferences held with designers and operators to analyze the results and modify construction to achieve the best results should also be included as a record for future generations.

The operational phase (Fig. 5), is, for the most part, the agency's chapter. The A/E consults on the initial start-up and as requested by the Agency during the life of the facility. With the transition to full production, years of developmental effort come to fruition. If the need for the hatchery has been realistically established, if the preliminary studies and project design have been meticulous, thorough, and candid, if construction has been faithful to the project concept, then the finished hatchery will bring great benefits to the agency and its constituents.

Biological Considerations in Hatchery Design for Coolwater Fishes

HARRY WESTERS

Michigan Department of Natural Resources
Stevens T. Mason Building
P.O. Box 30028,
Lansing, Michigan 48909

Abstract

The evolution from extensive to intensive fish rearing is described. Intensive rearing techniques require a knowledge of the nutritional, physiological, physical and ethological requirements of the fish. Specific biological parameters in each of these areas are identified and a model for an imaginary coolwater fish, called "walmusk", is used to illustrate the application of these parameters to the design of an intensive fish production facility.

Originally, and during its centuries-long development, controlled fish production took place in static water environments in ponds. The production capacity of such ponds depends on the species reared and the abundance of fish food produced by the pond. Annual production can amount to several tons per hectare. The Chinese have developed pond culture into a fine art of balanced polyculture, using different species, each feeding on a different trophic level. Thus, a fuller utilization is made of the production capacity of the pond. In pond culture, the function of the pond is two-fold. It provides a residence for the fish, and it is the medium in which the food organisms are produced. On a quantitive basis, these two functions limit the productive capacity of a fish pond.

As fish density increases, normally the second function will become the first limiting factor, since the pond will be unable to meet the demand for food organisms for the fish. To alleviate this limitation, the pond can be enriched with organic or inorganic fertilizers, thus increasing the production of food organisms.

However, if further increase in fish production is to be attained, food from outside the pond environment must be introduced. At relatively low fish densities, the supplemental feeding constitutes only a small percentage of the total food requirements of the fish, and, therefore, incomplete diets, such as cheap cereals, can be used. As fish densities increase, the proportion of food required shifts more towards those introduced from the outside, and ultimately, nutritionally complete diets are needed.

When the pond is managed this way, it no longer serves the function of the production of fish food, but only as a residence for the fish. As fish production continued to increase, the residence function will become a limiting factor. This occurs when the pond no longer meets the environmental requirements of the fish, when the metabolic activity of the fish outstrips the pond's ability to take up oxygen and purify the waste products produced. As oxygen levels drop, and excretory products accumulate, stresses imposed on the fish will manifest themselves in reduced growth rates, reduced resistance to diseases, etc., and, finally, in increased mortalities. Artificial aeration can often be helpful, since oxygen depletion is usually the first environmental problem in the pond. Such intensification in pond management (e.g., fertilization, supplemental feeding, and aeration), is found with more or less sophistication throughout the world. Production levels attained vary greatly with species produced, techniques applied, and climatological conditions.

When the residence function of the stagnant pond has reached its endpoint, further increases in production can only be realized by providing "additional" residence, through supplying the pond with additional water. Table 1 shows how fish production can be influenced by the type of pond management used.

Fish culture operations in running water systems have already been practiced for a long time, primarily with salmonids, but lately with other species as well. In a flow-through system, the pond serves only as residence, while the water is the vehicle that delivers oxygen and removes waste products. Fish culture in running water systems has much merit, since it can be carried out in compact units that require little space. Environmental conditions can be controlled, especially when rearing units are installed in buildings. Water delivered to the units may be processed (heated, filtered, aerated, etc.,) and production can be mechanized and automated. Production capacity depends largely on the flow rates, or water exchange rates, in the rearing units, rather than on the size or surface area of the units. Figure 1 illustrates the direct relationship of flow rates (exchange rates) and rearing density for salmonids.

Fish culture conducted under intensive conditions requires a knowledge of the nutritional, physiological, physical, and ethological requirements of the fish. With present-day technology, fish can be produced without concern for seasonal or climatological effects on growth or reproduction.

Hatchery Design for Intensive Culture of Coolwater Fish

Nutritional Requirements

As stated, intensive culture in flow-through systems demands a knowledge of various biological requirements of the species produced.

Nutritional requirements are of foremost importance. The Subcommittee on Fish Nutrition of the National Research Council of the National Academy of Sciences has reviewed and evaluated published research concerning the nutritional needs of trout, salmon and catfish (1973) and the nutrient requirements of warmwater fish (1977). Relatively little is known about the specific nutritional requirements of coolwater fish. Orme (1978, this volume) has reported on the status of coolwater fish diets. Intensive culture cannot be successful without artificial diets which are nutritionally complete. In addition, physical and behavioral characteristics of fish will influence the preparation of feeds. For instance, the very small size of walleye fry requires microscopic nutritionally complete food particles which should be water-stable so they will not quickly loose specific nutrients through leaching after contact with water.

TABLE 1.—*Production levels of carp under various methods (modified from Bardach et al. 1972).*

Method	Production kg/hectare	Estimated production kg/m^3
Natural ponds	100–200	0.02–0.04
Fertilized ponds	500–1,000	0.10–0.20
Fertilized ponds with supplemental feeding ponds	2,000–5,000	0.40–1.00
Ponds with running water and supplemental feeding	400,000–2,000,000	80–400

Physiological Requirements

The metabolic activity of fish removes dissolved oxygen from, and adds metabolites to, water. If the biomass of fish in a pond is too high, oxygen depletion and the accumulation of metabolites will create an intolerable environment.

In order to determine the capacity of a pond for the production of any species of fish, we must know the concentration of dissolved oxygen and other chemical constituents in the water, rates of oxygen consumption and waste production under specific rearing conditions, the minimum safe concentration of dissolved oxygen, and the maximum allowable level for specific metabolites. Information is needed for each phase of the hatchery life cycle of each species.

Knowledge of the following parameters is particularly useful in designing and operating intensive fish culture facilities (Westers and Pratt 1977): (1) the level of

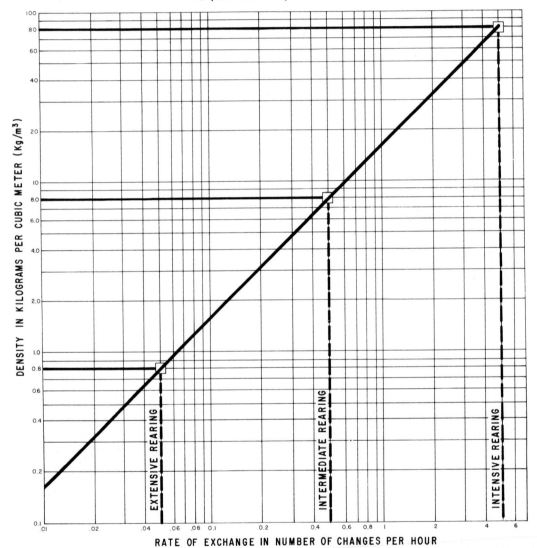

FIGURE 1.—*Relationship of water exchange rate to fish density for extensive, intermediate, and intensive culture of salmonids. Standard conditions: production level, 1 kg/lpm; fish length, 15–25 cm; temperature, up to 20 C.*

oxygen in the influent; (2) the rate of oxygen consumption; (3) the rate of ammonia production; (4) the minimum allowable level of oxygen in the effluent; (5) the maximum allowable level of ammonia in the effluent.

Although quite a lot of information about these parameters is available for certain salmonids (Beamish 1962; Brett and Zala 1975; Burrows 1964; Hartman 1977; McLean and Fraser 1974; Smart 1976; Smith and Piper 1975; Willoughby et al. 1972), very little is known as yet for coolwater fish. Pecor (1977) found that oxygen consumption and ammonia production rates for tiger muskellonge (northern pike × muskellunge) reared under intensive condition were 55% and 82% less, respectively, than corresponding rates for coldwater species. Tiger muskellunge requirements for water are less than half of that of salmonids, based on oxygen consumption, and less than one-fourth, based on ammonia production, if tolerance levels for these two parameters are the same as for salmonids. However, more data on the metabolic rates of these fish and their tolerance to ammonia and low oxygen levels under hatchery conditions are needed.

Physical Requirements

Overly high rearing densities of fish will result in stress. Stress indicators may be poor growth rate, poor conversion, reduced disease resistance, physiological and physical changes, and consequent higher mortality rates. It is important, therefore, to know what the maximum permissible rearing density is for each species cultivated. This "maximum permissible density," under hatchery conditions, is the level at which these parameters (growth rate, etc.) fall within the acceptable range. This presents a problem since "standards" are unavailable for most hatchery-reared fish.

Wedemeyer and Yasutake (1977) offer various clinical methods to measure effects of environmental stress on fish health, but do not give a simple method to measure the quality of the hatchery product.

Very high densities have been reported in intensive salmonid culture. Experimentally, densities as high as 540 kg/m³ for rainbow trout (Buss et al. 1970) and 335 kg/m³ for coho salmon fingerlings (Westers 1964) were obtained. Such extremely high densities are in great contrast with those generally considered acceptable for hatchery production. Recommended maximum rearing densities for salmonids produced under intensive rearing conditions vary from 32 kg/m³ (Burrows and Combs 1968) to 50 kg/m³ (Piper 1970), to 96 kg/m³ (Westers 1970). These values are for salmonids of 15 to 20 cm, produced under a temperature range of 10–15 C.

Under experimental conditions—rainbow trout reared in small circular containers (volumes of 8.8, 12.5, 17.1, and 37.5 liters)—Kincaid et al. (1976) found that growth rate was the single trait most sensitive to overcrowding. The authors found that optimum rearing densities, from a production point of view, ranged from 50 to 80 kg/m³, depending on type and size of container. Incidentally, this observation gives support to the notion that the design of a rearing container is important.

Wedemeyer (1976) found measurable physiological changes (hyperglycemia) in 10–13-cm coho salmon when the fish were held at densities of just over 16 kg/m³, but the condition was severe only in those fish held at densities of 96 and 192 kg/m³. In smolting coho salmon of 15 to 20 cm, densities greater than 16 kg/m³ caused severe blood chemistry disturbances.

Rainbow trout fingerlings produced in circular rearing tanks at 11.9 C, did not show any growth depression until a density of 26 kg/m³ was reached (Brauhn et al. 1976). Although the investigators expressed "density" in terms of flow (kg/liter per min), they provided data to calculate the exchange rates. Their findings indicated no growth depression at or below "densities" of 1.0, 1.5, and 3.0 kg/lpm at inflow rates of 12, 8, and 4 lpm, respectively. Corresponding exchange rates (R) were 1.56, 1.04, and 0.52. The result was a space-related density of 26 kg/m³ for all three conditions.

Little is known about the effects of density on coolwater fish. Pecor (1978, this volume) speculates that tiger muskullunge fingerlings may show reduced growth rates at a density of 32 kg/m³. However, he points out that the growth rate comparisons involved fish reared at two different locations.

Considerable attention has been given to this subject of density, since space is a most important factor in the design of intensive fish rearing facilities. Because rearing density and carrying capacity per unit flow are complementary, these two criteria provide the most significant design information for rearing units for intensive facilities irrespective of whether they are to produce coldwater, coolwater, or warmwater fish. In all cases, maximum use must be made of both water and space.

The significance of water velocity in hatchery design has two aspects. It must meet the requirements of the fish and it must allow the solids to settle (Westers and Pratt 1977). To meet both criteria satisfactorily in rearing units with circulating water is difficult, and nearly impossible when relatively fast exchange rates are desired. Rectangular, flow-through rearing units are preferred.

Ethological Requirements

A knowledge of feeding behavior is important. Such things as striking distance, movement of the food particles, sinking rate, shape, color preference, light conditions, feeding locations (surface, water column, bottom), etc., are all factors which could influence certain facets of hatchery design. Some of the factors may be quite critical, others very flexible, depending on species and/or fish developmental stage in the hatchery life cycle. For instance, peculiarities in feeding behavior could dictate specific design requirements for automatic feeders (% area coverage, minimum particle size, minimum amounts, frequencies, etc.). Factors in the design and operation of rearing units, such as optimum water depth relative to surface area, water velocities, and even color preference (background), can be influenced by feeding behavior. Hatcheries must be designed and operated with fish behavior in mind.

Fright reactions to various activities (people movement, sounds, light variations) may stress fish. Cannibalism, bunching, piling into corners, jumping and other "abnormal" distribution or movement patterns, could require some specific design feature or operational mode. Any one or more of the behavior patterns mentioned could occur during particular phases of the hatchery rearing cycle. The ideal hatchery is designed accordingly.

Application of Design Criteria to Intensive Fish Production Facilities

An exercise in the application of design parameters will follow. Since sufficient data are unavailable for any particular species of coolwater fish, hypothetical data for an imaginary coolwater fish, called walmusk, will be used.

In this example, facilities have to be designed for the intensive production of walmusk fry, reared to 20-cm fingerlings.

Diets and Feeding

Nutritionally complete dry diets have been developed and are available in pellet form. They are readily acceptable to the fry and fingerlings when dispersed by automatic feeders. The following feeding regimes are needed to prevent cannibalism.

Fry to 2.0 cm: 5 min interval, 18 h per day.
2.0–5.0 cm: 10 min interval, 18 h per day.
5.0–10.0 cm: 15 min interval, 18 h per day.
10.0–20.0 cm: 30 min interval, 12 h per day.

These feeding regimes require automatic feeders which can dispense very small particles of food in accurate amounts at frequent intervals. An automatically controlled lighting system is required to extend feeding beyond normal daylight hours. Since walmusk do not show search or pursuit behavior, the feeders must distribute feed evenly over 90% or more of the surface area.

Metabolic Characteristics—Loading Levels

Walmusk have the following metabolic characteristics: (1) oxygen consumption rate, 115 g O_2 per kg food; (2) ammonia production rate, 11.0 g NH_3 per kg food; (3) minimum oxygen tolerance level, 5.0 mg/liter in effluent; (4) maximum ammonia tolerance level, 0.025 mg/

liter un-ionized ammonia. The pH of the water is 7.6. The maximum rearing temperature is 20 C. Under these conditions, 1.5 mg/liter total ammonia will produce 0.025 mg/liter un-ionized ammonia (Trussell 1972). The incoming water has 9.0 mg/liter oxygen. The flow requirements, based on oxygen, can be determined with:

$$\text{lpm/kg food} = \frac{\text{oxygen consumption rate per kg food}}{1.44 \times \text{available } O_2 \text{ in mg/liter}}.$$

The factor 1.44 represents the grams of O_2 per one lpm over a 24-hour period, for each mg/liter O_2. The specific flow requirements in lpm per kg food is:

$$\text{lpm/kg food} = \frac{115}{1.44(9-5)} = 20.$$

This can be expressed in lpm per kg fish as follows:

$$\text{lpm/kg fish} = \frac{115}{1.44(9-5)} \times \frac{\% \text{ feeding level}}{100}.$$

Feeding levels in percent body weight are determined with the hatchery constant method of Buterbaugh and Willoughby (1967). The walmusk has a hatchery constant of 100 for these particular water characteristics. [H.C. = 10 × conversion (1.7) × monthly growth rate (5.9 cm) = 100.] The flow requirements are as follows.

2-cm fish: 20 × 50 × 0.01 = 10.0 lpm/kg fish.
5-cm fish: 20 × 20 × 0.01 = 4.0 lpm/kg fish.
10-cm fish: 20 × 10 × 0.01 = 2.0 lpm/kg fish.
20-cm fish: 20 × 5 × 0.01 = 1.0 lpm/kg fish.

The same approach can be used with ammonia.

$$\text{lpm/kg food} = \frac{NH_3 \text{ production per kg food}}{1.44 \times \text{allowable level in mg/liter in effluent}}.$$

For walmusk:

$$\text{lpm/kg food} = \frac{11.0}{1.44 \times 1.5} = 5.0.$$

Expressing this in lpm/kg fish:

$$\text{lpm/kg fish} = \frac{11.0}{1.44 \times 1.5} \times \frac{\% \text{ feeding level}}{100}.$$

Flow requirements are as follows.

2-cm fish: 5.0 × 50 × 0.01 = 2.5 lpm/kg fish.
5-cm fish: 5.0 × 20 × 0.01 = 1.0 lpm/kg fish.
10-cm fish: 5.0 × 10 × 0.01 = 0.5 lpm/kg fish.
20-cm fish: 5.0 × 5 × 0.01 = 0.25 lpm/kg fish.

As noticed, the maximum flow requirements in terms of oxygen is four times that in terms of ammonia. Ammonia toxicity is therefore of no consequence until the water is reused four times.

Physical Requirements—Density Levels, Etc.

The maximum allowable rearing density levels for walmusk are as follows.

2-cm fish: 10 kg/m³.
5-cm fish: 30 kg/m³
10-cm fish: 60 kg/m³.
20-cm fish: 100 kg/m³.

The maximum allowable water velocities are:

Fry: 0.0025 m/s.
2-cm fish: 0.005 m/s.
5-cm fish: 0.010 m/s.
10-cm fish: 0.020 m/s.
20-cm fish: 0.030 m/s.

The relationship between production in terms of loadings (kg/lpm) and density (kg/m³) is illustrated with this equation (Westers 1970):

$$\text{kg/lpm} = (0.06/R) \times \text{kg/m}^3.$$

Flow requirements (lpm/kg fish) and their reciprocals (kg/lpm) are:

2-cm fish: 10.0; reciprocal = 0.10.
5-cm fish: 4.0; reciprocal = 0.25.
10-cm fish: 2.0; reciprocal = 0.50.
20-cm fish: 1.0; reciprocal = 1.0.

By knowing the maximum permissible loadings (kg/lpm) and densities (kg/m³) for the various phases of the hatchery life cycle, we can use the equation $\text{kg/lpm} = (0.06/R) \times \text{kg/m}^3$ to determine the optimum exchange rate (R); $R = (\text{kg/m}^3 \times 0.06)/\text{kg/lpm}$. Thus, we have these ex-

TABLE 2.—*Application of design data to rearing units for walmusk fingerlings.*

Item	Fish length (cm)			
	Fry–2	2–5	5–10	10–20
Fish weight (g)[a]	0.005–0.04	0.04–0.625	0.625–5.0	5.0–40.4
Unit capacity				
Loading (kg/lpm[b])	0.05–0.10	0.10–0.25	0.25–0.50	0.50–1.00
Density (kg/m³)	2–10	10–30	30–60	60–100
Hourly exchange rate (R)[c]	3.0–6.0	1.8–3.6	3.6–7.2	3.0–6.0
Maximum water velocity (V, m/s)	0.0025–0.005	0.005–0.010	0.010–0.020	0.020–0.030
Feeding rate (% of body rate)[d]	100–50	50–20	20–10	10–5
Rearing unit dimensions				
Length (L, m)	3.0	10.0	10.0	20.0
Width (m)	0.5	1.0	1.0	2.0
Depth (m)[e]	0.20 + 0.04	0.6 + 0.1	0.6 + 0.2	0.75 + 0.4
Water volume (m³)	0.3	6.0	6.0	20.0
Maximum number of fish per rearing unit (thousands)	120–75	1,500–288	288–72	360–74

[a] Weight (w) is calculated from length (l) and a condition factor (C) of 0.005: $w = Cl^3$.
[b] lpm = liters per minute.
[c] R is calculated from: kg/lpm = 0.06R × kg/m³.
[d] Feeding rate is based on a hatchery constant of 100.
[e] The number after the + sign gives the required freeboard.

change rates, in changes per hour:

  2-cm fish: 10 × 0.06/0.10 = 6.0.
  5-cm fish: 30 × 0.06/0.25 = 7.2.
  10-cm fish: 60 × 0.06/0.50 = 7.2.
  20-cm fish: 100 × 0.06/1.00 = 6.0.

Knowing the maximum allowable velocities, we can now determine the best length of the rearing unit. Velocity in a rectangular flow-through rearing unit can be expressed as

$$V = \frac{R \times L}{3{,}600},$$

where V is velocity in m/s, L is the length of the rearing unit in m, and 3,600 is the number of seconds per hour. This equation can be re-written:

$$L = \frac{V \times 3{,}600}{R},$$

and the length can now be determined:

  Fry: .0025 × 3,600/3.0 = 3.0 m.
  2-cm fish: .005 × 3,600/6.0 = 3.0 m.
  5-cm fish: .010 × 3,600/7.2 = 5.0 m.
  10-cm fish: .020 × 3,600/7.2 = 10.0 m.
  20-cm fish: .030 × 3,600/6.0 = 18.0 m.

The exchange rate (R) was reduced to 3.0 or (one-half of R = 6) to reduce velocity to 0.0025 m/s. Tanks selected are 3.0 m for early rearing (up to 2 cm); 10 m for intermediate rearing (up to 10 cm) and 20 m for final rearing to 20 cm or more. Table 2 summarizes the design data and its application.

Rearing tanks should always have a glass-like finish to reduce abrasiveness.

Behavioral Requirements

Walmusk fingerlings have a habit of suddenly jumping out of the water, especially along the perimeter of the rearing unit. It is unknown what triggers this behavior. They will jump nearly twice their length.

Another troublesome behavior is occasional crowding at the outlet screen. When this occurs, walmusk swim downstream and push their noses into the outlet screen. This can cause skin erosion, followed by fungus infection. Again, it is unknown what causes the fish to do this.

It has been observed that walmusk feed best and appear most "relaxed" when they are placed under relatively constant and uniform light conditions of about 260 lumens.

When hungry, these fish show cannibalistic tendencies, especially during the early part of the rearing cycle (2 to 10 cm). Feed should not be withheld for more than 6 consecutive hours during a 24-h period.

The rearing units should have a freeboard of twice the length of the fish produced to prevent the fish from jumping out. To avoid the problem of "nose-poking" into the outlet screen, a solid baffle (curtain) should be hung in front of the outlet screen at a distance of 10 cm, approximately 20 cm off the bottom. Additional baffles placed in the rearing unit will result in an efficient, self-cleaning tank, that constantly removes the solids from the rearing environment (Westers and Pratt 1977).

References

BARDACH, J. E., J. H. RYTHER, AND W. O. MCLARNEY. 1972. Aquaculture: the farming and husbandry of freshwater and marine organisms. John Wiley, New York.

BEAMISH, F. W. H. 1964. Seasonal changes in the standard rate of oxygen consumption of fishes. Can. J. Zool. 42:189–194.

BRAUHN, J. L., R. C. SIMON, AND W. R. BRIDGES. 1976. Rainbow trout growth in circular tanks: consequences of different loading densities. U.S. Fish Wildl. Serv. Tech. Pap. 86. 12 pp.

BRETT, J. R., AND C. A. ZALA. 1975. The daily pattern of nitrogen excretion and oxygen consumption of sockeye salmon (*Oncorhynchus nerka*) under controlled conditions. J. Fish. Res. Board Can. 32(12):2479–2486.

BURROWS, R. E. 1964. Effects of accumulated excretory products on hatchery reared salmonids. Fish Wildl. Serv. (U.S.) Res. Rep. 66. 12 pp.

———, AND B. E. COMBS. 1968. Controlled environments for salmon propagation. Prog. Fish-Cult. 30(3):123–136.

BUSS, K., D. R. GRAFF, AND E. R. MILLER. 1970. Trout culture in vertical units. Prog. Fish-Cult. 32(4): 187–191.

BUTERBAUGH, G. L., AND H. WILLOUGHBY. 1967. A feeding guide for brook, brown, and rainbow trout. Prog. Fish-Cult. 29(4):210–215.

HARTMAN, J. 1978. Ammonia production and Oxygen consumption of brown trout (*Salmo trutta faris*) in a three pass serial reuse system. Submitted to Prog. Fish-Cult.

KINCAID, H. S., W. R. BRIDGES, A. E. THOMAS, AND M. J. DONAHOO. 1976. Rearing capacity of circular containers of different sizes for fry and fingerling rainbow trout. Prog. Fish-Cult. 38(1): 11–17.

MCLEAN, W. E., AND F. J. FRASER. 1974. Ammonia and urea production of coho salmon under hatchery conditions. Environ. Prot. Serv. Pac. Region, Surveillance Rep. EPS S-PR-74-5. 61 pp.

NATIONAL ACADEMY OF SCIENCES—NATIONAL RESEARCH COUNCIL. 1973. Nutrient requirements of trout, salmon and catfish. N.A.S., Washington, D.C.

———. 1977. Nutrient requirements of warmwater fishes. N.A.S., Washington, D.C.

ORME, L. E. 1978. The status of coolwater fish diets. Am. Fish. Soc. Spec. Publ. 11:167–171.

PECOR, C. H. 1978. Preliminary trials with intensive culture of a coolwater fish in a three-pass water reuse system. Submitted to Prog. Fish-Cult.

———. 1978. Intensive culture of tiger muskellunge in Michigan during 1976 and 1977. Am. Fish. Soc. Spec. Publ. 11:202–209.

PIPER, R. G. 1970. Know the proper carrying capacity of your farm. Am. Fishes U.S. Trout News. May–June.

SMART, G. 1976. The effects of ammonia exposure on gill structure of the rainbow trout (*Salmo gairdneri*). J. Fish Biol. 8:471–475.

SMITH, C. E., AND R. G. PIPER. 1975. Lesions associated with chronic exposure to ammonia. Pages 497–514 *in* W. E. Ribelin and G. Migaki, eds. The pathology of fishes. Univ. Wis. Press, Madison.

TRUSSEL, R. P. 1972. The percent un-ionized ammonia in aqueous ammonia solutions of different pH levels and temperatures. J. Fish. Res. Board Can. 29(10):1505–1507.

WEDEMEYER, G. A. 1976. Physiological response of juvenile coho salmon (*Oncorhynchus kisutch*) and rainbow trout (*Salmo gairdneri*) to handling and crowding stress in intensive fish cultures. J. Fish. Res. Board Can. 33(12):2699–2702.

———, AND W. T. YASUTAKE. 1977. Clinical methods for the assessment of the effects of environmental stress on fish health. U.S. Fish Wildl. Serv. Tech. Pap. 89. 18 pp.

WESTERS, H. 1964. A density and feeding study with coho salmon. Mich. Dep. Nat. Resour., Lansing. 10 pp. (Unpublished.)

———. 1970. Carrying capacity of salmonid hatcheries. Prog. Fish-Cult. 32(1):43–46.

———, AND K. M. PRATT. 1977. The rational design of fish hatcheries for intensive culture based on metabolic characteristics. Prog. Fish-Cult. 39(4):157–165.

WILLOUGHBY, H., H. H. LARSEN, AND J. T. BOWEN. 1972. The pollutional effects of fish hatcheries. Am. Fishes U.S. Trout News.

CONSIDERATIONS IN PERCID MANAGEMENT

Case Histories of Stocking Walleyes in Inland Lakes, Impoundments, and the Great Lakes— 100 Years with Walleyes

Percy W. Laarman

Michigan Department of Natural Resources
Fisheries Division
Ann Arbor, Michigan 48109

Abstract

Evaluations on stocking walleyes in 125 bodies of water during the last 100 years were reviewed. Walleye stockings were separated into three categories: (1) introductory plants where walleyes were absent; (2) maintenance plants where natural reproduction was absent or very limited; and (3) supplemental plants where efforts were made to augment natural reproducing walleye populations. About 48% of the introductory plants were successful. Approximately 32% of maintenance plants were successful. Only about 5% of the supplemental plants were considered to be successful. Success or failure of walleye stocking appeared to depend more on environmental and biological conditions of individual bodies of water than on the number and size of walleyes that were stocked.

Billions of walleyes have been stocked in North American waters during the last century. In Michigan, about 2.5 billion fry have been stocked in inland waters since 1882 and more than 3 million fingerlings since 1952. During the last decade, from 4 to 6 million fingerlings have been stocked annually in Minnesota waters (Johnson 1971). Forney (1975) reports the current annual production of walleye fry exceeds 300 million at Oneida Hatchery, New York. Since 1924, 842 million walleyes have been stocked in waters of Saskatchewan (Marshall and Johnson 1971). In the Midwest alone, over 400 million fry are stocked annually (Klingbiel 1971).

In this report I attempt to distinguish between introductory, maintenance and supplemental stocking. Introductory plants are made into new impoundments and in natural waters where the species is absent. In many bodies of water where natural reproduction does not occur, periodic maintenance stocking is required to provide a fishery. Supplemental stocking is practiced in an attempt to increase the abundance of walleyes in naturally reproducing populations. Some plants are based on biological principles and others are motivated by political and public pressures.

The task of evaluating walleye stocking programs is difficult and expensive. Evaluations can be based on survival rates of stocked fish, contributions to abundance of year classes or contributions to the harvest. Expenses involved in determining contributions of stocked fish to the harvest may exceed initial costs of raising and planting the fish. The purpose of this report is to review case histories of stocking walleyes and the evaluations of these stockings.

Inland Lakes

The relationships between stocking of fry and the commercial catch of walleyes in two large lakes in Minnesota were investigated. Lake of the Woods, 384,613 hectares in size, was stocked with fry beginning in 1926 (Carlander 1945). From 1931 to 1939, 62 to 164 fry per hectare were planted annually. The smallest number of fry was stocked in 1936 and that year produced the most abundant year class. Lack of a positive correlation between numbers of fry planted and abundance of walleyes was evident in other years also.

Walleye fry were first planted in Red Lakes (111,288 hectares) in 1919 (Smith and Krefting 1954). From 1925 to 1953, stocking rates ranged from 98 to 1,080 fry per hectare. No significant relationship could be shown between numbers of fry stocked and abundance of walleyes. Also, the relative strengths of eight year classes were not significantly correlated with numbers of fry planted.

Stocking of fry at relatively high densities appeared to supplement walleye populations in Clear and Spirit lakes in Iowa (Carlander et al. 1960; Rose 1955). Clear Lake is a 1,474-hectare eutrophic lake with a maximum depth of 5 m. The lake contains a mixed population of fishes with the walleye being the principal game fish. Walleye fry have been stocked periodically since 1920. From 1948 to 1958, fry were stocked in even-numbered years at the rate of 12,350 to 23,465 per hectare. No stocking was done in the odd-numbered years. The contribution of fry stocking was evaluated by seining and examination of the relative abundance of various year classes in the population. More fingerlings were collected by seine in years of stocking than in years of no stocking. The mean number of walleyes caught in gill nets from year classes produced when stocking occurred was a little over three times greater than from year classes when fry were not stocked. Stocking of 23,000 fry per hectare gave no detectable advantage over the lower rate of 12,000 per hectare.

Spirit Lake is 2,137 hectares in size with a maximum depth of 7.3 m. In 1947, the number of adult walleyes was estimated at about 30,000 by the tag-and-recapture method (Rose 1947). Walleyes were not stocked from 1944 to 1948. From 1949 to 1954, about 7,400 fry per hectare were stocked (Rose 1955). Evaluations of these plants were made from seine hauls of shoal areas after 1 July each year. During the 6 years of stocking fry, the mean catch per seine haul of young-of-year walleyes was about 7.5 times greater than in the 5 years of no stocking. Abundance of yearlings was about 15 times greater and adults 11 times greater during years of stocking. In 1954, a population estimate of about 43,000 adult walleyes supported the survey evidence of an increase in the adult population.

The increase in walleye abundance during years of stocking coincided with a low population density of the predacious white bass. This factor complicated interpretation of the data on walleye stocking.

From 1964 to 1969, fingerling walleyes were marked and stocked in Spirit Lake (McWilliams 1975). Samples collected by seining showed that stocked fingerlings contributed an average of 2% to the respective year classes. Only the 1965 planting contributed a significant amount (12.9%) to that year class.

Oneida Lake in New York is another large lake that has been stocked with walleye fry. The lake is 20,700 hectares with a maximum depth of 17 m. Walleye and yellow perch are the predominant species, comprising about 70% of the biomass of fish captured in experimental gill nets, trap nets, and trawls (Forney 1976).

Contribution of stocked fry to the fry population during 1968 to 1973 was evaluated in Oneida and nine other lakes (Forney 1975). Density of fry was estimated from catches in a high-speed sampler. The other nine lakes ranged in size from 62 to 4,444 hectares. Four of the lakes supported substantial walleye fisheries and six lakes provided limited or sporadic fishing for walleyes. Oneida Lake was stocked in even-numbered years at a rate of 9,000 to 15,000 fry per hectare. The other lakes were stocked in odd-numbered years at a rate of 12,000 fry per hectare.

The density of fry in Oneida Lake was 3–5 per 100 m^3 in the years of stocking and 1 or less per 100 m^3 in years of no stocking. Stocking also appeared to increase the density of fry in most of the other nine lakes.

In another study on Oneida Lake, Forney (1976) reported the initial size of the year class was strongly influenced by stocking fry, but the difference did not persist at age I. An increase in density of fry from stocking did not necessarily mean an increase in year-class strength in subsequent years.

A walleye fishery was established with the introduction of fry in Gogebic Lake, Michigan. The lake is 5,982 hectares with a maximum depth of 11.3 m. During the late 19th century, the lake supported a good smallmouth bass fishery. In 1897, 84 adult northern pike were introduced. This introduction resulted in a northern pike fishery that lasted until the mid-1920's and then declined.

From 1904 to 1907, a total of 1.2 million walleye fry were introduced. Another 27 million fry were stocked from 1933 to 1940. Coinciding with the decline of the northern pike fishery, walleye became the predominant species in the harvest. A partial creel census, conducted intermittently from 1928 to 1963, showed that the walleye comprised from 82 to 92% of the anglers' creel and northern pike comprised from 2 to 10%. A special creel census was conducted during 1940, 1941, and 1947 (Eschmeyer 1950). Walleyes constituted 81 and 89% of the catch in 1940 and 1941. In 1947, northern pike were better represented and walleyes decreased to 54% of the total catch.

Despite the predominance of walleyes in the fishery, public sentiment has been strong in recent years to increase the walleye population. From 1972 to 1975, 4.1 million fry were stocked.

Another successful introduction of walleyes was made in Seagull Lake in Minnesota (Micklus 1961). The lake is 1,512 hectares in size with a maximum depth of 39.6 m. A creel census in 1937 revealed that 100% of the catch consisted of northern pike and lake trout. Walleyes were first planted in 1938 and various times in subsequent years. Another census in 1957 showed that walleyes comprised 60.9% of the catch by weight.

An evaluation of stocking walleye fingerlings in Many Point Lake, a 694-hectare lake in Minnesota, was made by conducting censuses of the anglers' catch (Olson and Wesloh 1962) from 1955 to 1960. In 1956 and 1957, walleye fingerlings were stocked at the rate of 143 and 168 per hectare. No appreciable increase was evident in the catch from those 2 years of stocking.

Kempinger and Churchill (1972) evaluated the stocking of walleyes in Escanaba Lake, Wisconsin. This lake is 119 hectares in size with a maximum depth of 7.6 m. Walleyes were not native to the lake. From 1933 to 1942, about 8,400 fry per hectare were stocked annually which resulted in a reproducing walleye population. Four experimental plants of 7.6-cm marked fingerlings were made in 1954, 1958, 1959, and 1961, at the rates of 94 to 410 per hectare. The contribution of these experimental plantings were determined from a creel census. Thirteen percent of the fish planted in 1954 were caught by anglers. Plantings in the other 3 years contributed less than 1% to the fishery.

Pike Lake, a 211-hectare lake in Wisconsin, was stocked with fry and fingerling walleyes from 1933 to 1954 to augment the walleye population (Mraz 1968). From 1959 to 1961, fin-clipped fingerlings, 8.9 to 10.7 cm in length were stocked at the rate of 20 to 25 per hectare. Population studies conducted from 1959 to 1962 revealed that stocked fingerlings contributed little to the total walleye population.

Two consecutive plantings of fingerlings at the rate of 38 to 96 per hectare, did not create a good walleye fishery in Lake Ripley, a 173-hectare lake in Wisconsin (Threinen 1955).

In Lake Francis, Minnesota, where natural reproduction is lacking, Groebner (1960) reported the walleye harvest increased from 0.2 to 5.3 kg per hectare following 6 consecutive years of stocking fingerlings at an average annual rate of 69 per hectare.

From 1924 to 1969, walleyes were stocked in 286 waters in Saskatchewan (Marshall and Johnson 1971). About 60% of these waters were shallow lakes subject to winterkill. Many of these waters maintained walleyes a sufficient length of time to support periodic sport fishing. Most of the other lakes contained natural walleye populations. Studies on seven of the lakes indicated that year classes in years of stocking were no more abundant than year classes dependent on natural reproduction.

Johnson (1971) reported on the evaluation of stocking walleye fingerlings in 41 Minnesota lakes. Twenty-eight of the lakes contained reproducing walleye populations. Bass and panfish were predominant in the other 13 lakes. Walleye populations in these lakes were usually low with sporadic natural reproduction. From 1961 to 1969, all lakes had from two to four plants of fingerlings at intervals of 1 to 4 years. The mean stocking rates were 333 per littoral hectare in the walleye lakes and 550 per littoral hectare in the bass-panfish type lakes. Total length of the fingerlings averaged about 3.6 cm and ranged from 2.5 to 6.6 cm. Evaluation of the stocking program was done by comparing relative year-class abundance as indicated from experimental gill net catches in 1969 and 1970.

In the walleye lakes, stocking did not make significant contributions to the abundance of year classes. In the bass-panfish lakes, walleye year classes were significantly higher in 50% of the years when stocking occurred. When stocking apparently made a contribution, the stocked fingerlings were larger than in years when stocking did not make a contribution.

From 1951 to 1963, fingerling walleyes were stocked in 60 to 70 lakes in Michigan (Schneider 1969). Plants were for introductory purposes, maintenance, and to supplement existing walleye populations. The schedule for stocking varied from consecutive years to every third year or longer. Production of fingerlings was irregular so that many lakes were not stocked according to the designed schedule. Stockings resulted in a contribution to the fishery in 4 lakes, a very limited contribution in 20 lakes, and negative results in 17 lakes. The remaining lakes were not evaluated.

Lake Charlevoix, 6,988 hectares in size, received from 3.5 to 5 fingerlings per hectare in 1955, 1957, and 1958. A spawning run up the Jordan River was established and a small fishery was developed. Three other lakes, ranging in size from 152 to 394 hectares, provided returns to the angler when stocked at rates of 10 to 79 fingerlings per hectare; however, naturally reproducing populations were not established.

Fife Lake, 233 hectares in size, was one of the lakes that contained a naturally reproducing population. In 1961 and 1962, fingerlings were stocked at the rate of 92 and 100 per hectare, respectively. Comparison of the catch of year classes in the 2 years of stocking with years of no stocking showed the rate of return of stocked fish was 1% and 0.1%. From 1969 to 1974, fingerlings were stocked at the rate of 57 to 114 per hectare (Pettengill 1975). No relationship between year-class strength and stocking could be shown from a census conducted in 1974. The stocking program has not significantly changed the walleye population in Fife Lake.

Impoundments

Walleyes were successfully introduced into Canton Reservoir, Oklahoma. The impoundment was built in 1948 and is 3,035 hectares in size, with a maximum depth of 11.9 m (Grinstead 1971). The walleye stocking program was started in 1950 when 75 fingerlings were stocked. The following year 198 fry per hectare were stocked. Extensive trap netting in 1960 failed to take walleye.

In 1961, a new stocking program was started when 264 fry per hectare and a total of 2,900 fingerlings were stocked. During 1962 and 1963, 226 and 82 fry per hectare were stocked. These plants established the first natural reproducing walleye population in Oklahoma.

Lake Meredith in Texas was impounded in 1965 and, at full capacity, has an area of 6,684 hectares with a maximum depth of 37.5 m. Following impoundment several species of fish were introduced. In 1965 and 1966, a total of 2.5 million walleye fry were stocked. A study conducted from 1968 to 1974 was designed to investigate walleye life history factors and evaluate the introduction of walleyes (Kraai and Prentice 1974). Walleyes comprised 15.6 to 42.2% of the catch in gill nets from 1966 to 1972. Fry stocking resulted in a self-sustaining walleye population.

Quabbin Reservoir in Massachusetts reached spillway level in 1946, with a size of 10,125 hectares and a maximum depth of 45.7 m (Bridges and Hambly 1971). Two-thirds of the volume of water provided habitat for coldwater species. During 1952 to 1960, a lake trout-walleye management program was initiated and a total of 20,670,000 fry, 5,500 yearlings, and 2,000 adult walleyes were introduced. Walleyes first entered the catch in 1960 but have contributed little to the harvest and have failed to established themselves.

McConaughy Reservoir in Nebraska was impounded in 1941. It is 14,000 hectares in size, with a maximum depth of 43 m, and supports self-sustaining warm- and coldwater fisheries (McCarraher et al. 1971). Beginning in 1943, walleyes were stocked for 10 consecutive years and then intermittently until 1962. The plantings resulted in a walleye fishery.

Elephant Butte Reservoir in New Mexico was impounded in 1916. From 1964 to 1970, the size has varied from 2,040 to 8,000 hectares, due to release of water for irrigation purposes. During this period maximum depth has varied from 10 to 35 m (Jester 1971). Walleye eggs or fry were planted periodically since 1935, but not in consecutive years until 1959 and 1960. Little success was experienced prior to the 1959 and 1960 plantings. From 1963 to 1967, 3,214,000 fry were stocked. The program was discontinued after 1967 to determine whether a self-sustaining walleye population would become established. In years when natural reproduction was not confirmed, stocking did not result in strong year classes.

Walleyes did produce a strong year class in 1964 when gizzard shad spawned early in response to rapid warming of the water. It appeared that success of walleyes was dependent upon availability of prey within about 1 month after hatching, when young walleyes changed to a fish diet.

Lake McBride, an impoundment in Iowa, has a surface area of 384.5 hectares and a maximum depth of 14.6 m. Fry were stocked from 1958 to 1962 and in 1964 at the rate of 2,470 per hectare. From 1965 to 1969, stocking rates varied from 4,940 to 10,500 per hectare (Mitzner 1971). Due to lack of suitable spawning habitat there was little or no survival from natural reproduction. Population density and angler catch were significantly correlated with stocking rates.

Belleville Lake, a 506-hectare impoundment in Michigan, was treated with rotenone in 1973 to remove a predominant carp, *Cyprinus carpio*, population. From 1973 to 1975 a total of 4,350,000 walleye fry and 16,568 fingerlings were stocked along with several other species of fish. A creel census in 1975 showed that 16,436 or 10.9% of the total number of fish creeled were walleyes; however in 1976, only 1,550 walleyes were harvested. This represented 1.9% of the total number of fish harvested. Evaluation of this project is still in progress.

Erickson and Stevenson (1972) investigated environmental factors in relation to the success of stocking walleyes in more than 100 impoundments in Ohio. In 24 impoundments survival from stocking produced catch rates of at least 1 fish per 100 hours of netting or angling. Even less survival was found in 20 other waters. Complete lack of survival was evident in the remainder of the impoundments. Best survival was found: (1) where populations of smallmouth bass and white bass were present; (2) in the larger reservoirs; (3) where water levels receded slowly; (4) in waters with low turbidities; and (5) where rip-rap or gravel shoals were present.

Great Lakes

Objective evaluations of stocking walleyes in the Great Lakes are few. Hile (1937) examined the relationship between stocking walleye fry in Saginaw Bay from 1924 to 1930 and the corresponding abundance of walleyes from 1929 to 1935 when the stocked fish would have entered the fishery. Although an increase in abundance of walleyes occured in Saginaw Bay during the period investigated, an increase also occurred in southern Lake Huron, southern Lake Michigan, and Green Bay waters where stocking was not a factor. The conclusion was that the general increase in abundance in Saginaw Bay was not the result of stocking, but rather represented a natural fluctuation in abundance.

Summary

After a century of stocking walleyes and less than half that time spent in evaluations, results have varied considerably; however, some general patterns are evident. A summary of successes and failures obtained

TABLE 1.—*Degree of success, expressed in percentages, of stocking walleyes for introductory, maintenance, and supplemental purposes from published and unpublished data.*

Stocking purpose	Total number of lakes	Degree of success (%)		
		Poor	Limited	Good
Introductory	27	29.6	22.2	48.2
Maintenance	10	35.0	32.5	32.5
Supplemental	58	86.2	8.6	5.2

from published and unpublished data is shown in Table 1. Introductions of walleyes into newly created impoundments or rehabilitated lakes have resulted in established populations in many waters. Usually fry have been used for introductory purposes; however, introduction of fingerlings or a combination have been successful in some cases. Stocking fry at rates as low as 100 per hectare have been successful; although in other cases stocking rates greater than 10,000 per hectare have failed.

Maintenance stocking of fingerlings in lakes where natural reproduction is very limited or absent appears to be more successful than stocking in lakes containing good natural populations. Good returns have been realized from stocking rates of 15 to 130 fingerlings per hectare while poor returns have been evident from similar stocking rates. There is some indication, although certainly not conclusive, that survival rate of larger fingerlings was better than of smaller fingerlings. More information is needed on relative survival rates of stocking different size fingerlings under similar biological and environmental conditions.

Ideally, supplemental stocking should be used to augment weak natural year classes. However, this goal has been difficult to attain. Supplemental stockings of fry at relatively high densities of at least 7,400 per hectare have been positively correlated with year-class strength (Rose 1955; Carlander et al. 1960). In several other cases, higher stocking rates of fry have failed to increase year-class abundance. Attempts to strengthen naturally produced year classes by stocking fingerlings have also met with very limited success (Olson and Wesloh 1962; Mraz 1968; Johnson 1971). Stocking rates as high as 250 fingerlings per hectare have not provided significant returns to the harvest.

The return of stocked walleyes to the harvest will depend not only on survival due to environmental conditions, but also on fishing pressure and harvest regulations. Lower returns can certainly be expected when regulations designed to allow walleyes to reach maturity before being harvested are in effect.

Walleye stocking programs should be adjusted to fit each individual body of water and not be based on a set number or size of fish. The ultimate question to be answered is whether the cost of raising and stocking the fish is commensurate with the benefits obtained.

References

BRIDGES, C. H., AND L. S. HAMBLY. 1971. A summary of eighteen years of salmonid management at Quabbin Reservoir, Massachusetts. Am. Fish. Soc. Spec. Publ. 8:243–254.

CARLANDER, K. D. 1945. Age, growth, sexual maturity and population fluctuations of the yellow pikeperch, *Stizostedion vitreum vitreum* (Mitchill), with reference to the commercial fisheries, Lake of the Woods, Minnesota. Trans. Am. Fish. Soc. 73:90–107.

———, R. R. WHITNEY, E. B. SPEAKER, AND K. M. MADDEN. 1960. Evaluation of walleye fry stocking in Clear Lake, Iowa, by alternate-year planting. Trans. Am. Fish. Soc. 89(3):249–254.

ERICKSON, J., AND F. STEVENSON. 1972. Evaluation of environmental factors of Ohio reservoirs in relation to the success of walleye stocking. Dingell-Johnson Final Rep., Ohio F-29-R-11, VI-6.

ESCHMEYER, P. 1950. The life history of the walleye, *Stizostedion vitreum vitreum* (Mitchill) in Michigan. Mich. Dep. Conserv. Inst. Fish. Res Bull. 3. 99 pp.

FORNEY, J. L. 1975. Contribution of stocked fry to walleye fry populations in New York lakes. Prog. Fish-Cult. 37(1):20–24.

———. 1976. Year-class formation in the walleye (*Stizostedion vitreum vitreum*) population of Oneida Lake, New York, 1966–73. J. Fish. Res. Board Can. 33(4, 1):783–792.

GRINSTEAD, B. G. 1971. Reproduction and some aspects of the early life history of walleye, *Stizostedion vitreum* (Mitchill) in Canton Reservoir, Oklahoma. Am. Fish. Soc. Spec. Publ. 8:41–51.

GROEBNER, J. F. 1960. Appraisal of the sport fishery catch in a bass-panfish lake of southern Minnesota Lake Francis, Le Suer County, 1952–1957. Minn. Dep. Conserv. Invest. Rep. 225. 17 pp. (Mimeo.)

HILE, R. 1937. The increase in the abundance of the yellow pikeperch, *Stizostedion vitreum* (Mitchill), in lakes Huron and Michigan, in relation to the artificial propagation of the species. Trans. Am. Fish. Soc. 66(1936):143–159.

JESTER, D. B. 1971. Effects of commercial fishing, species introductions, and drawdown control on fish populations in Elephant Butte Reservoir, New Mexico. Am. Fish. Soc. Spec. Publ. 8:265–285.

JOHNSON, F. H. 1971. Survival of stocked walleye fingerlings in northern Minnesota lakes as estimated from the age-composition of experimental gill net catches. Minn. Dep. Nat. Res. Sect. Fish. Invest. Rep. 314, 12 pp. (Mimeo.)

KEMPINGER, J. J., AND W. S. CHURCHILL. 1972. Contribution of native and stocked walleye fingerlings to the anglers' catch, Escanaba Lake, Wisconsin. Trans. Am. Fish. Soc. 101 (4):644–649.

KLINGBIEL, J. 1971. Management of walleye in the upper midwest. Pages 151–163 in R. J. Muncy and R. V. Bulkley, eds. Proceedings of north central warmwater fish culture-management workshop. Iowa Cooperative Fish. Unit, Ames.

KRAAI, J. E., AND J. A. PRENTICE. 1974. Walleye life history study. Final rep. Tex. Fed. Aid Proj. F-7-R-22. 28 pp.

MARSHALL, T. L., AND R. P. JOHNSON. 1971. History and results of fish introductions in Saskatchewan 1900–1969. Sask. Dep. Nat. Resour. Fish Wildl. Branch Fish. Rep. 3, 30 pp.

McCARRAHER, D. B., M. L. MADSEN, AND R. E. THOMAS. 1971. Ecology and fishery management of McConaughy Reservoir, Nebraska. Am. Fish. Soc. Spec. Publ. 8:299–311.

McWILLIAMS, D. 1975. 0-age fish production and survival in Spirit Lake. Study 101-3. Pages 1–4 in A compendium of fisheries research in Iowa. Iowa Conserv. Comm. Fish. Sect., Des Moines.

MICKLUS, R. C. 1961. A comparison of sport fisheries of 1937 and 1957 on Seagull Lake, Cook County. Minn. Fish Game Invest. Fish. Ser. 3:43–50.

MITZNER, L. 1971. Iowa walleye stocking in impoundments. Pages 130–133 in R. J. Muncy and R. V. Bulkley, eds. Proceedings of north central warmwater fish culture-management workshop. Iowa Cooperative Fish. Unit, Ames.

MRAZ, D. 1968. Recruitment, growth, exploitation, and management of walleyes in a southeastern Wisconsin Lake. Wis. Dep. Nat. Resour. Tech. Bull. 40.

OLSON, D., AND M. WESLOH. 1962. A record of six years of angling on Many Point Lake, Becker County, Minnesota, with special reference to the effect of walleye fingerling stocking. Minn. Dep. Conserv. Res. Planning Invest. Rep. 247. 5 pp. (Mimeo.)

PETTENGILL, T. D. 1975. Evaluation of a walleye, *Stizostedion vitreum vitreum* (Mitchill), stocking program, Fife Lake, Michigan. M.S. Thesis. Cent. Mich. Univ., Mt. Pleasant. 36 pp.

ROSE, E. T. 1947. The population of yellow pikeperch (*Stizostedion v. vitreum*) in Spirit Lake, Iowa. Trans. Am. Fish. Soc. 77:32–42.

———. 1955. The fluctuations in abundance of walleyes in Spirit Lake, Iowa. Proc. Iowa Acad. Sci. 62:567–575.

SCHNEIDER, J. C. 1969. Results of experimental stocking of walleye fingerlings, 1951–1963. Mich. Dep. Nat. Res. Res. Dev. Rep. 161. 31 pp.

SMITH, L. L., JR., AND L. W. KREFTING. 1954. Fluctuations in production and abundance of commercial species in the Red Lakes, Minnesota, with special reference to changes in the walleye population. Trans. Am. Fish. Soc. 83:131–160.

THREINEN, C. W. 1955. What about walleye stocking? Wis. Conserv. Bull. 20(11):20–21.

Reestablishment of Sauger in Western Lake Erie

MICHAEL R. RAWSON AND RUSSELL L. SCHOLL

Ohio Department of Natural Resources, Division of Wildlife,
Lake Erie Fisheries Unit
Sandusky, Ohio 44870

Abstract

The sauger was reintroduced into western Lake Erie by stocking 882,000 fry and fingerlings into lower Sandusky Bay between 1974 and 1976. Recaptures in research and commercial fishing gear have indicated good survival from each of these annual stockings. Their distribution ranged from Maumee Bay east to Cleveland, Ohio, and north to the Canadian shoreline. Areas of concentration occurred in Sandusky Bay and along the southern shore of western Lake Erie. Growth has exceeded by far the historical growth of native saugers. Back-calculated lengths at the end of years of life 1 through 3 were 200, 381, and 433 mm. All of the males and most females were mature as 2-year olds. Fecundity estimates for 2-year-old females averaged 65,000 eggs per fish. Mature saugers migrated up the Sandusky and Maumee Rivers to spawning areas during March and April. Natural recruitment has occurred. Sauger fed primarily on freshwater drums and emerald shiners. Sport catches of saugers have occurred at spawning areas in tributary rivers and along the southern shoreline of Lake Erie.

For many years a great number of fish have been stocked in western Lake Erie for purposes of either augmenting the natural recruitment or introducing exotic species. Scientists have documented the demise of various indigenous fishes and generally attributed it to excessive fishing and inferior environment. The sauger was the first of the three varieties of *Stizostedion* in Lake Erie to decline in abundance (Hartman 1972). The absence of reported captures up to the early 1970's led to suggestions that the sauger had approached extinction in the lake (Applegate and Van Meter 1970). This report describes efforts by the Ohio Department of Natural Resources to reestablish this formerly abundant species.

The Historical Fishery

The sauger was most abundant in the shallow and turbid waters of western Lake Erie (Deason 1933; Doan 1941; Trautman 1957; Van Meter and Trautman 1970). The early-spring (spawning) distribution was mainly along the south shore and tributaries of the Western Basin (Fig. 1), particularly in Maumee Bay (Hartman 1972). Moore (1894, unpublished) reported early spring movements of saugers near shore in western Lake Erie (often in advance of ice break-up) and documented the sauger fishery in the region west of Vermilion, Ohio during winter, early spring, and fall. The most important sauger fishery was at Toledo, Ohio, although catches were common along the entire south shore of the Central Basin. In 1883, saugers were described as so abundant in the spring and fall at Cleveland, Ohio that "scarcely any other fish can exist" (United States Fish Commission 1883).

Initial commercial harvest reached 2.4 million kg in 1885 and peaked at 2.8 million kg in 1916 (Ohio Department Natural Resources 1956) (Fig. 2). Moore (1894, unpublished) observed that saugers seemingly filled a void in the fisheries created by the decline of walleyes and blue pike. Regier et al. (1969) attributed the increase of saugers in the fishery to increased turbidity and fertility of the lake along southern Lake Erie shores. Since that time, commercial harvest regularly decreased to commercial extinction in the 1960's. Combined effects of environmental degradation and heavy fishing pressure have been suggested as limiting factors (Applegate and Van Meter 1970; Hartman 1972; Regier et al. 1969; White 1975).

The sauger was also an important sport fish. An estimated 4,000 saugers were harvested in one day from Cleveland Harbor (United States Fish Commission 1883). Doan (1944) reported a winter sport harvest of 3,600 kg of saugers by ice fisher-

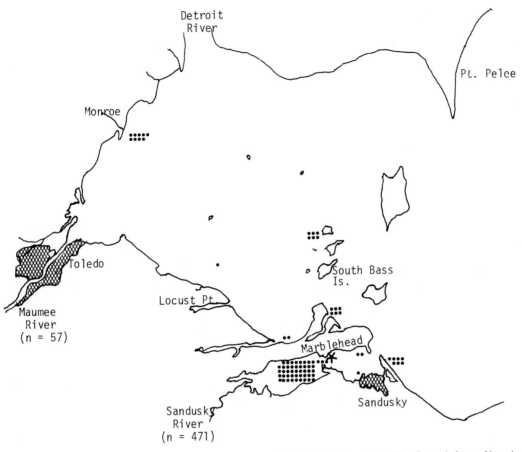

FIGURE 1.—*Western Lake Erie; locations of sauger captures by Ohio Division of Wildlife (●) and the stocking site (★) in Sandusky Bay.*

men at South Bass Island. However, by 1963, saugers had been unimportant in the sport fishery for fifteen years (Keller 1964).

Historically, native sauger were relatively short lived and growth was poor. Moore (1894, unpublished) observed that a 330-mm commercial harvest size and intense fishing pressure in the 1880's prevented the sauger from attaining a total length much over 380 mm. A comprehensive study indicated less than 10% of saugers in the 1927–1928 commercial trap net fishery were older than age III at 264 mm (Deason 1933). Observations by Trautman (1957) and Doan (1941, 1942) support this slow growth.

Successful walleye hatches in 1970 and 1972 suggested that similar historical sauger spawning and nursery areas in the Western Basin would be productive. The Ohio Department of Natural Resources, Division of Wildlife, began the reintroduction of sauger in the spring of 1974 when 229,000 fingerlings and 330,000 fry were planted in the lower Sandusky Bay. In 1975 and 1976, 213,000 and 90,000 fingerlings, respectively, were planted at the same location.

Recaptures

Saugers were first recaptured in August, 1974 in lower Sandusky Bay. Trawl and gill net surveys in Lake Erie captured 85 saugers through the fall of 1977. Saugers from each of the three plantings were represented. Sauger captures were distributed along the shoreline of the Western Basin between Monroe, Michigan and Sandusky Bay (Fig. 1).

Other research organizations reported 60 additional sauger captures in the past four years (C. E. Herdendorf and D. R. Wolfert, personal communications; Rawson, unpublished). Captures occurred primarily at sites in Sandusky Bay, Maumee Bay, and Sandusky River. The Ontario Ministry of Natural Resources reported a September 1977 capture of a 1977 year class sauger (S. J. Nepszy, personal communication).

Captures of saugers in commercial fishing gear have been reported since October 1974. More than 2,000 captures have been reported from trap nets, gill nets, and seines. Sandusky Bay and the southern shoreline of the Western Basin near Locust Point were sites of major concentrations.

Early-spring spawning concentrations of saugers in the Sandusky and Maumee Rivers were sampled by electrofishing in 1976 and 1977. There was a greater concentration of sauger captures in the Sandusky River ($n = 467$) than in the Maumee River ($n = 83$).

Sport-caught saugers were taken primarily from the Sandusky and Maumee Rivers and Sandusky Bay. The 1977 harvest from the rivers were estimated at 1,900 saugers. Two of the river saugers were recorded as new sport records for Ohio.

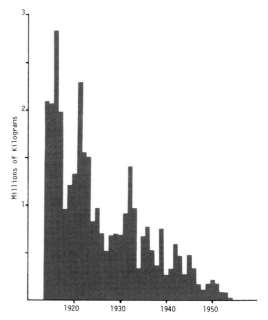

FIGURE 2.—*Ohio commercial harvest of saugers, 1914–1958.*

Growth

Limited growth analyses (31 fish, sexes combined) suggest a faster growth by saugers now in Lake Erie than formerly. Back-calculated lengths at the end of years of life one through three averaged 200, 381, and 433mm, respectively. Sauger weights at the end of years one through four averaged 0.16, 0.53, 1.25, and 1.36 kg, respectively. In October 1976, a two-year-old sauger was captured which measured 520 mm and weighed 1.7 kg, exceeding the 490-mm maximum reported for Ohio waters by Trautman (1957). Deason (1933) reported average lengths for native Lake Erie saugers at end of years one to six were 99, 201, 264, 310, 345, and 401 mm, respectively. In other areas, back-calculated lengths of saugers at the end of 2 years were 338–396 mm in Tennessee Reservoirs (Stroud 1949; Hassler 1958) 325 mm in Lewis and Clark Lake, South Dakota (Nelson 1968).

Food Preference

Stomachs were examined from 21 saugers collected during summer and fall from Sandusky Bay and the open lake. Fish was the major food item in 19 stomachs which contained food. Prey species and their overall percent of food volumes were: freshwater drum (35%); emerald shiners (21%); gizzard shad (15%); alewife (5%); channel catfish (<1%); invertebrates (<1%). Unidentified fish remains constituted 24% of the total contents.

Current food preferences are similar to reported preferences of native saugers (Doan 1942) except that more alewives and gizzard shad are eaten now; this reflects the increased abundance of these prey species during the last 20 years in Lake Erie.

Fecundity

Estimated fecundity of five mature two-year-old saugers (373–395 mm total length) collected in March and April, 1976 before spawning ranged from 58,000 to 77,400 and averaged 64,859 eggs per female. Eggs per fish kg averaged 105,800. This compares with 15,871 eggs per female in Lake Winnebago (Priegel 1969) and 58,250 in Norris Reservoir (Smith 1941). Many estimates averaged 45,000 eggs per fish or 65,000 eggs per kg (Carlander 1950; Carufel 1963; Hassler 1958).

Spawning Areas

Electrofishing surveys were conducted between 15 March and 1 May in 1976 and 1977 in the Maumee and Sandusky Rivers at walleye spawning sites. Concentrations of mature saugers were discovered along shale bedrock ridges and sand-gravel substrate in the Sandusky River and along cobble-boulder riffles in the Maumee River. Few saugers were captured where bottom substrate was predominantly mud. The highest capture rates occurred during periods of warming water in late March and early April.

In 1976, all the saugers captured were mature 2-year olds ($n = 158$). Most of the sauger were males (87%). In 1977 68% of the 3-year olds ($n = 259$) and 88% of the 2-year olds ($n = 130$) were males. Females flowing ripe eggs were caught on 12–15 April.

Mature sauger were also captured in early spring commercial trap nets off of Locust Point.

Movements

To further define sauger movements and distribution, 235 saugers were tagged from their spawning concentrations in the Sandusky and Maumee Rivers in 1977 with Floy FD-68 anchor tags. Six were recaptured within four days of tagging and 8 others were recaptured after spending considerably more time (averaging 35 days) in the river after tagging. One fish remained in the Sandusky River spawning area for 51 days before it was removed by an angler.

Four saugers were recaptured in commercial fishing gear within a month after tagging, three in Sandusky Bay and one near Monroe, Michigan. Angler recaptures were reported at three lake locations: the mouth of Sandusky Bay; the east side of Catawba Island; and from a hot-water discharge near Lorain, Ohio.

Saugers are now distributed over most of their original range in Lake Erie: along the Michigan shoreline; north to the Canadian shoreline; along the southwest shoreline as far as Cleveland. Saugers have remained especially abundant in Sandusky Bay, the original stocking site. Sauger recaptures have occured primarily at near-shore areas of high turbidity. Sport catches of saugers were also near shore, but none occurred at walleye reef areas with low turbidity.

Interactions with Walleye

The absence of saugers on walleye reefs suggests segregation of saugers and walleyes during most of the summer and fall. Differences in food preference in the summer and fall suggested that saugers and walleyes select different prey species even when both species are in the same area. Saugers preyed primarily on freshwater drum, but also ate emerald shiners, gizzard shad, and alewifes. Walleyes preyed primarily on shiners.

Mature saugers concentrated at the same spawning areas as walleyes in the Sandusky and Maumee Rivers. However, saugers were much less abundant than walleyes during the peak walleye run in March and April. Saugers were sometimes captured at a greater frequency in certain areas, but segregation between species was not apparent at these spawning areas. Individual saugers remained in the spawning area for an extensive time, whereas Van Vooren (1976) documented that tagged walleyes showed little tendency to remain in the rivers for more than four days.

Spring gill net sampling of walleye spawning populations at Western Basin reefs captured 3,700 walleyes, but only one sauger. However, both mature saugers and walleyes were abundant in early spring commercial trap nets fished at near-shore areas adjacent to these reefs. Sauger utilization of this area for spawning was not documented.

Successful natural reproduction has been

documented by limited capture of 1977 year class saugers.

Future Programs

The evaluation of this stocking success has led to two future management considerations. The Sandusky River spawning stock can be developed by continuing annual plants of between 250,000 to 750,000 fingerlings, or planting 2.0 million fingerlings at eight locations to establish a basin-wide sauger population.

References

APPLEGATE, V. C., AND H. D. VANMETER. 1970. A brief history of commercial fishing in Lake Erie. U.S. Fish Wildl. Serv. Fish. Leaflet 630. 28 pp.

CARLANDER, K. D. 1950. Growth rate studies of saugers, *Stizostedion canadense*, and yellow perch, *Perca flavescens* (Mitchill) from Lake of the Woods, Minnesota. Trans. Am. Fish. Soc. 79:30–42.

CARUFEL, L. H. 1963. Life history of sauger in Garrison Reservoir. J. Wildl. Manage. 27(3):450–456.

DEASON, H. J. 1933. Preliminary report on the growth rate, dominance, and maturity of the pike-perches (*Stizostedion*) of Lake Erie. Trans. Am. Fish. Soc. 63:345–360.

DOAN, K. H. 1941. Relation of sauger catch to turbidity in Lake Erie. Ohio J. Sci. 41(6):449–452.

———. 1942. Some meteorological and limnological conditions as factors in the abundance of certain fishes in Lake Erie. Ecol. Monogr. 12(3):294–314.

———. 1944. The winter fishery in western Lake Erie with a census of the 1942 catch. Ohio J. Sci. 44(2):69-74.

HARTMAN, W. L. 1972. Lake Erie: effects of exploitation, environmental changes, and new species on the fishery resources. J. Fish. Res. Board Can. 29:899–912.

HASSLER, W. W. 1958. The fecundity, sex ratio and maturity of the sauger, *Stizostedion canadense* (Smith) in Norris Reservoir, Tennessee. J. Tenn. Acad. Sci. 33(1):32–38.

KELLER, M. 1964. Lake Erie sport fishing survey. Ohio Dep. Nat. Resour. Div. Wildl. Publ. 316. 19 pp.

NELSON, W. R. 1968. Reproduction and early life history of sauger, *Stizostedion canadense* in Lewis and Clark Lake. Trans. Am. Fish. Soc. 97(2):167–174.

OHIO DEPARTMENT OF NATURAL RESOURCES. 1956. Summary of the Ohio Lake Erie commercial fish catch 1885–1955. Ohio Dep. Nat. Resour. Div. Wildl. 65 pp.

PRIEGEL, G. R. 1969. The Lake Winnebago sauger; age, growth, reproduction, food habits, and early life history. Wis. Dep. Nat. Resour. Tech. Bull. 43. 63 pp.

REGIER, H. W., V. C. APPLEGATE, AND R. A. RYDER. 1969. The ecology and management of the walleye in western Lake Erie. Great Lakes Fish. Comm. Tech. Rep. 15. 101 pp.

SMITH, C. G. 1941. Egg production of walleyed pike and sauger. Prog. Fish Cult. 54:32–34.

STROUD, R. H. 1949. Rate of growth and condition of game and pan fish in Cherokee and Douglas Reservoirs, Tennessee, and Hiwassee Reservoir, North Carolina. J. Tenn. Acad. Sci. 24(1):60–74.

TRAUTMAN, M. B. 1957. The fishes of Ohio. Ohio State Univ. Press, Columbus. 686 pp.

UNITED STATES FISH COMMISSION. 1883. On the impropriety of depositing white-fish minnows off the harbor of Cleveland, Ohio—fishing for saugers. Bull. U.S. Fish Comm. 3. 2 pp.

VAN METER, H. D., AND M. B. TRAUTMAN. 1970. An annotated list of the fishes of Lake Erie and its tributary waters exclusive of the Detroit River. Ohio J. Sci. 70(2):65–79.

VAN VOOREN, A. R. 1976. Characteristics of walleye spawning population. Annu. Rep. Dingell-Johnson Proj. F-35-R-14, Study IV. Ohio Dep. Nat. Resour. Div. Wildl., Sandusky.

WHITE, A. M. 1975. Water quality baseline assessment for the Cleveland area—Lake Erie—Vol. 11. The fishes of the Cleveland metropolitan area including the Lake Erie shoreline. U.S. Environ. Prot. Agency Proj. GOO5107. 182 pp.

Management for Walleye or Sauger, South Basin, Lake Winnipeg

R. O. SCHLICK

Fisheries and Wildlife Branch
Department of Renewable Resources and Transportation Services
Winnipeg, Manitoba R3H 0W9

Abstract

The South Basin of Lake Winnipeg just prior to 1970, when the lake was closed due to mercury contamination in fish, produced mainly sauger. From 1972 onwards, the harvest in the fall fishery, which is the main seasonal fishery, was predominately of walleye. However, over 42 seasons, total sauger and walleye production has been almost equal (only 1% difference). Dominance of either species in the catch has varied not only from year to year but sometimes from season to season within the same year. These shifts in dominance do not appear related to market conditions but are probably the result of environmental factors or migrations. This indicates that both species are important to the South Basin fishery. Maturity studies indicate a change to a larger gill net mesh size would be desirable, especially for walleye management. Economically, fishermen would suffer by an increased mesh size, especially in the winter fishery, although a continuing price differential in favour of walleye over sauger would lessen this economic impact. Speculatively, environmental factors (e.g., water transparency) are likely to favour the sauger over the walleye in the future.

The South Basin of Lake Winnipeg (Fig. 1) is commercially fished in the summer, fall, and winter. The fall fishery is the most important, both in terms of amount of fish harvested and of numbers of fishermen involved. The South Basin, for this study, is all of Lake Winnipeg south of Grindstone Point. This fishery was closed because of mercury pollution in 1970 and 1971. When the fall fishery reopened in 1972, walleye was the main commercial species harvested, whereas prior to 1972 sauger was the main species in the fall fishery. Walleye continued to be the dominant quota species (among walleye, sauger, and lake whitefish) in the fall fisheries from 1972 through 1974. Therefore the question arose "should we manage the South Basin for walleye or sauger or both"? Management for walleye would require all gill nets to have a minimum stretch measure mesh size of 108 mm. To answer this question we must examine past harvest records, economic return to the fishermen, and environmental factors.

The Past Fish Harvest

The summer fishery has been predominantly a sauger fishery except in 1961, 1973, and 1974 (Table 1). Sauger catches in the South Basin varied from 31 to 89% of the total sauger–walleye catches (1958–1974) and averaged 63%. The summer fishery averaged 45,284 kg of sauger, 26,855 kg of walleye.

In the fall fishery, walleye harvest in the South Basin exceeded that of sauger over the period 1945–1953 (Table 2). From 1954 to 1955 the fishery harvested more sauger than walleye. From 1956 to 1958, harvest records did not distinguish the two species. From 1959 to 1969 the fishery harvested mainly sauger. This area of the lake was closed to commercial fishing in 1970 and 1971 due to mercury contamination. Upon reopening in 1972 through to 1974, this fishery harvested mainly walleye again. Thus, over the 25 fall seasons from 1945 to 1974 for which there are reliable records, walleye was the main species in 12 seasons and sauger was dominant in 13 seasons. The fall fishery in these 25 seasons has averaged 139,340 kg of walleye and 120,150 kg of sauger. Therefore, sauger averaged 46% of the total walleye-sauger catch over this period.

Winter catch records of walleye and sauger from the South Basin (Table 3) could only be separated from the Channel Area for two winter seasons, 1973–1974 and 1974–1975. In each of these two seasons sauger production exceeded walleye production. The winter fishery averaged 119,800 kg of sauger and 58,600 kg of walleye, with sauger accounting for 67% of the winter sauger-walleye catches for the two seasons. These two winter seasons cover

TABLE 1.—Summer walleye and sauger catches, South Basin, Lake Winnipeg.

Year	Walleye kg	Sauger kg	% sauger
1958	56,744	114,396	66.8
1959	3,402	25,719	88.0
1960	12,565	19,822	61.2
1961	5,171	5,035	49.3
1962	14,696	15,604	51.5
1963	10,115	74,752	88.1
1964	3,357	27,714	89.2
1965	7,484	16,329	68.6
1966	3,946	17,146	81.3
1967[a]	1,996	4,309	68.3
1968	12,292	75,568	86.0
1969	31,132	86,613	73.6
1970–1971	Closed due to mercury pollution		
1972	64,804	85,856	57.0
1973	107,967	49,211	31.3
1974	67,149	61,188	47.7
Total	402,820	679,260	
Average	26,855	45,284	62.8

[a] The poor production of both species in 1967 was due to poor prices set by the fish companies, since there was a glut of fish on the market.

TABLE 2.—Fall walleye and sauger catches and prices South Basin, Lake Winnipeg.

Year	Walleye Kilograms	¢/kg	Sauger Kilograms	¢/kg	% sauger
1945	192,958	31	123,150	22	39.0
1946	200,987	31	179,940	24	47.2
1947	249,249	31	129,501	22	34.2
1948	301,367	31	104,699	22	35.3
1949	277,553	20	215,955	9	43.8
1950	275,648	29	97,613	29	26.2
1951	137,030	37	44,770	26	24.6
1952	169,417	33	102,829	20	37.8
1953	282,815	29	90,129	18	24.2
1954	86,686	31	133,356	24	60.6
1955	227,749	29	246,981	22	52.0
1956		31		29	
1957		29		24	
1958		40		37	
1959	15,925	53	19,150	49	54.6
1960	20,870	55	25,373	49	54.9
1961	30,679	33	32,507	33	51.4
1962	29,267	42	93,596	37	76.2
1963	9,185	55	140,550	49	93.9
1964	13,543	55	122,084	37	90.0
1965	19,709	68	153,485	62	88.6
1966	27,038	90	133,981	66	83.2
1967	18,740	46	45,265	26	70.7
1968	30,292	57	95,238	31	75.9
1969	55,057	73	175,753	62	76.1
1970–1971	Closed due to mercury pollution				
1972	305,500	112	147,433	97	32.6
1973	377,881	126	151,940	110	28.7
1974	309,037	128	138,471	112	30.9
Total	3,483,482		3,003,749		
Average	139,339		120,150		46.3

the same years during which the summer and fall seasons produced predominantly walleye. Thus, it obviously would not be economical for fishermen to fish only for walleye, by using 108 mm or larger mesh nets. In addition, winter fishing involves greater operating expenses per gross income to fishermen (Richard Peters, personal communication), so the amount of sauger caught is very important to the economics of the winter fishery.

At present time, walleyes predominate in both summer and fall fisheries while saugers dominate the winter fishery. Historically, sauger was the most important commercial species in the South Basin in the summer and in the winter while walleye was the most important species in the fall fishery. However, walleye harvest ex-

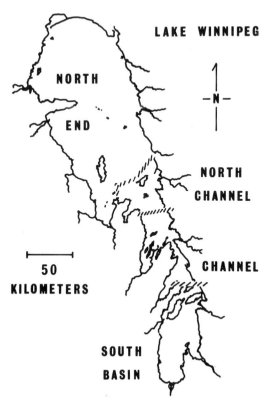

FIGURE 1.—Major areas of Lake Winnipeg.

TABLE 3.—Winter walleye and sauger catches, South Basin, Lake Winnipeg.

Year	Walleye kg	Sauger kg	% sauger
1973–1974	56,866	100,777	63.9
1974–1975	60,332	138,847	69.7
Total	117,198	239,624	
Average	58,599	119,812	67.2

ceeded that of sauger by only 1% over 42 seasons (Tables 1, 2 and 3). Therefore, it must be concluded both species are very important to the South Basin fishery. The prevalence of one species over another not only varies from year to year but can also vary from season to season within the same year. These shifts in varying dominance of sauger and walleye in the catch suggest the possibility, among other factors, of significant movements of fish populations into or out of the South Basin. Thus, there is no guarantee that a species dominant one season or year will remain dominant in subsequent years.

Economic Considerations

The average price difference fishermen received for walleye over sauger has varied up to 26¢ per kg (Table 2) and averaged 10¢ higher for walleye in the years 1945–1974. Presently (1977) the price difference is about 40¢ per kg.

The South Basin is being primarily managed as a sauger fishery because the minimum legal mesh size is 76 mm stretch measure, which is a more suitable mesh size for the smaller-maturing sauger, although present maturity data suggests that the sauger are also being caught too early in life to reproduce successfully (Davidoff 1978, this volume). Managing the South Basin for a walleye fishery (e.g. changing the minimum mesh size to 108 mm) would result in a prolonged fishery in an attempt to reach the existing quota which is based on an aggregate weight of walleye, sauger, and lake whitefish. This would increase the fishermen's operating costs and place an extra strain on the present walleye population. Of course, with a reduced quota, an increase in the minimum allowable mesh size would mean more young walleye would escape, thus increasing the walleye spawning stock in the South Basin.

In the fall fishery of 1973 there were 233 fishermen involved in the South Basin. A change in mesh size to 108 mm mesh would require each fisherman to obtain about 10 nets each at a cost of about $70.00 per net completely corked and leaded. This changeover would cost about $163,00.00.

Experiments with gill net mesh sizes involving 76 mm, 83 mm, and 95 mm stretch measure (Davidoff 1973) indicated that the 76-mm mesh would yield the greatest gross income to fishermen at Victoria Beach and Gimli, both areas being within the South Basin. This would indicate that a 76-mm mesh fishery is probably better suited to the South Basin economically, though, not biologically, than a larger minimum mesh size fishery.

Water Transparency

Doan (1942) demonstrated that average April-May turbidity values of Lake Erie at Cleveland were significantly related to Ohio's commercial sauger catch three years later. Increased spring turbidities had no effect on the state's yellow perch catches. Small walleye catches resulted in years of high precipitation in the spring.

Both walleye and sauger exhibit negative phototropism, but the sauger more so than the walleye. As a result, in clearer water bodies, the walleye is a twilight feeder, while the sauger, which inhabit more turbid waters feed more actively throughout the day (Ali and Anctil 1968). In Alberta, walleyes are more widely distributed than saugers, suggesting a wider water transparency tolerance than sauger. Saugers occur in the larger rivers of Alberta such as the Milk River and the lower Red Deer River (Paetz and Nelson 1970). "In very turbid water, saugers usually succeed over walleye. In Ombabika Bay, Lake Nipigon, a predominantly walleye dominance changed to a sauger dominance in a period of a few years following diversion into the bay of turbid water" (Scott and Crossman 1973). "Optimum transparency in a shallow lake, which will allow walleye to feed intermittently throughout the day is in the order 1–2 m Secchi disc" (Ibid). Regier et al. (1969) stated that walleyes prefer water having Secchi disc readings of roughly 1.2–3 m and a bottom that contains areas of clean rock and sand.

In Moose Lake, Manitoba, which had a mean Secchi disc reading of 1.49 m, few saugers were found with the walleyes and northern pike, but in Cedar Lake, Manitoba, with a mean Secchi disc reading of 0.55 m, a substantial sauger population coexisted with the walleye and northern pike (Garside et al. 1973). Secchi disc readings in the Grand Rapids area of Lake Winnipeg in August, 1974, varied from 0.91 to 2.13 m. This area is historically a walleye producing area whereas the Pine Dock area of Lake Winnipeg is a sauger producing area. Ward (1975) found Secchi disc readings of 0.30 m in the Pine Dock vicinity. Thus, from the literature review it would appear that waters with transparency readings of 0.9 to 3.0 m are best suited for walleye while those with Secchi disc reading of less than 0.9 m are better suited for sauger. From 1962 to 1970 Secchi disc readings in the South Basin have varied from 0.3 to 1.2 m and averaged 0.5 m. These transparency readings should favour the sauger over the walleye.

References

ALI, M. A., AND M. ANCTIL. 1968. Corrélation entre la structure rétinienne et l'habitat chez *Stizostedion v. vitreum* et *S. canadense*. J. Fish. Res. Board Can. 25(9):2001–2003.

DAVIDOFF, E. B. 1973. Lake Winnipeg gill net mesh size experiment. Manit. Dep. Mines, Resour. Environ. Manage. Res. Branch Rep. 73–12.

———. 1978. The Matheson Island sauger fishery of Lake Winnipeg, 1972–1976. Am. Fish. Soc. Spec. Publ. 11:308–312.

DOAN, K. H. 1942. Some meteorological and limnological conditions as factors in the abundance of certain fishes in Lake Erie. Ecol. Monogr. 12: 293–314.

GARSIDE, E. T., A. J. DERKSEN, AND W. H. HOWARD. 1973. Summer food relations and aspects of the distribution of the principal percid fishes of Saskatchewan River delta prior to 1965 impoundment. Manit. Dep. Mines, Resour. Environ. Manage. Res. Branch Rep. 73-18.

PAETZ, M. J., AND J. S. NELSON. 1970. The fishes of Alberta. Queen's Printer, Gov. Alberta 282 pp.

REGIER, H. A., V. C. APPLEGATE, AND R. A RYDER. 1969. The ecology and management of the walleye in western Lake Erie. Great Lakes Fish. Comm. Tech. Rep. 15. 101 pp.

SCOTT, W. B., AND E. J. CROSSMAN. 1973. Freshwater fishes of Canada. Bull. Fish. Res. Board Can. 184. 966 pp.

WARD, N. J. R. 1975. Effects of gill net colour, twine size and twine structure on Lake Winnipeg commercial fishing income. Practicum, Nat. Resour. Inst., Univ. Manitoba.

Food, Growth, and Exploitation of Percids in Ohio's Upground Reservoirs[1]

KENNETH O. PAXTON AND FREDERICK STEVENSON

*Ohio Department of Natural Resources, Division of Wildlife
Findlay, Ohio 45840*

Abstract

Yellow perch grew rapidly and reached mean total lengths in excess of 200 mm at age I+ or II+. Growth abruptly declined in subsequent years, and was related to the absence of piscivorous forage for older and larger yellow perch. Yellow perch fed throughout life on zooplankton and aquatic insects, primarily chironomid larvae and pupae. Young yellow perch grew too rapidly to be utilized by slower growing age 0+ walleyes. Yellow perch were also unimportant as forage for age I and older walleyes, and growth of walleyes was poor. Walleye forage was primarily zooplankton and aquatic insects through age II; fish and crayfish became important at ages III and IV. Though growth was slow, walleyes were readily harvested and along with yellow perch accounted for 44% of the total angler harvest from 1972 to 1977.

Northwestern Ohio has only one natural lake and low relief of the terrain does not permit construction of onstream impoundments. Consequently, upground reservoirs have been developed for water storage and are the principal inland waters available to area anglers. These impoundments, somewhat similar to pumped-storage reservoirs, are constructed proximal to their source stream by a combination of excavation and dike building. When completed, an earthen dike encompasses the impoundment and rises 4–8 m above the surrounding terrain. Water is then pumped into the reservoir from the source stream. Burgess and Niple, Limited (1967) have provided detailed descriptions of the morphometry and uses of upground storage reservoirs.

Ohio's Division of Wildlife has managed the sport fisheries of upground reservoirs since 1946, but management objectives have been based on limited studies conducted during routine fisheries surveys. Clark (1965) determined that 13 of 23 Ohio inland waters supporting walleye populations yielding an annual minimum catch per effort of 0.01 fish were upground reservoirs. Fishery records also indicated that yellow perch was a successful and important species in many upground reservoirs. A comprehensive study of upground reservoir fisheries initiated in 1971 is the basis for this paper.

Study Areas

Studies were conducted on Ferguson and Killdeer Reservoirs from 1972 to 1977. Both are located in the Lake Erie drainage of northwestern Ohio, Ferguson Reservoir in Allen County and Killdeer Reservoir in Wyandot County.

Ferguson Reservoir, constructed in 1960, is a 123-hectare impoundment with a mean depth of 6.8 m and a maximum depth of 12.0 m. The shoreline is very regular with a shore development of 1.43. Killdeer Reservoir, impounded in 1972 and similar in size at 115.3 hectares, has a mean depth of 7.1 m and a maximum depth 12.8 m. It is nearly square with a shore development of 1.05.

Fish populations of both reservoirs are similar; yellow perch, walleyes, bluegills, and smallmouth bass are important species. White bass are also abundant in Ferguson, but are not present in Killdeer Reservoir. Several other centrarchid and ictalurid species inhabit the reservoirs, but occurrence of forage species is extremely limited.

Methods

Young-of-the-year yellow perch and walleyes were sampled with a 3.7-m otter trawl of 5.1-cm stretch mesh with a codend liner of 6-mm mesh. Age I and older yellow perch were captured with gill nets of 2.5-, 3.8-, 5.1-, and 7.6-cm stretch mesh. All trawling and gill netting occurred during daylight from 0900 to 1600 hours. Electro-

---

[1] A contribution from Federal Aid in Fish Restoration Project F-29-R, Ohio.

fishing was conducted at night from 2000 to 0300 hours to collect age I and older walleyes.

Stomach contents of all fish were examined under a binocular microscope. All organisms were identified and counted; identifiable fragments were counted as whole organisms. Organisms were identified to family or genus. Results were expressed as percentage occurrence and as the average number of organisms per stomach for each stomach containing a given organism.

Subsamples of walleye and yellow perch caught in trap nets during April of each year were used for age–growth determinations. Scales were removed from an area immediately below the lateral line and beneath the insertion of the spinous dorsal fin. The anterior edge of each scale was considered an annulus since spring annulus formation had not occurred. The relationship of yellow perch body and scale lengths was linear and is expressed by the equation $L = 27.6061 + 1.4022 S$ for Ferguson Reservoir and $L = 56.4243 + 1.9588 S$ for Killdeer Reservoir, where L equals total fish length in mm and S equals the anterior scale radius in mm. Empirical growth data were utilized for 1972 year class walleyes at ages I through IV.

Angler harvest data were obtained from 1972 to 1976 for Ferguson Reservoir. Killdeer Reservoir was closed to angling for two years subsequent to impoundment in 1972, thus harvest data were determined from 1974 to 1977. A random stratified census was scheduled in all years from either April or May through September or October. Total angler harvest and effort were calculated by direct proportion for each month censused.

A random creel census was conducted on Ferguson Reservoir during winter ice cover from 1974 to 1976. Effort and total harvest were estimated by direct proportion. No winter creel census was conducted on Killdeer Reservoir due to minimal angling pressure.

Results

Food of Yellow Perch

Food of age 0+ yellow perch in Ferguson Reservoir was determined in both 1972 and 1975. Food of fry less than 25 mm was not determined, but young yellow perch

TABLE 1.—*Food of 0+ yellow perch, 1972 and 1975, from Ferguson Reservoir in percentage occurrence and (in parentheses) average number of organisms per stomach.*

Date	Mean total length (mm)	Number of stomachs Examined	Number of stomachs With food	Copepods	Cladocerans	Chironomids	Corixids	Trichopterans	Amphipods	Other invertebrates
					1972					
June 12	28	30	30	100 (29)	87 (5)					
June 30	47	30	30	100 (165)	100 (18)					17 (1)
July 18	61	30	30	100 (238)	100 (74)					30 (2)
July 31	71	30	30	100 (74)	100 (260)	50 (5)	7 (1)			77 (5)
August 15	83	30	29	97 (391)	100 (107)	28 (10)				69 (5)
September 19	115	30	30	53 (11)	90 (193)	80 (18)				20 (3)
October 26	132	30	30	27 (2)	100 (140)	93 (25)				7 (1)
November 16	132	30	29	3 (1)	93 (57)	83 (9)				7 (1)
March 14, 1973	134	30	29	35 (5)	100 (151)	90 (14)				24 (1)
					1975					
June 9	29	30	30	100 (55)	60 (4)					
June 25	45	30	30	100 (145)	100 (18)	20 (3)	27 (2)		3 (1)	11 (1)
July 16	68	30	30	97 (137)	93 (37)	57 (11)	60 (7)		53 (8)	50 (3)
July 30	82	10	10	40 (2)	100 (86)	90 (2)	100 (8)	20 (1)	100 (31)	90 (5)
August 13	100	28	28		46 (63)	64 (3)	64 (6)	64 (9)	93 (36)	21 (2)
September 22	138	16	16		63 (39)	88 (70)	63 (13)	19 (4)	94 (10)	31 (4)
October 22	146	20	20		95 (153)	95 (28)	50 (3)	5 (2)	100 (28)	5 (2)
November 19	150	22	17	4 (6)	70 (31)	78 (6)	4 (1)		9 (50)	
March 19, 1976	153	22	8		100 (28)	25 (1)				

TABLE 2.—*Food of adult yellow perch, 1972–1976, from Ferguson Reservoir in percentage occurrence and (in parentheses) average number of organisms per stomach.*

Month	Number of stomachs Examined	Number of stomachs With food	Chironomids	Zooplankton	Other invertebrates	Crayfish
1972						
April	31	11	91 (18)	9 (1)	18 (2)	9 (1)
June	37	37	92 (116)	68 (123)	92 (56)	3 (1)
August	47	46	98 (33)	94 (137)	26 (5)	2 (3)
October	17	17	100 (80)	82 (141)	41 (1)	
1973						
April	30	24	38 (16)	96 (94)		
June	6	6	83 (305)	83 (1196)	50 (3)	17 (1)
August	18	17	94 (138)	88 (228)		12 (2)
October	30	30	97 (142)	87 (171)	40 (2)	
1974						
April	49	45	92 (19)	87 (201)	2 (1)	
June	52	48	19 (15)	75 (2073)	2 (1)	33 (1)
August	54	54	87 (46)	100 (239)	19 (1)	
October	53	51	84 (62)	88 (1623)	18 (3)	
1975						
April	52	42	50 (16)	95 (269)	5 (1)	
June	51	46	98 (257)	67 (138)	33 (1)	13 (1)
August	51	51	98 (141)	100 (121)	16 (2)	8 (2)
October	40	38	100 (130)	84 (118)	13 (1)	
1976						
June	30	30	80 (37)	97 (555)	17 (7)	17 (8)
August	33	32	91 (54)	84 (195)	22 (43)	3 (1)
October	5	5	60 (35)	100 (164)	40 (8)	

greater than 25 mm had become demersal and were readily sampled with trawls. These fish fed initially on zooplankton, primarily small *Daphnia*, *Alona*, *Bosmina*, and copepods (Table 1). The diet rapidly changed to include aquatic insects, with a corresponding decline in utilization of copepods as forage. Cladocerans remained an important part of the diet throughout the year.

Food habits of age 0+ yellow perch differed during 1972 and 1975. The 1975 year class began consuming chironomids in mid-June, but the 1972 year class consumed only zooplankton until late July. Corixids, amphipods, and Trichoptera naiads were also important forage items in 1975, but were not utilized in 1972. Growth of the 1975 year class was faster; fish reached a mean length of 153 mm by 19 March 1976, compared to 134 mm on 14 March 1973 for the 1972 year class. The difference in first-year growth of the two year classes was statistically significant (Student's t-test, $P < 0.05$).

Food of adult yellow perch in Ferguson Reservoir was remarkably consistent during the five year period 1972–1976. Cladocerans, primarily *Daphnia* species, and chironomids provided most of the forage (Table 2). *Chaoborus* were important forage items only in 1972, and other aquatic insects always were relatively unimportant in the diet. Crayfish were utilized as forage principally after cohort production when young crayfish were small. Fish were never observed in the diet of adult yellow perch from April–October.

Feeding patterns were not readily discernible, although adult yellow perch did not feed heavily during spawning concentrations. Greatest occurrence of nonfeeding and minimum stomach content volumes were generally observed during the April spawning period. Winter feeding of adult yellow perch was examined and indicated no appreciable changes from summer feeding. Chironomid larvae and *Daphnia* remained the primary forage items, but were consumed in lesser amounts.

The diet of Killdeer yellow perch during 1975–1977 was basically similar to that of yellow perch from Ferguson Reservoir. Immature chironomids were important forage items, with peak utilization occurring during June or July (Table 3). *Leptodora* and *Eurycercus* were the most heavily consumed cladocerans. *Daphnia* and copepods were relatively unimportant. Zooplankton utilization was greatest during May or June and then declined through September. Other invertebrates consumed sporadically and in varying amounts included ostracods, gastropods, and immature culicids, ceratopogonids, and trichopterans. Crayfish consumption, greatest during early summer, coincided with annual production of crayfish cohorts. Increased crayfish utilization during 1976 was a valid indicator of a large 1976 cohort. Fish were eaten by adult yellow perch, but did not contribute a major portion of the total yearly diet. Species identified from yellow perch stomachs included green sunfish, white crappie, and yellow perch.

TABLE 3.—Food of adult yellow perch, 1975–1977, from Killdeer Reservoir in percentage occurrence and (in parentheses) average number of organisms per stomach.

	Number of stomachs		Food items in stomachs containing food				
Month	Examined	With food	Chironomids	Zooplankton	Other invertebrates	Crayfish	Fish
			1975				
June	4	4	75 (18)	75 (272)	25 (148)	25 (1)	
July	16	15	93 (200)	60 (122)	47 (57)	7 (2)	
August	26	25	36 (28)	92 (96)	12 (2)	4 (1)	12 (1)
September	8	8	25 (9)	13 (1)	25 (2)		75 (2)
			1976				
June	31	26	19 (3)	27 (186)	39 (3)	73 (2)	4 (1)
July	46	29	55 (89)	35 (28)	51 (7)	38 (1)	10 (1)
August	20	17	59 (10)	59 (86)	71 (63)		6 (1)
September	2	2			100 (2)	100 (1)	
			1977				
May	24	24	100 (99)	92 (1156)	54 (15)		4 (1)
June	30	29	97 (257)	59 (63)	41 (6)	14 (3)	3 (1)
July	19	13	85 (68)	15 (1)	54 (1)	23 (1)	
August	7	7	86 (43)		71 (7)	29 (1)	
September	17	16	100 (48)		94 (8)		

Growth of Yellow Perch

Scale analysis indicated nine year classes, 1966–1972 and 1974–1975, in Ferguson Reservoir; strong 1968 and 1972 year classes accounted for 91% of all fish analyzed. Annulus formation was evident for most fish by mid-May and completed by mid-June. Average water temperatures ranged from 15 C in mid-May to 20 C in mid-June. Yellow perch from year classes prior to 1970 grew at a slower rate and females grew faster than males (Table 4). First-year growth of all year classes was somewhat comparable. Greatest annual growth increments occurred in the first year of life and decreased thereafter. Growth of 1972 year class fish exceeded

TABLE 4.—Mean calculated total lengths (mm) at each annulus of yellow perch from Ferguson and Killdeer Reservoirs. Only data based on at least 10 fish are shown.

Year class	Sex	Number of fish	Year of life							
			1	2	3	4	5	6	7	8
			Ferguson Reservoir							
1967	Male	31	117	167	204	234	252	267		
1968	Male	378	115	181	217	239	253	261	267	271
	Female	14	115	187	234	268				
1969	Male	10	118	190	220					
1972	Male	538	120	217	243	254	264			
	Female	123	126	235	273	284	316			
1974	Male	35	124	220						
1975	Male	88	154	231						
	Female	42	164	258						
			Killdeer Reservoir							
1970	Male	52	116	165	192	228	242	254		
1971	Male	101	118	170	219	238	250			
	Female	12	117	195						
1974	Male	242	124	196						
	Female	54	120	197						
1975	Male	24	110	160						
	Female	56	109	157						

TABLE 5.—Food of 1972 year class walleyes, 1972–1975, from Ferguson (F) and Killdeer (K) Reservoirs in percentage occurrence and (in parentheses) average number of organisms per stomach.

Month	Number of stomachs				Food items in stomachs containing food									
	Examined		With food		Chironomids		Zooplankton		Other invertebrates		Fish		Crayfish	
	F	K	F	K	F	K	F	K	F	K	F	K	F	K
1972														
June–July	44	76	44	76	50 (7)	74 (27)	84 (80)	74 (78)	7 (1)	8 (2)	32 (2)			
Aug.–Oct.	52	33	39	18	62 (9)	89 (10)	97 (224)	28 (46)	18 (8)	6 (1)	8 (1)			
1973														
April–May	20	25	18	25	89 (12)	100 (270)	89 (91)	88 (9)	22 (2)	36 (2)	22 (1)			
June–July	25	10	23	9	74 (8)	100 (38)	78 (97)	11 (2)	13 (3)	89 (3)	30 (11)	11 (1)	4 (1)	
Aug.–Oct.	51	13	24	9	25 (6)	78 (149)	38 (110)	33 (2)	29 (2)	78 (18)	63 (1)	22 (6)		
1974														
April–May	47	59	44	57	75 (24)	91 (60)	50 (54)	60 (17)	20 (2)	79 (10)	34 (1)	4 (1)	2 (1)	
June–July	53	50	41	41	24 (12)	83 (15)	39 (378)	49 (42)	5 (1)	73 (7)	32 (1)	49 (3)	46 (2)	7 (1)
Aug.–Oct.	73	75	60	65	88 (19)	11 (17)	60 (495)	5 (8)	48 (3)	23 (20)	3 (1)	91 (3)	7 (1)	
1975														
April–May	51	51	50	38	56 (12)	63 (34)	24 (191)		32 (2)	50 (2)	12 (1)	45 (1)	50 (3)	
June–July	48	47	40	39	28 (34)	36 (14)	15 (5)	5 (34)	3 (4)	64 (4)	20 (1)	46 (1)	73 (2)	8 (1)
Aug.–Oct.	57	60	36	41	25 (3)	2 (1)	6 (46)		17 (2)	5 (1)	64 (1)	100 (2)	31 (4)	2 (1)

that of the 1968 year class at all ages. All year classes after 1970 grew rapidly and were recruited to the harvestable population (arbitrarily >200 mm) as age I+ fish during August, or approximately 16 months of age. Growth slowed appreciably after surpassing 200 mm. Male yellow perch attained a maximum mean length of 271 mm at age VIII. Females grew faster and attained larger size; age VII females reached 341 mm in length.

The adult yellow perch population in Killdeer Reservoir was composed of five year classes, 1970–1972 and 1974–1975. Yellow perch from the 1970–1972 year classes, part of the initial stocking effort in spring 1973, were relocated from Lake Erie, thus only the last four completed years of growth occurred in Killdeer Reservoir. The 1974–1975 year classes resulted from a combination of natural reproduction and supplemental stocking of fingerling yellow perch. All year classes attained the greatest annual increment during the first year. Growth of all fish was similar by age I (Table 4). Growth increments declined in all succeeding years except 1973, when 1970 year class fish exhibited an increased growth rate corresponding to their relocation from Lake Erie to the reservoir. Yellow perch did not attain mean lengths greater than 200 mm until age II+. Growth of both sexes was similar through age II, but females exceeded males in growth thereafter and attained greater mean lengths.

Food of Walleyes

Food habits of 1972 year class walleyes in both Ferguson and Killdeer Reservoirs were examined from 1972 through 1975. Both reservoirs were stocked with walleye fry during April 1972.

Young-of-the-year walleyes in Ferguson Reservoir fed on cladocerans, primarily *Daphnia* spp. and *Leptodora*, immature chironomids, and fish (Table 5). *Chaoborus* larvae and copepods were consumed in limited amounts, but were relatively unimportant in the diet. Cladocerans contributed the major portion of the diet throughout the year, with fish being important during June and July.

In 1973 food habits were similar to, but fish became more important than, in 1972. In 1974–1975 crayfish were important forage from May–August. Cladocerans remained a basic food in 1974, but were relatively unimportant in 1975. Immature chironomids, particularly pupae, were also important in 1974. However, by 1975 fish and crayfish constituted the majority of the diet. Most fish in walleye stomachs were unidentifiable, but those identified included bluegill, white crappie, bluntnose minnow, white sucker, and channel catfish. Only one yellow perch was identified from a walleye stomach during the four years. Walleye growth was poor in all years and was unquestionably related to low availability of forage fish.

Food of age 0+ walleyes in Killdeer Reservoir was almost entirely zooplankton and immature chironomids (Table 5). Both chironomid larvae and pupae were consumed, with pupae primarily being important during June. Copepods were important only in early June, but *Daphnia* spp. were utilized throughout the year. No fish were observed in the diet of young walleyes during 1972. In 1973, zooplankton was relatively unimportant as walleye forage, but chironomid pupae were heavily utilized. Other aquatic insects were also eaten, especially the mayfly naiad, *Caenis*. Young-of-the-year bluegills were also eaten as fish entered the diet. Fish became the most important forage source in 1974–1975; bluegill remained the primary species utilized. Other species consumed included smallmouth bass, white crappie, green sunfish, rock bass, *Ictalurus* spp., and yellow perch. Chironomid pupae remained important forage in 1974 and through June 1975, but zooplankton was not important. Other aquatic invertebrates also supplied a significant portion of the walleye diet in 1974–1975. Included among these were mayfly naiads, ceratopogonid larvae, adult caddisflies, and the amphipod *Hyalella azteca*.

Growth of Walleyes

Growth of 1972 year class walleyes, similar in both Ferguson and Killdeer Reservoirs, was slower than normal for Ohio waters. Tucker and Taub (1970) found male walleyes in Hoover Reservoir were 254 mm at age I and 369 mm by age II; females were slightly larger than males at each annulus. Age–growth data from

TABLE 6.—*Mean empirical total lengths (mm) of 1972 year class walleyes from Ferguson and Killdeer Reservoirs.*

Reservoir	Year of life			
	1	2	3	4
Ferguson	181	234	293	330
Killdeer	173	265	288	337

Ohio Division of Wildlife files (unpublished) indicate these lengths are normal for many Ohio waters. Empirical growth of the 1972 year class indicated Killdeer walleyes were only 265 mm and Ferguson walleyes 234 mm at age II (Table 6). Walleyes were not separated by sex as only males were mature prior to age IV. However, at age IV some females were mature and exceeded males in mean total length: 388 mm to 321 mm in Ferguson Reservoir, and 341 mm to 331 mm in Killdeer Reservoir. Growth fluctuated yearly in both reservoirs, but Killdeer walleyes exceeded Ferguson walleyes by only 7 mm in length at age IV.

Angler Harvest

Yellow perch and walleyes accounted for 44% of the April–October angler harvest from Ferguson Reservoir during the five year period 1972–1976 (Table 7). Percids were harvested in lowest numbers during 1972 (11%) and greatest in 1974 (79%). Harvest of both species increased in 1973 and 1974 after recruitment of the 1972 year classes. Yellow perch harvest declined thereafter, reflecting minimal contributions of year classes from 1973 to 1975. Harvest of walleyes did not decline as rapidly, indicating continued recruitment of 1974 and 1975 year classes.

Winter ice fishing contributed significantly to the total annual harvest from Ferguson Reservoir. During January 1974, an estimated 8,816 yellow perch were caught in an 18-day period, approximately 50% of the total annual yellow perch harvest from April 1973–March 1974 (Paxton 1975). In succeeding years, winter harvest of yellow perch contributed 25–35% of the annual total. Walleyes were not as vulnerable to winter angling, and ice fishing ac-

TABLE 7.—*Estimated total angler harvest by species from Ferguson and Killdeer Reservoirs, 1972–1977, from April through October.*

Year	Angler effort (total hours)	Species							
		Yellow perch	Walleye	Bluegill	Crappie	Black bass	White bass	Others	Total
Ferguson Reservoir									
1972	17,577	3,295	53	10,735	15,633	5	40	1,189	30,950
1973	15,831	8,885	940	1,739	1,755		242	853	14,414
1974	16,110	9,751	2,761	2,041	18	16	208	961	15,756
1975	17,978	1,956	2,567	647	180	186	343	2,632	8,511
1976	16,834	1,701	1,689	714	17	1,158	1,117	242	6,638
1972–1976	84,310	25,558	8,010	15,876	17,603	1,365	1,950	5,877	76,269
Killdeer Reservoir									
1974	22,071	175	5,782	2,635	81	7,353		442	16,468
1975	17,398	2,571	2,147	2,592	453	138		1,347	9,248
1976	18,851	2,811	893	2,471	66	151		837	7,229
1977	19,288	7,569	636	5,054	158	161		1,742	15,320
1974–1977	77,608	13,126	9,458	12,752	758	7,803		4,368	48,265

counted for only 2–8% of the annual walleye harvest.

Fish populations in Killdeer Reservoir were highly variable from 1974 to 1977. Fish stocked in 1972 were heavily exploited when the reservoir opened to angling in 1974, and there were immediate large reductions in walleye and black bass populations (Table 7). Walleyes continued to decline in abundance as no reproduction occurred prior to 1976. However, yellow perch increased in both population abundance and angler harvest from 1974 through 1977. Yellow perch and walleyes contributed only 37% of the total harvest in 1974, but slightly more than 50% each year from 1975 through 1977. Percids contributed 44% of the total angler harvest during 1974–1977.

Discussion

While percid species provided quality angling, yearly harvest fluctuated greatly in response to relative strength of successive year classes. Recruitment of yellow perch occurred as age I+ or II+ fish, and all but the strongest year classes were virtually eliminated from the harvestable population by age III or IV. Walleyes were also rapidly exploited.

Growth of yellow perch was initially rapid on a diet composed primarily of zooplankton and chironomids. Yellow perch exceeded 200 mm as age I+ or II+ fish, but growth declined rapidly in years after 200 mm was attained. Large yellow perch did not switch to a piscivorous diet as such a forage source was not present. They were thus forced to exist on essentially the same diet throughout their life. Schneider (1972) hypothesized that a yellow perch fishery is limited by the forage available to large yellow perch, but that forage for smaller yellow perch is usually not limiting. Pycha and Smith (1955) and Buck and Thoits (1970) also concluded growth of young yellow perch was not related to growth of larger and older yellow perch within the population. Our findings indicated this to be true for upground reservoirs.

Growth of 1972 year class walleyes was slow, to mean lengths less than 350 mm at age IV. There was no apparent relationship between walleyes and yellow perch, as young yellow perch were not an important part of the walleye diet. Although yellow perch and walleye year classes occurred synchronously in 1972 and 1975 (Ferguson Reservoir), yellow perch grew too rapidly to be utilized as forage by age 0+ walleyes. Young yellow perch were 61 mm on 18 July 1972, while age 0+ walleyes were only 85 mm. Although fish did become important forage for walleyes at ages II–IV, yellow perch were not consumed despite year-class production in both 1974 and 1975.

In summary, yellow perch grew well and were rapidly recruited and removed from the population through heavy exploitation. Walleye growth was poor, but small walleyes did provide angling opportunity and were readily harvested. There was no predator-prey relationship between walleyes and yellow perch, which accounted at least partially for poor walleye growth.

References

Buck, D. H., and C. F. Thoits III. 1970. Dynamics of one-species populations of fishes in ponds subjected to cropping and additional stocking. Ill. Nat. Hist. Surv. Bull. 30(2):69–165.

Burgess and Niple, Limited. 1967. The northwest Ohio water development plan. Columbus, Ohio. 299 pp.

Clarck, C. F. 1965. Walleye in Ohio and its management. Ohio Dep. Nat. Resour. 29 pp. (Unpublished.)

Paxton, K. O. 1975. Effects of ice fishing on yellow perch in an Ohio upground reservoir. Ohio Fish Wildl. Rep. 5. 11 pp.

Pycha, R. L., and L. L. Smith, Jr. 1955. Early life history of the yellow perch, *Perca flavescens* (Mitchill), in the Red Lakes, Minnesota. Trans. Am. Fish Soc. 84:249–260.

Schneider, J. R. 1972. Dynamics of yellow perch in single-species lakes. Mich. Dep. Nat. Resour. Res. Dev. Rep. 184. 47 pp.

Tucker, T. R., and S. H. Taub. 1970. Age and growth of the walleye, *Stizostedium vitreum vitreum*, in Hoover Reservoir, Ohio. Ohio J. Sci. 70(5): 314–318.

Effects of Water Level Management on Walleye and Other Coolwater Fishes in Kansas Reservoirs

CALVIN L. GROEN AND TROY A. SCHROEDER

*Kansas Fish and Game Commission
Box 54A, Pratt, Kansas 67124*

Abstract

The environmental and biological effects of water level management are monitored on certain Kansas reservoirs. The basic water level management plan consists of a rising water level in the spring to improve fish spawning and nursery conditions, followed by a mid-summer drawdown for revegetation of the fluctuation zone and to increase forage availability for piscivorous fishes. An improved forage base, increased walleye growth, increased walleye recruitment and harvest, enhanced survival of stocked walleye fry and northern pike fingerlings, an improved fish population structure, and improved water quality are attributed to water level management. No documented sustained natural northern pike recruitment has occurred. Manipulation of water levels has proven to be a valuable tool for the fisheries manager.

A major fishery management problem is the decline in fishing success after the high initial fishery production typical of newly impounded waters. Most Kansas reservoirs are generally windswept and turbid, and have fluctuating water levels due to their operation as flood control structures. In the past, water level variations were often untimely, with regard to fish life histories, and resulted in few benefits to the sport fishery. Beyond their basic hydrologic functions, such reservoirs can be managed to sustain a productive sport fishery by recreating, to some extent, the conditions of a newly impounded reservoir. Typical effects of a newly impounded reservoir on fish populations are increased reproductive rates, high survival of offspring, rapid growth and recruitment, and increased fishing success (Keith 1975). Through cooperative efforts between the Kansas City and Tulsa Districts of the U.S. Army Corps of Engineers, the Kansas Water Resources Board, the Kansas Parks and Resources Authority, and the Kansas Fish and Game Commission, mutually acceptable water level management plans have been developed and are presently being implemented on thirteen Kansas reservoirs.

Water Management Plans

The first water level management plans were implemented in Kansas in 1970 on Council Grove Reservoir, a 1,215-hectare impoundment on the Neosho River with an average depth of 4.0 m, and in 1972 on Milford Reservoir, a 6,478-hectare reservoir on the Republican River with an average depth of 7.6 m, and an average annual fluctuation of 1.6 m. Due to uncontrollable weather conditions and inflows, a plan may have to be implemented for several years to achieve the desired results. Many of the effects of a plan on the fishery may not be noticeable until a year or more after its implementation. Also, individual bodies of water may have different management objectives requiring a unique water level management plan for optimum fisheries benefits. However, water level management plans on Kansas reservoirs basically contain the following similarities (Fig. 1).

(1) Gradually rising water level in the spring inundates terrestrial vegetation and rocky areas, which enhances spawning and nursery habitat (Wood 1951; Walburg and Nelson 1966; Johnson and Andrews 1973; Bross 1969; Hassley 1970). Increased area and volume also encourages increased production of forage fishes. Any sudden drawdown should be avoided to prevent interruption of spawning activity or stranding and desiccation of eggs and fry.

(2) A relatively stable or slightly rising water level through late spring and early summer provides additional favorable spawning and nursery conditions. Flooded terrestrial vegetation provides additional protective cover for certain species. Decom-

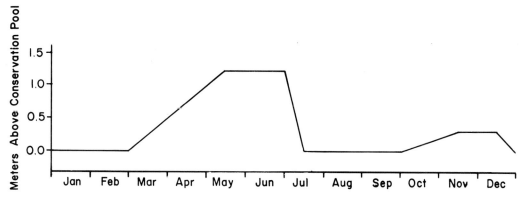

FIGURE 1.—*A typical Kansas water level management plan.*

posing vegetation adds nutrients to the lake and reduces colloidal clay turbidity.

(3) Around mid-summer a steady drawdown is initiated to expose previously inundated areas for revegetation, make forage fish more available to predator fish, and control rough fish.

(4) In the fall, inflows are retained in the reservoir to inundate the lower band of terrestrial vegetation. This enhances waterfowl habitat.

(5) Drawdown in the winter sets the stage for favorable fish spawning conditions the next spring. This drawdown reduces ice and wave damage to existing shoreline vegetation. It also provides additional water storage capacity to serve as a buffer against large inflows, thereby reducing chances for untimely spring drawdowns.

Success of a water level management plan depends upon the type and amount of vegetation that can be grown in the drawdown zone. Flooded vegetation is valuable as fish spawning habitat, as a nursery area providing protective cover and an abundant food supply for young fish, for increasing lake productivity, for turbidity control, and for waterfowl habitat.

Wood (1951) reported that productivity of a lake declines when the originally flooded vegetation has decomposed. He added that productivity of large impoundments increases when the water level is lowered, vegetation invades the drawdown zone, and the water level subsequently rises to inundate the terrestrial growth. Fish grew faster in Norris Reservoir, Tennessee, following a period of dry years when water levels receded and vegetation grew on the exposed shoreline. Hulsey (1957) reported that the fishery at Nimrod Reservoir, Arkansas, was greatly improved by a fall drawdown that exposed about 80% of the lakes original area.

Kansas water level management plans have attempted to recreate "new reservoir" conditions on a smaller scale. Drawdown zones generally are 10 to 20% of the total reservoir surface area. In Kansas reservoirs, more vegetation is established after a mid-summer drawdown than after a late summer drawdown. Most of the recent water level management plans have incorporated a mid-summer drawdown in an effort to grow a lush stand of vegetation in the drawdown zone.

Excellent stands of annual nutsedge (*Cyperus* spp.) and smartweed (*Polygonum* spp.) occurred naturally in the drawdown zone at Council Grove Reservoir following the first mid-summer drawdown. In addition to natural vegetation, 40.5 hectares, about one-fourth of the drawdown zone, was hand seeded with Japanese millet (*Echinochloa* sp.) in early July. Much of the millet reached 150 cm in height and matured within 60 days. Similar stands of natural and seeded vegetation were established at other Kansas reservoirs following mid-summer drawdowns. A total of 1,620 hectares were seeded to Japanese millet in conjunction with water level management plans at Kansas reservoirs during 1977.

After two or more years of successive mid-summer drawdowns, different types of vegetation were concentrated in zones. The lower portion of the drawdown zone consisted mostly of annual nutsedge, smartweed, and Japanese millet. The upper portion of the drawdown zone consisted predominately of cockleburs (*Xanthium* spp.), cottonwoods (*Populus* spp.), and willows (*Salix* spp.). Millet was present in the upper zone but it was somewhat stunted due to competition from other species. It was further noted that areas of smartweed, cockleburs, cottonwoods, and willows increased with repeated early drawdowns.

The type of vegetation to be planted in conjunction with a water level manipulation plan depends upon time of drawdown and intended use of the vegetation. For early drawdowns, Japanese millet and hybrid sudan-sorghum have produced lush stands valuable for fisheries habitat, turbidity control, and waterfowl food. In Kansas, these species are planted prior to August at rates of 6 to 12 kg/hectare. For late drawdowns, annual ryegrass (*Lolium* sp.), wheat (*Triticum* sp.), or rye (*Secale* spp.) can be seeded during September or early October. Ryegrass is usually planted at 11 kg/hectare while wheat and rye are planted at 34 to 68 kg/hectare.

Water clarity has improved in Kansas reservoirs in years when vegetation in the drawdown zone was flooded. Improvement was most evident prior to the mid-summer drawdown, and was usually greatest in reservoirs that had the largest drawdown zones. Rooted plants stabilize the shoreline and reduce turbidity from wave action. Decomposing vegetation entering the lake helps to precipitate colloidal clay. Bennett (1970) and Hulsey (1957) found that water transparency increased in years preceded by lowered water levels and results were more dramatic when lakes had been lowered to less than one-half their original area or held low for one or more complete growing seasons. By contrast, improved water quality in Kansas reservoirs was achieved by yearly drawdowns that were small enough not to interfere with any intended purpose of the reservoirs. Greater water clarity improves sport fish spawning success, sport fish growth, and angler success, and makes the lake more attractive for all recreational users.

Fisheries

Northern Pike

Although good northern pike fisheries have occurred in some Kansas reservoirs for a short period of time, all resulted from fry or fingerling stockings of new impoundments, or from one natural year class associated with the initial filling of new impoundments. Sustained natural recruitment has not occurred in any Kansas reservoir. When a reservoir is four or five years old, angling success for northern pike falls to unacceptable levels due to decreasing populations.

The first natural northern pike reproduction in Kansas occurred at Norton Reservoir in 1966. In 1965, about 5,000 northern pike fry per hectare were stocked. During the autumn and winter of 1965, the water level rose approximately 3 m into previously unflooded vegetation, much of which was short grass prairie dominated by buffalo grass (*Buchloë dactyloides*) and blue grama grass (*Bouteloua gracilis*). The water level remained stable through the 1966 spawning season and the year class of northern pike produced was larger than the stocked 1965 year class. Limited northern pike spawning success occurred at Marion and Perry reservoirs under similar conditions.

In an attempt to prolong the new reservoir conditions, Melvern Reservoir was filled in stages over a three-year period. Northern pike (1973 year class) were stocked at Melvern in 1973. In March of 1974, during the second year of filling, the reservoir water level rose to 60 cm into previously unflooded terrestrial vegetation, much of which was native prairie grasses. As a result, a good year class of northern pike was produced.

All instances of Kansas northern pike recruitment occurred in association with newly flooded grassland and all were produced by age 1 fish. Hassler (1970) also reported that successful northern pike reproduction is usually associated with newly flooded vegetation. This flooded vegetation provides a spawning substrate, a nursery area with protective cover, and an abundant food supply of invertebrates.

Kansas' first water level manipulation plan was implemented at Council Grove Reservoir mainly to create northern pike spawning habitat. This plan included a spring rise in water level between February and May to flood terrestrial vegetation which was established after the late summer drawdown. However, during the three years the plan was implemented (1970–1972), no documented northern pike recruitment occurred. A combination of the following factors was thought to have been responsible for this lack of recruitment: adequate vegetation could not be established in the short growing season after the September drawdown date; not enough early runoff was available to flood the existing terrestrial vegetation by spawning time; large spring runoffs usually carried a high silt load; and only a small breeding population of northern pike existed in the reservoir. The entire population resulted from a 1965 fry stocking, and many of these fish had been harvested or were near the end of their natural life span.

Due to poor natural recruitment, northern pike fisheries have had to be maintained by stocking. Until 1976, most maintenance stockings were made with intermediate size, 15.2 to 25.4-cm fish. Due to the difficulty in raising the northern pike to this size, stocking rates were limited to about 2.5 fish per hectare.

In 1976, small fingerlings, 3.8 to 7.6 cm, were stocked in Council Grove Reservoir at the rate of 37 per hectare. Later sampling indicated good survival of these smaller fish. This stocking success is attributed to excellent habitat created by water level management. Between April 15 and May 1, the water level rose nearly 1.5 m and flooded much terrestrial vegetation in the drawdown zone.

In 1977, another fingerling stocking was made at Council Grove Reservoir. The fish were of a similar size and stocked at the rate of 25 per hectare. However, due to a lack of spring rainfall, no flooded vegetation was available and the water was turbid. No evidence of survival was found in later fish samples. Northern pike fingerling stockings during the 1977 spring rise at Tuttle Creek Reservoir were successful. Therefore, it appears that to be successful, fingerling stocking should be made only when favorable water conditions and flooded vegetation exist.

Although no significant change in northern pike growth has occurred as a result of the water level management plan, growth rates have been very good at Council Grove Reservoir. With all year classes and both sexes combined, the mean length of northern pike at annuli 1–6 are, respectively: 459, 650, 789, 913, 969, and 999 mm.

Walleye

Downward trends in the relative abundance of walleye have been monitored in many Kansas impoundments. During late March and April, walleyes concentrate along the dams for spawning. Tag returns and angler surveys have indicated a high rate of walleye loss from reservoirs at this time, especially during high water releases. Walleyes are lost from Pomona Reservoir at discharges beginning at 56.6 m^3/s; 60% of the tagged walleyes caught came from the Pomona tailwaters. Walburg (1971) reported that large numbers of young saugers and walleyes were lost during discharges from Lewis and Clarke Reservoir in early June after spawning. High discharge rates during walleye spawning seasons can result in a significant portion of the brood stock and age 0 fish being discharged from a reservoir.

This situation occurred at Milford Reservoir during the 1973 walleye spawning season. Due to large inflows, high discharges of up to 283.2 m^3/s occurred. Many walleyes were lost from the reservoir; subsequent fishing was good in the tailwater, but poor in the reservoir.

An objective of Milford's 1974 water level management plan was to rebuild the walleye population in the reservoir. The plan called for a rise in the water level beginning in April, the month walleyes usually spawn in Milford Reservoir. The bene-

TABLE 1.—*Angling and harvest trends at Milford Reservoir from 1974 through 1976.*

	1974	1975	1976
Fishermen	111,520	160,431	172,533
Catch rate (fish/hour)	0.49	0.91	0.82
Total fish	176,630	397,635	425,572
Total kilograms	94,374	205,922	223,904
Fish/fisherman	1.6	2.5	2.5
Walleye harvest			
Reservoir	1,254	12,382	17,272
Tailwater	3,707	329	1,014

ficial effects desired from the 1974 plan were (1) inundated rocky areas free of algae, a desirable spawning habitat for walleye; (2) additional reservoir storage capacity to buffer any sudden inflows, and reduce the chances of high releases and drawdowns during walleye spawning season; (3) inundated terrestrial vegetation to stimulate production throughout the food chain and rapid growth of young fish; (4) an expanded habitat for the prolific spawning of forage fishes.

Since 1974, three strong walleye year classes have been added to the Milford population in conjunction with stable or rising water levels during walleye spawning season. Catch of walleyes per effort in standard gill nets tripled from 1973 to 1976. Creel surveys have revealed a dramatic increase in the number of walleyes harvested from Milford Reservoir, from 1,254 in 1974 to 17,272 in 1976 (Table 1). Conversely, the tailwater fishery has declined since 1973. Discharge rates from the reservoir since 1973 have been below 56.6 m³/s during the time walleyes congregate off the dam for spawning. This is a result of the extra capacity for spring runoff created by the winter drawdown. The abundance of walleyes at Council Grove has also increased since implementation of water level management. Walleyes caught per modified fyke net-night during the spring spawning season has steadily increased from five in 1974 to 16.5 in 1977. Altogether, strong walleye recruitment and the development of a walleye sport fishery has occurred on seven reservoirs having water level management programs. Similar trends in other reservoirs are anticipated in the future.

Since initiation of water level management, an increase in forage fish has been observed. Multiple gizzard shad spawns have consistently occurred on Kansas reservoirs operating with water management plans. Late gizzard shad spawns seem to be associated with rising water levels and expanding habitat. September seining samples at Milford have demonstrated a wide length range of shad (50–152 mm). A substantial increase in the number of emerald shiners and red shiners has also occurred at Milford Reservoir since 1973.

Growth rates of young-of-the-year walleyes and other sport fish are increasing in reservoirs with a water level plan. More forage fish, better water quality, and the summer drawdown are contributing factors.

Fish Population Trends

Perhaps the greatest effect of water level management can be seen in changes in the fish population structure of certain reservoirs. Results of the Council Grove and Milford Reservoir plans are increased numbers and weights of harvestable game and panfish caught per unit of standard sampling effort. This increase was greatest in harvestable sport fish and least in rough fish. Data collected from several Tennessee Valley Authority impoundments show those with the largest percentage of predator species were those having the widest fluctuation in water level (Bennett 1970). Wood (1951) stated that bottom feeders (many rough fish) may be limited by a decreased food supply during the drawdown period. Declining water levels can reduce littoral zone production of invertebrate fauna (Bennett 1954).

Test-netting results at Council Grove Reservoir show the portion of the total fish population composed of harvestable game and panfish increased from 15.5 to 52.9% by number and from 30.3 to 44.1% by weight between 1974 and 1976. During this time period the portion of the population composed of rough fish declined from 30.3 to 28.1% by number and from 66 to 54.2% by weight. The number of harvestable sport fish caught per net-night of sampling effort increased from 3.7 in 1974 to 18.4 in 1976, while the number of rough fish caught per-net night of sampling effort increased only from 7.2 to 9.8 during the same period.

Milford Reservoir test-netting results also show an increase in the number of fish caught per unit of effort. The catch per fyke net-night increased from 23.2 fish in 1973 to 42.4 fish in 1976. The catch per experimental gill net night increased from 30.3 to 42.8 fish during the same period. Since water level management began, the relative abundance of walleye, white bass, white crappie, river carpsucker, and gizzard shad has increased at Milford Reservoir.

Desired fishery benefits have been achieved on Kansas multipurpose reservoirs through water level management. It has been possible to fluctuate water levels above or below the planned conservation pool levels. The timing of these fluctuations is determined by the desired results and by other reservoir purposes. Although water level management plans may not be considered management "cure-alls", they have proven to be valuable fishery management tools. Benefits produced by water level management are an improved forage base, increased walleye recruitment and harvest, enhanced survival of stocked northern pike fingerlings, improved predatory fish growth rates, improved water quality, and an improved fish population structure.

References

BENNETT, G. W. 1954. The effects of a late-summer drawdown on the fish population of Ridge Lake, Coles County, Illinois. Trans. N. Am. Wildl. Nat. Resour. Conf. 19:259–270.

———. 1970. Management of lakes and ponds. Van Nostrand Reinhold, New York. 375 pp.

BROSS, M. G. 1969. Fish samples and year-class strength (1965–1967) from Canton Reservoir, Oklahoma. Proc. Okla. Acad. Sci. 48:194–199.

HASSLER, T. J. 1970. Environmental influences on early development and year-class strength of northern pike in Lake Oahe and Sharpe, South Dakota. Trans. Am. Fish. Soc. 99(2):369–375.

HULSEY, A. H. 1957. Effects of a fall and winter drawdown on a flood control lake. Proc. Annu. Conf. Southeast Assoc. Game Fish Comm. 10:285–289.

JOHNSON, J. N., AND A. K. ANDREWS. 1973. Growth of white crappie and channel catfish in relation to variations in mean annual water level of Lake Carl Blackwell, Oklahoma. Proc. Annu. Conf. Southeast Assoc. Game Fish Comm. 27:767–776.

KEITH, W. E. 1975. Management by water level manipulation. Pages 489–497 in H. Clepper, ed. Black bass biology and management. Sport Fishing Institute, Washington, D.C.

WALBURG, C. H. 1971. Loss of young fish in reservoir discharge and year-class survival, Lewis and Clarke Lake, Missouri River. Am. Fish. Soc. Spec. Publ. 8:441–448.

———, AND W. R. NELSON. 1966. Carp, river carpsucker, smallmouth buffalo and bigmouth buffalo in Lewis and Clarke Lake, Missouri River. Fish Wildl. Serv. (U.S.) Res. Rep. 69. 30 pp.

WOOD, R. 1951. The significance of managed water levels in developing the fisheries of large impoundments. J. Tenn. Acad. Sci. 26(3):214–235.

DEVELOPMENT OF ESOCID FISHERIES

Management Evaluation of Stocked Northern Pike in Colorado's Small Irrigation Reservoirs[1]

STEPHEN A. FLICKINGER

Department of Fishery and Wildlife Biology
Colorado State University
Fort Collins, Colorado 80523

JOHN H. CLARK

Alaska Department of Fish and Game
Division of Commercial Fisheries
333 Raspberry Road, Anchorage, Alaska 99502

Abstract

Northern pike of two size groups were stocked into several small irrigation reservoirs in Colorado. Small northern pike approximately 50-mm total length were stocked at an approximate rate of 62/hectare, whereas large northern pike averaging 377-mm total length were stocked at an approximate rate of 25/hectare. Out of 18 introductions of 50-mm northern pike, nine (50%) resulted in no observable northern pike populations, seven (39%) resulted in northern pike populations of too low a density to interest fishermen, and two (11%) resulted in northern pike populations of a density which would attract fishermen. Survival of 50-mm stocked northern pike was dependent upon large numbers of small forage fishes. In the two instances when 50-mm fingerlings became established in sufficient numbers to interest fishermen, cost per surviving catchable-sized northern pike averaged $0.16. All four introductions of 377-mm northern pike resulted in northern pike populations of a density which would interest fishermen. Survival of stocked 377-mm northern pike averaged 35% and cost per surviving fish averaged $4.62. Few statistical differences in average size and number of resident fish populations were found within a year after northern pike were introduced into study reservoirs; however, the general trend was an increase in average total length and a decrease in number of resident fishes following introduction of northern pike.

Colorado and several other states and provinces have many publicly and privately owned small plains reservoirs, built principally for irrigation purposes, which could provide public or private fishing. However, in Colorado most such reservoirs are dominated by rough fish populations. Introduction of a predacious game fish such as northern pike can fulfill two discrete management functions in this kind of fishery: (1) provide fishing for northern pike where there is little to attract fishermen now; and (2) improve existing fishing by reducing numbers and thereby increasing average size through predation upon overabundant rough-fish and stunted panfish populations. To become a useful management tool and fulfill the function of providing direct fishing, stocked northern pike must: (1) be inexpensive so that a combination of initial cost, stocking rate, and survival does not make the surviving fish more expensive than a fishing license or more expensive than a private pond owner is willing to pay; (2) survive in sufficient numbers for fishermen to have a reasonable chance of catching them; (3) grow large enough to be attractive to fishermen; and (4) survive consistently reservoir to reservoir and year to year so that the fishery management biologist can develop a dependable fishery. To become a useful management tool and fulfill the function of indirectly improving fishing, stocked northern pike must: (1) be inexpensive so that a combi-

---

[1] Financial assistance was provided by the U.S. Department of Commerce, National Oceanic and Atmospheric Administration, National Marine Fisheries Service under the Commercial Fisheries Research and Development Act (P. L. 88-309), Projects 6-11-D and 1-100-R and by the Colorado Division of Wildlife.

nation of initial cost, stocking rate, and survival does not make the surviving northern pike more expensive than alternative methods of fish population control such as chemical reclamation, or this method of fish population control must be longer lasting and more feasible than alternative methods of fish population control; (2) survive in sufficient numbers to have a potential influence on the resident fish community; (3) prey upon undesirable portions of the fish community to the extent that size composition, or species composition, or both will become more desirable to the angler; and (4) demonstrate reservoir-to-reservoir and year-to-year consistency in survival and utilization of forage fish populations so that the fishery management biologist can consistently use stocked northern pike as a management tool. Management success of stocked northern pike is highly dependent upon survival which, in turn, is highly dependent upon size of the introduced fish.

Survival of stocked northern pike fry has been estimated at one% or less, and frequently no northern pike can subsequently be found after a waterway has been stocked (Royer 1971; Schryer et al. 1971; Hicks 1972; Powell 1973; Vasey 1974). However, fry are commonly stocked because they can be propagated economically and with consistency (Hiner 1961; Sorenson et al. 1966). Northern pike fingerlings exhibit higher survival than fry (Gasaway 1971; Beyerle and Williams 1972; Beyerle 1973). Rearing northern pike fingerlings larger than 50 mm is expensive because live forage fish must be provided. Additionally, there is uncertain hatchery production of fingerlings (Clark 1974). Age I northern pike (200 to 400 mm) can be captured from winterkill lakes and stocked elsewhere (Williams and Jacob 1971). Survival of these larger fish should be better than that of fingerlings, but supply is limited and cost is high.

The objective of this research was to evaluate the management success of stocking 50-mm fingerling and 377-mm yearling northern pike in several of Colorado's small irrigation reservoirs containing resident fish communities. Fifty-millimeter fingerlings were selected because that size is near the maximum that can be reared without live forage fish and hence it is relatively inexpensive, and 377-mm yearlings were selected because the Colorado Division of Wildlife was using that size in their management program. The criteria of evaluation included cost, survival, and growth of stocked fish as well as the impact of stocked northern pike upon average total length and number of fish in the resident fish populations.

General Procedures

Eleven privately owned reservoirs in the vicinity of Fort Collins, Colorado, were selected as study sites. Physical, chemical, and preliminary fish population data were gathered for background information.

All study reservoirs were stocked in June, 1973, and seven study reservoirs were stocked in May, 1974, with northern pike averaging about 50 mm total length. The general stocking rate was 62 fish/hectare. These fingerling northern pike were obtained from fish research ponds located on the Foothills Campus of Colorado State University. Egg source was Pishkun Reservoir, Montana. Williams and Jacob (1971) listed the hatchery production cost of 100-mm northern pike at $0.03 each. We felt that 50-mm fingerlings should have higher yields and would spend less time in rearing ponds than 100-mm fingerlings and therefore used a cost of $0.02 each for the 50-mm fingerlings stocked in this study.

Four study reservoirs were stocked during the winter of 1973–1974 with northern pike averaging 377 mm in total length. The general stocking rate was 25 fish/hectare. These yearling northern pike were purchased by the Colorado Division of Wildlife from Mr. Jim Cody, who had captured the fish in the Red Lake region of Minnesota. Yearling northern pike were purchased for $1.41 each; this figure was used to generate cost estimates for 377-mm northern pike introductions.

Several weeks after northern pike were introduced into study reservoirs, population estimates of northern pike and major

TABLE 1.—Survival and cost of northern pike stocked at an approximate size of 50 mm and at an approximate rate of 62/hectare.

Reservoir	Year of northern pike introduction	No. of weeks from stocking to evaluation	No. of northern pike/hectare resulting from introduction	Survival of stocked northern pike	Cost per surviving northern pike	Total cost of stocked northern pike/hectare
Baker Lake	1974	6	0.0	0.0%		$1.43
Baldridge Lake	1974	49	0.0	0.0%		$1.18
Duck Lake	1973	97	0.0	0.0%		$1.26
Launer's Lake	1974	7	0.0	0.0%		$1.23
Mahood Reservoir	1973	9	0.0	0.0%		$1.25
Mahood Reservoir	1974	10	0.0	0.0%		$0.93
Park Creek Reservoir	1973	12	0.0	0.0%		$1.24
Roulard Lake	1974	15	0.0	0.0%		$1.24
Swanson Lake	1974	12	0.0	0.0%		$1.23
Average		24	0.0	0.0%		$1.22
James Lake	1973	97	0.1[a]	0.2%[a]	$19.60	$1.24
Smith Reservoir	1974	13	0.3	0.5%	$ 7.37	$1.24
Reservoir No. Nine	1973	97	0.3[a]	0.5%[a]	$ 5.49	$1.23
Swanson Lake	1973	60	0.4	0.7%	$ 4.74	$1.23
Baker Lake	1973	54	0.5	0.8%	$ 4.21	$1.23
Baldridge Lake	1973	97	0.8[a]	1.3%[a]	$ 2.53	$1.24
Launer's Lake	1973	57	1.0	1.6%	$ 2.03	$1.23
Average		68	0.5	0.8%	$ 6.56	$1.23
Smith Reservoir	1973	10	6.6	10.7%	$ 0.19	$1.24
Roulard Lake	1973	14	10.2	16.4%	$ 0.12	$1.24
Average		12	8.4	13.6%	$ 0.16	$1.24
Grand average		40	1.1	1.9%		$1.23

[a] These estimates are based upon the number of northern pike gillnetted in 1975.

age I and older resident fish populations were conducted. Fish were captured with a 2.5-m by 61-m beach seine (13-mm stretched mesh). Captured fish were measured, marked, and released. Subsequently, fish were recaptured, and Schnabel or Petersen population estimates were computed. During April and May, 1975, stocked northern pike were sampled with 2.5-m by 91.5-m gill nets (50-mm bar mesh) for estimates of average size and growth.

Study Areas

Elevations of study reservoirs were within 230 m of each other, giving similar climates. Study reservoirs averaged 6.6 hectares (range of 2 to 15 hectares) and had mean depths ranging from 0.9 to 2.6 m. Dissolved solids concentrations (range of 520 to 8,000 micromhos/cm) in study reservoirs were intermediate compared to typical lakes (Reid 1961). The fishes most frequently encountered were carp, black bullheads, white suckers, black crappies, white crappies, and yellow perch. All study reservoirs seemed reasonably comparable, and there is no reason to think that they were different from other small, plains irrigation reservoirs.

Survival and Cost of Stocked Northern Pike

In this study, 18 introductions of northern pike averaging approximately 50 mm total length which had been stocked into 11 small irrigation reservoirs were evaluated. With only two exceptions, survival of 50-mm northern pike was low, and cost per surviving northern pike was high (Table 1). From the standpoint of fisherman appeal, it was felt that at least 2 northern pike/hectare must survive before the introduction could be termed successful. Fifty percent of the introductions were complete failures, producing no observable northern pike populations. Thirty-nine percent of the introductions resulted in unacceptably small northern pike populations. Survival from stocking to evaluation of these small northern pike populations averaged 0.8%, and cost per surviving fish averaged $6.56. From a fishery management standpoint,

TABLE 2.—*Survival and cost of northern pike stocked at an average size of 377 mm and at an approximate rate of 25/hectare.*

Reservoir	Year of northern pike introduction	No. of weeks from stocking to evaluation	No. of northern pike/hectare resulting from introduction	Survival of stocked northern pike	Cost per surviving northern pike	Total cost of stocked northern pike/hectare
Swanson Lake	1974	18	4.5	19.3%	$7.32	$33.05
Mahood Reservoir	1973	32	7.2	29.0%	$4.87	$34.83
Baker Lake	1973	28	9.9	40.1%	$3.51	$34.85
Launer's Lake	1973	32	12.7	51.0%	$2.76	$34.89
Average		27	8.6	35.8%	$4.62	$34.40

the introductions which resulted in small populations of northern pike were also failures. Eleven percent of the introductions resulted in northern pike populations of an acceptable density. Survival from stocking to evaluation of these northern pike populations averaged 13.6%, and cost per surviving fish averaged $0.16. From a management standpoint, only 11% of the introductions of 50-mm northern pike were successful. Lack of consistency of management results may preclude 50-mm northern pike from being a useful management tool in small irrigation reservoirs. However, management success of stocking 50-mm northern pike appeared to be directly related to availability of small forage fish existing in the reservoir at time of stocking. For example, the fathead minnow population in Smith Reservoir appeared high when the successful 1973 stocking was made, but appeared low when the unsuccessful 1974 stocking was made. (A cursory sampling in 1975 indicated a high population of fathead minnows again and also a good survival of stocked 50-mm northern pike). Fifty-millimeter northern pike might demonstrate more favorable management results if resident fish populations were sampled prior to stocking and only bodies of water containing adequate sizes and amounts of forage were stocked. As a single test of this recommendation, Reservoir Number Nine, which contained only an insignificant amount of small forage in 1973 when an unsuccessful stocking of 50-mm northern pike was made, was stocked again with 50-mm northern pike in 1976 when samples revealed abundant fathead minnows and unusually small (60 mm) age I carp. When the reservoir was intentionally poisoned in 1977, a standing crop of 3.4 northern pike/hectare with an average size of 350 mm was ennumerated, making this second stocking successful as defined earlier. At stocking rates used, the average total cost of stocking 50-mm northern pike was only $1.23/hectare (Table 1). Because of the low cost involved in introducing 50-mm northern pike, they are an economical management tool in improving the fisheries of small irrigation reservoirs where their survival is likely.

In this study, four introductions of northern pike averaging 377 mm were evaluated. All of the introductions resulted in northern pike populations of an acceptable density (2 northern pike/hectare). Survival from stocking to evaluation of these introductions averaged 35.8%, and cost per surviving fish averaged $4.62 (Table 2). From a management standpoint, 377-mm introduced northern pike were consistently successful, but cost per surviving northern pike approached the 1975 cost of a Colorado fishing license ($6.00). At stocking rates used, the total cost of stocking 377-mm northern pike was $34.40/hectare (Table 2). Cost may preclude introduced 377-mm northern pike from being a useful management tool to provide northern pike fishing in small irrigation reservoirs. Where cost is justified (when northern pike are used to control overabundant rough fish and panfish), 377-mm northern pike introductions may provide a relatively economical means of improving the fisheries of small irrigation reservoirs. Chemical rehabilitation is commonly used to correct overabundant fish populations and cost for shallow reservoirs like the ones in this study would be approximately $40/hectare. Furthermore, this method of fish population control usually needs to be repeated every few

TABLE 3.—*Average total length and average growth of northern pike residing in the study reservoirs from stocking to spring, 1975.*

Reservoir	Date of introduction	Average total length when stocked (mm)	Average total length during spring 1975 (mm)	Average increment of growth (mm)
Roulard Lake	June 1973	49	533	484
Baldridge Lake	June 1973	49	600	551
James Lake	June 1973	49	602	553
Reservoir No. Nine	June 1973	49	634	585
Smith Reservoir	June 1973	49	646	597
Mean of means		49	603	554
Smith Reservoir	May 1974	49	363	314
Mahood Reservoir	December 1973	377	574	196
Baker Lake	December 1973	377	589	212
Mean of means		377	581	204
Swanson Lake	March 1973	377	542	164

years. Stocking northern pike after introduction may or may not be needed depending on spawning habitat.

This study has demonstrated that stocking 50-mm northern pike resulted in inexpensive, but inconsistent, management results and that stocking 377-mm northern pike resulted in consistent, but expensive management results. The question that arises is: what size northern pike should be stocked in small irrigation reservoirs? The best size would be that which demonstrates consistently favorable management results at the lowest possible cost. Management based upon stocking northern pike fry has generally been unsuccessful (Royer 1971; Schryer et al. 1971; Hicks 1972; Powell 1973; Vasey 1974). Our data indicate that the most appropriate size for stocking lies between 50 mm and 377 mm but do not indicate precisely what this size should be. A review of the literature also failed to identify the optimum size but did provide some insight into the question. Gasaway (1971) estimated that survival through the first year for fingerling northern pike stocked in Great Plains reservoirs was 1–10%. Beyerle and Williams (1972) estimated that survival through the fall of the first year for stocked 60- to 90-mm northern pike fingerlings in Long Lake, Michigan was 28%. Beyerle (1973) estimated that survival through the second year for stocked 90- to 120-mm northern pike fingerlings ranged from 4.4 to 54.7% for a series of six introductions in Daggett and Emerald Lakes, Michigan. Needham (1973) found that 150-mm northern pike fingerlings survived 7.5 times as well as 60- to 70-mm northern pike fingerlings when both sizes were stocked into Fort Peck Reservoir, Montana. Vasey (1974) stated that 150-mm northern pike stocked into Thomas Hill Reservoir, Missouri, survived better than other sizes of northern pike which were stocked. Snow (1974) found that only 6.6% returned to the creel when 250- to 550-mm northern pike were stocked into Murphy Flowage, Wisconsin. Perhaps survival to catchable size for stocked 150-mm northern pike would range from 25 to 50%. Williams and Jacob (1971) estimated that their 180-mm northern pike might be provided for a cost of $0.25 per fish. It would seem that 150-mm northern pike could be reared for a similar, or even lower, cost. On the basis of these predictions, 150-mm northern pike might be established in small irrigation reservoirs with consistency for a cost of $0.50 to $1.00 per surviving fish. However, fish of this size would be fairly hard to rear, and hatchery production would be uncertain. Our speculation on management success also needs verification.

Growth of Stocked Northern Pike

In this study, growth of northern pike in nine small Colorado irrigation reservoirs was evaluated. Although growth varied among reservoirs (Table 3), average increment of growth in all reservoirs was above that of most northern pike populations reviewed by Carlander (1969). In 1975 the legal minimum size limit for northern pike in Colorado was 610 mm. With just a little more growth during the summer of 1975, all populations would reach legal size within two years of stocking. Introduction of both 50-mm and 377-mm northern pike in Colorado's small, plains irrigation reservoirs were successful from the standpoint of growth of individual fish.

Impact on Resident Fish

Chemical eradication of resident fish communities and subsequent restocking of game fishes has been the typical remedy for reservoirs with unwanted fish populations. However, reservoirs used for irrigation receive incidental rough-fish and panfish introductions through ditch systems; consequently, chemical eradication of fishes has to be repeated every few years. A management alternative to chemical eradication is the stocking of predacious game fishes.

Few statistical differences in average size and number of resident fish populations were found in this study after northern pike were introduced into study reservoirs (Tables 4 and 5). However, the general trend was an increase in average total length of resident fishes and a decrease in population number of resident fishes following introduction of northern pike.

Doxtater (1967) and McCarraher (1959) found that northern pike were successful in preventing overpopulation and stunting of bluegills. On the contrary, Beyerle (1971) found that northern pike were not able to control numbers of bluegills. The number of bluegills in Roulard Lake drastically decreased after northern pike were stocked in our study. Powell (1973) found that the number of white crappies decreased but that rate of growth did not increase after northern pike were introduced into Carbody Reservoir, Colorado. In contrast, numbers of black crappies in Smith Reservoir and in Roulard Lake and number of crappies (mixed species) in Mahood Reservoir did not change significantly after northern pike were introduced in our study. On the other hand, average size of black crappies in Smith Reservoir and in Roulard Lake and average size of crappies in Mahood Reservoir did not change significantly after northern pike were introduced, which agrees with Powell's (1973) findings. Beyerle and Williams (1968) and Coble (1973) found that northern pike selected soft-rayed forage fishes over spiny-rayed forage fishes. Mauck and Coble (1971) found that carp and fathead minnows were more vulnerable to predation than were other forage species investigated. In our study, number and average size of carp in Smith Reservoir, Roulard Lake, Mahood Reservoir, and Launer's Lake did not change significantly after northern pike were introduced. However, the fathead minnow population in Smith Reservoir drastically decreased after northern pike were introduced. Mauck and Coble (1971) found that white suckers and yellow perch were intermediate in vulnerability to northern pike predation. In our study,

TABLE 4.—*Change in average size of rough fish and panfish after northern pike were introduced in the study reservoirs.*

Species	Reservoir	Average total length in 1973 (mm)	Average total length in 1974 (mm)	Change in total length from 1973 to 1974 (mm)
Carp	Roulard Lake	333	334	+1
Carp	Mahood Reservoir	324	326	+2
Carp	Smith Reservoir	286	315	+29
Black bullhead	Smith Reservoir	203	245	+42
White sucker	Roulard Lake	321	312	−9
White sucker	Smith Reservoir	217	250	+33
White sucker	Mahood Reservoir	247	288	+41
White sucker	Launer's Lake	289	335	+46
Black crappie	Roulard Lake	185	173	−12
Crappie	Mahood Reservoir	163	173	+10
Black crappie	Smith Reservoir	191	210	+19
Bluegill	Roulard Lake	122	173	+51[a]
Green sunfish	Roulard Lake	151	173	+22
Yellow perch	Mahood Reservoir	178	162	−16
Yellow perch	Smith Reservoir	161	167	+6
Yellow perch	Launer's Lake	147	158	+11
Yellow perch	Roulard Lake	87	105	+18[b]

[a] Statistically significant at $\alpha = 0.05$.
[b] Statistically significant at $\alpha = 0.15$.

TABLE 5.—*Change in resident fish population numbers after northern pike were introduced in the study reservoirs.*

Species	Reservoir	Fish/hectare in 1973	Fish/hectare in 1974	Changes in fish/hectare from 1973 to 1974
Carp	Mahood Reservoir	1,273	695	−578
Carp	Launer's Lake	4,301	1,813	−2,488
Black bullhead	Launer's Lake	4,440	2,359	−2,081
White sucker	Mahood Reservoir	47	7	−40[a]
White sucker	Smith Reservoir	198	130	−68
Crappie	Mahood Reservoir	750	1,005	+255
Black crappie	Smith Reservoir	13	49	+36
Black crappie	Roulard Lake	116	67	−49

[a] Statistically significant at $\alpha = 0.05$.

TABLE 6.—*Fulfillment of two management functions by stocked 50-mm and 377-mm northern pike in study lakes.*

Attribute	Performance of 50-mm northern pike	Performance of 377-mm northern pike
Ability to provide northern pike fishing		
Cost/surviving fish	$0.16	$4.62
Attractive numbers	11% of stockings	100% of stockings
Attractive size	Yes	Yes
Survival consistency	Poor	Excellent
Ability to improve overall fishing through predation		
Cost/surviving fish	$0.16	$4.62
Longer lasting and more feasible	Perhaps	Perhaps
Influential numbers	11% of stockings	100% of stockings
Alter size composition	Perhaps	Perhaps
Alter species composition	Perhaps	Perhaps
Consistency in survival	Poor	Excellent
Consistency in forage fish utilization	Poor	Perhaps

numbers of white suckers and yellow perch decreased in Mahood Reservoir after northern pike were introduced. Mauck and Coble (1971) found that bluegills and black bullheads were least vulnerable to northern pike predation. In our study, number of black bullheads in Launer's Lake did not change significantly after northern pike were introduced. However, bluegills in Roulard Lake nearly disappeared, but it is presumed that they were the most appropriately sized forage in the reservoir. In general, results of our study are in agreement with the literature regarding the impact of stocked northern pike upon resident fish populations. Overall, during their first year of reservoir residence, the impact of northern pike upon resident fish populations of Colorado's small, plains irrigation reservoirs was not great.

Management Functions

It was hypothesized that northern pike could fulfill two fishery management functions in small plains reservoirs. These functions were: (1) to provide fishing for northern pike and (2) to improve fishing overall through predation upon overabundant and stunted resident fish populations. Attributes of stocked fish which would determine whether or not stocked 50-mm and 377-mm northern pike could fulfill the two management functions were discussed in the introduction of this paper. Introduction of 50-mm northern pike resulted in the combination of low initial cost coupled with stocking rate and survival to provide an economical surviving fish, and additionally growth was good to excellent (Table 6). However, in only a few cases did enough northern pike survive to attract fishermen and certainly survival was too inconsistent among reservoirs and years to provide the basis for a dependable fishery. Likewise, with regard to the second function, cost was competitive with other methods of population control, but survival in magnitude and consistency were inadequate for 50-mm northern pike to be a useful management tool. On the other hand, introduction of 377-mm northern pike resulted in consistent survival, attractive numbers, and good to excellent growth which would make them potentially useful in both functions (Table 6). However, cost per surviving fish for stocked 377-mm northern pike was close to the 1975 cost of a fishing license in Colorado, making their use in the first function questionable. Although cost was more favorable with regard to the second function, not enough time had elapsed (one year) to document whether or not the generally favorable trends in alteration of size and species composition of resident fish communities would continue until reservoirs were providing attractive non-northern pike fishing.

Conclusions

It is concluded that stocking 50-mm and 377-mm northern pike into small plains irrigation reservoirs has merit only in special cases, and that overall, neither size adequately fulfills fishery management functions. It is recommended that intermediate sizes of northern pike be investigated in hope that a size can be found that demonstrates more consistent fishery management results than have 50-mm northern pike and that will cost considerably less than have 377-mm northern pike.

References

BEYERLE, G. B. 1971. A study of two northern pike-bluegill populations. Trans. Am. Fish. Soc. 100(1):69–73.

———. 1973. Growth and survival of northern pike in two small lakes containing soft-rayed fishes as the principal source of food. Mich. Dep. Nat. Resour. Fish. Div. Res. Rep. 16 pp.

———, AND J. E. WILLIAMS. 1968. Some observations of food selectivity by northern pike in aquaria. Trans. Am. Fish. Soc. 97(1):28–31.

———, AND ———. 1972. Contribution of northern pike fingerlings raised in a managed marsh to the pike population of an adjacent lake. Mich. Dep. Nat. Resour. Res. Dev. Rep. 274. 20 pp.

CARLANDER, J. D. 1969. Handbook of freshwater fishery biology. Iowa State Univ. Press, Ames. 752 pp.

CLARK, J. H. 1974. Variability of northern pike pond culture production. M.S. Thesis. Colo. State Univ., Fort Collins. 35 pp.

COBLE, D. W. 1973. Influence of appearance of prey and satiation of predator on food selection by northern pike (*Esox lucius*). J. Fish. Res. Board Can. 30(2):317–320.

DOXTATER, G. 1967. Experimental predator-prey relations in small ponds. Prog. Fish-Cult. 29(2): 102–104.

GASAWAY, C. R. 1971. Estimating the costs of sustained stocking of northern great plains reservoirs. Pages 65–83 *in* R. J. Muncy and R. V. Bulkley, eds. Proc. N. Central Warmwater Fish Cult.-Mange. Workshop, Iowa State Univ., Ames.

HICKS, D. 1972. Introductions. Prog. Rep., Dingell-Johnson Proj. F-22-R-5, Job 1. Okla. Dep. Wildl. Conserv., Oklahoma City. 3 pp.

HINER, L. E. 1961. Propagation of northern pike. Trans. Am. Fish. Soc. 90(3):298–302.

MAUCK, W. L., AND D. W. COBLE. 1971. Vulnerability of some fishes to northern pike (*Esox lucius*) predation. J. Fish. Res. Board Can. 28(7):957–969.

MCCARRAHER, D. B. 1959. The northern pike-bluegill combination in north-central Nebraska farm ponds. Prog. Fish-Cult. 21(4):188–189.

NEEDHAM, R. G. 1973. Inventory of waters of the project area. Prog. Rep. Dingell-Johnson. Proj. F-11-R-19, Job I-a. Mont. Dep. Fish Game, Helena. 11 pp.

POWELL, T. G. 1973. Effect of northern pike introductions on an overabundant crappie population. Colo. Div. Wildl. Fish. Res. Sect. Spec. Rep. 31. 6 pp.

REID, G. K. 1961. Ecology of inland waters and estuaries. Reinhold, New York. 375 pp.

ROYER, L. M. 1971. Comparative production of pike fingerlings from adult spawners and from fry planted in a controlled spawning marsh. Prog. Fish-Cult. 33(3):153–155.

SCHRYER, F., V. EBERT, AND L. DOWLIN. 1971. Determination of conditions under which northern pike spawn naturally in Kansas reservoirs. Final Res. Rep. Dingell-Johnson Proj. F-15-R, Job C-3. Kan For. Fish Game Comm., Pratt. 37 pp.

SNOW, H. E. 1974. Effects of stocking northern pike in Murphy Flowage. Wis. Dep. Nat. Resour. Tech. Bull. 79. 20 pp.

SORENSON, L., K. BUSS, AND A. D. BRADFORD. 1966. The artificial propagation of esocid fishes in Pennsylvania. Prog. Fish-Cult. 28(3):133–141.

VASEY, F. W. 1974. Life history of introduced northern pike in Thomas Hill Reservoir. Final Rep. Dingell-Johnson Proj. F-1-R-23, Job 2. Mo. Dep. Conserv., Jefferson City. 14 pp.

WILLIAMS, J. E., AND B. L. JACOB. 1971. Management of spawning marshes for northern pike. Mich. Dep. Nat. Resour. Res. Dev. Rep. 242. 21 pp.

An Evaluation of the Muskellunge Fishery of Lake Pomme de Terre and Efforts to Improve Stocking Success

Lawrence C. Belusz

Missouri Department of Conservation
1104 S. Grand
Sedalia, Missouri 65301

Abstract

Two independent creel census methods (roving and probability) were used to measure fishing pressure and angler success for muskellunge at Lake Pomme de Terre. The two methods provided significant differences in estimated muskellunge harvest and angler effort. The probability method was considered to be more accurate since it utilizes only completed trip information for calculating angler effort. Angler acceptance of muskellunge as a trophy fish has increased since this species was first stocked in 1966. Heavy exploitation of initial releases of muskellunge by anglers in 1972, coupled with limited hatchery production of fingerlings, has resulted in a limited adult population that provides an annual harvest of 100 to 200 muskellunge per year. Stocking mortality of fingerling muskellunge can be very high and delayed releases of muskellunge held in isolation coves showed that the period of greatest mortality occurred within 48 hours. It is suggested that fingerlings be released in areas near aquatic vegetation and that releases be made after dark to reduce stress. Delayed mortality of muskellunge may be caused by latent pathogens under stress conditions associated with transportation and handling.

A quantitative, random creel census was initiated in 1965 to measure fishing pressure, harvest, and angler success by season on Lake Pomme de Terre. In 1966, the Missouri Department of Conservation initiated a muskellunge stocking program to establish a trophy fishery in the lake, which is the only muskellunge lake in Missouri. Modifications in the creel census were made in 1967, 1972, and 1976 to more reliably measure angler success and harvest of all species. Attempts to monitor the adult muskellunge population with the use of nets and electrofishing gear have been conducted yearly since 1970. In addition, several studies to measure short-term stocking success and methods of reducing stocking mortality have been conducted.

Study Area

Lake Pomme de Terre is a 3,167-hectare Corps of Engineers flood control reservoir located in the Ozark Uplands region of west-central Missouri (Fig. 1). Completed in 1961, the lake impounds portions of the Pomme de Terre River and Lindley Creek which are tributaries of the Osage River. Multipurpose pool is at elevation 256 m above mean sea level, and flood pool of 6,480 hectares is at elevation 266 m. Total length of shoreline varies from 174 km at multipurpose pool to 188 km at flood pool. Maximum reservoir depth is 27.4 m and averages 9.1 m at multipurpose pool. Since Ozark streams drain shallow, stoney soils, water clarity in the lake is exceptional with Secchi disc readings of 1.8 m or more being common. Mean surface temperatures by season are 13.3 C in spring (March–May), 26.7 C in summer (June–August), and 17.8 C in fall (September–October). Total hardness is 188 mg/liter.

Methods

From 1965 through 1973 a creel census clerk surveyed anglers on portions of the two major arms of the lake on a one-third time basis (seven days per arm each month) from March through November. The areas surveyed were on the Pomme de Terre Arm (895 hectares) and the Lindley Creek Arm (827 hectares), 55% of the total lake area (Fig. 2). In 1976 the survey was modified to census each creel area eleven days per month. The clerk, operating from a boat, interviewed fishermen during an 8-h day which included all combinations of sunrise to sunset hours. Angler counts within the creel area were made twice daily; all possible combinations of morning and afternoon hours were sampled. Work days were scheduled on a random basis, and weekends and holidays were weighted twice the value of weekdays. Data from weekdays and weekends were calculated separately and then combined.

Total hours fished in the census area were estimated by multiplying mean daily fisherman counts (hourly counts), average length of the census day (daylight hours), and number of days in the sampling period. Catch rates for each species were determined by dividing the total number of fish caught by the total hours fished by all anglers actually interviewed by the clerk. The estimated total catch by species was calculated as the product of the catch rate and estimated total hours.

In 1973, the creel survey was modified to gain additional information on species selection by anglers. This modification, called Angler Preference, indicated what species was fished for regardless of the composition of the catch. This enables the calculation of more precise catch rates for individual species by using data only from anglers who fished for that particular species. Angler harvest data for the period 1966–1973 were taken from Hanson (1974).

During 1976 a second method of creel survey was initiated at the lake. This method, probability sampling, was described by Fleener (1971) and is based on the unbiased allocation of sampling effort. All sampling factors, time of day, day of week, and sampling location, are weighted according to expected use and sampled according to assigned probabilities. Sampling probabilities were modified on a monthly basis as necessary to insure that sampling effort was allocated according to fisherman use. With this method, a clerk surveys all anglers exiting a single access point during an 8-h time period 22 days per month from March through November and these data (all completed trips) are expanded for all other access points on the lake according to their assigned weights. Total fishing hours were estimated by multiplying actual fishing hours (completed trips) by the reciprocal of the product of day, site, and time probabilities and dividing by site sampling frequency. Catch rates for all species were calculated by dividing the estimated catch (actual catch at a single site expanded for all other sites) by the total estimated fishing hours by species.

A study to determine the relative value and accuracy of both census methods in estimating fisherman effort and harvest parameters is currently underway. Preliminary observations indicate that since

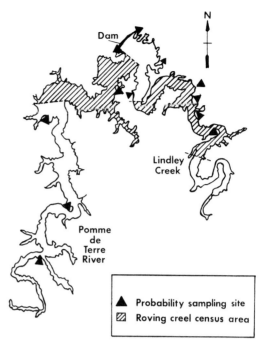

FIGURE 2.—Location of probability sampling sites and roving census areas, Lake Pomme de Terre.

FIGURE 1.—Location of Lake Pomme de Terre and associated drainage pattern.

TABLE 1.—*Muskellunge stocking record, Lake Pomme de Terre, 1966–1977.*

Year	Fry[a] number	Fingerlings Number	Fingerlings Length (mm)
1966	36,000	57[b]	178–381
1966	15,000	1,450	178–381
1967	12,850	835	229–280
1968	7,700	798	229–280
1969	11,200	370	229–280
1970		298	229
1971			
1972		419	229–280
1973			
1974		207	229–280
1975		298	229–280
1976		950	254–330
1976		275	127
1977		2,214	282
Totals	82,750	8,171	

[a] Fry size 38–51 mm.
[b] Great Lakes Strain. Al other releases were of Ohio strain.

the probability method utilizes only completed trip information, estimates of effort and harvest are more accurate than those calculated by roving census methods. However, disregarding relative degree of accuracy, both census methods provide usable trend data which reflect fluctuations in angler effort and harvest.

Muskellunge Stocking History

Beginning with their introduction in 1966, muskellunge have been stocked yearly through 1977 except in 1971 and 1973 (Table 1). Initial releases in 1966 included both the Great Lakes strain (*Esox masquinongy masquinongy*) and Ohio strain muskellunge (*E. m. ohioensis*) but only the Ohio strain thereafter. In early years both fry and fingerlings were stocked, but fry releases were discontinued after 1969 when the decision was made to produce a smaller number of larger fish which could have greater survival rates as indicated in studies by Oehmcke (1969) and Hacker (1966). It was felt that predation on fry by the abundant centrarchid population and lack of sufficient aquatic vegetation for fry escapement would nullify the effect of releasing large numbers of small muskellunge. Even though the four-year total of fry releases amounted to over 25 muskellunge per hectare it is doubtful if significant survival occurred. Fry releases as practiced at that time consisted of scattering fry along shallow areas devoid of vegetation. Natural reproduction has not been documented in the reservoir.

Growth of stocked muskellunge has been good, with legal size (762 mm total length) reached during their fourth summer of life at age III+ (Table 2). A length–weight relationship was calculated from legal fish, of both sexes, caught by fishermen during the fall months (September–November). The calculated values were $\log W = -2.8735 + 2.2344 \log L$, where L is mm and W is g. The calculated r^2 value was 0.99.

Harvest

Harvest of the first legal muskellunge was authenticated in 1968 two years after stocking was initiated. The first creel observation and harvest estimate was made in 1970 (Table 3). In 1972, an estimated 1,238 muskellunge were harvested in the roving creel area, which includes all of the productive muskellunge fishing areas. This harvest represents a 39% return of all large fingerlings stocked from 1966 to 1968. However, the estimated catch in 1973 of 112 muskellunge was less than 10% of the 1972 catch, indicating heavy exploitation of the legal muskellunge population in 1972. Estimated harvest in 1976 (no census in 1974–1975) of 251 muskellunge in the roving census area and 133 muskellunge for the entire lake by probability methods continued this trend of low muskellunge harvest. As a check on both census methods, information on muskellunge harvest was provided by each of the four marinas at the lake. This voluntary information showed that 83 legal muskellunge were brought in

TABLE 2.—*Average calculated length and weight at annulus formation for muskellunge from Lake Pomme de Terre.*

Age group	Length mm	Weight g
I	310	493
II	561	1,857
III	739	3,437
IV	894	5,260
V	932	5,773
VI	1,008	6,878
VII	1,072	7,829
VIII	1,156	9,341

TABLE 3.—*Muskellunge catch as measured by roving and probability census methods, Lake Pomme de Terre, 1970–1977.*

	Census method						
	Roving[a]					Probability[b]	
Parameter	1970	1972	1973	1976	1977	1976	1977
Estimated catch							
Legal muskellunge	128	1,238	112	251	53	133	63
Sublegal muskellunge	0	0	0	85	41	122	112
Average length (mm)							
Legal muskellunge	845	965	1,067	959	851	991	841
Sublegal muskellunge	0	0	0	635	457	548	477
Muskellunge/hectare							
Legal	0.07	0.70	0.10	0.20	0.03	0.04	0.02
Sublegal	0	0	0	0.05	0.02	0.04	0.04

[a] Both creel areas combined, 1,722 hectares.
[b] Total lake area, 3,167 hectares.

by fishermen during 1976. Angler counts show that an average of 55% of all fishing trips occurred from one of these marinas from March through November. Therefore, the adjusted estimate of muskellunge catch was 151 fish, which compares favorably with the probability estimate. Harvest in 1977 continued to decline with 53 muskellunge estimated by roving census and 63 muskellunge estimated by probability methods even though angler effort for muskellunge was 20% greater than in 1976.

The low angler harvest at Pomme de Terre from 1970–1977 produced a return of 0.20 muskellunge per hectare. This return is comparable with muskellunge harvest at other lakes including Lake St. Clair, Michigan, 0.62/hectare (Schrouder 1973); Lake Chautauqua, New York, 0.99/hectare (Mooradian and Shepherd 1973); and Cave Run Lake, Kentucky, 0.16/hectare (James Axon, personal communication). Catch rate by muskellunge fishermen at Lake Pomme de Terre averaged 0.015 muskellunge per hour from 1973–1977 (Table 4). By comparison, catch rate at Lake Chautauqua was 0.013 muskellunge per hour from 1965 to 1973 (Mooradian and Shepherd 1973) and at Cave Run Lake in 1973 averaged 0.02 muskellunge per hour (James Axon, personal communication).

Stocking Mortality

Miles et al. (1974), have provided considerable insight into the problems of lactic acid buildup in muskellunge under stress conditions. However, methods of reducing stress and associated mortality have been less than encouraging. For states relying on a limited supply of muskellunge fingerlings each year, the number of muskellunge surviving the stress period associated with transportation and handling becomes increasingly important. Efforts to determine short-term mortality and attempts to reduce the effects of stocking stress by using isolation coves was previously reported by Belusz (1975). Further studies using this technique were made in 1975, 1976, and 1977 to gather additional information on stocking survival (Table 5).

In October, 1975, 656 muskellunge were released in a 0.24-hectare cove sealed off from the lake with a block net. Muskellunge were held in this area for a period of 240 hours and provided with forage. Twenty-four hours after release several muskellunge showed dark coloration on the head and from the dorsal fin to the tail. In addition, many muskellunge had red areas at

TABLE 4.—*Muskellunge angler effort as measured by roving and probability census methods, Lake Pomme de Terre, 1973–1977.*

	Census method				
	Roving[a]			Probability[b]	
Parameter	1973	1976	1977	1976	1977
Estimated total trips	1,134	3,011	1,843	1,532	3,181
Estimated total hours	2,042	8,972	5,530	7,048	14,634
Percent of anglers	1.5	3.5	2.3	2.2	7.6
Percent of hours	1.2	4.5	4.0	2.4	13.0
Hours per hectare	1.2	5.2	3.2	2.2	4.6
Catch rate (fish/h)	0.013	0.011	0.016	0.025	0.014

[a] Both creel areas combined 1,722 hectares.
[b] Total lake area, 3,167 hectares.

TABLE 5.—*Survival of stocked muskellunge fingerlings as determined by isolation cove release method, Lake Pomme de Terre.*

Year	Number stocked	Transit time (h)	Isolation period (h)	Number released	% survival[a]
1972	420	2	53	419	99.8
1974	223	12	7[b]	207	92.8
1975	656	2	240	298	46.8
1976	450	18	56	450	100.0
1976	500	18	56	500	100.0
1977	2,224	2	132	2,214	99.9

[a] Fish used for bacterial examination not included in calculation.
[b] Study terminated prematurely.

the base of the fins (particularly the dorsal) and in the mouth. Within 24 hours of release, 45% were dead. At 72 hours after release, 51% were dead. No additional mortality was observed after 96 hours; total mortality was 53% of the number released. The isolation area was checked a minimum of twice daily with scuba to locate dead muskellunge. It appears that the first 48 hours after release is the critical period for muskellunge survival.

The isolation release method was again used in 1976 to determine short-term mortality. Two groups of muskellunge were transported for 18 hours in furacin and salt and released after dark. Muskellunge released after dark were noticeably less active than those released during daylight hours in previous years. It is likely that prolonged inactivity during the night was beneficial in reducing stress associated with transportation and handling. Both groups of muskellunge were held in the isolation cove for 56 hours with no mortality (Table 5).

In 1977, muskellunge were again released after dark and held in the isolation cove for 132 hours. Two muskellunge were found dead within 48 hours of release with no additional mortality thereafter. Muskellunge were provided with forage during the holding period and they fed actively throughout the period of confinement. The resumption of active feeding is particularly important in helping acclimate stocked muskellunge to their new environment.

During the 1975 and 1977 isolation studies, kidney stabs were taken from muskellunge prior to transportation and throughout the holding period. Samples taken prior to transportation did not indicate the presence of disease organisms on either TSA or Ordahl media. However, samples taken during the isolation period did show the presence of several species of typically nonpathogenic bacteria. This suggests that handling stress could trigger outbreaks of latent pathogens causing delayed mortality of stocked muskellunge.

Information collected from the isolation cove studies demonstrates the great variability of stocking success that can be anticipated with muskellunge. Release of a large number of fingerlings can be offset by reduced survival, as in 1975 (Table 5), so that investment return as measured by angler success may be only as high as in years of limited release with high survival. The implication here is that an apparently low angler harvest may in reality be a significant return in terms of stocked fish surviving to legal size.

Conclusion

The low harvest of muskellunge during the 1976 and 1977 census period is a direct result of weak muskellunge year classes from 1969 to 1973 when only 1,087 fingerlings were released. Average length of muskellunge harvested steadily increased from 1970 through 1973 indicating continuing growth but little recruitment to the population. However, in 1976 and 1977, the average length of muskellunge harvested decreased, indicating some recruitment of the 1972 and 1974 year classes into the legal population. A corresponding increase in the catch of sublegal muskellunge was documented by probability census methods in 1976 (estimated catch of 122 sublegal muskellunge) and 1977 (estimated catch of 112 sublegal muskellunge), indicating significant survival of recent muskellunge releases.

It is obvious that the presence of muskellunge has produced an added dimension to the quality of the fishing experience in Lake Pomme de Terre. This new dimension of a "trophy fish" has met with wide angler acceptance and is verified by increases in angler trips, particularly for muskellunge. Due to recent advances in intensive culture of hybrid esocids, which have higher production rates and lower cost, questions have been raised concerning the continued stocking of muskellunge. However, the decision to maintain at least one pure muskellunge lake in Missouri has been reaffirmed. Current policy of releasing 2,000 fingerlings per year coupled with increases in stocking survival should dramatically increase muskellunge harvest within 2–3 years. The long-term result will be the establishment of a stable trophy fishery for muskellunge in Lake Pomme de Terre.

References

BELUSZ, L. C. 1975. The use of isolation coves in assessing muskellunge stocking mortality. Proc. Annu. Conf. Southeast. Assoc. Game Fish Comm. 29:251–253.

FLEENER, G. G. 1971. Recreational use of the Platte River, Missouri. Pages 63–78 in Stream channelization: a symposium. Am. Fish. Soc. N. Central Div. Spec. Publ. 2.

HACKER, V. A. 1966. An analysis of the muskellunge fishery of Little Green Lake, Green Lake County, Wisconsin 1957–65. Wis. Dep. Nat. Resour. Fish Manage. Rep. 4. 22 pp.

HANSON, W. D. 1974. Harvest of fish in Pomme de Terre Reservoir, Missouri. Mo. Dep. Conserv. Dingell-Johnson Proj. F-1-R-23, Study I-3, Job 1.

MILES, H. M., S. M. LOEHNER, D. T. MICHAUD, AND S. L. SALIVAR. 1974. Physiological responses of hatchery reared muskellunge (*Esox masquinongy*) to handling. Trans. Am. Fish. Soc. 103(3):336–342.

MOORADIAN, S. R., AND W. F. SHEPHERD. 1973. Management of muskellunge in Chautauqua Lake. N.Y. Fish Game J. 20(2):152–157.

OEHMCKE, A. A. 1969. Muskellunge management in Wisconsin. Wis. Dep. Nat. Resour. Fish Manage. Rep. 19. 22 pp.

SCHROUDER, J. D. 1973. Muskellunge management in Michigan. Mich. Dep. Nat. Resour. Tech. Rep. 73-31. 21 pp.

Evaluation of Esocid Stocking Program in Wisconsin

LEON D. JOHNSON

Wisconsin Department of Natural Resources
Spooner, Wisconsin 54801

Abstract

There was great variability in the range of survival of stocked esocids in Wisconsin waters. Generally muskellunge and northern pike survived in the range of zero to 60% over short-term intervals. Despite high variations, the stocked fingerlings added to the lake populations. There was an overall tendency for hybrids of these two species, to survive at higher values, up to 85%. Dry-diet-fed hybrids, however, tended toward lower survival than minnow-fed hybrids.

Many states stock esocids to augment the fish populations of their waters. Such stocking is an old program in Wisconsin. Webster (1929) reported that muskellunge fry were stocked in lakes as early as 1899. This stocking program has been expanded since then to include fingerlings, not only of the muskellunge but the northern pike and the hybrid obtained by cross-fertilization of these two species. This paper reports on the survival evaluation of these fish after stocking, as reported by fish managers located throughout Wisconsin.

Methods

All esocid fingerlings stocked in the various waters were marked by a fin clip or an internal tag. No marks were used where it was previously established that there had been no natural reproduction. The size of fish in individual introductions ranged from 13-mm fry to fingerling groups that approached 400 mm total length. The stocking was replicated in many of the lakes. Recaptures were made by one or more methods including electrofishing, fyke netting, and creel census. Petersen or Schnabel fish population estimates (Ricker 1975) were used for most of the survival evaluations.

Muskellunge Survival

There was great variation in the consistency of muskellunge fingerling survival among similar as well as different size groups. Survival varied among those stocked in different lakes and even among replicate stockings within the same lake. The range extended from zero to 60% for the 25 stocking efforts made in 11 lakes (Table 1). Of this total, there were 11 complete stocking failures where no muskellunge could be found, or where there were too few fish captured to make population estimates.

Four lakes, Arrowhead, Brandy, Johnson, and Sparkling, were studied over a short term in an attempt to evaluate successive stocking of small (99-mm), medium (203-mm), and larger (295-mm) fingerlings within the same lake. These fish were stocked as they became available during the same summer. The larger fingerlings survived better in three of the lakes, but there was zero survival for all sizes in Sparkling Lake.

Over long-term studies, few 38-mm fingerlings survived in Murphy Flowage, but survival was also low (1.0%) for the 229-mm fingerlings that were stocked later. In contrast, high survival of 305-mm fingerlings (31%) was observed in Escanaba Lake; however, the later stocking of 203–279-mm size groups failed.

Among the various sizes of muskellunge stocked in all lakes combined, the survival of the larger size fingerlings was significantly greater than the small sizes at the 90%, but not the 95% level. Short-term survival was significantly less over four months than over shorter intervals at the 95% level of significance. The number of fingerlings stocked per hectare, however, showed no relationship to the survival.

Contributions of Stocked Muskellunge

Despite the high variations in survival, the stocked muskellunge contributed to their population in many of the lakes (Ta-

TABLE 1.—Survival of muskellunge stocked in Wisconsin waters.

Lake	Surface hectares	Average total fish length (mm)	No. muskellunge stocked per hectare	Interval to estimate	Percent survival	Investigator
Murphy Flowage	73	38	14	1–10 yr	[a]	Snow
		229	27	1–10 yr	1	Snow
Clear	31	203	91	4–7 yr	7	Snow (1968)
Winter	274	333	2	3.6 yr	[a]	Pratt
		279	2	1.6 yr	[a]	Pratt
Round	1,236	298	1	1 mo	59	Pratt
Arrowhead	40	99	10	4 mo	23	Serns
		203	10	2.5 mo	31	Serns
		295	10	1 mo	41	Serns
Brandy	46	99	10	4 mo	0	Serns
		203	10	2.5 mo	7	Serns
		295	10	1 mo	60	Serns
Escanaba	119	305	2	2–10 yr	31	Kempinger (1971)
		203	2	2–10 yr	[a]	Kempinger (1971)
		253	1	1–6 yr	0	Kempinger (1971)
		279	1	1–5 yr	[a]	Kempinger (1971)
Fishtrap	133	256	2	5–8 yr	1	Klingbiel
High	297	262	2	5–9 yr	17	Klingbiel
Johnson	32	99	10	4 mo	7	Serns
		203	10	2.5 mo	37	Serns
		295	10	1 mo	53	Serns
Sparkling	51	99	10	4 mo	0	Serns
		203	10	2.5 mo	0	Serns
		295	10	1 mo	0	Serns

[a] Insufficient captures prohibited estimation of population size.

ble 2). Although there are insufficient data for many of our waters, there is considerable information on a number of them. Some of these data were extracted from Klingbiel (1966). No actual population estimates were made, but these data show not so much minimum survival rates, but contributions to the populations. This is perhaps one of the most important objectives of stocking. From known numbers of muskellunge caught at subsequent dates, the catch contained from zero to 67.6% stocked muskellunge in 18 study lakes.

Long-Range Muskellunge Survival

Few studies have contained estimates of muskellunge survival over many years. Some of these data were obtained from 2,000-hectare Lac Courte Oreilles in Sawyer County (Johnson 1975) (Table 3). For every 1,000 muskellunge stocked at the age of 6 months, only 114 reach one year of age. Each of these may be expected to live 7.1 additional years. For muskellunge from 1 through 5 years, the life expectancy ranges between 5 additional to 6.4 additional years of life. After 5 years, we would expect to have only 99 muskellunge remaining from the original 1,000 stocked finger-

TABLE 2.—The contribution of stocked fingerlings (180–300 mm total length) to the muskellunge populations of various Wisconsin lakes.

Lake	Surface hectares	No. fingerlings stocked	No. muskellunge captured	% caught that were stocked
East Fork Chippewa	81	420	27	0.0
Gallilee	87	300	62	4.8
Upper Clam	79	135	37	8.1
Holcomb Flowage	1,720	4,670	34	67.6
Amnicon	171	170	139	0.7
Bone	678	1,670	597	54.1
Butternut	389	2,300	37	13.5
Musser Flowage	206	150	23	0.0
Wilson Flowage	123	365	21	9.5
Island L. Chain	472	1,165	56	5.4
Barber	103	250	73	1.4
Connors	165	615	40	7.5
Grindstone	1,337	1,650	145	24.8
Lac Courte Oreilles	1,953	8,700	26	38.5
Lost Land	519	1,025	185	2.2
Lower Clam	82	100	21	0.0
Mason	73	160	111	2.6
Shell	984	2,700	241	46.9

TABLE 3.—Life table for three year classes (1955–1956–1957) of muskellunge based on returns of 3,300 muskellunge stocked as 200–275-mm finclipped fingerlings. Returns are based on muskellunge captured in fyke nets each spring during the years 1956 through 1973 in Musky Bay of Lac Courte Oreilles. Mean length of life, 5.22 years.

Age, years	No. alive at beginning of year	No. dying during year	No. dying per 1,000 alive at beginning of year	Average no. of years of life remaining at beginning of year
0.5–1	1,000	886	886	1.4
1–2	114	5	44	7.1
2–3	109	5	46	6.4
3–4	104	5	48	5.7
4–5	99	5	51	5.0
5–6	94	29	308	3.2
6–7	65	13	200	3.4
7–8	52	16	307	3.2
8–9	36	10	277	3.3
9–10	26	7	269	3.2
10–11	19	5	263	3.2
11–12	14	1	71	3.2
12–13	13	4	307	2.5
13–14	9	2	222	2.3
14–15	7	2	285	1.8
15–16	5	3	600	1.3
16–17	2	0	0	1.5
17–18	2	0	0	0.5

lings. Five-year-old muskellunge have usually attained the 76-cm legal size limit.

For each year from 5 through 10, the life expectancy is only 3.2 additional years at any attained birthday. More numbers of muskellunge die each year, now partly due to fish caught by anglers. For fish that live to become 10 years old, we may expect only 19 left from the 1,000. There will only be 2 left of those that have attained an age of 18 years. Older muskellunge live a tenuous existence, with an additional life expectancy of only one-half year at any attained birthday past 18 years. The maximum known age of an individual fish at this writing is 26 years.

Northern Pike Survival

Percent survival of northern pike stocked in lakes was generally low except for Murphy Flowage, where one cohort was followed beginning at 4.7 months (76%) through 4.3 years (0.4%) (Table 4). These were large fish stocked at an apparent optimum cool water temperature in mid-December. The subsequent decline in this population was attributed to disease and migration from the flowage.

Most other survival figures for lakes were low, and yet these stocked fish usually added to the northern pike population. The stocking of 2,500 fry per hectare in Lamereau Lake during each of 2 years resulted in zero survival after the first summer season and 3% survival after the second season. The apparent low second-year value for stocking nonetheless provided 75 northern pike per hectare. In a similar example, the 1% survival in Eagle Lake at its stocking rate of 4,890 fingerlings per hectare contributed 42 large (381 mm) yearlings per hectare, a satisfactory figure. Krohn (1969) reports many additional variations of stocking success.

Our data (Table 4) are insufficient for correlations of survival with size of the northern pike at the time stocking, or with the interval from stocking to the estimate.

TABLE 4.—Survival of northern pike stocked in Wisconsin waters.

Lake	Surface hectares	Average total fish length (mm)	No. northern pike stocked per hectare	Interval to estimate	Percent survival	Investigator
Lamereau	4	13	2,500	4.6 mo	0	Beard
		13	2,500	5.6 mo	3	Beard
Eagle	210	13	4,890	1 yr	1	Schumacher
Murphy Flowage	73	392	117	4.7 mo	76	Snow (1974)
				10.7 mo	15	Snow (1974)
				1.4 yr	9	Snow (1974)
				1.8 yr	4	Snow (1974)
				2.4 yr	1	Snow (1974)
				3.4 yr	0.6	Snow (1974)
				4.3 yr	0.4	Snow (1974)
Nebish	38	330	14	4 yr	3	Krohn (1969)
		330	13	3 yr	2	Krohn (1969)

TABLE 5.—Survival of hybrids (muskellunge × northern pike) stocked in Wisconsin waters.

Lake	Surface hectares	Average total fish length (mm)	No. hybrids stocked per hectare	Interval to estimate	Percent survival	Investigator
Fish	144	292[a]	25	1.7 mo	42	Snow
		292[a]	25	7.7 mo	17	Snow
Bullhead	27	208[b]	5	0.7 mo	22	Belonger
		315[a]	5	1.4 mo	65	Belonger
Wilke	39	243[a]	5	6.0 mo	[c]	Belonger
		221[b]	5	6.0 mo	[c]	Belonger
Eagle	210	381[a]	2	6.0 mo	85	Schumacher and Tills
		203[b]	2	6.0 mo	0	Schumacher and Tills
Potters	64	292[a]	44	6.0 mo	47	Tills and Schumacher
		292[a]	18	6.0 mo	51	Tills and Schumacher
		203[b]	18	6.0 mo	[c]	Tills and Schumacher

[a] Minnow diet.
[b] Dry food diet.
[c] Insufficient captures prohibited estimation of population size.

It is likely that the variability of survival following stocking follows the same trends observed for muskellunge.

Hybrid Survival

The studies of hybrid stocking success were all short-term evaluations over 1.7 to 6 months. The survival varied from zero to 85%, but values were generally higher than those for either the muskellunge or northern pike.

There was an apparent tendency for low survival of dry-diet-fed hybrids, as demonstrated by no or insufficient captures that prohibited estimation of population size. Minnow-fed hybrids stocked on the same dates in Eagle and Potter lakes survived at 85% and 51%, respectively, compared to the nonsurvival of the dry-diet fingerlings. Variability was again noted, however, in Wilke Lake, where few fish survived of either the dry-diet- or minnow-fed hybrids.

Among the hybrids stocked in all lakes combined there was an apparent significant relationship of fingerling survival to size, at the 95% level of significance. This significant value includes dry-diet-fed fish, which were smaller and generally had poorer survival. There were no other significant relationships of survival to the time interval in the lake or to the number of fingerlings stocked per hectare.

Despite these initial studies showing lower survival of dry-diet-fed hybrids, further studies are warranted in view of the ease with which this fish feeds on pellets in the hatcheries.

Acknowledgment

My appreciation is extended to Thomas Beard, Brian Belonger, James Kempinger, John Klingbiel, Frank Pratt, E. Randy Schumacher, Steven Serns, Howard Snow, and Donald Tills for permitting the use of unpublished data, and to Ronald Masterjohn for compiling tables.

This study was supported in part by Federal Aid to Fish Restoration Funds under Wisconsin Dingell-Johnson Project F-83-R.

References

JOHNSON, L. D. 1975. How many muskies aren't there anymore? Wis. Cons. Bull. 40(5):20–21.

KEMPINGER, J. J. 1971. Muskellunge fishing in Escanaba Lake with liberalized regulation, 1946–1971. Wis. Dep. Nat. Resour. Informal Ser. Rep. 6 pp.

KLINGBIEL, J. 1966. An evaluation of stocking large muskellunge fingerling. Wis. Dep. Nat. Resour. Fish Manage. Div. Rep. 3. 11 pp.

KROHN, D. C. 1969. Summary of northern pike stocking investigations in Wisconsin. Wis. Dep. Nat. Resour. Tech. Bull. 44. 35 pp.

RICKER, W. E. 1975. Computation and interpretation of biological statistics of fish populations. Bull. Fish. Res. Board Can. 191. 382 pp.

SNOW, H. E. 1968. Stocking of muskellunge and walleye as a panfish control practice in Clear Lake, Sawyer County. Wis. Dep. Nat. Resour. Res. Rep. 38. 18 pp.

———. 1974. Effects of stocking northern pike in Murphy Flowage. Wis. Dep. Nat. Resour. Tech. Bull. 79. 20 pp.

WEBSTER, B. O. 1929. Propagation of muskellunge. Trans. Am. Fish. Soc. 59:202–203.

Management Implications of Hybrid Esocids in Pennsylvania

ROBERT B. HESSER

Fisheries Managememt Section
Pennsylvania Fish Commission
Bellefonte, Pennsylvania 16823

Abstract

In Pennsylvania, only the sterile tiger muskellunge (northern pike × muskellunge) has been utilized to any great extent among esocid hybrids, and has proven superior to both parents and other hybrids in many hatchery categories. Tiger muskellunge have been planted in 109 waters in Pennsylvania. All new impoundments over 28 hectares are stocked with tiger muskellunge fry at the rate of approximately 250/hectare unless they occur in a parent species' natural range. Following initial introductions of fry, only 18–20-cm fingerlings are stocked subsequently, at the rate of 1 to 2/hectare, usually biennially. Stream stocking rates are 625 fry/km or 20 fingerlings/km for streams less than 100 m wide and 1,000 fry/km or 30 fingerlings/km for wider streams. Fry are stocked in streams only as introductions to rehabilitated waters. Fingerlings are stocked thereafter usually on a biennial basis. Available information indicates the tiger muskellunge is difficult to sample; grows rapidly, particularly in new impoundments; returns better to the angler than the muskellunge; and is readily accepted by a growing segment of anglers as a valued trophy.

Fisheries biologists have stated many justifications for use of esocid hybrids over purebreds in Pennsylvania. Examples are: (1) hybrids are more adaptable to hatchery techniques because they hatch better, grow faster, are more disease resistant and will more readily accept dry food; (2) most hybrids are sterile, which permits their introduction where purebred esocids would have caused problems; and (3) following stocking, hybrids are fast growing, attractive sport fishes which offer better returns on investment.

These justifications are much too generalized and hatchery-oriented for considering hybrids over purebreds. It is better to first identify comprehensive goals for managing large purebred esocids, excluding chain pickerel, and then determine the benefits of hybrids over parental species and which hybrids produce these benefits.

Role of Hybrids in Meeting Management Goals

There are no known natural populations of large esocid hybrids in Pennsylvania. Consequently, all hybrid esocid populations are established by fishery managers with hatchery-produced fish. The present rationale for stocking hybrids is to provide a trophy fishery where none exists (and frequently where no large esocids at all are present), and to add a predator in a niche which other predators are not adequately filling. These goals must be reached without adversely affecting an existing fishery. If establishment of a trophy fishery, which in Pennsylvania usually involves a relatively small but growing segment of the angling public (Heasley and Cawley 1977), adversely affects a historically important fishery serving the larger segment of our angling public, the optimum fishery objective is not being met.

Evaluation of hybrid esocids must consider both hatchery and post-stocking factors. Hatchery considerations include difficulty in obtaining sperm and eggs of both parent species simultaneously, egg hatchability, growth, disease resistance, acceptance of dry diet, ability to reproduce, and production economics. Post-stocking considerations include adaptability to a variety of water-quality conditions, feeding habits, growth, ability to reproduce and recruit sufficiently to produce self-sustaining wild populations, size and frequency of stockings, angler acceptability, angling mortality, angling regulations, and promotion of the best hybrid(s).

All factors must be weighed and compared, particularly hatchery versus post-stocking benefits.

Hatchery versus Management Benefits

In the past, hatchery benefits not necessarily concrete management objectives, influenced the choice of hybrids to be stocked. This was due to the heavy emphasis placed on hatchery production and research benefits, while too few fisheries managers were employed to conduct the field research necessary for elevating management programs to the sophistication of hatchery programs. It seems so obvious that a better approach is a cooperative venture between the fish culturist and fishery manager. The question is: At what point do the hatchery benefits tip the scales over the post-stocking benefits, or vice versa?

The objective can only be the production of the best quality, most economical, most adaptable and persistent hybrid which grows rapidly, is an effective predator, and returns well under the simplest regulations to a satisfied angler. Angler satisfaction requires special emphasis and must be influenced by an active promotional effort which tells the angler exactly why a given hybrid is being utilized in a management program.

Historical Utilization of Esocid Hybrids

At least six esocid hybrids have been produced in Pennsylvania, and the majority have at least some potential management significance (Buss and Miller 1967). However, only the tiger muskellunge (northern pike × muskellunge) has been widely utilized in our statewide management program. A small number of other hybrids, primarily those involving the Amur pike (*Esox reicherti*), have been stocked in two waters (Meade 1976). These fish were planted primarily on an experimental basis, not through any well-defined management scheme. Both impoundments were selected because each is located on the clean headwaters of a stream grossly polluted by coal mine drainage. This condition provides a natural barrier against the spread of an exotic species, the implications of which could not be predicted with any certainty. Presently no significant fishery has been identified as having resulted from these introductions.

A few individual hybrids, including one very large specimen, have been taken from the Bald Eagle Creek drainage downstream from Commission hatchery installations on Spring Creek. One northern pike × Amur pike hybrid taken by an angler was 119.7 cm total length and weighed 11.9 kg. This and other individuals apparently escaped from the hatchery during Hurricane Agnes in June 1972, since none had ever been intentionally stocked.

Management Potential of Hybrids

Northern Pike × Amur Pike

The hybrid between northern and Amur pikes fares well in the hatchery and readily accepts dry diet. Compared to other large esocids and hybrids, it is highly excitable in the hatchery and is a good fighter on hook and line. It is characterized by varied unusual markings and it is difficult to identify by the angler.

Unfortunately, no data are available on this hybrid's growth after stocking except for the one specimen from the Bald Eagle Creek basin, but there is reason to believe that it is a viable hybrid with some management significance if a well-designed program could be established for its use. It is fertile, but we do not know if it will reproduce in the wild or what characteristics future generations will exhibit if reproduction is successful.

Northern Pike × Muskellunge

The tiger muskellunge has proved hatchery benefits which consistently equal or exceed those of both parents, and it is sterile. Its sterility poses no problem in management, since no significant reproduction of the large esocids or their hybrids have ever been verified in state waters outside their natural ranges. The natural range for northern pike and muskellunge is considered to be the upper Ohio River and Lake Erie basins in western Pennsylvania. In waters outside their natural range, populations of large esocids and their hybrids are completely dependent upon periodic maintenance stocking.

Sterility of the tiger muskellunge offers at least one significant advantage. One can guarantee that its planting will not lead to its spread throughout a river basin, such as interstate water, where a joint management program between states cannot be derived due to administrative apprehension over the introduction and spread of these exotics.

Tiger Muskellunge in Management Programs

Since the tiger muskellunge is the only hybrid now being utilized in Pennsylvania's statewide management program, the remainder of this paper will be devoted to discussion of its use.

Habitat Requirements

The tiger muskellunge appears to thrive in a variety of habitats as does the purebred muskellunge. Insufficient information is available to differentiate significantly between the preferred habitats of these two fishes in Pennsylvania. Scott (1964) found that tiger muskellunge can tolerate higher water temperatures than the muskellunge. It also appears that the tiger muskellunge may tolerate a wider range of water conditions, including more productive waters than the muskellunge. In general, the tiger muskellunge appears to have habitat requirements which lie either midway between those of its parents or which tend to be more like those of the northern pike.

Tiger muskellunge have been planted in a wide variety of waters throughout Pennsylvania with varying physical and chemical characteristics. For lakes, areas vary from 30 to 3,359 hectares, maximum depths from 5 to 54 m, pH from 5.9 to 8.6, and total alkalinity from 2 to 110 mg/liter. Stream widths vary from a few hundred meters to a kilometer or more on large rivers. Stream chemical parameters are similar to those of lakes.

Stocking

Tiger muskellunge were first stocked in Pennsylvania in 1965 when fingerlings were planned in three lakes. Since the initial stockings, tiger muskellunge have been introduced into a total of 109 waters or segments of waters. The number of waters stocked annually varied from lows of one in 1966 to 50 in 1975.

Tiger muskellunge plantings from 1965 through 1969 involved only fingerlings. In 1970, the first fry planting of 22,000 was made in Brady's Lake, Monroe County. Since then, fry have been planted in 75 waters, including new impoundments and rivers reclaimed from coal mine pollution.

All new major impoundments over 28 hectares, unless they occur in the parent species' natural range or unless otherwise proscribed by the area fisheries manager, are presently stocked with tiger muskellunge fry. Following initial introductions of fry, 18–20-cm fingerlings are stocked from late August to early October, usually on a biennial basis. The only exception to this procedure occurs when fry or fingerlings become available in quantity over and above the annual allocations requested. Area fisheries managers maintain contingency lists of waters where these fish may be planted with reasonable expectation of maximum survival and harvest.

Pennsylvania's fry and fingerling stocking rates for tiger muskellunge have been adapted from rates for muskellunge from other states, primarily Michigan. Little data exist from our plantings which can be used to document the validity of the present stocking rates. Although the tiger muskellunge program indicates that the present stocking rates are reasonably successful, many hard data are needed for streamlining the program.

Approximate fry stocking rates are 250/hectare for impoundments, 625/km for streams less than 100 m wide, and 1,000/km for wider streams.

Approximate fingerling stocking rates are 1 to 2 hectare for impoundments, 20/km for streams less than 100 m wide, and 30/km for wider streams.

Growth and Feeding Habits

Tiger muskellunge consistently grow fast, particularly in new impoundments. Weithman and Anderson (1977) found tiger muskellunge in small Missouri ponds grew at a significantly higher rate than northern pike and muskellunge. In a new Pennsylvania impoundment, Beltzville Lake, tiger muskellunge planted as fry on May 10, 1971 yielded an angler-caught specimen of 62 cm on May 27, 1972—only one year and 17 days from planting. One month later a 65-cm hybrid was collected in the lake with Commission trap nets. The popularity of the tiger muskellunge with anglers grew as rapidly as the fish, and by May of 1974 many catches of fish exceeding 100 cm and weighing up to 9 kg were verified.

Nockamixon State Park Lake, a 586.8-hectare lake with a maximum depth of 16.8, was first stocked with fry and fingerlings in 1974; the average length of several specimens collected in October 1974 was 36 cm. During a trapnetting survey on August 11–15, 1975, 28 tiger muskellunge were taken ranging from 20 cm to 73 cm in length. Considering the difficulty which we have experienced in trapnetting tiger muskellunge, we concluded from this sampling that there is indeed a significant population of fast-growing tiger muskellunge of varying year classes in Nockamixon Lake.

The largest specimen of tiger muskellunge recorded in Pennsylvania, either from angler catch or from surveys, was taken in April 1977 from Brady's Lake—one of the first lakes to be stocked with this hybrid. This specimen exceeded 120 cm and weighed 10.9 kg. A second large specimen measured just over 110 cm and also weighed 10.9 kg.

Tiger muskellunge are reported to be voracious feeders (Schrouder 1973) and are far less sedentary than muskellunge. Commission staff have observed tiger muskellunge cruising extensively along lake shorelines. We have also noted that upon being placed in tubs after removal from trapnets, the tiger muskellunge is quite likely to respond to the movement of persons by jumping from the tubs. These observations suggest that if tiger muskellunge are more active than the muskellunge they also would feed more often. Therefore, its feeding habits should cause it to return better to the angler. Although these hybrids are less vulnerable to angling than the northern pike (Beyerle 1973; Weithman 1975), they offer a very favorable alternative to both parents since the muskellunge is much more difficult to catch. The tiger muskellunge is less likely to be cropped down to existing size limits, as has been the case in most of our northern pike populations, and yet will return to the angler better than the muskellunge.

Angler Acceptability

The tiger muskellunge has been readily accepted by anglers in Pennsylvania. This was particularly noted in Beltzville Lake where instant tiger muskellunge anglers were created. The stimulation of 100-cm fish, weighing 9 kg in three years after planting as fry, has created positive feelings with anglers. A few, however, dispaired of such a voracious, "fish-eating beast" being introduced into local waters and readily started rumors that all the other game fishes would soon be eaten.

In the Pocono Mountains area of northeastern Pennsylvania, where the majority of our natural lakes are found, some opposition to tiger muskellunge introductions has arisen because of the regional popularity of the chain pickerel, particularly for ice fishing. However, on one Pocono water, Bradys Lake, the tiger muskellunge has been taken through the ice in significant numbers and little opposition has arisen to its continued stocking there. Despite this particular case, continued introduction of tiger muskellunge into waters with highly acceptable, naturally reproducing chain pickerel fisheries raises a serious question that must be properly addressed.

There is little doubt that because of the prestige of the muskellunge name, the hybrid name of tiger muskellunge is more impressive to the fisherman than would be names such as tiger pike or norlunge. We feel that anglers consider the tiger muskellunge to be nothing more than striped muskellunge and it should be here to stay.

Angling Regulations

Since esocid hybrids are being managed to provide a trophy fishery, a conservative policy concerning size and creel regulations is desirable. For those hybrids which reproduce, size at maturity becomes a significant factor. But the tiger muskellunge is not capable of reproduction; therefore, spawning size is not a factor.

The present 76.2-cm size regulation, for both tiger muskellunge and muskellunge, corresponds to a sexually mature female muskellunge of at least age III+ and in some cases IV+. A 76.2-cm specimen is generally felt to represent a trophy-sized fish; consequently the present size regulation will be maintained because it is realistic for both muskellunge and hybrid.

Two additional advantages in maintaining a trophy-size regulation are that more large and efficient predators will be assured as will a larger catch of sublegal fish. There is really little justification for the present closed season for either hybrid or muskellunge since the tiger muskellunge fisheries are totally supported by stocking and the muskellunge populations, even in their natural range, are supplemented by periodic stockings.

Some difficulty has been encountered in the enforcement of regulations pertaining to northern pike, Amur pike, muskellunge, and hybrids. The difficulty has arisen from the two existing size regulations and improper identification by anglers of these fishes in waters where more than one purebred and/or hybrid is found. The present size and creel regulations on Pennsylvania's inland waters are 61.0 cm and two fish daily for northern pike and Amur River pike, and 76.2 cm and two fish daily for muskellunge and muskellunge hybrids. Since Amur pike have been stocked in only one lake and anglers are becoming more familiar with tiger muskellunge, enforcement difficulties have recently abated. Suffice it to say that responsible management must seek realistic regulations not only to benefit the resource, but also to preclude complicated and difficult situations for the angler in obeying those regulations.

Evaluation and Sampling

Evaluations of tiger muskellunge introductions were attempted at Beltzville Lake and Lake Wilhelm, the first new impoundments in Pennsylvania stocked with tiger muskellunge. Both received fry in 1971, fingerlings in 1972, fry in 1973, and fingerlings in 1974. Beltzville Lake has a surface area of 383 hectares and Lake Wilhelm's area is 704 hectares. Beltzville is fed by a good trout stream containing chain pickerel and Wilhelm is fed by a low gradient, meandering stream containing wild northern pike.

Both lakes were sampled during the summers of 1972 and 1973 with trapnets and experimental gill nets, and through the use of day and night electrofishing. Only six tiger muskellunge were taken at Lake Wilhelm. In fact, by April 1977, only 15 tiger muskellunge had been taken by staff and students of the Biology Department, Grove City College, who have sampled the lake annually since its construction (F. J. Brenner, personal communication).

On the other hand, during the 1973 survey alone 42 northern pike were captured from Lake Wilhelm, indicating to us that this indigenous species had adapted readily in the new lake, possibly at the expense of the tiger muskellunge. A few chain pickerel, which is one of two species of pickerel occurring naturally in the area, were collected from Beltzville Lake, but unlike the northern pike at Lake Wilhelm, they appeared to offer no serious competition to the tiger muskellunge.

Reports from Commission enforcement personnel at Lake Wilhelm indicate very few catches of tiger muskellunge have been made by anglers and the largest measured was a 39–cm 6–kg specimen. On the other hand, catch reports from Beltzville Lake, as reported earlier in this paper, were substantial and the lake fast became a popular fishery for tiger muskellunge.

Two significant points arose from these evaluations. Tiger muskellunge populations are difficult to sample by most conventional methods in sufficient numbers to give reliable statistical data. Second, although insufficient data were obtained to conclusively prove our contention, it appears that the tiger muskellunge did not compete well with the wild northern pike population in Lake Wilhelm.

Surveys conducted during the spring of 1977 on lakes containing tiger muskellunge indicated no activities such as concentration of individuals at prospective spawning sites which could be construed as spawning-related. Despite the use of all available conventional sampling techniques, very limited random catches of tiger muskellunge were made.

Summary

Esocid hybrids, especially the tiger muskellunge, appear to offer some significant benefit over purebreds in producing an acceptable trophy fishery which must be sustained through stocking. However, the present need for producing additional hybrids is not great in Pennsylvania.

The present and future fisheries management thrust must be directed toward better understanding and enhancement of both the existing esocid purebred and hybrid programs. It is obvious that better methods must be derived for sampling sterile hybrids. A far better knowledge and understanding of all the fisheries management implications of the use of hybrids must be sought. These implications include stocking rates for both impoundments and streams; stocking rates versus survival, growth, and harvest rates, competition with purebred esocids, other hybrids, and other game fishes; effect upon the overall fish community; program economics; and acceptance by anglers.

References

BEYERLE. G. B. 1973. Comparative growth, survival and vulnerability to angling of northern pike, muskellunge, and their hybrid tiger muskellunge stocked in a small lake. Mich. Dep. Nat. Resour. Fish Res. Rep. 1779. 13 pp. (Mimeo.)

BUSS, K., AND J. MILLER. 1967. Interspecific hybridization of esocids: hatching success, pattern development, and fertility of some F_1 hybrids. U.S. Bur. Sport Fish. Wildl. Tech. Pap. 14. 30 pp.

HEASLEY, D. K., AND M. E. CAWLEY. 1977. Activity and preference of Pennsylvania fishermen: 1974. Pa. State Univ. 48 pp.

MEADE, J. W. 1976. The propagation, hybridization and experimental introduction of Amur pike, *Esox reicherti*, in the United States. Pa. Fish. Comm. Benner Spring Fish. Res. Stn. 19 pp. (Mimeo.)

SCHROUDER, J. 1973. Muskellunge management in Michigan. Mich. Dep. Nat. Resour. Fish Div. Tech. Rep. 73-31. 21 pp.

SCOTT, D. P. 1964. Thermal resistance of pike (*Esox lucius* L.), muskellunge (*E. masquinongy* Mitchell), and their F_1 hybrid. J. Fish. Res. Board Can. 21:1043–1049.

WEITHMAN, A. S. 1975. Survival, growth, efficiency, preference and vulnerability to angling of esocidae. M.S. Univ. Mo., Columbia. 71 pp.

———, AND RICHARD O. ANDERSON. 1977. Survival, growth, and prey of Esocidae in experimental systems. Trans. Am. Fish. Soc. 106(5):424–430.

MANAGEMENT CASE HISTORIES

The Matheson Island Sauger Fishery of Lake Winnipeg, 1972–1976

E. B. DAVIDOFF

Fish and Wildlife Branch
Manitoba Department of Renewable Resources
and Transportation Services
Winnipeg, Manitoba R3H 0W9

Abstract

Excess fishing pressure and the use of small-mesh nets have caused severe problems for populations of lake whitefish, sauger, and walleye in Lake Winnipeg, and resulted in declining catches prior to 1970. Following the introduction of a quota system in the Matheson Island fishery of Lake Winnipeg in 1972, sauger and walleye catches have increased. In 1972, when this fishery reopened following a two-year closure due to mercury pollution, the sauger catch consisted of three main age groups (III–V). In 1973, the modal age of sauger decreased to III and continued at this age through 1976. Age III sauger became the mainstay of the fishery in 1974 and subsequent years. The contributions of age IV and older sauger to the fishery have decreased since 1972. Age II sauger comprised only 1% of the total catch in 1972, increased to 27% in 1974, and comprised 15 and 17% of the total catch in 1975 and 1976, respectively. Since 1973, the commercial fishery has relied increasingly upon immature sauger, mainly females, for the bulk of its catch. There is evidence that the continued good catches of sauger at Matheson Island have been supported by stocks of fish from other areas that contain mature individuals.

Lake Winnipeg is Manitoba's single most important lake in terms of fishery resources. The three most valuable species present in order of value prior to 1976–1977 were sauger, walleye, and lake whitefish. The main sauger and walleye fisheries are conducted in the southern half of the lake during the summer, fall, and winter commercial fishing seasons, the fall being the most important season.

Until recently, the pattern of fisheries production on Lake Winnipeg has been boom or bust. Excess fishing pressure and the use of small-mesh nets have caused severe problems for fish populations and resulted in declining catches prior to 1970.

As a direct result of mercury pollution, a two-year fisheries closure was implemented in 1970 on walleye and sauger in Lake Winnipeg. This closure allowed these species a respite from fishing and enabled them to recover. With the regeneration of fish stocks came a unique opportunity to monitor production levels, and maintain fish stocks at a viable level. In order to do this job effectively, a workable quota system was developed. Quotas were introduced in 1971 (1) to conserve fish stocks and eliminate severe peaks and depressions in catch and income by spreading the benefits of good year classes over a period of several years; (2) to maintain optimum sustainable fish population levels, and optimize fishermen's income; and (3) to increase the number of major age groups and number of spawners in the commercial fishery to attain three main year classes, two of them mature.

Since the introduction of the quota system in 1971 both walleye and sauger catches in Lake Winnipeg have increased. Walleye catches increased from 389,137 kg in the year 1969–1970 to 1,212,933 kg in 1976–1977, a 212% increase. Sauger catches increased from 922,742 kg in 1969–1970 to 1,130,188 kg in 1976–1977, a 22% increase. Total catches of walleye and sauger from Lake Winnipeg in 1976–1977 amounted to 2,343,121 kg valued at 3.1 million dollars to fishermen. Sauger weights and values were 1,130,188 kg and 1.3 million dollars, respectively. Quotas have been partly successful in conserving fish stocks, in maintaining sustainable fisheries, and in aiding the elimination of severe peaks and depressions in fishermen's catch and income.

An important commercial fishery for sauger is located at Matheson Island, just north of the narrows in the Channel Area of Lake Winnipeg (Fig. 1). This commercial fishery uses gill nets as the principal gear. Fishing nets used can legally vary from 76 to 140 mm stretched mesh. However, those most commonly used are 83-mm stretched mesh. In addition, many 89-mm and 95-mm stretched mesh nets are used. Fishing is conducted from fiberglass yawls 6.7 to 7.3 m in length with pointed bottoms powered by 50-horsepower outboard motors (Fraser, unpublished data). Fishermen in this area have elected to fish on individual quotas instead of an area quota.

The purpose of this report is to document the changes in age composition that have occurred in the Matheson Island sauger fishery of Lake Winnipeg following its reopening in 1972, to interpret the effect of these changes, and to suggest measures for sound management of this important fishery.

Methods

Commercial catches were routinely sampled to monitor changes in length, weight, age composition, growth, survival, and year-class strength. Length, weight, and scale samples for age determination were collected weekly from Matheson Island commercial saugers by departmental personnel during the summer and fall commercial fisheries beginning in 1972. Samplers were instructed to randomly sample a maximum of 250 saugers/week from as many fishermen as possible, but no more than 50 saugers were to be sampled from any one fisherman's catch. Each fisherman's catch was sampled in proportion to the species and size category of the fish caught. Annual sample sizes during a season ranged from a low of 247 in the fall of 1976 to a high of 1,609 in the fall of 1973 (Table 1). The small number of fish sampled in 1976 resulted from walleyes replacing saugers in the commercial catch at the time samples were collected. Average fork length was highest when the fishery

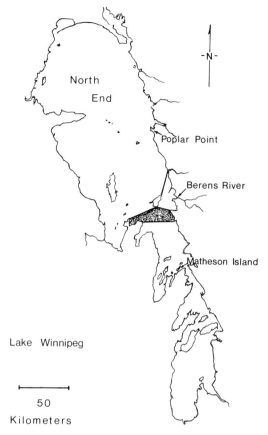

FIGURE 1.—Map of Lake Winnipeg showing localities mentioned in the text. The shaded area is closed to fishing during the summer season.

reopened in the summer of 1972 and lowest in the fall of 1976.

Age Composition

The accuracy of age determination decreases with the increasing age of the fish. Relatively few saugers were older than age V; therfore the ageing is believed to be relatively accurate.

When the fishery reopened in 1972 following a two-year closure due to mercury pollution, three main ages, III, IV, and V supported the sauger fishery and accounted for 97% of the total catch (Table 1). The modal age of saugers in the catch was age IV. Age V and older saugers comprised 25% of the catch while age III saugers made up 21% of the catch and age II saugers only 1%. A transition in age

TABLE 1.—Age composition, average length, and average weight, of sauger from commercial catches at Matheson Island by year and season, 1972–1976.

Age group	Year and season of capture									
	1972		1973		1974		1975		1976	
	Summer	Fall	Summer	Fall	Summer	Fall	Summer	Fall	Summer	Fall
0							1			
I					7	6	25			
II	9	8	21	59	182	481	27	325	166	42
III	176	217	166	891	566	817	352	851	505	163
IV	308	858	444	561	252	208	249	288	175	34
V	137	351	248	79	39	21	62	63	64	6
VI	7	36	44	14	4	2	13	9	17	2
VII	2	5	4	3			2		6	
Unknown				2	1					
Total	639	1,475	927	1,609	1,051	1,535	705	1,562	933	247
Average fork length (mm)	340	338	318	328	330	322	325	315	319	312
Average weight (g)	418	461	377	415	367	367	365	386	314	347

composition occurred in 1973; the percentage of age IV saugers in the catch decreased while the percentage of age III saugers increased and exceeded that of age IV saugers. The modal age of saugers in the 1973 catch decreased by one year to age III and continued at this age through 1976. Age III saugers became the mainstay of the fishery in 1974 and accounted for 54% of the total catch by number. This age group accounted for 53% and 62% of the catch in 1975 and 1976, respectively.

The contribution of age IV saugers to the fishery decreased from 53% in 1972 to 17% in 1974 and 16% in 1976. In addition, the percentage by number of age V and older saugers in the catch decreased from 25% in 1972 to 2% in 1974. Percentages in 1975 and 1976 were 7 and 6%, respectively.

The dramatic shift in age composition resulting from the increased exploitation of younger age groups is clearly shown by the increased percentages of age II saugers in the catch. This age group comprised only 1% of the total catch in 1972, reached a peak of 27% in 1974, and comprised 15 and 17% of the total catch in 1975 and 1976, respectively.

Since 1973, the commercial fishery has relied increasingly upon immature saugers, mainly females, for the bulk of its catch. Limited maturity data (C. G. Prouse, Department of Renewable Resources and Transportation Services, unpublished) indicate that practically all female saugers, age II and younger, caught by experimental fishing south of the Channel Area in 1976 were immature; 33% of age III, and 15% age IV females were immature. Male saugers were 87% mature at age II and 98% mature at age III. The increased reliance of the commercial fishery upon immature saugers does not ensure the future reproductive potential of the stock.

Whether or not these recent reductions in spawning stock combined with lower lake water levels in 1977 will have an additional adverse effect on the year class produced during this year will not be known until some years from now. How much longer the reproductive potential of this sauger stock can continue to be lowered before a severe drop in catch occurs is hard to predict. The consistent capture of large numbers of immature female saugers by the commercial fishery can lower the future reproductive potential of the stock. Total egg deposition is considerably reduced as younger fish are smaller and contain fewer eggs than older, larger fish. In addition, eggs from medium age fish may show greater viability than eggs from first-time spawners (Nikolskii 1965; Bagenal 1973).

Stock Maintenance

The present age structure of the Matheson Island sauger population suggests that this stock is in an unsatisfactory state and below the level capable of maintaining itself. However, this age structure is not strikingly different from that found by Rybicki (1961, 1964) and Howard (1967, 1971) in years prior to the closure of the commercial fishery in 1970. Their limited data suggest that age II and III saugers were important components of the commercial catch during the years 1961–1967. Total annual mortality rates were also generally high (Howard 1967). It is evident that despite the large catches of immature saugers taken by the fishery during these years the sauger population has maintained itself. There are several possible reasons for this. One possible explanation is that Matheson Island serves as a feeding area (Howard 1971) and that the sauger population in this area is maintained by sauger stocks which spawn in the North Channel and adjacent areas in the North End of Lake Winnipeg. Certainly the larger minimum mesh size of 95 mm in use in the North Channel would serve to protect age II saugers from being heavily fished. For example, age composition data from the Berens River sauger fishery, a fishing area located in the North Channel shows an older age composition than that at Matheson Island with relatively few age II saugers comprising the catch.

Recent tagging data (R. W. Moshenko, Department of Renewable Resources and Transportation Services—and, currently, Freshwater Institute—unpublished) indicate that saugers spawning at Berens River and Poplar Point in the North End of Lake Winnipeg contributed to the fishery at Matheson Island. Thus, there is some support for the hypothesis that the continuing good catches at Matheson Island are supported by stocks of fish from other areas that contain mature individuals. In fact, the stocks at Berens River and areas further north probably help maintain the reproductive potential of sauger stocks south of these areas.

Another possible reason for the maintenance of sauger stocks is the closure of the area directly south of the Berens River fishery during the summer fishing season. This area serves as a protected zone for any saugers which enter it from commercial fishing areas north or south of it. In addition, the complete cessation of commercial fishing between seasons provides a respite from fishing and provides an opportunity for stock(s) of sauger to mix and migrate partially or completely out of commercial fishing areas. Depending upon the location, timing and direction of their movements, some saugers could be completely unprotected from commercial fishing, or achieve partial or complete protection from fishing. For example, saugers moving into the North End of Lake Winnipeg would be completely protected from commercial fishing during September and October as fishing is not permitted in this area during the fall.

The presence of walleyes in the Matheson Island area serves to reduce fishing pressure on saugers, since fishermen commonly fish several large mesh nets to capture the more highly valued walleyes.

The above factors offer possible explanations for the continued success of the Matheson Island fishery. They also illustrate the need to further study the delineation of sauger stocks and their relationship to one another, and to consider them in managing the commercial fishery. The capture of young immature saugers must be curtailed to increase the number of potential spawners, maintain high stable catches, and provide for the possibility of increased catches in future years.

Recent tag return data are presently being analyzed to provide further insight into the delineation of sauger stocks and their relationship to one another.

Mesh size experiments have been conducted during the past two years to determine the best mesh size for walleyes and saugers in the commercial fishery. These experiments may lead to possible changes in mesh size regulations to curtail the capture of immature saugers and walleyes. The results from the tagging and mesh size experiments should provide the information required for better management of sauger, and provide for increased fishing quotas in future years.

Acknowledgments

I wish to thank the many fishermen who kindly permitted their catches to be sampled, and to acknowledge the effort made by various personnel of the Manitoba Department of Renewable Resources and Transportation Services in collecting data for this study. Appreciation is expressed to R. Janusz and J. Beyette who assisted in the computer processing, analysis, and drafting of data, and to H. Metheringham for reading the ages of the scale samples.

References

BAGENAL, T. B. 1973. Fish fecundity and its relations with stock and recruitment. Rapp. P.V. Reun. Cons. Int. Explor. Mer. 164:186–198.

HOWARD, W. 1967. A comparison of some Lake Winnipeg sauger statistics. Manit. Dep. Renewable Resour. Transp. Serv., Winnipeg. 15 pp.

———. 1971. Fall sauger sampling in 1967 and assorted sampling of pickerel and sauger—1964 to 1967. Manit. Dep. Renewable Resour. Transp. Serv. Res. Branch, MS Rep. 70-17. 16 pp.

NIKOLSKII, G. V. 1965. [Theory of fish population dynamics as the biological background for rational exploitation and management of fishery resources.] Nauka Press, Moscow. 382 pp. English edition, Oliver and Boyd, Edinburgh, 1969. 323 pp.

RYBICKI, R. W. 1961. Lake Winnipeg walleye–sauger investigation fall, 1960 and winter, 1961. Manit. Dep. Renewable Resour. Transp. Serv., Winnipeg. 10 pp.

———. 1964. Tabulations of sauger data collected from Lake Winnipeg in 1961, 1962 and 1963. Manit. Dep. Renewable Resour. Transp. Serv., Winnipeg. 17 pp.

Northern Pike, Tiger Muskellunge, and Walleye Populations in Stockton Lake, Missouri: A Management Evaluation

JOHN A. GODDARD AND LEE C. REDMOND

Division of Fisheries
Missouri Department of Conservation
Jefferson City, Missouri 65101

Abstract

A three-stage filling of 10,072-hectare Stockton Lake coupled with spring releases of walleye and northern pike fry and fingerlings into the flooded terrestrial vegetation in 1970, 1971, and 1972 resulted in good survival, exceptional growth, and a quality fishery. Angler interest in these species was high and accounted for 20% of the fishing trips in 1972. With minimum size limits of 457 mm on walleyes and 762 mm on northern pike, harvest reached a high of 1.9 and 1.2 fish per hectare (2.4 and 4.5 kg per hectare) in 1972 and averaged 1.0 and 0.4 fish per hectare (1.5 and 1.5 kg per hectare). The size limits provided many additional hours of quality angling; in 1971 nearly 31,000 sublegal walleyes and northern pike were caught and released to be caught again (10.3 per hectare). In addition, the size limits permitted a very respectable harvest of quality size fish, averaging 1.4 fish and 3 kg per hectare. While remaining in the lake, these species effectively preyed on forage fishes, primarily gizzard shad. Tiger muskellunge were released in 1975 to replace the northern pike, many of which succumbed to natural mortality.

Fisheries managers in Missouri and elsewhere have demonstrated the value of stocking small impoundments early with selected sizes and species of forage and predator organisms to fill primary niches. This approach usually results in a fish population containing all sizes of primary predator species, usually largemouth bass in Missouri, within two years after original stocking (Hoey and Redmond 1974). These predators can utilize the wide range in sizes of established forage organisms.

Overharvest of largemouth bass in small impoundments also has been well documented. The angling harvest of adult largemouth bass by fishermen in small Missouri lakes has been as high as 69% during the first four days of fishing at 12.5-hectare Jo Shelby Lake (Redmond 1974), and 72% of adult largemouth bass during the first season (May 28–October 15) at 83-hectare Little Dixie Lake (Turner 1963). The result was poor quality fishing in succeeding years. Preventing overharvest of largemouth bass has been effective in maintaining good quality fishing, and a minimum length limit for regulating bass harvest is our best technique for maintaining good quality fishing past the first season (Redmond 1974). The early stocking with various sizes of predator and forage organisms, and the prevention of overharvest of the predators by appropriate size limits, is the basis of Missouri's fisheries management program for new small impoundments.

These principles were applied to planning development of the fishery at Stockton Lake, a new large reservoir, except that coolwater species were used in addition to largemouth bass. An outstanding fishery has been maintained there through the first seven years of impoundment. Northern pike and walleye were introduced to fill the role of top trophic level predators which could utilize larger forage fishes than would the three black basses native to the impounded stream. Equally important was the establishment of a more diverse fishery and a trophy fishery in Stockton Lake.

The Study Area

Stockton Lake is a 10,072-hectare impoundment in west-central Missouri completed by the Corps of Engineers in 1969 for flood control and power generation. The reservoir has a drainage area of 300,000 hectares. The shoreline is approximately 400 km long; maximum depth is 29.6 m at the dam; mean depth is 10.9 m

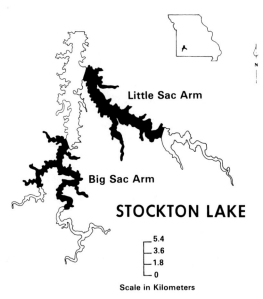

FIGURE 1.—*Stockton Lake, Missouri. Creel census areas are in black.*

at conservation pool elevation of 264 m above mean sea level. At flood stage the reservoir has a surface area of 15,500 hectares and a mean depth of 13.3 m. The reservoir basin was totally cleared of timber except for the upper ends of the arms. There was some regrowth of trees and shrubs in the three years before total filling in early 1972.

The reservoir began filling in December, 1969 and reached an elevation of 256 m above mean sea level by August, 1970. A two-stage filling plan was recommended to provide at least a second year of flooding terrestrial vegetation for maximum survival of stocked fish as well as good reproduction, survival, and growth of native fishes. Three separate years of spring flooding of vegetation was actually achieved as the lake level reached elevations of 260 m in late February, 1971; 263 m in September, 1971; and full pool elevation of 264 m in January, 1972.

A chemical analysis of the reservoir waters by Jones (1978), revealed alkalinity readings of 115.9 mg/liter; total hardness, 125.5 mg/liter; calcium hardness, 98.0 mg/liter; chloride, 5.4 mg/liter; specific conductance, 266 micromhos per cm.

Methods

Initial fish releases into the reservoir were made during April, 1970, and included the stocking of 2,000,000 northern pike fry, 4,220,000 walleye fry, and 10,000 northern pike fingerlings 35 to 50 mm in length. During April, 1971, an additional 675,000 northern pike fry and 5,000,000 walleye fry were stocked. A release of 35,000 25- to 50-mm walleye fingerlings in May of 1972 completed the coolwater species stocking program.

To reduce competition for forage between the two species, the northern pike fry and fingerlings were released among extensive stands of recently flooded terrestrial vegetation in the upper one-half of the reservoir and the walleye fry and fingerlings were released over flooded vegetation, primarily fescue fields, in the lower one-half of the reservoir. Numbers of naturally produced young largemouth bass were very high in the reservoir the first few years.

Northern pike fry were made available from Gavins Point National Fish Hatchery at Yankton, South Dakota, while most of the walleye fry were produced at the Missouri Department of Conservation's Chesapeake Hatchery. Brood stock walleye at Chesapeake Hatchery were from the Osage River in Missouri.

Beginning in April, 1971, a daytime quantitative creel census as described by Kathrein (1953) and modified by Fry (1962) and by Hanson (1969) was conducted from March 1 to November 30 each year from 1971 to 1976 on two areas of the reservoir to evaluate success of early fisheries management efforts. The creel census areas on Big Sac Arm (1,070 hectares) and Little Sac Arm (1,928 hectares) comprise about 30% of the reservoir area and 30% of the shoreline length (Fig. 1).

Boat-mounted 230-volt DC electrofishing gear was used to collect fish for age, growth, and length-frequency determinations.

During the spring of 1973, 91 walleyes and 82 northern pike greater than 406 mm and 508 mm, respectively, were tagged with numbered Floy FD68B anchor tags. Tags were attached to the fish at the posterior end of the dorsal fin near its base.

Stocking Success

Growth of the two species was excellent through 1970; walleyes averaged 302 mm in total length and northern pike averaged 493 mm by mid-October, only six months after their introduction.

Forage fishes, consisting primarily of 50- to 100-mm gizzard shad, were very abundant throughout 1970, even into the late fall and early winter.

Growth of both walleyes and northern pike continued to be good throughout the next several years. Northern pike attained mean lengths of 515 mm by March of 1971 (age I), 726 mm by March, 1972 (age II), and 830 mm by March, 1973 (age III). Walleyes reached mean lengths of 310 mm in March, 1971 (age I), 452 mm in March, 1972 (age II), 485 mm in March, 1973 (age III), 599 mm in March, 1974 (age IV), and 632 mm by March, 1975 (age V). Collections by electrofishing during March of 1972 revealed that 20% of the northern pike and 35% of the walleyes exceeded the minimum harvestable lengths of 762 and 457 mm, respectively, for these species.

Successful reproduction by the northern pike in Stockton Lake was anticipated but never verified although 5,000,000 young were successfully hatched at a nearby hatchery from eggs and milt collected from Stockton Lake fish in the spring of 1973.

The first successful walleye spawning occurred during the spring of 1973, and additional year classes were produced through 1976. Beginning in early March of each year many sexually mature walleyes migrated several kilometers up the major tributary streams of the reservoir to spawn in the gravel riffles, which provided a substantial springtime river fishery for large walleye up to 5 kg in size.

In 1973, anglers voluntarily returned 7.7% of walleye tags and 6.1% of northern pike tags. A study comparing voluntary and reward-stimulated returns from tagged black basses showed that rewards doubled the rate of return. Thus, we assume that the harvest of legal walleyes and northern pike during 1973 was about 12 to 15 percent of the population.

A chronic die-off of northern pike began during the summer of 1971 and continued through 1973. All moribund northern pike examined had large hemorrhagic areas around the mouth, eyes, vent, and at the base of the fins. Secondary fungal infections were noted in most fishes examined. The bacteria *Aeromonas liquifaciens* was isolated and identified from all kidney samples tested. *A. liquifaciens* is a common soil and water bacterium which can become a systemic pathogen in fishes weakened by stress. The stress, in this instance, was thought to result from a combination of the reservoir's high summertime water temperatures, which approached 32 C near the surface, and infection by the protozoan parasite *Trichodina* sp. which was found in large numbers on the gills and over the bodies of those northern pike examined for parasites. It is believed that a substantial number of northern pike succumbed during 1971 and 1972.

In an effort to replace the northern pike, the hybrid muskellunge or "tiger muskellunge" (northern pike male × muskellunge female) was introduced into the reservoir in September, 1975 with the release of 2,510 250-mm fish. These fish were obtained as fry from the state of Michigan. Additional releases of 10,000 200-mm and 2,600 225-mm tiger muskellunge were made during October, 1976. The tiger muskellunge, like the northern pike, is protected under a statewide 762-mm length limit and the daily creel limit for both the tiger muskellunge and the northern pike was reduced from 2 to 1 in January, 1976. Growth of the tiger muskellunge was excellent; fish from the 1975 release averaged 394 mm in March, 1976. Those released during the fall of 1976 averaged 376 mm when captured in March, 1977. Some of the 1975 release exceeded the 762-mm length limit in May, 1977. The largest of these measured 815 mm and weighed 4.8 kg. No mortalities of tiger muskellunge have been reported.

Harvest

Angler harvest of the Stockton Lake fishes began in earnest during the early fall months of 1970. The northern pike was protected under existing statewide regulations that included a minimum harvestable length limit of 762 mm and a daily creel limit of two fish. The walleye harvest was restricted only by a daily creel limit of four fish.

Periodic interviews of Stockton Lake anglers during early fall of 1970, prior to the creel census, revealed that large numbers of these small 250- to 300-mm walleyes were being caught and removed. It was expected that a large population of gizzard shad would be available for forage, and a minimum harvestable length of 457 mm was established in April, 1971 for Stockton Lake walleyes. This was done to build and maintain sizable numbers of large walleyes in the reservoir able to both provide quality fishing and maintain heavy predation on the forage fish population. A 305-mm minimum length for black basses was also established for the same reason.

Angling for northern pike and walleyes was substantial during the early years following impoundment. Fishing trips for these species varied from 16.0% of the total trips in 1971, to 20.4% in 1972, and 10.6% in 1973 (Table 1). The substantial amount of fishing for walleyes and northern pike during 1971 indicated angler interest in catching, but not necessarily keeping, a large fish since only 6.0% of the northern pike and 4.9% of the walleyes caught then were of legal size.

Trips by anglers specifically seeking the northern pike declined steadily from a high of 8.7% of the total trips in 1971 to zero in 1976. This decline in popularity of the northern pike fishery corresponds closely to the rapid decline of the northern pike population. Estimates of the angler catch of the northern pike ranged downward from 12,350 (4.1 fish and 7.2 kg per hectare) in 1971 to only 220 in 1975. In 1975 all were caught by anglers fishing for other species. The harvest of legal northern pike ranged from 1.2 to 0.1 fish per hectare (Table 2) and averaged 0.4 per hectare (1.5 kg per hectare) for the six-year period.

Anglers specifically seeking walleyes accounted for 7.3% of the estimated angling trips in 1971 and 12.6% in 1972, but this proportion declined thereafter to 1.6%. This decline apparently was not associated with a declining walleye population, but probably indicated a change in angler preference toward large and readily available white and black crappie populations. Both the estimated total catch and harvest of walleyes increased during 1975 and 1976 from those in 1973 and 1974. Catches of walleyes ranged from a low of 2,960 (1.0 fish and 1.0 kg per hectare) in 1974 to a high of 20,295 (6.7 fish and 3.4 kg per hectare) in 1971 (Table 2).

Angler harvests of 3,248 legal-size walleyes during 1975 and 3,970 legal-size walleyes during 1976 were exceeded only by the harvest of 5,785 walleye in 1972, which were fish from the original stocking. Average harvest of legal-size walleyes for the six-year period was 1.0 fish and 1.5 kg per hectare.

Tiger muskellunge releases into the reservoir in October of 1975 were followed by a catch of 329 sublegal fish in 1976. Verified catches of four tiger muskellunge exceeding the 762-mm minimum harvestable length were made during the spring of 1977.

The catch of predator fishes, black

TABLE 1.—*Fishing trips in the Stockton Lake creel census areas 1971–1976, daytime only, March–November.*

	1971[a]	1972	1973	1974	1975	1976
Total trips	55,422	84,354	79,949	102,130	124,730	117,387
Northern pike fishing trips	4,822	6,548	4,095	1,477	105	0
(% of total trips)	8.7	7.8	5.1	1.4	0.1	0
Walleye fishing trips	4,046	10,646	4,422	5,878	2,096	1,913
(% of total trips)	7.3	12.6	5.5	5.8	1.7	1.6

[a] March not censused.

TABLE 2.—*Catch of northern pike and walleyes from Stockton Lake creel census areas 1971–1976, daytime only, March–November. Minimum legal sizes were 762 mm for northern pike and 457 mm for walleye.*

	1971	1972	1973	1974	1975	1976
Northern pike						
Numbers						
Legal size	740	3,602	799	654	220	451
Sublegal size	11,610	631	0	0	0	0
Numbers/hectare						
Legal size	0.2	1.2	0.3	0.2	0.1	0.2
Sublegal size	3.9	0.2	0	0	0	0
Kilograms/hectare						
Legal size	0.8	4.5	1.3	1.0	0.4	1.0
Sublegal size	6.4	0.4	0	0	0	0
Walleye						
Numbers						
Legal size	989	5,785	3,085	1,524	3,248	3,970
Sublegal size	19,306	1,775	216	1,436	1,746	626
Numbers/hectare						
Legal size	0.3	1.9	1.0	0.5	1.1	1.3
Sublegal size	6.4	0.6	0.1	0.5	0.6	0.2
Kilograms/hectare						
Legal size	0.3	2.4	1.7	0.9	1.7	2.0
Sublegal size	3.1	0.5	0.04	0.1	0.4	0.1

basses, northern pike, walleyes, and crappies, in the creel census areas for the years 1971 through 1976 ranged from a low of 36.5 fish (14.7 kg) per hectare in 1973, to 147.4 fish (37.6 kg) per hectare in 1976; the average annual catch was 86.5 (28.0 kg) per hectare (Table 3). The catch of most species of fish was depressed on all large Missouri reservoirs censused during 1973 due to extremely high water levels. Stockton Lake remained more than 3 m above conservation pool level from late March through early June.

Discussion

Survival and growth of the stocked fry of both species appeared to be very good, probably because of the abundance of flooded terrestrial vegetation in the three succeeding spring periods. This "three-stage" filling of the basin provided a fertile environment as the flooded plants slowly decayed. Zooplankton was plentiful for young fishes as the aquatic environment continued to expand. Small forage fishes, primarily gizzard shad, were also abundant every year and small gizzard shad up to 75 mm were available throughout the growing period each year. The continuing rise of the water into terrestrial vegetation provided protection and an important feeding area for the young fishes. Aggus and Elliott (1975) correlated high survival and good growth of young basses in Bull Shoals Lake to the amount and duration of flooding of terrestrial vegetation. Apparent high survival and good growth of the cool-water predators, particularly the northern pike, in the early years at Stockton Lake probably can be attributed to the same factors.

Growth rate of walleyes in Stockton Lake through age V exceeded that of 22 other walleye populations in the United States and Canada (Table 4) by 10 to 50%. Growth of northern pike was also exceptionally good, surpassing that of good populations occurring well within the natural range for the species, and much better than the typical growth to 203 mm, 406 mm, and 584 mm at ages I, II, and III, respectively, cited by Buss (1961).

Early management efforts, including stocking of selected predator species and harvest control through minimum size limits on the northern pike, walleyes, and black basses, resulted in the development of a high quality large reservoir fishery that yielded an average annual harvest of 65.6 predator fish per hectare (northern pike, walleyes, black basses, and crappies) weigh-

TABLE 3.—*Catch and harvest of predator fishes (northern pike, walleyes, black basses, crappies) in Stockton Lake creel census areas 1971–1976, daytime only, March–November.*

	1971	1972	1973	1974	1975	1976	Average 1971–1976
Catch							
Numbers/hectare	77.1	40.8	36.5	80.6	136.8	147.4	86.5
Kilograms/hectare	35.0	24.5	14.7	21.7	34.7	37.6	28.0
Harvest							
Numbers/hectare	18.5	29.5	31.2	72.7	119.2	122.5	65.6
Kilograms/hectare	11.9	20.3	13.2	20.1	28.4	33.0	21.2

TABLE 4.—Mean total length (mm) at each annulus for walleyes from Stockton Lake and 22 other populations (from Colvin 1975).

	Age I	Age II	Age III	Age IV	Age V
Stockton Lake, Missouri	310	452	485	599	632
Red Cedar River, Wisconsin	162	264	328	373	419
Pike and Round Lakes, Price County, Wisconsin	140	208	264	307	333
Lake Winnebago, Wisconsin	147	256	330	376	411
Red Cedar Lake, Wisconsin	130	246	320	373	427
Pike Lake, Washington County, Wisconsin	175	290	367	423	469
Big Eau Pleine Reservoir, Wisconsin	180	299	393	458	498
Three Mile Lake, Ontario	122	183	236	292	315
Red Lakes, Minnesota	140	211	267	312	347
Oneida Lake, New York	157	237	303	353	387
Lac la Ronge, Saskatchewan		234	282	328	378
Lake of the Woods, Minnesota	175	234	295	350	386
Lake Gogebic, Michigan	117	239	307	361	401
Many Point Lake, Minnesota	127	218	310	376	411
Leech Lake, Minnesota				361	409
Spirit Lake, Iowa	147	248	318	377	428
Clear Lake, Iowa	182	287	369	429	471
Saginaw Bay, Lake Huron	163	305	408	470	518
Lower Bay Quinte, Lake Ontario		342	425	492	549
Utah Lake, Utah	205	353	404	456	489
Current River, Missouri	207	330	406	456	496
Norris Reservoir, Tennessee	262	417	475	505	528
Canton Reservoir, Oklahoma	310	422	484	536	571

ing 21.2 kg per hectare. The coolwater species accounted for 8.1% by number and 14.2% by weight of the average annual predator harvest. Since the minimum legal sizes were 762 mm and 457 mm, these were quality-size fish, some of them truly trophies. Three state record northern pike weighing 8.2, 8.3, and 8.4 kg were caught at Stockton Lake in the springs of 1973, 1974, and 1975, respectively.

The six-year average harvest of walleyes at Stockton Lake is higher than any previously documented in Missouri. It compares well with that reported for 14 populations by Colvin (1975). Numbers and weights of walleyes harvested ranged from 0.3 fish (0.4 kg) per hectare for Pool 11 of the Mississippi River upward to 30.5 fish (16.9 kg) per hectare for Little Cut Foot Sioux Lake, Minnesota (Table 5). Mean total length of walleyes harvested from these 14 populations ranged from 317 mm to 406 mm; that of walleyes harvested from Stockton Lake during 1971–1976 ranged from 465 mm to 569 mm and averaged 513 mm. Although the annual harvest of walleyes larger than 457 mm from Stockton Lake is the highest yet documented for Missouri lakes, it is low compared to that of many northern lakes. However, the estimated average annual catch of both legal and sublegal walleyes of 2.4 fish per hectare averaging 432 mm long certainly places Stockton Lake among the better walleye fisheries.

The catch-and-release fishing that resulted from the sublegal northern pike and walleyes added significantly to the angling experience by providing an additional average annual catch of 2.1 fish per hectare.

Though the data are limited, returns of tags from walleyes and northern pike indicated only 12 to 15% of the legal-size

TABLE 5.—Number, weight, and mean length of walleyes harvested from Stockton Lake and 14 other walleye populations (from Colvin 1975).

	Walleye harvest		Mean length in creel (mm)
	No/ hectare	Kg/ hectare	
Stockton Lake (6-yr average harvest)	1.0	1.5	513
Stockton Lake (6-yr average catch)[a]	2.4	2.2	432
Red Cedar River, Wisconsin	11.9–21.9	4.7–9.2	351–362
Pool 4, Mississippi River	1.2	1.2	
Pool 5, Mississippi River	0.9	1.0	
Pool 7, Mississippi River	1.0	0.6	
Pool 11, Mississippi River	0.3	0.4	
Big Eau Pleine Reservoir, Wisconsin	2.5	0.7	317
Laura Lake, Wisconsin	3.5		356
Leech Lake, Minnesota	3.3–3.6	2.0–2.3	
Many Point Lake, Minnesota	4.9–7.2	2.8–4.1	
Cut Foot Sioux Lake, Minnesota	19.7	10.3	366
Little Cut Foot Sioux Lake, Minnesota	30.5	16.9	
Lake Winnibigoshish, Minnesota	1.8–2.8	1.2–1.7	381–406
Lake of the Woods, Minnesota, Manitoba, and Ontario	0.6–0.9	0.3–0.5	
Splithead Lake, Minnesota	5.8–6.6	2.7–3.0	

[a] Legal and sublegal sizes combined.

fish in the population were harvested in 1973. This agrees with observations of the walleye harvest in some other Missouri reservoirs. Although substantial populations appear to be present during netting and electrofishing surveys, limited numbers appear in the anglers' creels. Although the harvest rate at Stockton Lake appears to be lower than desired, the coolwater species contributed substantially to the anglers' interest, added diversity to the fishery, and provided a trophy dimension as well as quality food fish.

References

AGGUS, L. R., AND G. V. ELLIOTT. 1975. Effects of cover and food on year-class strength of largemouth bass. Pages 317–322 in H. Clepper, ed. Black bass biology and management. Sport Fishing Institute, Washington, D.C.

BUSS, K. 1961. The northern pike. Spec. Purpose Rep., Penn. Fish Comm., Benner Spring Fish Res. Stn. 58 pp. (Mimeo.)

COLVIN, M. A. 1975. The walleye population and fishery in the Red Cedar River, Wisconsin. M.S. Thesis. Univ. Wis., Stevens Point. 97 pp.

FRY, J. P. 1962. Harvest of fish from tailwaters of three large impoundments in Missouri. Proc. Annu. Conf. Southeast. Assoc. Game Fish Comm. 16:405–411.

HANSON, W. D. 1969. Harvest of fish in Pomme de Terre Reservoir and its tailwater. Prog. Rep. Dingell-Johnson Proj. F-1-R-18, Work Plan 18, Job 3, Mo. Dep. Conserv., Jefferson City 7 pp.

HOEY, J. W., AND L. C. REDMOND. 1974. Evaluation of opening Binder Lake with a length limit for bass. Pages 100–105 in J. L. Funk, ed. Symposium on overharvest and management of largemouth bass in small impoundments. Am. Fish. Soc. N. Central Div. Spec. Publ. 3.

JONES, J. R. 1978. Chemical characteristics of some Missouri Reservoirs. Trans. Mo. Acad. Sci. (In press.)

KATHREIN, J. W. 1953. An intensive creel census on Clearwater Lake, Missouri, during its first four years of impoundment, 1949–1952. Trans. 18th N. Am. Wildl. Nat. Resour. Conf.:282–295.

REDMOND, L. C. 1974. Prevention of overharvest of largemouth bass in Missouri impoundments. Pages 54–68 in J. L. Funk, ed. Symposium on overharvest and management of largemouth bass in small impoundments. Am. Fish. Soc. N. Central Div. Spec. Publ. 3.

TURNER, J. L. 1963. A study of a fish population of an artificial lake in Central Missouri with emphasis on *Micropterus salmoides* (Lacépède). M.S. Thesis. Univ. Mo., Columbia. 59 pp.

Responses of Northern Pike to Exploitation in Murphy Flowage, Wisconsin

HOWARD E. SNOW

Wisconsin Department of Natural Resources
Box 309, Spooner, Wisconsin 54801

Abstract

The harvest and population dynamics of the fish populations in Murphy Flowage were studied for 15 years. A complete creel census and annual population estimates were utilized to describe the relationship between exploitation and other selected statistics of the northern pike, which was the major predator species present. The entire study was conducted under liberalized fishing conditions (no bag, season, or size limits). Annual densities of native northern pike over 356 mm ranged from 5.8 to 40.6/hectare (mean = 20.8) and the biomass averaged 14.7 kg/hectare. The addition of stocked northern pike (117.2/hectare) in December of the ninth year of the study temporarily increased the standing stock to 121.6/hectare the following spring. The annual harvest of northern pike averaged 6.9 fish/hectare (6.7 kg/hectare) and they were caught at an average annual rate of 3.8/100 hours of angling. Annual exploitation rate (u) averaged 26%; natural mortality rate (v), 40%; and total mortality rate (A), 66%. Total mortality was not density dependent within the range of population density observed for native northern pike but became density dependent only when the numbers of fish were increased to high levels by stocking. I conclude that except when their densities were artificially high, the most important determinant of northern pike population density in Murphy Flowage was the availability of suitable forage (small bluegills). The results suggest that forage may contribute to predator abundance only above some threshold density of prey. Conversely, at lower forage levels predators may control the abundance of forage species.

There is very little quantitative information in the literature concerning the response of northern pike populations to exploitation. There is an even greater gap in our knowledge when it comes to subdividing total mortality for this species into exploitation and natural causes.

The present work draws upon a 15-year study (1955–1970) of Murphy Flowage which included compulsory creel census, liberalized fishing regulations (no bag, season, or size limits), and annual population estimates. It also discusses the impact of three management techniques (northern pike stocking, panfish removal, and winter drawdown), on the native northern pike population.

Study Area

Formed in 1937 by the impoundment of Hemlock Creek, a trout stream, Murphy Flowage was located in northwestern Wisconsin in the headwaters region of the Red Cedar River. The 73-hectare flowage had a maximum depth of 4.3 m, an average alkalinity of 37 mg/liter and a mean annual flow at the outlet of 0.52 m³/s. A large percentage of the total area was covered by dense aquatic vegetation until 1967. Thereafter, winter drawdown reduced the amount of aquatic plants for the remainder of the study (Beard 1973). On June 1, 1970, high water destroyed the Murphy Flowage Dam and terminated the study.

The total biomass of fish in Murphy Flowage averaged in spring approximated 28 kg/hectare of game fish and 336 kg/hectare of panfish. Northern pike accounted for approximately 65%, and the largemouth bass about 35%, of the biomass of game fish. Bluegills comprised about 80% of the total biomass of panfish.

Methods

Exploitation rates and total harvest of northern pike were determined from the registration-type compulsory creel census conducted throughout the study. Petersen mark-and-recapture estimates and Bailey's modified formula were made in 100-mm size groups for fish over 356 mm each spring (mainly age II and older). All estimates of mortality were obtained from Petersen estimates of the marked (fin-clipped) segment of the northern pike population surviving from the previous year. All notations used for exploitation and mortality are those of Ricker (1975). Further details on methods used have been described previously (Snow 1971, 1974, 1978; Churchill and Snow 1964).

The Northern Pike Population

The spring standing stock and biomass estimates of the native northern pike populations larger than 356 mm total length averaged 20.8 fish/hectare weighing 14.7 kg/hectare. Numbers per hectare increased (with variation) from 5.8 in 1955 to 40.6 in 1965, then declined to 7.0 in 1970 (Fig. 1). Biomass followed a similar pattern, increasing from 5.0 kg/hectare in 1955 to 25.1 kg/hectare in 1964 and then declining to 8.2 kg/hectare in 1970. The addition of stocked northern pike (117.2/hectare and mean total length of 391 mm) in December 1963 increased the maximum values of numbers and biomass to 121.6 fish/hectare and 61.1 kg/hectare in May, 1964, the first estimate after stocking. The numbers of stocked northern pike declined to 10.1/hectare by May 1965 and 0.5/hectare in May 1968 after which the numbers of stocked northern pike were too low to estimate (Snow 1974). Annual average numbers and biomass including stocked northern pike were 27.0 fish/hectare and 17.5 kg/hectare.

Most of the variations in numbers of northern pike throughout the study occurred among northern pike under 556 mm total length rather than among larger northern pike (Table 1). Some variations were due to various experimental management techniques. Panfish removal in 1960 and 1961 was followed by numerical increases of small panfish and northern pike under 556 mm in 1962 and 1963. Northern pike stocking in 1963 was followed by an increase of small native northern pike in 1965. Winter drawdown in 1967, 1968, and

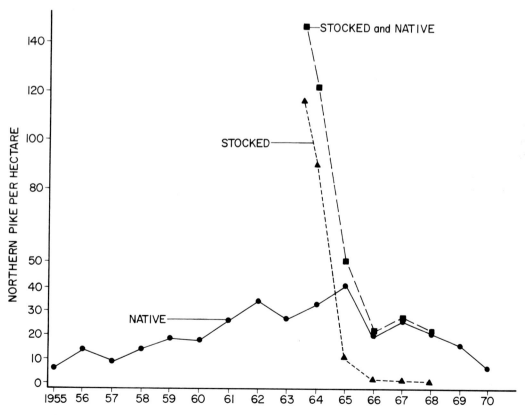

FIGURE 1.—*Estimated number of northern pike in Murphy Flowage for fish 356 mm and larger.*

TABLE 1.—*Estimated number of native northern pike per hectare each spring in Murphy Flowage.*

Year	Total length, mm			
	356–455	456–555	556–655	>655
1955	2.5	2.7	0.5	0.1
1956	5.1	4.3	3.6	0.9
1957	1.0	4.6	2.8	0.5
1958	5.8	3.2	3.5	1.5
1959	10.9	5.0	1.4	1.1
1960[a]	9.4	4.6	2.1	1.5
1961[a]	14.8	7.1	2.4	1.1
1962	23.4	7.0	2.4	0.9
1963[b]	20.0	6.6	2.0	0.9
1964	11.9	14.0	4.5	1.9
1965	34.6	3.8	2.1	0.1
1966	9.8	7.4	1.9	0.4
1967[c]	15.1	5.7	3.7	2.0
1968[c]	11.2	5.1	3.3	1.9
1969[c]	3.0	4.9	5.3	2.0
1970	1.7	1.4	2.1	1.8

[a] Panfish partially removed.
[b] Northern pike stocked in December.
[c] Water drawdown (October through March).

1969 was accompanied by a decrease in abundance of small northern pike to low levels by 1970.

Harvest

Northern pike comprised 2.1% of the total number of fish and made up 19.6% of the total biomass taken by anglers (Snow 1978). Annually, the harvest of northern pike per hectare averaged 6.9 fish and varied from a low of 2.2 in 1966 and 1969 to a high of 11.1 in 1956 (Table 2). The most notable trend during the study was a decline in annual harvest after northern pike were stocked in December 1963. Before stocking (1955–1963) the average annual harvest of northern pike per hectare was 8.7 compared to 4.1 after stocking (1964–1969). Stocked northern pike comprised a major part of the northern pike harvest only during 1964.

The mean length of all native northern pike was 538 mm and of all stocked northern pike, 447 mm (Table 3). If a 456-mm size limit on northern pike had been in effect, as there was in several areas of Wisconsin, 22% of the native fish and 71% of the stocked fish harvested would have been illegal.

The harvest rate of northern pike per 100 hours averaged 3.8 (winter, 8.8; summer, 3.1). The summer harvest rate in 1964, the first season after stocking, was the only one that exceeded the following winter rate. The maximum (18.5) and minimum (1.0) harvest per 100 hours occurred in winter. Harvest rate presented here is based on the total hours fished by all anglers whether or not they were fishing for northern pike.

Exploitation and Mortality

Natural mortality (v) averaged 40% and varied from 9 to 86%, while exploitation (u) averaged 26% and varied from 3 to 50% (Fig. 2). Total mortality (A), which is equal to $v + u$, averaged 66% and varied from 47 to 90% throughout the 15-year study period.

The largest variation in mortality occurred after stocking 117 northern pike

TABLE 2.—*Annual angler harvest of northern pike by anglers in Murphy Flowage.*[a]

Year	Fish/hectare[b]	Kg/hectare[b]
1955	4.9	5.2
1956	11.1	10.9
1957	8.9	10.1
1958	10.6	12.9
1959	6.9	7.4
1960	6.2	6.5
1961	9.9	9.3
1962	10.6	9.2
1963	9.4 (4.9)	8.6 (2.8)
1964	10.6 (60.5)	6.2 (48.5)
1965	3.5 (14.0)	2.5 (14.2)
1966	2.2 (9.8)	2.0 (14.6)
1967	3.5 (2.8)	4.4 (4.4)
1968	2.7 (0.4)	2.8 (1.6)
1969	2.2	2.8
Average	6.9	6.7

[a] Percentages of the total catch which were stocked fish are shown in parentheses.
[b] 117 northern pike per hectare were stocked in December 1963.

TABLE 3.—*Percentage distribution of the harvest, total harvest, and average length of northern pike caught in Murphy Flowage, 1955–1969.*

Total length (mm)	Native			Stocked		
	Summer	Winter	Total	Summer	Winter	Total
254–355	2.3	0.5	1.8	1.4	48.2	6.1
356–455	25.1	8.9	20.5	69.3	25.0	64.9
456–555	40.5	37.7	39.7	24.9	16.1	24.0
556–655	23.0	38.5	27.4	3.8	7.1	4.1
655–755	7.7	12.8	9.1	0.6	3.6	0.9
756+	1.4	1.8	1.5			
Total fish	4,987	1,979	6,966	505	56	561
Average length	526	569	538	439	505	447

per hectare in December, 1963. The stocking program was followed by an increase in total and natural mortality and a decrease in exploitation rate during the first 2 to 3 years. After 3 years, total mortality was at pre-stocking levels; however, exploitation remained low and was compensated for by increased natural mortality.

Based on a comparison with other waters (Snow and Beard 1972; Groebner 1964; Johnson and Peterson 1955), annual exploitation in Murphy Flowage could be considered above average during the first four years of this study (1955–1958), average or slightly above during the next five years (1959–1963), and well below average after northern pike were stocked (1964–1969). The variation in exploitation is closely related to the variation in fishing pressure and standing stock.

Except for 1964–1966, when high natural mortality occurred after the stocking of northern pike, u and v were linearly and negatively correlated ($p < 0.01$; Fig. 2). The coefficient of determination (r^2) indicates that exploitation accounts for 76% of the variability in natural mortality.

The v intercept ($u = 0$) on Figure 2 is 60%, which is an estimate of average natural mortality in the absence of angling. If the years after winter drawdowns (1968 and 1969) are ignored as possible sources of abnormal mortality (which seems to be the case because of lower fishing pressure and exploitation), the revised regression (1955–1963) has a v intercept of 54%. These two figures—54% and 60%—bracket a best estimate of natural mortality in the absence of angling for northern pike larger than 356 mm. The estimation of natural mortality in the absence of fishing (in addition to that reported for brook trout by McFadden [1961]) has also been determined for the smallmouth bass population in Oneida Lake, New York (Youngs 1972; and Forney 1972). To my knowledge, this is the first such estimate for northern pike.

These data do not indicate the maximum allowable mortality that can occur without fear of depleting the stock. However, during most years of study the standing stock of fish over 356 mm (Fig. 1) increased,

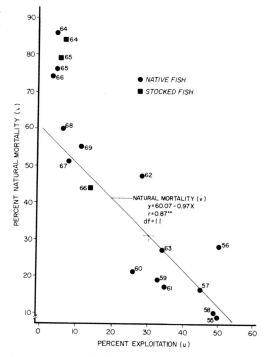

FIGURE 2.—*Relationship between annual exploitation rate and natural mortality of northern pike in Murphy Flowage. (The points in the upper left corner, which include native fish 3 years, and stocked fish 2 years, after stocking 117 fish per hectare, are not considered in the correlation calculations.)* **$P < 0.01$.

particularly during and immediately following the years of highest exploitation (1955–1963); total mortality during this time of increase averaged 60% and reached as high as 78% in one year (1956).

Angling Parameters and Stock Density

The regression of harvest rate of northern pike on stock density indicates that there was an intensive relationship between these parameters (Fig. 3). The rate of increase of fishing success decreased with increasing stock density. At low levels of stock density an increase or decrease in density had a much greater effect on fishing success than an equivalent change at high densities. These results point to the need for knowledge of the density of a fish population before any management program aimed to improve fishing success is undertaken. For example, stocking of catchable size fish may result in little in-

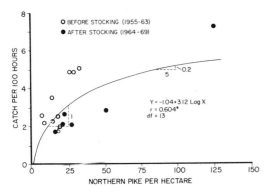

FIGURE 3.—*Relationship between population density in spring and the harvest rate of northern pike during the open water season immediately following.* *P < 0.05.

This relationship implies that at any given level of fishing pressure exploitation rate is higher at low stock densities than high stock densities. This is partially a consequence of the decreasing fishing success rate with increasing stock density shown in Figure 3. It theoretically could also be a consequence of the physical limitations of fishing. It takes time to bait hooks, cast and retrieve, fight a fish to the creel, to unhook it and bag it, and time to move from one fishing spot to another, etc. There is a maximum number of fish that can be landed in a given time. Therefore if there are more fish present a smaller percent of them will be landed in that time and exploitation rates will be proportionately lower.

crease in catch rate in high population densities, while at low densities there could be a vast improvement.

There was an inverse density-dependent relationship of exploitation to population density and fishing pressure (Figure 4).

Based on these data it was concluded that stock density is an important determinant of harvest rate and that stock density and fishing pressure are important determinants of exploitation.

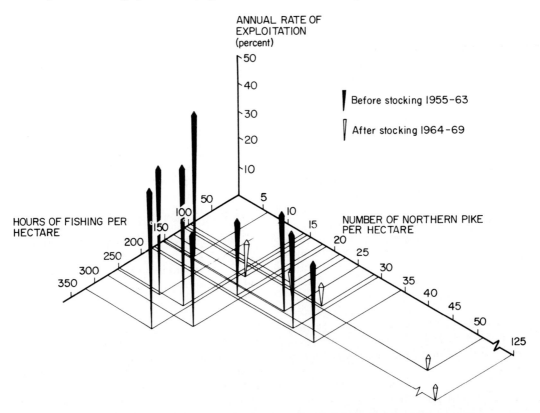

FIGURE 4.—*A three-dimensional portrayal of the inverse density-dependent relationship of annual exploitation to angling intensity and density of northern pike. Observations are based on the 1955–1969 seasons when there were no bag, season, or size limits in effect. Stocked and native fish are combined.*

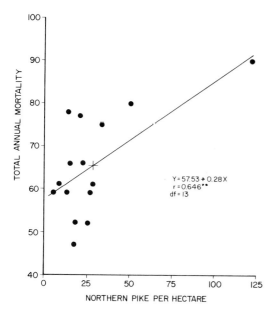

FIGURE 5.—*Relationship between total mortality and population density of northern pike in Murphy Flowage 1955–1969.* **$P < 0.01$.

The relationships shown in Figures 3 and 4 are to my knowledge the only examples of commonly accepted theoretical situations for a recreational coolwater fishery. Although the relationship in Figure 4 has been shown for a stream trout fishery (McFadden 1961; Hunt et al. 1962) this is the first time it has been shown for a northern pike population.

Population Density and Mortality

It has previously been shown that losses of northern pike from disease and emigration were density dependent and positively correlated the first three to five years after stocking northern pike in Murphy Flowage (Snow 1974). The regression of total mortality (losses from all causes) on population density for all 15 years of study was significant at the 0.01 level (Fig. 5). However, when data from 1964, the first year after stocking, were excluded, the relationship was not significant (r values declined from 0.646 to 0.407). Therefore I concluded that total mortality was not density dependent within the range of population density observed for native northern pike, and that it was the high mortality occurring the first year after stocking 117 northern pike per hectare that made the relationship significant.

Stock Density and Prey

Between 1955 and 1970, the standing stock of adult native northern pike varied from 6 to 41 per hectare. Many factors could be related to these variations of stock density; however, one of the most important is the available food supply.

Population estimates of all major fish species each spring and intensive electrofishing surveys in spring and fall showed that bluegills were the dominant prey species. The percentage occurrence of bluegills in northern pike stomachs was 50% in Murphy Flowage and could be as high as 72% if fish remains are considered as bluegills (Johnson 1969). The average size of bluegills consumed was 70 mm and 84% were within the 50–80-mm size range. There was an average of 3 years difference in age between bluegills of this preferred size and bluegills of the smallest size group for which population estimates were calculated (115–140 mm). The relationship of northern pike abundance to bluegill abundance 3 years later was significant at the 0.01 level (Fig. 6). However, if the three or four years of highest abundance of bluegills were not included, the relationship shown in Figure 6 would not be significant. Therefore, prey may contribute to preda-

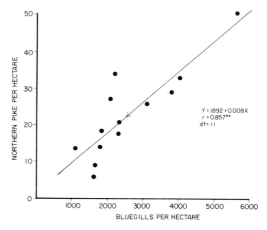

FIGURE 6.—*Regression of northern pike population density on density of 115–140 mm bluegills three years later.*

tor abundance only above some threshold density of prey. Conversely, at lower prey levels, predators may control the abundance of prey species. Since northern pike are often stocked with the intention of controlling panfish, knowledge of the size distribution and relative abundance of both prey species and native predator species is extremely important.

The maximum population density that a northern pike population can attain before factors other than available forage become limiting is not entirely known. However, it is likely below the maximum density attained in this study. At the time of northern pike stocking in December 1963, the standing stock of northern pike was increased from 30 to an estimated 148 fish per hectare. Prey abundance increased to maximum levels from 1962 to 1965. If northern pike were regulated entirely by available forage, they should have remained more abundant after stocking than they did. However, the estimated population of stocked and native northern pike declined by 58% in 12 months. Anglers only accounted for 5% of the decline. These results suggest that the maximum carrying capacity was about 40 to 50 northern pike/hectare (20 kg/hectare for fish over 356 mm). Among the homeostatic mechanisms seemingly involved were territorial behavior (a space factor), which resulted in downstream migration mainly of stocked fish, and a parasitic outbreak affecting stocked and native fish; both of these mechanisms have been shown to be density dependent (Snow 1974).

Another consideration concerning the abundance of the northern pike population is the presence of predators other than northern pike. Numerical abundance of largemouth bass over 200 mm was about equal to or slightly less than that of northern pike, but the biomass was about one-half that of northern pike. About two-thirds of the diet of largemouth bass between 1959 and 1964 was crayfish (Snow 1971). The only other predator has been a very small population of stocked muskellunge. Therefore northern pike was the only major predator of bluegills in Murphy Flowage.

I conclude that except when northern pike densities were artificially high, the availability of suitable forage was the most important determinant of the population density of northern pike in Murphy Flowage.

Acknowledgments

I am indebted to the following individuals who contributed directly to the study: Thomas D. Beard; Leon D. Johnson; Lyle Groth; Donald Stafford; Ronnie Masterjohn; Alvin Johnson; Ingvald Tronstad; and Jon Peterson. Thanks are also due to many other individuals in Research, Fish Management, and Law Enforcement, and Rusk County officials who provided equipment and assisted at various times. Warren Churchill and Donald Thompson advised on statistical procedures.

The manuscript was critically reviewed by Gordon R. Priegel, Lyle M. Christenson, Robert Hunt, and Ruth Hine.

This research was supported in part from funds supplied by the Federal Aid in Fish Restoration Act under the Dingell-Johnson program.

References

BEARD, T. D. 1973. Overwinter drawdown: impact on the aquatic vegetation in Murphy Flowage, Wisconsin. Wis. Dep. Nat. Resour. Tech. Bull. 61. 14 pp.

CHURCHILL, W., AND H. E. SNOW. 1964. Characteristics of the sport fishery in some northern Wisconsin lakes. Wis. Dep. Nat. Resour. Tech. Bull. 32. 48 pp.

FORNEY, J. L. 1972. Biology of smallmouth bass in Oneida Lake, New York, N.Y. Fish Game J. 19(2):132–154.

GROEBNER, J. F. 1964. Contributions to fishing harvest from known numbers of northern pike fingerlings. Minn. Dep. Nat. Resour. Sect. Fish. Invest. Rep. 280. 24 pp.

HUNT, R. L., O. M. BRYNILDSON, AND J. T. MCFADDEN. 1962. Effects of angling regulations on a wild brook trout fishery. Wis. Dep. Nat. Resour. Tech. Bull. 26. 58 pp.

JOHNSON, F. H., AND A. R. PETERSON. 1955. Comparative harvest of northern pike by summer angling and winter darkhouse spearing from Ball Club Lake, Itasca county, Minnesota. Minn. Dep. Nat. Resour. Sect. Fish. Invest. Rep. 164. 14 pp.

JOHNSON, L. D. 1969. Food of angler-caught northern pike in Murphy Flowage. Wis. Dep. Nat. Resour. Tech. Bull. 42. 26 pp.

McFADDEN, J. T. 1961. A population study of the brook trout, *Salvelinus fontinalis*. Wildl. Monogr. 7. 73 pp.

Ricker, W. E. 1975. Computation and interpretation of biological statistics of fish populations. Bull. Fish. Res. Board Can. 191. 382 pp.

Snow, H. E. 1971. Harvest and feeding habits of largemouth bass in Murphy Flowage, Wisconsin. Wis. Dep. Nat. Resour. Tech. Bull. 50. 25 pp.

———. 1974. Effects of stocking northern pike in Murphy Flowage, Wisconsin. Wis. Dep. Nat. Resour. Tech. Bull. 79. 20 pp.

———. 1978. A 15-year study of the harvest, exploitation, and mortality of fishes in Murphy Flowage, Wisconsin. Wis. Dep. Nat. Resour. Tech. Bull. 103. 22 pp.

———, and Thomas D. Beard. 1972. The northern pike of Bucks Lake, Wisconsin, including evaluation of an 18.0 inch size limit. Wis. Dep. Nat. Resour. Tech. Bull. 56. 20 pp.

Youngs, W. D. 1972. Estimation of natural and fishing mortality rates from tag recaptures. Trans. Am. Fish. Soc 101:542–545.

An Evaluation of the Muskellunge Fishery in Cave Run Lake, Kentucky

JAMES R. AXON

Division of Fisheries
Department of Fish and Wildlife Resources
Frankfort, Kentucky 40601

Abstract

Cave Run Lake, 3,347 hectares, was impounded in 1974 on the Licking River. The river had a native population of muskellunge, but an additional 0.3 fish/hectare were stocked above the dam in 1973. Since then, annual stockings of the lake have occurred at the rates of 1.1–3.1 fish/hectare; lengths of stocked fish have been 102–356 mm. The largest planting was in 1974: 10,445 fish of 102–305 mm length. Yearly standing crops of muskellunge in coves have ranged between 0 and 0.7 fish (0.04–0.7 kg)/hectare. In 1975, anglers took 56 muskellunge (214 kg) of legal size (762 mm minimum length) at a rate of 1 fish/58 h. In 1976, these statistics improved to 1,029 fish, 4,140 kg, and 1 fish/48 h. Muskellunge provided 21% by weight of the total angler harvest that year. The 1977 muskellunge take was 478 fish (2,300 kg) at a rate of 1 fish/82 h. The 1974 year class provided 68% and 78% of the muskellunge harvest in 1976 and 1977, respectively. Muskellunge reach legal length between ages II and IV in Cave Run Lake, compared with ages III–VI in Kentucky streams. Carp and gizzard shad were the only food items identified in stomachs of muskellunge.

Cave Run Lake, a 3,347-hectare U.S. Army Corps of Engineers flood-control reservoir, was created in the spring of 1974 (Fig. 1). Part of the Licking River and tributaries in Bath, Menifee, Morgan, and Rowan counties in Northeast Kentucky was impounded by the lake. Seasonal pool fluctuation in the lake is 1.83 m; maximum and mean depths at normal summer pool are about 19.81 and 4.57 m. Thermal stratification occurs throughout the lake from June through September each year. Lake transparency is rather good during this time of year, as Secchi disk readings were recorded between 1.5 and 5.2 m in 1974–1976.

A native population of muskellunge inhabited the Licking River drainage where Cave Run Lake is now located, prior to impoundment in 1974. Natural reproduction is assumed to still occur above the lake in North Fork Creek and Licking River. The standing crop of muskellunge in 1967–1969, in the area where the lake was later formed, ranged from 1.2 kg/hectare in Beaver Creek to 3.9 kg/hectare in North Fork Creek (Brewer 1975). Total fish standing crop fluctuated between 40.0 kg/hectare in North Fork Creek and 53.2 kg/hectare in the Licking River (Turner 1964).

In anticipation of Cave Run Lake, the Kentucky Department of Fish and Wildlife Resources designated this lake to be managed for a muskellunge fishery. The Minor E. Clark Fish Hatchery, located immediately below Cave Run Dam, was the site for native broodfish spawning and fingerling muskellunge production as of 1973. The first stocking of muskellunge above Cave Run Dam was in 1973, and included 936 fish that measured 76–279 mm total length (Table 1).

Methods

Fish population and standing crop data were obtained by sampling 0.4–12-hectare cove areas, which were partitioned by a block net and treated with 1 mg/liter of rotenone (2.5% solution). Two coves, with a total area of 1.5 hectare, were sampled in 1974–1975; 2.1 hectares were sampled in 3 coves during 1976; 2.8 hectares were sampled in 3 coves during 1977. All fish were identified, measured to the nearest 2.45-cm group according to total length, and weighed during the first day of the 3-day cove studies. Weights were not taken during the second and third days. Scales were taken from muskellunge, which were measured to the nearest 0.25 cm total length, to determine age and growth.

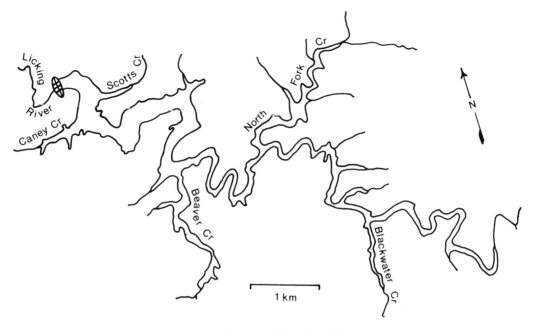

FIGURE 1.—*Map of Cave Run Lake.*

Muskellunge were captured in Cave Run Lake by fishing four gill nets in October 1974, and in mid-summer and October during the three following years. Each net was 91.4 m long, 2.4 m deep, and contained a bar mesh of either 25, 38, 51, or 64 mm. These nets were usually fished during a 3-day period in the Caney Creek arm of the lake. Nets were positioned perpendicular to the shoreline and immediately above the thermocline during the mid-summer netting period. In October, nets that were at a depth less than 6 m were fished on the lake bottom since the lake was no longer thermally stratified. Stomachs from muskellunge that did not survive being netted were visually observed for food items; muskellunge total length, weight, and scale samples were also obtained.

A year-round mail-in survey was implemented in May 1975 that rewards anglers with a fish pin and certificate when they submit a self-addressed questionnaire envelope with 4 post-pectoral scales from each muskellunge they harvest. Envelopes are available to anglers at boat docks, tackle shops, and by request to the Department. The self-addressed envelope has a set of questions about the fish length and weight, date and location of the catch, fishing intent and method, and name and address that the angler must answer.

A non-uniform probability creel survey (Pfeiffer 1966) was utilized at Cave Run Lake for estimating fish harvest and other characteristics of the entire fishery and the muskellunge fishery. The survey was conducted between 7 a.m. and 7 p.m. from March through October, 1975–1977. Cave Run Lake was divided into 2 areas that were simultaneously surveyed by a creel clerk in each area. The duration of a creel survey period was 6 h, from 7 a.m.–1 p.m. or 1 p.m.–7 p.m., and occurred once each week. Four of 6 h in a period were desig-

TABLE 1.—*Muskellunge stocked in the Cave Run Lake area since Cave Run Lake Dam was constructed in 1973. The lake, having a surface area of 3,346.7 hectares at summer pool, was impounded in 1974.*

Year of stocking	Year class	Number of muskellunge stocked	Total length (mm)	Number of fish stocked per hectare
1973	1973	936	76–279	0.3
1974	1973	235	254–356	0.1
	1974	10,445	102–152	3.1
1975	1974	73	305	
	1975	3,589	152–203	1.1
1976	1976	7,949	152–178	2.4
1977	1977	7,990	171	2.4

nated for interviewing anglers and 2 h were allowed to make an instantaneous count of all anglers.

Muskellunge age and growth were determined from scales mailed in by anglers. The Dahl-Lee direct proportion method was used to determine growth of muskellunge.

Fish Population Studies

Annual fish population and standing crop estimates in cove sampling areas at Cave Run Lake during 1974–1977 varied from 2,845 to 18,667 fish/hectare and 94 to 176 kg/hectare, respectively (Table 2). Fish standing crop in the lake is similar to that in other eastern Kentucky flood-control reservoirs (90-250 kg/hectare) but lower than western Kentucky reservoirs (220–440 kg/hectare). Muskellunge were collected in cove fish population samples at Cave Run Lake in 1974–1975, all of which belonged to the 1974 year class. During each year, the muskellunge population was 0.7 fish/hectare; standing crop was 0.04 kg/hectare in 1974 and 0.73 kg/hectare in 1975.

Cave Run Lake provides a good population of forage fishes for muskellunge. Certain species of fish, that were classified as potential muskellunge forage, had standing crops of 68–142 kg/hectare during 1974–1977. Gizzard shad and carp ranked 1 and/or 2 among the forage fish group in 1974–1976; Bigmouth buffalo was second to gizzard shad in standing crop during 1977. Gizzard shad and carp were the only fish identified in stomachs of muskellunge.

The 1974 year class dominated gill net catches of muskellunge during 1974–1977; 18 of 24 muskellunge caught were of this year class. Catch frequency of this year class increased during each netting period through 1975, after which the catch decreased. By the first netting period in 1976, these age II+ fish were near or had surpassed the minimal creel length of 762 mm.

TABLE 2.—Fish population and standing crop data obtained from cove-rotenone studies at Cave Run Lake in 1974–1977.

Measurement	Year			
	1974	1975	1976	1977
Total cove area sampled (hectares)	1.53	1.53	2.09	2.81
Fish population (no./hectare)				
Total	18,667	3,738	2,845	6,266
Muskellunge				
0–761 mm	0.7	0.7	0	0
>761 mm	0	0	0	0
Standing crop (kg/hectare)				
Total	176.5	94.2	104.4	159.6
Muskellunge				
0–761 mm	0.0[a]	0.7	0	0
Gizzard shad	38.0	16.1	22.1	39.9
Carp	18.8	14.9	32.9	33.7
Commercial fishes[b]	29.9	19.5	18.0	54.1
Panfishes[c]	31.3	17.8	15.5	14.0

[a] 0.04 kg/hectare.
[b] Commercial fishes include smallmouth and bigmouth buffalo; highfin carpsucker; golden, river, and shorthead redhorse; and spotted sucker.
[c] Panfishes include rock bass, bluegill, green sunfish, longear sunfish, and warmouth.

TABLE 3.—Year-class distribution of muskellunge from mail-in survey returns and creel survey data in 1975, 1976, and 1977 at Cave Run Lake.

Year class	Mail-in returns				Creel survey results			
	1975	1976	1977	Total	1975	1976	1977	Total
1967	1			1				
1968	1			1				
1969	1	2	1	4				
1970	2	3	3	8				
1971		7	7	14			20	20
1972	13	21	20	54				
1973	26	92	68	186	56	328	85	469
1974		259	309	568		701	372	1,073
Total	44	394	408	846	56	1,029	478	1,563

Mail-in Survey

Muskellunge returns from the mail-in survey in 1975–1977 were not indicative of the total harvest in Cave Run Lake, but the survey did provide data on year class distribution, age and growth, lake location of catch, and method of fishing. Muskellunge returns increased from 1975 through 1977. Six year classes were represented in each year's angling returns (Table 3). A sudden improvement of the muskellunge fishery in August and September, 1975, was attributed to the 1972 and 1973 year classes, many of which became legal size (762 mm) during these months. In 1975, 26 of 44 returns were of the 1973 year class. This was the first contribution to the fishery of stocked muskellunge.

The muskellunge fishery vastly improved

in 1976 due to the availability of many legal-size fish from a strong 1974 year class, which provided 65% of the total returns. Thirty-eight percent of the estimated muskellunge harvest from creel survey data in 1976 (399 of 1,029 fish) were recorded during the mail-in survey, which was considered an excellent response by anglers.

Muskellunge returns in 1976 steadily increased from 3 in March to 140 in July. During this time, the 1974 year class was beginning to enter the fishery. This year class made up a larger percent of returns each month from April (5%) through August (83%). The second largest contribution to the 1976 mail-in survey returns was the 1973 year class.

The 1974 year class muskellunge continued to dominate mail-in survey returns, 76% of the total, in 1977. The 1975 year class was expected to enter the catch in 1977, but none of this age group was among the mail-in survey returns, possibly because the 1975 stocking was relatively small (Table 1). Eighty-five percent of the estimated muskellunge harvest from the 1977 creel survey (408 of 478 fish) were recorded during the mail-in survey.

Area distribution of catch for 819 muskellunge returns in 1975–1977 varied only slightly among the 4 major areas of Cave Run Lake: North Fork, 225 fish; Main lake, 208 fish; Licking River, 196 fish; Beaver Creek, 190 fish.

Muskellunge Fishery

The muskellunge fishery in Cave Run Lake began to develop in 1975, the second year of impoundment, and continued to progress through 1976 before declining in 1977 according to creel survey statistics (Table 4). During 1975, a harvest of 56 muskellunge that weighed 214 kg was

TABLE 4.—*The muskellunge fishery at Cave Run Lake (3,346.7 hectares) during March 1 through October 31, 1975–1977, based on a non-uniform probability creel survey.*

Date of creel survey		Hours fished by muskellunge anglers	Number of muskellunge fishing trips	Muskellunge harvest		Number of fish harvested by muskellunge anglers	Hours fished per muskellunge harvested by muskellunge anglers	Percent success of muskellunge anglers
				Number	kg			
March	1975							
	1976	1,163	314	18	66	18	65.8	5.6
	1977	1,289	260	0				
April	1975							
	1976	3,531	811	114	498	0		
	1977	3,450	605	100	533	100	34.5	16.5
May	1975							
	1976	1,689	355	67	240	49	34.9	13.9
	1977	3,918	987	88	467	88	44.5	4.5
June	1975							
	1976	3,430	2,147	248	1,091	216	15.9	10.1
	1977	13,462	4,001	122	490	122	110.7	3.0
July	1975	65	65					
	1976	8,954	6,462	223	1,129	223	40.2	9.0
	1977	4,064	1,318	20	132	20	202.9	1.5
Aug.	1975							
	1976	16,002	4,947	205	696	205	78.2	4.1
	1977	3,928	1,098	99	459	99	39.6	9.0
Sep.	1975	2,223	578	56	214	56	57.5	9.7
	1976	2,631	799	108	372	97	27.2	12.1
	1977	3,814	1,148	49	221	49	78.5	4.2
Oct.	1975	929	263					
	1976	1,514	328	10	48	10	145.3	3.2
	1977	5,222	1,496	0				
Total	1975	3,217	905	56	214	56		
	1976	38,918	12,165	1,029	4,140	818		
	1977	39,148	10,912	478	2,301	478		
Mean	1975						57.5	6.2
	1976						47.6	6.7
	1977						82.0	4.0

estimated. All of these fish were caught in September and belonged to the 1973 year class. The average size of these fish was 813 mm and 3.83 kg. There were 905 fishing trips made by muskellunge anglers, of which 6.2% were successful. Harvest rate by muskellunge anglers was 1 fish/39.7 h in September and 1 fish/57.5 h for the year.

The 1974 year class was responsible for the great improvement in the muskellunge fishery in 1976. Anglers harvested 1,029 muskellunge that weighed 4,140 kg in 1976, which convert to a harvest rate of 0.3 fish and 1.23 kg/hectare. Weight of harvested muskellunge represented 21% of the total weight of fishes creeled in 1976. The mean length and weight of creeled muskellunge were 828 mm and 4.02 kg. The 1974 year class contributed 701 fish to the harvest; the remaining 328 fish were from the 1973 year class. Muskellunge were caught at a rate of 1 fish/47.6 h; the best monthly catch rate was 1 fish/15.9 h in June. Fishing pressure for muskellunge increased from 0.9 h/hectare in 1975 to 11.6 h/hectare in 1976. Of 12,165 muskellunge fishing trips in 1976, 6.7% were successful.

The 1973 year class represented the entire creel survey catch of muskellunge in March and April 1976. This year class continued to contribute to the fishery through July, but did not appear in the catch thereafter. The first fish from the 1974 year class entered the creel in May and sustained the fishery through October, at which time the 1976 creel survey ended. During May-October, a large number of 1974 year class fish became legal size (762 mm), which resulted in a harvest increase from 67 fish in May to 284 in June, 223 in July, 205 in August, and 108 in September. Seventy-eight percent of the muskellunge harvested from May-September were of the 1974 year class.

The 1977 muskellunge harvest in Cave Run Lake was less than in 1976; harvest of 478 muskellunge (0.14 fish/hectare) weighing 2,301 kg (0.69 kg/hectare) was estimated. This was considered an underestimate, since 408 fish were known to have been caught according to mail-in survey results. The mean length and weight of creeled muskellunge increased to 887 mm and 4.82 kg in 1977. Fishing pressure by muskellunge anglers was 11.7 h/hectare, only 0.1 h/hectare higher than in 1976. It took 82.0 h to harvest a muskellunge, compared to 47.6 h the previous year. Four percent of fishing trips were successful. The best monthly harvest rate and percent success were 34.5 h/fish and 16.5% in April.

The decline of creeled muskellunge in 1977 is primarily attributed to the lack of recruitment by the 1975 year class fish into the legal size group. The 1974 year class was still the major contributor to the catch in 1977: 78%. The percent of mail-in survey returns and harvest of muskellunge that belonged to the 1974 year class were within 2% of each other in 1976 and 1977.

A minimum survival estimate of 1974 year class muskellunge to age II was 10.2%, based on an accumulated harvest of this year class during 1976–1977. Any additional harvest of 1974 year class fish after 1977 would add to this value. The 1,073 muskellunge that were harvested from the 1974 year class were assumed to be of the 10,518 fish stocked in 1974 and 1975, not from natural reproduction.

One of the considerations of the muskellunge stocking evaluation in Cave Run Lake was the effect, if any, of such a large predator stocking on other fish populations and their fishery. The only noticeable effect was an indirect one on black bass. Fishing pressure by black bass anglers increased from 27.0 h/hectare in 1975 to 37.8 h/hectare in 1976, but fell in 1977 to 16.9 h/hectare. The decline of the bass catch rate from 0.35 fish/h in 1975 to 0.12 fish/h in 1976 was largely responsible for the decline in fishing pressure on bass during 1977. However, part of the reduction in pressure was influenced by the increased fishing interest for muskellunge. The population of black bass, 241 mm or longer, expanded from 7 fish/hectare in 1976 to 12/hectare in 1977 as bass fishing pressure decreased. Consequently, the catch rate for black bass improved from 0.12 fish/h in 1976 to 0.16 fish/h in 1977.

Muskellunge Age and Growth

Scales were examined from muskellunge recorded during the mail-in survey to

TABLE 5.—*Back-calculated total lengths (mm) for muskellunge of the 1967–1974 year classes, based on scales returned by anglers during the 1975–1977 mail-in survey.*

Year class	Number of fish	Age							
		I	II	III	IV	V	VI	VII	VIII
1974	568	330	587						
1973	185	287	574	744	981				
1972	64	260	451	654	813	981			
1971	14	243	422	554	780	893	996		
1970	8	198	376	535	671	779	921	1,054	
1969	4	223	442	586	746	859	936	1,017	
1968	1	228	495	528	661	803	922	1,019	
1967	1	243	488	589	691	790	907	963	1,021

determine age and growth in Cave Run Lake (Table 5). The 1967–1971 year-class fish grew to legal size of 762 mm between ages III and V, which is typical growth in Kentucky streams. The 1972 year class grew at a similar rate during the first 2 years of growth; but in 1974, the first year of impoundment for Cave Run Lake, growth accelerated. Fish from this year class became of legal length between ages III and IV. The length mode of muskellunge from the 1973–1974 year classes reached 762 mm between ages II and IV.

The first year growth of muskellunge from the 1974 year class was greater than for the previous year classes because of the newly impounded habitat. Increased growth rate for all year classes during 1974 demonstrates the impact of Cave Run Lake on muskellunge growth.

References

BREWER, D. L. 1975. Musky studies. Dingell–Johnson F-31-R(1–6). Ky. Dep. Nat. Resour., Frankfort. 105 pp.

PFEIFFER, P. W. 1966. The results of a non-uniform probability creel survey on a small state-owned lake. Proc. Annu. Conf. Southeast. Assoc. Game Fish Comm. 20:409–412.

TURNER, W. R. 1964. Pre- and post-impoundment studies. Ky. Dep. Fish Wildl. Resour. Job Prog. Rep., Dingell-Johnson F-16-R-5. Ky. Dep. Nat. Resour., Frankfort. 63 pp.

The Muskellunge in Lake St. Clair

ROBERT C. HAAS

Michigan Department of Natural Resources
Lake St. Clair Great Lakes Station
33135 South River Road, Mt. Clemens, Michigan 48045

Abstract

The population of muskellunge in Lake St. Clair has been exploited by intensive sport fishing for many years. Growth data from trap-netted muskellunge showed that females did not reach maturity until about 914 mm total length, which is equal to Michigan's size limit. Mean total lengths of females were greater than males after age V. Growth rates of Lake St. Clair muskellunge are very similar to those in the St. Lawrence River and the average size in the Lake St. Clair sport catch has apparently not changed in 40 years. A mail survey of ardent muskellunge fishermen in 1972 showed that 74 anglers caught 1,273 fish in 1,017 days of fishing. An analysis of the Michigan-Ontario Muskie Club catch and of tag returns from the general sport fishery showed substantial north to south movements during June of about 40 km. Tag recoveries also indicate that there are separate stocks of muskellunge inhabiting the east and west areas of the lake.

There have been several investigations of the muskellunge in Lake St. Clair (Williams 1948, 1961; Krumholz 1949). This species is relatively scarce, difficult to capture in survey gear, and dispersed over a large area. The majority of information has been collected by observing the sport fishery.

Lake St. Clair has a long history of excellent sport fishing for many species including yellow perch, walleye, smallmouth bass, northern pike, and muskellunge. An angler contact creel census conducted in 1966 and 1967 (Michigan Department of Natural Resources, unpublished) provided an estimate of 1,500,000 angler days per year and a catch rate of 3–4 fish per angler day for Lake St. Clair. Recent Michigan mail survey data provide confirming evidence that the annual effort on Lake St. Clair and the connecting rivers has been 1,000,000 angler days and that 3–4 million fish of various species have been caught each year by Michigan anglers (Jamsen 1973, 1976, 1977). Yellow perch and walleye contributed about 70% to this catch.

Study Area

Lake St. Clair and the St. Clair and Detroit Rivers make up the connecting waters between Lake Huron and Lake Erie (Fig. 1). Approximately two-thirds of Lake St. Clair is within Canada. The total length of this waterway, including the connecting rivers, is 137 km and the surface area is 1,110 km². Lake St. Clair has a maximum depth of 6.4 m, a volume of 3.4 km³ and a mean flow of about 5,350 m³/s (Leach 1973). The water level varies 0.3–0.9 m annually.

Methods

Trap nets were set in the north area of the lake during May and June from 1969 to 1974 to catch brood stock for potential hatchery production of muskellunge. The nets were set in a 0.5-km² area which was approximately 1.6 km from shore (Fig. 1).

Scale samples and data on total length and sex were taken from netted muskellunge. Scales were impressed in cellulose acetate and read on a microprojector. All fish in good condition were tagged with plastic dart tags and released at the net site. The tags were imbedded in the dorsal musculature and secured between pterygiophores of the dorsal fin. Records of the Michigan-Ontario Muskie Club for years 1968–1976 provided data on the size and areal distribution of muskellunge in the sport catch.

Growth

Table 1 gives age and length data for 329 muskellunge that were captured in trap nets. Females were significantly longer than

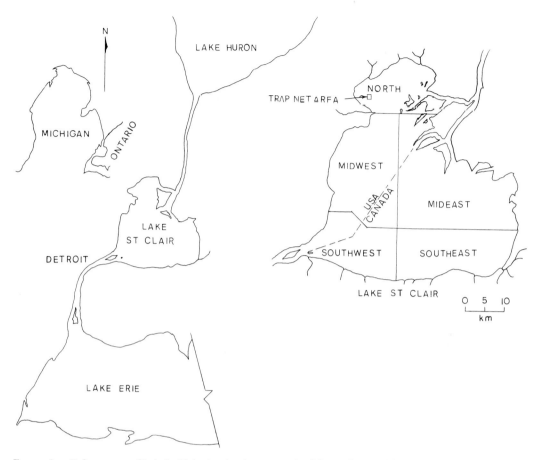

FIGURE 1.—*Reference map of Lake St. Clair, showing the areas used to delineate the sport fishery, and connections to adjacent Great Lakes.*

males beyond age VI ($P < 0.05$). The sample size for younger females was too small to determine whether a difference also exists for the younger age groups. The mean lengths, sexes combined, are very similar to Hourston's (1952) data on mean lengths of age groups (total lengths converted from fork lengths) for muskellunge from the eastern area of Canada (a majority of fish were from the St. Lawrence River). The Lake St. Clair fish are larger than those reported from the central and western

TABLE 1.—*Total length in mm of various ages of muskellunge caught in trap nets in Lake St. Clair.*

	Age group							
	III	IV	V	VI	VII	VIII	IX	X
Males								
Number		4	26	60	62	31	6	
Mean length[a]		853 ± 166	861 ± 25	907 ± 18	988 ± 16	1,003 ± 26	975 ± 76	
Females								
Number			6	4	18	53	18	3
Mean length[a]			927 ± 93	919 ± 66	1,059 ± 31	1,140 ± 19	1,176 ± 25	1,214 ± 95
Total								
Number	5	9	37	69	88	91	26	4
Mean length[a]	533 ± 100	749 ± 91	864 ± 25	907 ± 16	1,001 ± 14	1,087 ± 20	1,128 ± 41	1,171 ± 148

[a] Means ± 95% confidence limits.

areas of Canada (Hourston 1952) and for three inland lakes in Wisconsin (Johnson 1971).

The following length-weight regression was calculated for 296 muskellunge, sexes combined, captured in trap nets from 1969 to 1974 (W = weight in g; L = total length in mm): $\log W = 3.0162 \log L - 5.2377$. Williams (1961) determined the following length-weight relation for a large number of Lake St. Clair muskellunge caught during the period 1939 to 1960: $\log W = 2.9969 \log L - 5.2202$.

The more recent relationship implies that fish are slightly heavier, which may be due to spring sampling when the fish are in spawning condition or may reflect a real change in the length-weight relationship.

Maturity

A sample of 189 mature male muskellunge from trap nets averaged 949 mm total length and 102 mature females averaged 1,113 mm. Only 3% of the mature females were less than the present minimum size limit (in Michigan) of 914 mm. The smallest mature male was 737 mm and 34% were less than 914 mm. The males apparently reach maturity 2-3 years younger than females.

Water temperatures at the netting site varied between 11 and 16 C during the netting periods of 1969-1974. A total of 70 obviously ripe females were trapped and the weighted mean water temperature associated with the day of their capture was 14 C. A summary of daily water temperatures near the netting site from 1965 through 1974 (data for 1965-1967 are from City Water Department, New Baltimore, Michigan, unpublished) showed that 14 C is normally reached in the netting area between May 21 and May 27. Scott and Crossman (1973) reported that the optimum spawning temperature for muskellunge was 13 C.

The presence of numerous ripe females at the optimum spawning temperature is strong evidence that the netting area was actively being used for spawning. However, the physical characteristics of the area are quite different from the normal muskellunge spawning habitat described by Scott and Crossman (1973). The Lake St. Clair area is in open water with a minimum depth of 3 m and has very little or no vegetation until midsummer. Water- and wind-induced currents are quite strong.

Migration and Sport Fishery

The muskellunge sport fishery is very mobile, utilizes both Michigan and Ontario waters, and has well organized lines of communication. It had developed to the point in the late 1930's that interested boat liveries began compiling catch records. Williams (1961) summarized information on 3,000 fish that had been caught in the lake between 1939 and 1960. He demonstrated a difference in the average size of fish caught at the north and south ends of the lake and speculated that mature fish moved north during the spring to spawn, where they were quite vulnerable to sport fishing. He thought that the greater average size reflected a high percentage of mature (spawning) fish.

The Michigan-Ontario Muskie Club, which was formed in 1958, today probably represents the majority of ardent muskellunge fishermen. They have kept length, weight, and location records for all fish entered in their annual contest, which provide a comparison of the catch from year to year.

During the 1950's there was concern among Michigan biologists and some Muskie Club members that the liberal size limit and open season were not adequately protecting the fish (Williams 1961). In 1971 the Muskie Club voluntarily adopted a 914-mm minimum size restriction which raised the average size in their catch. (Michigan's statutory size limit was 762 mm until 1976 when it was also raised to 914 mm. Ontario's size limit remains at 762 mm.) The mean total lengths of the Muskie Club's registered catch for years 1968 through 1976 (except 1970 when fishing was banned due to mercury contamination) are shown in Table 2.

The size of muskellunge in the sport catch in Lake St. Clair has apparently not changed greatly since the late 1930's. The

TABLE 2.—Mean total lengths in mm of muskellunge caught in Lake St. Clair by members of the Michigan-Ontario Muskie Club.

	Year							
	1968	1969	1971	1972	1973	1974	1975	1976
Number caught	343	424	285	156	75	108	99	117
Mean length[a]	972 ± 15	968 ± 28	1,036 ± 12[b]	1,033 ± 25	1,064 ± 27	1,054 ± 20	1,063 ± 20	1,035 ± 18

[a] Means ± 95% confidence limits.
[b] Michigan-Ontario Muskie Club adopted a 914-mm size limit compared to the previous 762-mm limit.

average length in the north area was 1,013 mm during the period 1939–1954 and the average length in the south was 914 mm during the period 1956–1960 (Williams 1961).

The size of muskellunge in the Lake St. Clair fishery appears to resemble that of the St. Lawrence River more closely than other fisheries reported in the literature, which was expected since both groups belong to one semidiscrete Great Lakes population (Scott and Crossman 1973). Hasse (1976) examined 798 sport-caught muskellunge from New York's waters of the St. Lawrence River and found an average total length of 998 mm. Similar studies from New York's Chautauqua Lake (Heacox 1946; Mooradian and Shepherd 1973) found average lengths of sport-caught muskellunge of 897 mm and 856 mm, respectively. Hourston (1952) presents data from muskellunge caught by anglers in three areas of Canada also showing a larger average total length in the St. Lawrence River area (958 mm) compared to the other two areas sampled (central, 827 mm; western, 841 mm).

The 1973 through 1976 records of the Michigan-Ontario Muskie Club, which are the only data available on the muskellunge sport catch, are shown in Table 3. The mean total length of these muskellunge caught in each of the three areas did not vary significantly ($P > 0.05$) during this time period. However, the mean length of the fish caught during 1968 in the north area (1,090 mm) was significantly greater ($P < 0.05$) than the mean lengths for the southwest (954 mm) and southeast (991 mm) areas (Table 2). These 1968 data are similar to those of Williams (1961) which showed a larger mean size in the sport catch at the north end of Lake St. Clair. There are two probable reasons for no difference between north and south in the later catch records. In 1969 Michigan enacted a closed season between December 15 and the first Saturday in June of the succeeding year, which protected larger muskellunge at the north end from being caught before dispersing from the spawning area. Secondly, the adoption of the 914-mm size limit precluded registration of smaller fish being caught in the southwest and southeast areas.

A closer look at the distribution of the Muskie Club catch indicates that muskellunge have an extensive seasonal movement pattern, as had been suggested by Williams (1961). The specific locations of capture have been summarized by five lake areas to facilitate review (Fig. 1). The majority of fish (97%) came from three areas of the lake: north (16%); southwest (47%); and southeast (34%). The dates of individual muskellunge capture for each area were compared to see if fishing success varied with time. The mean day of capture by area varied significantly ($P < 0.05$), showing that the successful fishery was moving temporally from north to south, probably reflecting movements of both fish and fishermen. Since these club members

TABLE 3.—Total length in mm and chronology of capture of muskellunge caught in three areas of Lake St. Clair by members of the Michigan-Ontario Muskie Club during years 1973 through 1976.

	North area	Southwest area	Southeast area
Number of fish	63	190	137
Mean length[a]	1,084 ± 30	1,056 ± 15	1,037 ± 15
Mean day of capture[a]	July 9 ± 10 days	July 28 ± 5 days	Aug. 23 ± 4 days

[a] Means ± 95% confidence limits.

TABLE 4.—*Chronology of capture of tagged muskellunge by fishermen in three areas of Lake St. Clair during years 1969 through 1977.*

	North area	Southwest area	Southeast area
Tagged from trap nets[a]			
Mean day recaptured[b]	June 15 ± 5 days	July 24 ± 10 days	
Angler-tagged			
Mean day recaptured[b]	June 13 ± 10 days	Aug. 15 ± 28 days	Aug. 11 ± 31 days

[a] All fish tagged during second half of May.
[b] Means ± 95% confidence limits.

are well informed muskellunge fishermen, their pattern of fishing activity is likely to represent general fish distribution.

The recoveries of tagged muskellunge by anglers provide good evidence of considerable north-to-south movement following spring spawning. A total of 357 trap-netted fish were tagged in the north area and 50 have since been recovered, 43 by fishermen and 7 in trap nets, during subsequent years at the tagging site. Table 4 shows the mean day of recapture for the north and southwest areas. The majority (88%) of the sport recoveries came from these two areas which are separated by a distance of 42 km. The means for the two areas were significantly different ($P < 0.05$) indicating a general southward movement of the fish after the middle of June. Most angler recoveries of net-tagged fish in the north area (75%) occurred between June 6 and June 14 (none recovered after June 24) while the same percentage in the southwest area occurred between June 14 and September 12. The southward movement probably takes place very shortly after the usual mid-May through early June spawning period. None of the trap net tags were recovered in the southeast area in spite of the intensive sport harvest of muskellunge (Table 3). The 1966–1967 creel census showed that fishing effort for combined species was extensive in all five areas of Lake St. Clair from June through October (Michigan Department of Natural Resources, unpublished) so that tagged muskellunge would be vulnerable to capture in all areas throughout the fishing season.

A few ardent fishermen tagged and released 530 sport-caught muskellunge between 1969 and 1975, using the same type of tag as was used for netted muskellunge. Almost all of these fish were tagged in the north (28%), southwest (43%), and southeast (24%) areas of Lake St. Clair. There have been 32 (6%) recoveries of which 20 were recaptured in the tagging area. The mean day of recapture by area is shown in Table 4. The time of tagging and recapture were significantly earlier ($P < 0.05$) in the north area compared to either the southwest or the southeast areas, implying that during June both fish and fishermen involved in tagging were concentrated in the north area.

Since a number of these angler-tagged fish were marked and released in the southwest and southeast areas, they provide additional insight into muskellunge movement patterns. The fish tagged in the southwest area either were recaptured there during summer months or were taken in the north area during subsequent spawning seasons. None were taken in the southeast area. The muskellunge tagged in the southeast area were all recaptured in or very near the southeast area and none were taken in the north area. This evidence strongly suggests that there are at least two groups of muskellunge in Lake St. Clair: one inhabits the west side of the lake and utilizes the north area for spawning while another group probably inhabits the east side and possibly spawns in the mideast area. Ripe or spawning muskellunge have not been reported by members of the Muskie Club in areas of the lake other than the north.

Seasonal movements of this magnitude (40 km) in native muskellunge populations may be unique to Great Lakes waters. Crossman (1956) reported that tagged muskellunge in a river in Ontario moved very little, if at all, during summer and had highly variable movements not exceeding 1.6 km during the fall. Muir and Sweet (1964) reported that muskellunge netted in an Ontario river traveled mean distances of 2.9–4.3 km when transplanted to other provincial waters.

There is evidence that fishing pressure has substantially increased since the 1940's, as would be expected from human population growth and increased leisure time. A Lake St. Clair creel census run in 1942 and 1943 (Krumholz and Carbine 1943, 1945) provided an annual estimate of total fishing effort of about 300,000 angler days. The present level of effort is at least three times greater (Jamsen 1973, 1976, 1977).

A rigorous attempt to estimate total sport harvest of muskellunge from Lake St. Clair has never been made; however, in 1972 a mail survey was made of 240 members of the Michigan-Ontario Muskie Club. Only 50% of the members responded after receiving two reminders and 33% of the respondents did not fish during 1972. The 74 active respondents reported catching 1,273 muskellunge in 1,017 days of fishing.

Acknowledgments

I wish to thank all the members of the Michigan-Ontario Muskie Club for allowing me to use their catch records.

References

CROSSMAN, E. 1956. Growth, mortality, and movement of a sanctuary population of maskinonge (*Esox masquinongy* Mitchell). J. Fish. Res. Board Can. 13(5):599–612.
HASSE, J. J. [1976?]. A comparison of the St. Lawrence River muskellunge fishery to other muskellunge waters with a recommendation on regulation changes. N.Y. State Dep. Environ. Conserv. 13 pp. (Mimeo.)
HEACOX, C. 1946. The Chautauqua Lake muskellunge research and management applied to a sport fishery. Trans. 11th N. Am. Wildl. Conf.: 419–425.
HOURSTON, A. S. 1952. The food and growth of the maskinonge (*Esox masquinongy* Mitchell) in Canadian waters. J. Fish. Res. Board Can. 8(5): 347–368.
JAMSEN, G. C. 1973. Michigan's 1972 sport fishery. Mich. Dep. Nat. Resour. Surv. Stat. Serv. Rep. 122. 6 pp.
———. 1976. Michigan's 1975 sport fishery. Mich. Dep. Nat. Resour. Surv. Stat. Serv. Rep. 156. 7 pp.
———. 1977. Michigan's 1976 sport fishery. Mich. Dep. Nat. Resour. Surv. Stat. Serv. Rep. 165. 7 pp.
JOHNSON, L. D. 1971. Growth of known-age muskellunge in Wisconsin: and validation of age growth determination methods. Wis. Dep. Nat. Resour. Tech. Bull. 49. 24 pp.
KRUMHOLZ, L. A. 1949. Length-weight relationship of the muskellunge, *Esox m. masquinongy* in Lake St. Clair. Trans. Am. Fish. Soc. 77:42–48.
———, and W. F. CARBINE. 1943. The results of the cooperative creel census on the connecting waters between Lake Huron and Lake Erie in 1942. Mich. Dep. Conserv. Fish. Res. Rep. 879. 24 pp.
———, and ———. 1945. Results of the cooperative creel census on the connecting waters between Lake Huron and Lake Erie, 1943. Mich. Dep. Conserv. Fish. Res. Rep. 997. 24 pp.
LEACH, J. H. 1973. Seasonal distribution, composition and abundance of zooplankton in Ontario waters of Lake St. Clair. Proc. 16th Conf. Great Lakes Res. 1973:54–64.
MOORADIAN, S. R., AND W. F. SHEPHERD. 1973. Management of muskellunge in Chautauqua Lake. N.Y. Fish Game J. 20(2):152–157.
MUIR, B., AND J. SWEET. 1964. The survival, growth, and movement of *Esox masquinongy* transplanted from Nogies Creek Sanctuary to public fishing waters. Can. Fish Cult. (32):31–44.
SCOTT, W. B., AND E. J. CROSSMAN. 1973. Freshwater fishes of Canada. Bull. Fish. Res. Board Can. 184. 966 pp.
WILLIAMS, J. E. 1948. The muskellunge in Michigan. Mich. Conserv. 17(10):10–11, 15.
———. 1961. The muskellunge in Lake St. Clair. Mich. Dep. Nat. Resour. Instit. Fish. Res. Rep. 1625. 30 pp.

GENERAL MANAGEMENT CONCEPTS

The Incremental Method of Assessing Habitat Potential for Coolwater Species, with Management Implications

KEN D. BOVEE

Cooperative Instream Flow Service Group
U.S. Fish and Wildlife Service
Fort Collins, Colorado 80521

Abstract

The IFG incremental method assesses the effects of stream flow regimes on fish communities. It utilizes one or more hydraulic simulation procedures to determine the distribution of depth, velocity, and substrate within a channel at different discharges. A composite probability of use for each combination of depth, velocity, and substrate is determined for each life stage of each species under study. A weighted usable area, roughly equivalent to the physical carrying capacity of the stream reach, is then determined for each month of the year. The weighted usable area may then be used to interpret changes in both standing crop and species composition due to changes in the hydraulic features of the stream.

At first glance, there appear to be few common features among various riverine habitat alterations such as stream dewatering, flow augmentation, channelization, bank stablization, habitat improvement, or sedimentation. Each appears to be a unique problem, requiring a unique solution. However, each of these problems involves some alteration of river hydraulics, and the responses of different aquatic species to those changes. Thus, it is possible to utilize a standard methodological approach in the solution of any of these problems.

The Incremental Method was developed by personnel of the Cooperative Instream Flow Service Group, U.S. Fish and Wildlife Service, Fort Collins, Colorado. The IFG incremental method allows quantification of the amount of potential habitat available for a species and life history phase, in a given reach of stream, at different streamflow regimes with different channel configurations and slopes.

This method is composed of four components: (1) simulation of the stream; (2) determination of depths, velocities, substrates, and cover objects, by area; (3) determination of a composite probability of use for each combination of depth, velocity, substrate, and cover (where applicable) found within the stream reach, for each species and life history phase under investigation; and (4) the calculation of a *weighted usable area* (roughly a habitat's carrying capacity based on physical conditions alone) for each discharge, species, and life history phase under investigation.

Stream Reach Simulation

Several hydraulic simulation techniques, with varying input data requirements and levels of accuracy, are routinely used in assessment of instream flow requirements. However, the family of hydraulic simulations most promising in the assessment of channel manipulation is generally termed "backwater curve" calculation.

Several computer programs are available, which can predict the hydraulic parameters of depth, velocity, width, and stage for different discharges. The version utilized by the Bureau of Reclamation (Anonymous 1968) is termed "Pseudo," while the Corps of Engineers (Anonymous 1976) has a series of backwater curve programs titled "HEC."

Regardless of the title, all backwater curve or "water surface profile" calcula-

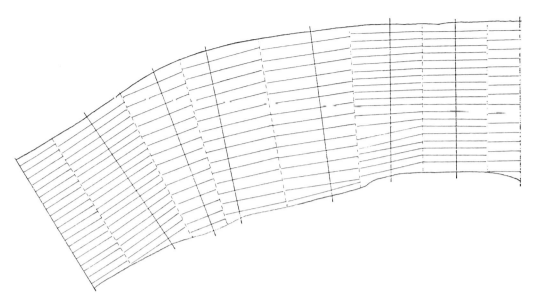

FIGURE 1.—*Computer conceptualization of a simulated stream reach. Hydraulic parameters of depth, velocity, and substrate for each major transect subdivision are assigned to the area of each subdivision block.*

tions utilize Manning's equation,

$$Q = N^{-1}R^{2/3}S^{1/2}A,$$

where

Q = discharge (m³/s);
R = hydraulic radius (m), or cross-sectional area divided by the wetted perimeter of the stream (roughly equivalent to mean depth);
S = energy gradient, assumed parallel to slope of the water surface;
A = cross-sectional area (m²);
N = roughness coefficient, which may be calculated from stream measurements, or estimated from a description of bed materials, channel uniformity, and channel shape.

Since $Q = VA$, Manning's equation may be restated as

$$V = N^{-1}R^{2/3}S^{1/2},$$

where V is mean velocity (m/s).

The stream reach simulation utilized by IFG uses several cross-sectional transects, each of which is subdivided into 9 to 20 subsections. The computer program then treats each subsection as an essentially separate channel. For any stage (water surface elevation), the mean depth and velocity of each subsection may then be calculated. An area represented by these values of depth and velocity is calculated by multiplying the width of the subsection by half the distance to the next transect upstream and the next downstream. This representation is illustrated in Figure 1.

The output of the stream reach simulation is in the form of a multidimensional matrix showing the surface area of stream having different combinations of hydraulic parameters (i.e., depth, velocity, substrate, and cover when applicable). Table 1 illustrates a depth-velocity matrix, although the computer is not limited to two dimensions. The numeral in the upper left-hand corner of the matrix refers to 585 m² per km of stream having a combination of depths less than 0.3 m *and* velocities less than 0.15 m/s. This is the total summation of areas within the stream reach with that combination of depths and velocities. These areas are not necessarily contiguous.

In order to evaluate the magnitude of impacts caused by changes in stream hydraulics, it is necessary to develop an information base for each species or group of species of interest. Such an information base is termed biological criteria.

Biological criteria are primarily aimed at those parameters affecting fish distribu-

TABLE 1.—Distribution of depth-velocity combinations, expressed as m² per km of stream, and (in parentheses) weighted usable areas (m²/km) for adult smallmouth bass.

Depth (m)	Velocity (m/s)								
	<0.15	0.15–0.30	0.30–0.45	0.45–0.60	0.60–0.75	0.75–0.90	0.90–1.05	>1.05	Totals
<0.30	585 (22)	78 (4)							663 (26)
0.30–0.45	270 (24)	141 (15)		123 (14)	51 (5)	18 (<1)	18 (0)	279 (0)	900 (58)
0.45–0.60	87 (11)	114 (17)	96 (15)	132 (21)	324 (38)	237 (5)	114 (0)	516 (0)	1,620 (107)
0.60–0.75	18 (3)	87 (17)	69 (15)	27 (6)	333 (53)	393 (11)	429 (0)	525 (0)	1,881 (105)
0.75–0.90	18 (4)	45 (11)	165 (44)	237 (63)	123 (24)	192 (7)	123 (0)	315 (0)	1,218 (153)
0.90–1.05	27 (7)	51 (15)	45 (15)	36 (9)	96 (23)	9 (<1)	447 (<1)		711 (69)
1.05–1.20	27 (9)	60 (23)		51 (21)	141 (43)	51 (3)	246 (3)		576 (102)
1.20–1.35		60 (29)		33 (17)	150 (58)	105 (7)	51 (1)		399 (112)
1.35–1.50		33 (22)		15 (11)	345 (189)	60 (6)			453 (228)
1.50–1.65				21 (20)	69 (50)	45 (6)			135 (76)
1.65–1.80		30 (27)			93 (68)	60 (8)			183 (103)
Totals	1,032 (80)	699 (180)	375 (89)	675 (182)	1,725 (551)	1,170 (53)	1,428 (4)	1,635 (0)	8,739 (1,139)

tion which are most directly related to streamflow and channel morphology: depth, velocity, temperature, and substrate. Cover, a habitat parameter of paramount importance to many species, is also indirectly related to streamflow. Cover may be incorporated into an assessment by evaluating the usability of available cover objects in reference to the flow parameters around them.

Species for which biological criteria are being developed are roughly divided into five classes.

(1) *Management-objective species* are sport and game fishes considered important and desirable by the public, and of importance to the objectives of the state management agencies.

(2) *Indicator species* are those with narrow habitat tolerances, which inhabit areas of streams which are particularly sensitive to changes in flow. It is assumed that if conditions are suitable for the indicator species, all other species will also have suitable habitat.

(3) *Rare and endangered species* are those which may be locally abundant, but with a highly restricted distribution, or those which occupy much of their former distribution, but in greatly reduced numbers.

(4) *Important non-game species* are those which may act in direct competition with game or sport species.

(5) *Forage species* are organisms occupying intermediate positions in the food chain, including both forage fish and aquatic invertebrates.

The criteria with which this method is concerned are termed probability criteria. The assumption is that the distribution and abundance of any species are not primarily influenced by any single parameter of stream flow, but related by varying degrees to all streamflow parameters. Furthermore, it is assumed that individuals of a species will tend to select the most favorable conditions in a stream but will also use less favorable conditions, with a lower probability of use as conditions become less favorable.

Given a sufficient number of observations and measurements, it is possible to determine a species' preferences within a certain parameter, such as depth. Based on this information, it is also possible to calculate the relative probability that the species will utilize some positive or negative increment of that parameter which falls outside of its preferred range.

Most flow assessment methodologies in current use address only one, or occasionally two life history stages. Frequently, a particular life history stage, or a certain time period is singled out as being critical for the continued well being of a fish population. For example, spawning success is commonly considered a critical factor in the maintenance of a fish population, but habitat evaluations for fry and juvenile fish are almost universally neglected. However, under the incremental method, probability criteria are developed for all life stages (Bovee and Cochnauer 1977).

Probability of Use

The composite use probability of any combination of hydraulic conditions encountered in the study reach may be determined from the individual probability-of-use curves for each species and life stage.

Figure 2 gives probabilities of use by adult smallmouth bass for depths and velocities. For a given increment of each parameter the use probability is read directly from the curve. For example, the use probability for the depth increment of 105 cm is 0.37. The use probability for the velocity increment of 15 cm/s is 0.81. The composite use probability for adult smallmouth bass for a depth of 105 cm *and* a velocity of 15 cm/s is 0.37 × 0.81 = 0.30. A composite probability is similarly calculated for each stream reach subsection.

Substrate or cover may also be incorporated into this determination of composite probability following the procedure detailed above. In the preceding example, if the substrate found with that combination of depth and velocity had a probability of use of 0.90, then the composite probability of use for that combination of depth, velocity, *and* substrate would be 0.37 × 0.81 × 0.90 = 0.27.

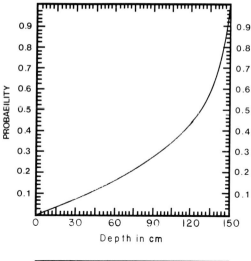

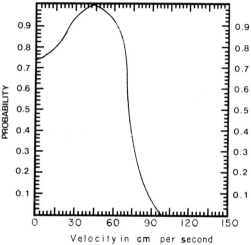

FIGURE 2.—*Probability-of-use curves for adult smallmouth bass.*

Weighted Usable Area

The weighted usable area is defined as the total surface area having a certain combination of hydraulic conditions, multiplied by the composite probability of use for that combination of conditions. This calculation is applied to each cell within the multidimensional matrix.

This procedure roughly equates an area of marginal habitat to an equivalent area of optimal habitat. For example, if 305 m² of surface area had the aforementioned combination of depth, velocity, and substrate it would have the approximate habitat value of only 82.4 m² of optimum habitat.

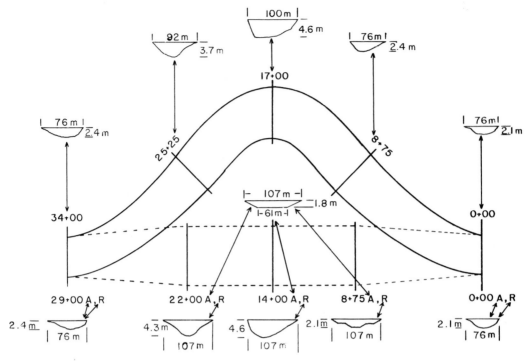

FIGURE 3.—*Plan and cross-sectional views of original (curved) channel, channelized reach, and rehabilitated reach. Station suffixes A and R refer to altered (channelized) and rehabilitated reaches, respectively. The figure is not drawn to scale.*

A worked example of an entire two-dimensional matrix (depth and velocity) is given in Table 1. In each cell of the matrix, the upper numeral refers to the surface area of stream having a certain depth-velocity combination. The numerals in parentheses refer to the weighted usable area. Due to the large number of calculations required for each matrix, a computer program has been developed to make all calculations for as many species as desired from a single set of hydraulic input data (Main 1978).

For each species and life stage, weighted usable area is plotted against various monthly flow regimes, such as median monthly flows or 1-in-10 year monthly low flows. Such plots can show critical time periods for a given life stage, the limiting habitat availability for each life stage (i.e., physical carrying capacity), and the limiting habitat availability for different species. Since changes in hydraulic characteristics will initiate differential species reactions, the method is particularly useful in predicting and quantifying changes in species composition.

Management Implications

Although the incremental method was designed primarily to assess changes in standing crop and species composition due to changes in flow regime, several other applications of the method have been identified. These applications include impact analysis for channel modification projects, cost-efficiency analysis for habitat improvement projects, stream classification and stocking programs, and negotiation of operating schedules for hydroelectric facilities. The example below shows the applicability to a stream channelization and rehabilitation project.

A typical stream meander and a by-pass channelized section, with representative cross-sections for both reaches, are shown in Figure 3. The original channel had an overall hydraulic gradient (water surface slope) of 0.00058. A large pool extended from station 12+25 to station 25+25 with the deepest portion crossed by station 17+00.

In contrast, the hydraulic gradient of the channelized reach is 0.00068. Channel cross-sections at stations 8+75A through 22+00A are uniform trapezoids with a top width of 107 m and a base width of 61 m. The channel is designed for a maximum discharge of 170 m^3/s.

The rehabilitation design for the channelized reach is shown in Figure 3 as sections 8+75R, 14+00R, and 22+00R. Station 8+75A was altered slightly by creating a notched control with raised berms near the outer margin of the cross-section. This modification allowed good water exchange in the pools at low flow, but slowed the pool current at high flows. Stations 14+00A and 22+00A were excavated to provide a pool of approximately the same dimensions as the original behind the new control feature at 8+75A. Because the banks were steepened to create the pool, rock was added to strategic locations to stabilize them.

Figure 4 shows the amount of weighted usable area in the original channel, the channelized reach, and the rehabilitated reach for 5 life stages of smallmouth bass. These areas were derived for a median-year flow regime (i.e., the monthly flow which would occur 50% of all years of record).

The weighted usable area is roughly equivalent to the carrying capacity of a stream reach, based on physical parameters alone. Therefore, it is assumed, that if there are no significant changes in non-physical parameters, there will be a one-to-one ratio between the weighted usable area and standing crop. Percent decreases or increases in weighted usable area may then be directly related to decreases or increases in standing crop. It is further assumed that the standing crop for any life stage of a species is set by the lowest monthly weighted usable area.

Figure 4 illustrates that the habitat area for adult smallmouth bass in the original channel remained fairly constant through the year, although the discharge ranged from less than 5.7 m^3/s to nearly 170 m^3/s. Habitat area for juveniles fluctuated somewhat more, being limited primarily by high discharges during spring and early sum-

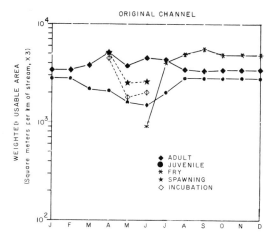

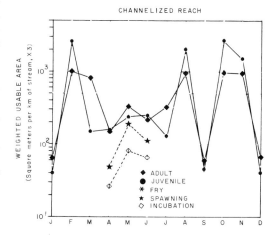

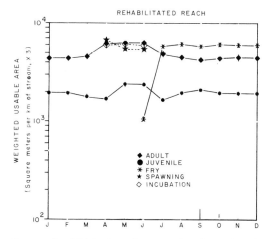

FIGURE 4.—*Weighted usable areas, in m^2 per km of stream, for five stages of smallmouth bass in the original channel, the channelized reach, and the rehabilitated reach.*

TABLE 2.—*Limiting weighted usable areas, and percent change from original conditions for five life stages of smallmouth bass.*

Life stage	Limited weighted usable area (m²)			Percent change	
	Original	Chan-nelized	Rehabil-itated	Chan-nelized	Rehabil-itated
Fry	9,455	0	9,242	−100	−2
Juvenile	15,982	354	15,302	−98	−4
Adult	34,221	531	37,515	−98	+10
Spawning	25,834	433	49,014	−98	+90
Egg incubation	18,666	225	45,811	−99	+75

mer. Black fry are most severely limited during June, when they are first congregating over the nests.

Diversion of the flow from the original channel into the channelized reach would reduce the adult weighted usable area from around 10,200 m² per km of stream to about 195 m² per km. This would relate to a decrease in standing crop of 98%. Juvenile standing crop would decline by 97% due to the channelization. Spawning and egg survival would decrease 98% and 99%, respectively. However, this loss would be immaterial because the habitat for young of the year would be completely eliminated in the channelized reach.

The success of the rehabilitation effort is also shown in Figure 4. Rehabilitation of the channelized section as demonstrated in Figure 3 would give an increase in adult standing crop of about 24% above that of the original channel. Juvenile carrying capacity would be nearly the same in the rehabilitated channel as in the original channel. Spawning and egg incubation would be enhanced considerably in the rehabilitated channel, although these gains would be off-set by a very small increase (16%) in the carrying capacity of fry.

In this example, the calculation of weighted usable area is not the final determination of the impact of channelization, nor the success of rehabilitation. Not only has the habitat per unit length (km) of stream been changed, but the total length of the reach has been decreased from 1,087 to 885 m. Total reach impacts, corrected for loss of reach length, are illustrated in Table 2.

The analysis in Table 2 illustrates that the rehabilitation of the channelized reach could improve the carrying capacity for adults about 10% above the carrying capacity of the original channel. While spawning and incubation would be greatly enhanced, overall recruitment in the rehabilitated channel would be slightly less than in the original channel due to loss of habitat for fry and juveniles.

Literally dozens of channel modifications could have been tested to determine the design which provided the most habitat for the least amount of money. Because of the cross-sectional design, and length between cross sections are known, it is possible to determine the approximate construction cost of the modifications. This method allows the planner to use cost-benefit analysis for habitat improvement projects.

Alternatively, the suitability of another desirable species, such as sauger or channel catfish, could have been tested in the rehabilitated reach. This procedure might have revealed a less expensive channel design which would support a good population of an alternative species. Such analytical capacity would allow the fisheries manager to modify the channel or the dominant species, or both, to arrive at the combination which meets his management goals for the least cost.

References

ANONYMOUS. 1968. Guide for application of water surface profile computer program. U.S. Bur. Reclamation, Sedimentation Section, Hydrology Branch, Denver, Colo. 17 pp. + exhibits.

———. 1976. HEC-2 water surface profiles. U.S. Army Corps of Engineers, Hydrologic Engineering Center, Davis, Calif. 17 pp. + exhibits.

BOVEE, K., AND T. COCHNAUER. 1977. Development and evaluation of weighted criteria—probability of use curves for instream flow methodologies. U.S. Fish Wildl. Serv. IFIP#3, Cooperative Instream Flow Group, Fort Collins, Colo.

MAIN, R. 1978. Guidelines for use of the HABTAT program in instream flow assessments. U.S. Fish Wildl. Serv. IFIP#9, Cooperative Instream Flow Group, Fort Collins, Colo. (In preparation.)

An Historical Review of the Commercial Fisheries of the Boreal Lakes of Central Canada: Their Development, Management, and Potential

George F. Adams

Fisheries Section
Manitoba Department of Renewable Resources and Transportation Services
Winnipeg, Manitoba R3H 0W9

Abstract

The fisheries resource of the numerous boreal lakes on the Laurentian Shield of central Canada traditionally have been important, both economically and socially, to the people in that region. These lakes continue to be valuable producers of highly preferred sport and commercial fish. Historically, the commercial fisheries have sought the dominant species, primarily walleye, lake whitefish, and northern pike. Serious declines in commercial fishing in this region, primarily since 1962 point to a need for more comprehensive fisheries policies and management strategies. Proposals for the management of these fisheries should be drawn first from a clearer understanding of past schemes and their reasons for failure. Effective fisheries development programs, and fisheries management in general, require: a recognition and respect for the value of fisheries resources and the economic implications of this value to fisheries and fishing communities; a sincere effort by fisheries institutions to eliminate the fragmented approach to new initiatives; a willingness to accept the adaptive or experimental approach; and an immediate transfer of insights and information directly to planning and policy-making.

The boreal lakes on the Uplands of the Canadian Shield, north and northwest of the upper Great Lakes, comprise a unique set of natural waters both from the standpoint of their morphological type and their history of commercial exploitation. The commercial fisheries, which have existed in this region since the 1880's, share the fisheries resources with numerous domestic fisheries, a substantial concentration of recreational fisheries (particularly in the accessible areas), and, in recent years, a growing bait-fish industry. Predominant in the commercial fisheries is the native fisherman. These fisheries play an important role in the lifestyle of the Ojibwa and Cree Indians and contribute substantially to their income (Rogers 1972).

Although management and management policy has been diverse over the years (Ryder 1970), the different uses of the fisheries resources have seldom conflicted with each other during their development. Initially, because the primary concern of governments was "economic growth and development," the role of resource management agencies was seen as that of offering initiatives to encourage the undertaking of commercial fishing and angler-outfitting activities. According to Whittington (1974), "the environment was viewed simply as a treasure chest which had to be opened as efficiently and quickly as possible." Because entry into the fisheries was largely uncontrolled, the industry is now facing a perpetuation of economic hardship. Since the 1960's, there has been a rapid decline in harvest and participation in the industry. For the most part, this decline in harvest was not brought about by overfishing and habitat modification, but the result of the following interrelated factors. More labor and capital is engaged in harvesting than is required to attain a given level of production. With fishing costs increasing more rapidly than the value of production, the erosion of profit margins in the industry has resulted in the stranding of labor and capital. The absence of lakeside managerial skills, primarily in the remote north, is an additional factor. Depleted fish stocks and mercury contamination in fish (from industrial and natural sources) are significant factors only to specific fisheries.

Thus, it is apparent that some management policies have been developed without a clear understanding of their implication to fishermen and the fisheries. Traditional harvest control measures, such as shortened seasons and aggregate quotas, have not contributed to the economic health of the fishing industry. Failure to control excess labor and capital can be attributed, in large measure, to public management policies that treat the fisheries resources as common property and fail, or are unable, to exercise proprietary rights (Cauvin 1978, this symposium). The increase, in recent years, of governmentally sponsored social-welfare programs for commercial fishermen (including direct and indirect subsidies), have inhibited natural adjustments and, in some cases, created a liability for fishing communities.

Although there are records of commercial catches since the early 1920's, knowledge of these lakes and rivers and the effects of exploitation on their fish fauna are based on relatively recent observations. Because commercial fishing is economically important to this region, fisheries management strategies must consider economics with the biological implications of exploitation. Before the planners and decision-makers can tell us where we are now and where we are going, an understanding of where we have been with these commercial fisheries and their management is in order. With this in mind, a review of the historical development of these fisheries and management trends should offer insights toward clearer policies and strategies for their management.

Boreal Lakes of Central Canada

The boreal region of Canada is characterized by its profusion of water. With some exceptions, the predominant feature of the landscape is a surface of low relief traversed by a vast number of highly dissected and usually shallow bodies of water. Because of the morphometry, drainages are meandering and broaden frequently into lakes.

Regional Geography

The boreal forested area of the Kazan, Severn, and Abitibi Uplands stretches from Saskatchewan to western Quebec and extends over 13 degrees of latitude (Fig. 1). The forest cover is typical northern conifers dominated by white spruce (*Picea glauca*), black spruce (*P. mariana*), jack pine (*Pinus Banksiana*), balsam fir (*Abies balsamea*) and tamarack (*Larix laricina*). White and red pine (*Pinus strobus* and *P. resinosa*) are characteristic species along the southern margin in Ontario—a region with extensive elements from the adjacent boreal and southern deciduous forests (Rowe 1972). White birch (*Betula papyrifera*) and trembling aspen (*Populus tremuloides*) occur in strong stands throughout the southern reaches of this area.

This entire area (Fig. 1) was overlain by the Laurentide ice sheet during the relatively recent continental glaciation. The lakes, which were formed in the wake of glacial retreat, are underlain by Precambrian rock of the Canadian Shield (primarily granitic gneiss), broken by volcanic and derived metamorphic rocks with strong pockets of sedimentary rocks and large magmatic intrusions. Loose ablation till and esker and morainic ridges were deposited by melting ice with some depressions filled in by outwash sand or by sediments of subsequent glacial lakes (Anonymous 1957).

Lying primarily in the humid microthermal climate zones, the lakes are in a region that annually experiences, from north to south, 1,000 to 3,000 degree-days over 5.6 C (120–180 days), 240 to 150 days of lake ice cover (Anonymous 1974), 400 to 800 mm and 120 to 180 days of precipitation, and 20 to 30 mm of runoff. The period of maximum flow is primarily from May to July.

Postglacial dispersal of species is probably accountable for the paucity of species in the more northern lakes as compared to the southern reaches of this region. Climate has also been shown to influence the northward distribution of species (Radforth 1944; Ryder et al. 1964). Unauthorized introductions through the careless dumping of the contents of angler bait buckets have added to the somewhat greater diversity of fauna found in these more southern waters. According to MacKay (1963), introductions in 1901 and subse-

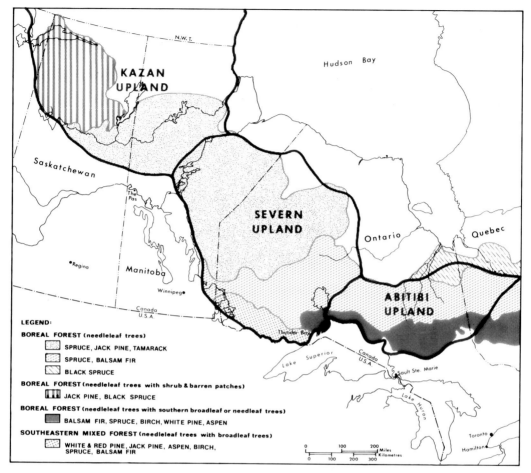

FIGURE 1.—Map of central Canada showing the zones of boreal forest and transition (southeastern mixed) forest on the Kazan, Severn, and Abitibi Uplands of the Laurentian Shield.

quent years by the Ontario Department of Game and Fisheries was the major influence on the dispersal of the small-mouth bass beyond its indigenous range. A greater depth of sedimentary deposits and a longer growing season in the warmer waters offer a habitat that is more suitable for many of the warm water species.

The underlying bedrock and soils and the severity of the climate greatly limit the ability of these lakes to produce fish (Ryder 1972). A feature of most of these lakes which redeems productivity is the extensive littoral zone, including irregular shorelines and great numbers of islands. The lakes appear to share a high degree of ecological similarity based on their geographic setting, their fish fauna, and man's impact upon them. Because of their geographic location, man's influence has been relatively minimal (Ryder and Johnson 1972) as compared to his impact on more fertile lakes to the south (United States). For the most part, the lakes in this region continue to be important producers of highly preferred sport and commercial fish.

The Commercial Fishery

Fishing was important to the native communities of this region even before it became an integral part of the settlement process. The role of the native peoples in this fishery was altered when expansion of the fur industry increased the demand for fish. The encroachment of civilization with the attendant decline in the fur trade forced native communities in the past 30 to 40 years to turn to other means of gaining a livelihood. This often was commercial fishing.

In response to market potential in the northern United States, the commercial fishery began in this region in the 1880's. The character of the harvest, primarily walleye, lake whitefish, and northern pike, was substantially determined by these export opportunities. Boreal lakes do not have a diverse fish fauna, a condition typical of Canadian freshwaters (Ryder et al. 1964; Scott and Crossman 1973). There are 80 species in the waters of this region, including the introduced rainbow trout, brown trout, arctic grayling, goldfish, and carp. These non-indigenous species were dispersed with varying degrees of success and influence on the existing biota. Of the 75 indigenous species, 29 are limited in their distribution. Catch reports by commercial fishermen (excluding bait fisheries) identify 25 species, but few play a major role in the harvest (Adams and Kolenosky 1974).

In general, it can be said, that this fishery expanded northward towards relatively unexploited lakes as reliable transportation permitted.

Historical Development of the Northern Ontario Commercial Fisheries

To develop an historical perspective that is representative to the many scattered and relatively small commercial fisheries on the boreal lakes of central Canada, I will review the development of this industry in northern Ontario. Within the boundaries of this region of Ontario, the myriad bodies of freshwater comprise more than 85,000 km² of surface area (not including Lake Superior), the majority of which is in the boreal forested Uplands. These lakes and rivers constitute 47% of the surface water area of Ontario, including the Great Lakes (Fig. 2).

The Period of Expansion

Fish from the waters of northern Ontario have long been important to the native peoples as a source of food both for themselves and their dogs and as bait for their traps. The first European explorers noted that the availability of preferred fish frequently determined the location of native villages (Judson 1961; Kennedy 1966).

Fishing effort increased materially in particular areas of this region with the expansion of trade routes. The vast majority of fish (primarily lake whitefish) were taken during the fall spawning runs and consumed during the winter months. Local fur trading posts often contracted with the native peoples for this supply of winter fish. With this growth in the fur trade came more efficient methods for taking fish (gill nets, hooks, etc.) and increased interest in summer fishing. While the methods employed were extremely crude, the effect upon many fish populations, especially the fall lake whitefish, lake trout, and brook trout runs (Anonymous 1905), was probably deleterious. However, it was only within the past century that an industry developed from fishing and within the past 30 to 40 years that remote native communities were encouraged to participate in this development.

The earliest report of commercial fishing activity from the northern lakes was 32 metric tons harvested from Lake Nipissing (Fig. 2) in 1885 (Anonymous 1886). Although pound nets were first reported on the Canadian side of the Lake of the Woods in 1893 (first report from the United States portion of this lake was in 1887), there is reason to believe a commercial fishery existed in these waters and in Rainy Lake (Fig. 2) prior to that date (Anonymous 1897; Evermann and Latimer 1910). Commercial fishing started on Lake Nipigon as early as 1898, but was not of any consequence until about 1917. In the Kenora district, harvests were first recorded from Rainy Lake, Eagle Lake, Wabigoon Lake, Butler Lake, and Lake Minnitaki in 1899.

As conditions for exporting improved, other lakes were fished commercially. In these developing fisheries, lake sturgeon initially was preferred (Evermann and Latimer 1910). However, early yields were not sustained because of this species' low reproductive potential, and its importance

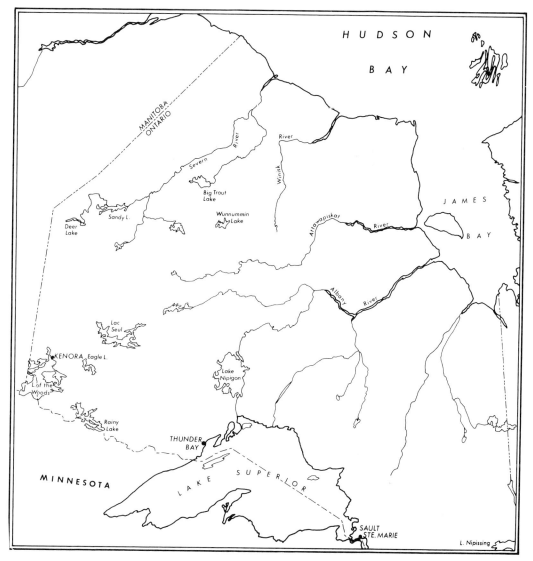

Figure 2.—Map of northern Ontario.

in the catch was short-lived. In 1895, lake sturgeon comprised nearly 50% of the 68.0-metric ton harvest from Lake of the Woods (Canadian waters only), while lake whitefish and walleye accounted for 30% and 14%, respectively. By 1899, slightly more than 20% of the harvest was lake sturgeon. Lake whitefish soon rose in dominance (about 1900), with walleye constituting the second-most important component in the catch.

By the turn of the century, fish buyers were active in the communities of Rat Portage (on Lake of the Woods) and Port Arthur (now Thunder Bay). A brisk trade soon developed, although there were few fulltime fishermen. With the continuance of favourable prices and the successful establishment of better markets in the early 1900's, the industry expanded considerably. Indicative of the active interest in commercial fishing at this time was a 20-year lease that the Canada Fish Company negotiated in 1902 with the Province of Ontario for rights to Lake Nipigon (Anonymous 1902). The fishing license in this lease allowed

454 metric tons annually for the first three years and twice that amount for each year thereafter. Although refrigerated storage and the practice of freezing whole or dressed fish was in widespread use before the turn of the century on the Great Lakes (Frick 1965) and Lake Winnipeg (Judson 1961), the lack of these facilities and the irregularity of train service did little to encourage summer fishing in the Kenora area even as late as 1911 (Anonymous 1911). Soon conditions for exporting the fish improved, and by 1913 an estimated 32 lakes were being fished by the industry in the Kenora and Thunder Bay districts (Fig. 2). References by the Ontario government to the benefits of commercial fishing (employment, and improved fishing by "thinning" of the coarse species) resulted in recommendations that more northern lakes be opened to the industry (Anonymous 1914). Originally, open-water commercial fishing was confined almost entirely to lakes near suitable transportation. The construction of the Trans Continental Railway in 1907 across the top of Lake Nipigon (Fig. 2) provided accessibility to lakes in this formerly remote area (Anonymous 1908).

Although the drive to develop more inaccessible fisheries in northern Ontario often met with failure (because of transportation difficulties, weather, poor handling, and lack of knowledge of the lakes), the number of men entering this fishery increased at a substantial rate. Expansion was not due to population growth, but was prompted by new transportation routes to United States markets. Concerns (Anonymous 1901) about the high fishing pressure on the more accessible lakes did not appear a deterrent to the growth of the commercial fisheries. Increases in licensing were partially based on the belief that World War I would cause food prices to rise (Anonymous 1917; Anonymous 1918). By the mid-1920's newly opened lakes, improvements in transportation and communication (radio), and the first marketing of fresh fish also contributed to this expansion. This decade also witnessed the first successful marketing of suckers (*Catostomus* spp. and *Moxostoma* spp.) and ciscos (primarily *Coregonus artedii*).

Production declines in the 1930's (Fig. 3) can be attributed directly to the decline in fish prices brought about by the worldwide economic depression. Returns to the fishermen were reduced further by the export market's rejection in 1930 of coregonid species for parasitism (*Triaenophorus crassus*) (Judson 1961). Uncertainties in fish marketing have long been a problem with the industry. As early as 1934, Ontario participated with the prairie provinces and the federal department of Fisheries and of Trade and Commerce in a conference solely concerned with the orderly marketing of fish as a means of providing better returns to fishermen. The prevailing uncertainties had encouraged a short-term point of view with most dealers. Although dealers emphasized quality (this was their greatest source of profit), this concern was often forgotten when the demand was strong. Volume output usually brought about severe waste problems because of quality and this usually resulted in losses to the fishermen and not the dealer. The interaction of a demand cycle in the competitive marketplace did not allow high quality to prevail. The result was deterioration in quality when the market was strong and a gradual recovery when the market was weak. Increased participation (Fig. 4), due to the loss of other employment in this region, compounded this economic low by leaving excess labor and capital (far more than required to satisfy the markets) stranded in the industry. Despite the introduction of winter tractors and airplanes, the northward thrust to remote lakes weakened during this decade because of the need for large capital investment. The growing war prosperity in the 1940's brought an increase in demand for freshwater fish, but export price ceilings on fish (1943) and certification requirements for lake whitefish (1944) deterred increases in production (Judson 1961). The Canadian inspection scheme with its emphasis on quality helped to force a longer-term viewpoint on the part of the industry. This, coupled with the growth of filleting, tended to stabilize the market.

Following World War II, higher prices and solid market expectations encouraged exporters to outfit more fishermen. This

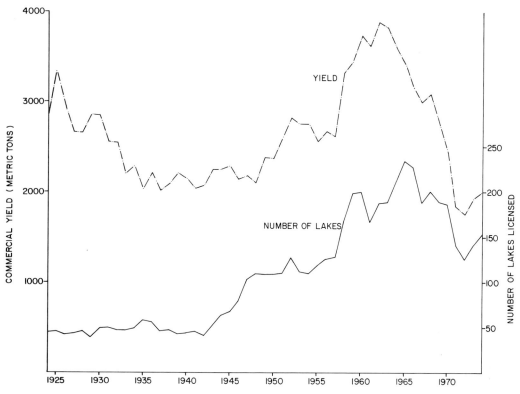

FIGURE 3.—*Comparison of commercial yield and the number of lakes licensed in the commercial fishing industry, northern Ontario, 1924–1974.*

period witnessed substantial increases in capital investment (Fig. 4), although production, at first, did not keep pace. Linen nets were replaced first by cotton nets and then by the more effective nylon nets (McCombie and Fry 1960; Pycha 1962). Gasoline-powered boats and snow toboggans, winter roads, and eventually airplanes speeded transportation of fish in the remote north during this time. Government efforts to assist expansion included programs to rehabilitate war veterans (provision of fishing licenses) and to encourage the native peoples to develop more successful fisheries.

The Native Fisheries

In the northward expansion of the fisheries, the last major area opened to exploitation was the Patricia District (basically north of Lac Seul). This region is inhabited primarily by Ojibwas and Crees, who traditionally moved from place to place throughout the year seeking game and good fishing. While it is true that "commercial fishing" began when natives first sold fish to the fur traders, there was no extensive effort to fish the Patricia lakes commercially until after the 1930's. The first commercial catches reported from this remote area were taken in Sandy Lake in 1935 (lake sturgeon only). It was not until a decade later that other lakes were opened, such as Wunnummin Lake and Deer Lake (Fig. 2). Harvest from the Patricia lakes was subject to greater fluctuations because efficiency was more dependent on transportation and weather conditions. Inadequate market development coupled with transportation costs, a condition which still exists in parts of this region, has resulted in unsold fish (northern pike, suckers, burbot, and sometimes lake whitefish) discarded at the lake.

During the past 30 years, the way of life in this remote region has changed significantly (Rogers 1972). Although native In-

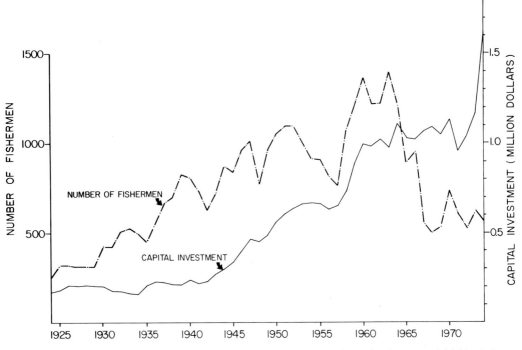

FIGURE 4.—*Comparison of the capital investment and the number of fishermen licensed in the commercial fishing industry, northern Ontario, 1924–1974.*

dians have always fished domestically, their perspective of commercial fishing often has not included a strong incentive toward development of a viable fishing enterprise. Even though present social values may be an inhibiting factor, the presence of a strong incentive is difficult to perceive when many of these fisheries are not within the geographic margin of economic viability and are heavily supported by government subsidies. Because these fisheries are still developing, they are considered to lack the "traditional" elements (experience) that are present in the southern commercial fisheries of this region. Often, commercial fishing by the native peoples is not simply a quest for employment. Fishing parties are usually family groups and the opportunity for the fishermen and his family to camp away from the community, enjoy fresh fish daily, hunt moose, and pick berries is considered an important event in the lives of the people. Because of the poor returns for effort, the younger people in communities where there are commercial fisheries usually have been reluctant to pursue this vocation. Until recently, the native, whenever he has sought a living from the commercial fishery, has been at the mercy of private dealers who have tended to be unscrupulous in their transactions. Government assistance to the native commercial fisherman has been spasmodic and as a rule relatively ineffective, although there have been significant exceptions. As early as 1939, the Indian Affairs Branch of the federal government encouraged the native peoples to produce more fish for sale (Martin 1962). Today, the Department of Indian and Northern Affairs together with the Ontario Ministry of Natural Resources and several other governmental agencies assist these remote fisheries to catch and market a quality product. Assistance has been in the form of fishing gear and equipment including shore installations, instruction in modern fishing and fish handling methods, and, more recently, new technology in processing, a new approach to marketing (creation of the Freshwater Fish

Marketing Corporation), and support programs to encourage marketing of lower-valued species.

The Decline in the Fisheries

Yield and participation in the industry continued to increase until the early 1960's (Figs. 3 and 4), at which time production declined (more than 50% within a decade). Fewer men are pursuing the livelihood and those who do fish, for the most part, do not utilize the harvest potential. In 1970, fishermen in northern Ontario landed only 29% of the allowable harvest of those species (lake whitefish, walleye, northern pike, lake trout, and goldeye) regulated under a catch-quota system (Adams 1972b). Based on lakes fished at that time, Adams (1972b) indicated a harvest potential for the above species of over 3,600 metric tons. There has not been a significant increase in this percentage of allowable harvest since 1970.

In looking at the reasons for decline, it is important to distinguish between the two types of fisheries: those that are economically accessible; and those that are located in the remote north (primarily the native fisheries) and, for the most part, outside the geographic margin of economic viability.

In *remote northern fisheries*, the reasons for declines in yield and participation are not related to depleted fish stocks, but rather to a pervasive tendency for the cost of production to exceed the value of production; to alternate employment opportunities, including government programs of welfare and assistance; to an absence of entrepreneurial skills; and, in some areas, to the discovery of unacceptable levels of mercury in fish.

The remote northern fisheries are characterized by low average incomes. Given the price that fishermen now receive for their fish, this value has not increased commensurate with the cost of operating, transporting, and purchasing new gear and equipment. As a result alternate employment opportunities have usually offered fishermen a better return for their labor.

Although many of these opportunities in remote areas are temporary (e.g. fire fighting, road construction, government tree-planting programs), they tend to be disruptive forces in these fisheries. Government programs of welfare and assistance for fishermen appear to have destroyed initiative in many cases by providing alternate sources of income. With increased costs of operating, fishing can be less remunerative than welfare.

With the discovery in 1970 of unacceptable levels of mercury contamination in fish (from natural and industrial sources), a number of fisheries in this region were closed. The impact of mercury contamination on employment and yield in the fishery was quite variable in this region (Adams 1972a). Some fisheries, such as portions of the Winnipeg, English, and Wabigoon river systems, Lac Seul, and Rainy Lake, faced serious economic circumstances. Many remote fisheries ceased because walleye (the most highly valued species) could not be marketed. According to Adams (1972a), mercury contamination was the major cause for a decline in yield in this region after 1970. The activities of man through the discharge of other deleterious effluents, logging operations, river diversions, water-level control systems, and poor land-use practices have all had varying effects on specific fisheries, but none were quite as noticeable as the effects of mercury contamination.

During this past decade of decline in the commercial fishery, the use of the fishery resource for recreational opportunities have increased, especially in the accessible lakes. While recreational uses have reduced the commercial fishing activities in some areas, the overall effect on the commercial harvest has been minimal in northern Ontario.

Finally, in regard to the native fishery, reliable local leadership with good business and managerial skills often is lacking. The inefficient use of capital in the fisheries is an added factor in destroying the profit margin. This is often compounded by poor handling techniques which reduce the value of the fish landed. Prior to the inception of the Freshwater Fish Marketing Corporation (1969), local fish companies provided this entrepreneurial force.

In *economically accessible fisheries*, declines in yield and participation have been brought about by a decreasing ability on the part of the industry to receive an economic return consistent with the productive potential of the resource. Excessive fishing effort, which has resulted in depleted fish stocks in specific areas (Adams and Olver, 1977), further erodes the profit margin in these fisheries. From an economic point of view, there appears to be too much labor and capital devoted to the production of fish. In other words, a reduction in the labor and capital engaged in harvesting today is unlikely to reduce aggregate production or the value of production.

Trends in Fisheries Management

When commercial fishing commenced in this region, jurisdiction for management was regarded to be the responsibility of the federal government. An officer was first appointed to Lake of the Woods in 1894 (Anonymous 1895). During the last quarter of the nineteenth century, attempts by Ontario to assert its proprietary right under the British North American Act (1867) resulted in the passage of The Fisheries Act (1885), which gave a larger share of fisheries jurisdiction to the province (Lambert 1967). Management prerogatives over the fisheries were not resolved until Ontario assumed management responsibility through the Act of 1907 (Thompson 1974).

As reported by Ryder (1970), early attempts at fisheries management were largely related to protective regulations, which were often implemented without any biological basis, subject to political whim, and sometimes injurious to the fisheries. Control measures placed on the commercial fishery have often increased the cost of fishing by attempting to control exploitation through the introduction of inefficiencies. Fisheries management agencies have traditionally promoted regulations with the fisheries resources in mind. This concern over fish conservation to the exclusion of labor and capital conservation has resulted in economic hardship for the commercial fishing industry. Inadequate protection caused the complete collapse of the lake sturgeon fishery in parts of this region and the serious exploitation of a number of other species, including the walleye (Wright 1892; Anonymous 1912) and lake trout (K. H. Loftus, personal communication).

Attempts to replenish depleted stocks popularized the development of a hatchery program. By 1926, Ontario had 15 hatcheries devoted to fish stocking (MacKay 1930). Unfortunately, enthusiastic support of fish planting resulted in occasional introductions of undesirable exotic species. A lake survey program was initiated in 1925, primarily to assess stocking potential. Despite the lack of positive evidence, accumulated in the late 1930's to the 1950's, to suggest that plantings of native species significantly affected production (Dymond 1956), stock replenishing programs for native species are still supported by governments in this region of central Canada.

Over time, the commercial fishery has become one of the industries most subject to government regulations. Since 1924, each holder of a commercial fishing license has been required to annually submit information on the quantity of fish taken and an estimate of investment in fishing equipment. Because this information did not adequately reveal fishing intensity, the industry was required, in 1945, to commence reporting daily catches (Anonymous 1948).

Quota control of harvest was slowly implemented in this region during the following decade. While proper application of catch-quota control was an important management step toward preventing overutilization by the commercial fishery in relatively infertile lakes, it failed to prevent overinvestment in the industry and decreasing profits to fishermen.

In response to declining conditions, in the 1960's, governments have designed a number of subsidy and incentive programs to enhance fishermen incomes. Although not readily apparent to the casual observer, these programs have further depressed the earning power of this industry and increased the tax burden of the resource owners (society at large). Because of excess labor and capital, contraction in participation would be considered a natural occurrence. Subsidy and incentive programs, however, prevent this attrition and can disguise unemployment in a "nonproductive" activity. This misallocation of labor and capital resources reduces returns from the potentially valuable fisheries resources and has the potential to place further stress on fish stocks. These programs fail to recognize that fish stocks are limited in their capacity to generate economic growth and development, and subsequently become an increasing liability to the fishing communities in this region.

Historically, development of the remote northern fisheries was based on welfare objectives rather than economics, and to this day little attempt has been made to separate the two. Because the native is not traditionally a "commercial fisherman," the initiative to develop has always come from an "outside" entrepreneur. Attempts to induce native fishermen into entrepreneurial roles and schemes involving the introduction of new fishing technology, new methods of processing, and new organizational structures, such as fishermen cooperatives, too often have met with failure because the existing way of life in native communities was neither recognized nor respected. In the long run, this entrepreneurial force would succeed only when it recognized and respected traditional values. Characteristically, these small fishing operations have low average incomes because of high operating and transporting costs. Consequently, their income is insufficient to provide a long-term return from labor and capital investments.

From a strict economic perspective, the fisheries resources in this region are being mismanaged under a policy that does not allow the fishing industry or the general public to realize a net return from harvesting fish. Overinvestment, decreasing returns to fishermen, increasing tax burden, and overexploitation in specific fisheries can be attributed to policies of unrestricted access to the fisheries resources. Resource management agencies display an apparent feeling of futility in overcoming the economic barriers to rational management. Rather than striving to remedy the root cause, the political arena will sell short long-term solutions with a series of "band-aid" attempts to protect and conserve the fisheries resources. For the most part, early stock-replenishing activities were one such attempt. Other efforts include traditional control measures, such as aggregate quotas, limited seasons, and closed areas. None of these measures specifically address the problem of excess labor and capital engaged in fishing and the resultant dissipation of profits to fishermen. Clearly, if policy is to continue to allow fisheries to be managed as common property, there will be a pervasive tendency for the cost of production to exceed the value of production.

Proposals for Future Management

In my opinion, this historical perspective is representative of fisheries development in the boreal region of central Canada and has a number of obvious implications for managing Canada's inland fisheries. These fisheries are important to the people and their economy (Adams and Kolenosky 1974; Regier 1976), and for this reason, the conditions which have precipitated this decline in the industry must be addressed.

Given an understanding of past deficiencies and the current problems facing the management of fisheries resources in this region, it is proposed that effective fisheries programs require (1) a recognition and respect for the value of fisheries resources and the economic implications of this value to fisheries and fishing communities; (2) a sincere effort by fisheries institutions to eliminate the fragmented approach to new initiatives; (3) a willingness to accept the adaptive or experimental approach; and (4) an immediate transfer of insights and information directly to planning and policy-making.

Resource Value

Fisheries resources are finite, and for this reason, there are limits to their capacity to provide economic returns. Fish are a valuable capital asset and as such, are capable of providing resource owners with a net dividend. A management policy which allows this net dividend to be dissipated is placing a zero value (or less than zero) on the primary asset of a commercial fishery—the fish. This approach neither recognizes nor respects the value of fisheries resources. It is proposed that efficiency can be attained in public resource management by development of a policy that will exercise proprietary rights to control fishing effort and will offer a strategy for allocating fish stocks based on the standard cost-benefit criterion of maximizing present values of net economic revenues. The economic implications of this strategy assure that the stock of resources being utilized in fishing (labor, capital, and fish) receive an economic return consistent with the productive potential of the fishery resource. For some fisheries in this region (those with excess labor and capital), this strategy will require a reduction in the number of fishermen.

The acceptance of this present-value strategy has serious implications to remote northern fisheries, where development costs are greater than the value of production. Without government assistance, commercial fishing in many parts of this region will diminish. Development strategies which seeks to build technical and organizational improvements into the way of life of the native (rather than to impose innovations in a revolutionary manner) and can overcome the barriers to labor mobility and adaptability are clearly needed. This is not an open plea for ill-conceived government programs that simply dump money and equipment at the fishermen's feet. Such programs, in effect, apply partial solutions based on an insufficiently comprehensive definition of the problems. The answer to problems associated with developing preconditions for growth must come from an assessment of all resources, the social infrastructure, and the opportunities to stimulate the formation of mobile and adaptable labor.

Fragmented Approach

Adjustments to the current industry infrastructure, which will be necessary over time, must minimize social hardship. Because the socioeconomic framework is likely to undergo continuous change, comprehensive plans and policy must be sufficiently flexible. For this reason, short-term as well as long-term objectives are required. Because traditional institutional arrangements in Canada often have kept fisheries management and research from successful application of their job, it is proposed that sincere efforts be made to jointly plan, initiate, and appraise management and development policy and programs. This is a formidable task when one considers that in Manitoba alone, there are 21 government agencies with programs that directly and indirectly affect fisheries management, many with divergent program objectives. The fragmented approach of a number of government agencies that indirectly involve themselves in fisheries management, through the provision of assistance and incentive programs to commercial fishermen, must be eliminated. There is potential for accomplishing this through joint fisheries management planning.

Experimental Management

The lakes in this region represent a major type of fish habitat capable of future development. Rather than searching for individual differences, management must develop harvest strategies built on indices that consider broad ecological relationships (Northcote and Larkin 1956; Ryder 1965), and must search for similarities in yield properties and fish community structure (Adams and Olver 1977). With an improved understanding of the responses of fish species and fish communities to exploitation, management should be prepared to test hypotheses and extend our knowledge of these systems. This adaptive (experimental) approach (Loftus 1976) may be labelled risky and probably not the best scientific procedure, but until our understanding is sufficient to measure the effects of fishing and to account for natural fluctuations, it should be considered (present economic conditions demand it). It is proposed that the long-term solution for fisheries management depends on enhancing natural production through experimental management programs and developing alternate management options. With this approach in mind, Adams and Olver (1977) suggested "pulse-fishing" as a harvesting strategy to be considered for percid lakes in this region. This practical management experiment, which allows heavy exploitation for one fishing season followed by a moratorium sufficient to allow populations to recover, is presently being conducted in remote boreal lakes in Manitoba. Fishermen operating in this experiment will fish new lakes on a rotational basis for up to 7 years.

Planning and Policy

From history, it is apparent that serious planning and clear policy is required and that reasons for the failure of past schemes offer insights for future public resource management. Management strategies drawn from several of these proposals have been tested here and there locally to meliorate conditions that contributed to earlier difficulties. Insights and information drawn from a reexamination of former and traditional fisheries protocols and development initiatives can be usefully applied to develop programs that would help fishery managers regulate the fisheries resources in the best interest of the commercial fishing industry and the resource owners.

Acknowledgments

I wish to express my appreciation to D. M. Cauvin, P. J. Colby, K. H. Doan, and W. F. Sinclair for their constructive review of the manuscript.

References

ADAMS, G. F. 1972a. Impact of mercury contamination on the Ontario commercial fishery. Ont. Minist. Nat. Resour. Fish. Branch. 31 pp. (Mimeo.)
———. 1972b. Fishery resources of inland northwestern Ontario. Ont. Minist. Nat. Resour. Fish. Branch. 11 pp. (Mimeo.)
———, AND D. P. KOLENOSKY. 1974. Out of the water: Ontario's freshwater fish industry. Ont. Minist. Nat. Resour. Toronto. 68 pp.
———, AND C. H. OLVER. 1977. Yield properties and structure of boreal percid communities in Ontario. J. Fish. Res. Board Can. 34:1613–1625.
ANONYMOUS. 1886. Annual report of the Department of Fisheries, Dominion of Canada, for the year 1885. MacLean, Rogers, Ottawa.
———. 1895. 28th annual report of the Department of Marine and Fisheries. S. E. Dawson, Ottawa.
———. 1897. Report of the Joint Commission relative to the preservation of fisheries and waters contiguous to Canada and the United States. S. E. Dawson, Ottawa.
———. 1901. 34th annual report of the Department of Marine and Fisheries. S. E. Dawson, Ottawa.
———. 1902. 4th annual report of the Department of Fisheries of the Province of Ontario, Toronto.
———. 1905. 7th annual report of the Department of Fisheries of the Province of Ontario, Toronto.
———. 1908. 1st annual report of the Game and Fisheries Department of the Province of Ontario, 1907. L. K. Cameron, Toronto.
———. 1911. 5th annual report of the Game and Fisheries Department. Prov. Ont., Toronto.
———. 1912. 6th annual report of the Game and Fisheries Department. Prov. Ont., Toronto.
———. 1914. 8th annual report of the Game and Fisheries Department. Prov. Ont., Toronto.
———. 1917. 11th annual report of the Game and Fisheries Department. Prov. Ont., Toronto.
———. 1918. 12th annual report of the Game and Fisheries Department. Prov. Ont., Toronto.
———. 1948. Report of the Minister of Lands and Forests of the Province of Ontario for the fiscal year ending March 31, 1947. Sessional Pap. 3. Baptist Johnston, Toronto.
———. 1957. Atlas of Canada. Dep. Mines Tech. Surv., Ottawa. 110 maps.
———. 1974. The national atlas of Canada, 4th ed. Dep. Energy, Mines Resour.; Inf. Can. MacMillan, Toronto. 267 pp.
CAUVIN, D. M. 1978. The allocation of resources in fisheries: an economic perspective. Am. Fish. Soc. Spec. Publ. 11:361–370.
DYMOND, J. R. 1956. Artificial propagation in the management of Great Lakes fisheries. Trans. Am. Fish. Soc. 86:384–392.
EVERMANN, B. W., AND H. B. LATIMER. 1910. The fishes of Lake of the Woods and connecting waters. Proc. U.S. Nat. Mus. 39:121–136.
FRICK, H. C. 1965. Economic aspects of the Great Lakes fisheries of Ontario. Bull. Fish. Res. Board Can. 149. 160 pp.
JUDSON, T. A. 1961. The freshwater commercial fishing industry of western Canada. Ph.D. Thesis. Univ. Toronto.
KENNEDY, W. A. 1966. A history of commercial fishing in inland Canada. Fish. Res. Board Can. Biol. Stn., London, Ont. 37 pp.
LAMBERT, R. S., with R. Pross. 1967. Renewing nature's wealth. Ont. Dep. Lands For. Hunter Rose, Toronto, Ont. 630 pp.
LOFTUS, K. H. 1976. Science for Canada's fisheries rehabilitation needs. J. Fish. Res. Board Can. 33:1822–1857.
MACKAY, H. H. 1930. The present status of fish culture in the Province of Ontario. Trans. Am. Fish. Soc. 60:33–44.
———. 1963. Fishes of Ontario. Ontario Dep. Lands For., Toronto. 292 pp.

MARTIN, H. B. 1962. Commercial fisheries in the Patricia District of Ontario. Vol. 1. Dep. Citizenship Immigration Can.; Indian Affairs Branch; Ont. Dep. Lands For. Res. Branch. 159 pp.

MCCOMBIE, A. M., AND F. E. J. FRY. 1960. Selectivity of gillnets for lake whitefish, *Coregonus clupeaformis*. Trans. Am. Fish. Soc. 89:176–184.

NORTHCOTE, T. G., AND P. A. LARKIN. 1956. Indices of productivity in British Columbia lakes. J. Fish. Res. Board Can. 13:515–540.

PYCHA, R. L. 1962. The relative efficiency of nylon and cotton gillnets for taking lake trout in Lake Superior. J. Fish. Res. Board Can. 19:1085–1094.

RADFORTH, I. 1944. Some considerations on the distribution of fishes in Ontario. Roy. Ont. Mus. Zool., Toronto 25. 116 pp.

REGIER, H. A. 1976. Science for the scattered fisheries of the Canadian interior. J. Fish. Res. Board Can. 33:1213–1232.

ROGERS, E. S. 1972. Ojibwa fisheries in northwestern Ontario. Ont. Min. Nat. Resour., Toronto. 49 pp.

ROWE, J. S. 1972. Forest regions of Canada. Dep. Environ. Can. For. Serv. Publ. 1300. 172 pp.

RYDER, R. A. 1965. A method for estimating the potential fish production in north-temperate lakes. Trans. Am. Fish. Soc. 94:214–218.

———. 1970. Major advances in fisheries management in North American glacial lakes. Am. Fish. Soc. Spec. Publ. 7:115–127.

———. 1972. The limnology and fishes of oligotrophic glacial lakes in North America (about 1800 A.D.). J. Fish. Res. Board Can. 29:617–628.

———, AND L. JOHNSON. 1972. The future of salmonid communities in North American oligotrophic lakes. J. Fish. Res. Board Can. 29:941–949.

———, W. B. SCOTT, AND E. J. CROSSMAN. 1964. Fishes of Northern Ontario, north of the Albany River. Roy. Ont. Mus. Zool., Toronto 60. 30 pp.

SCOTT, W. B., AND E. J. CROSSMAN. 1973. Freshwater fishes of Canada. Bull. Fish. Res. Board Can. 184. 966 pp.

THOMPSON, P. C. 1974. Institutional constraints in fisheries management. J. Fish. Res. Board Can. 31:1965–1981.

WHITTINGTON, M. 1974. Environmental policy. Pages 203–227 *in* G. B. Doern and V. S. Wilson, eds. Issues in Canadian public policy. Macmillan, Toronto.

WRIGHT, R. R. 1892. Preliminary report of the fish and fisheries of Ontario, 1891. Pages 419–476 *in* Commissioners report. Ontario Game Fish Comm. Warwick and Sons, Toronto.

The Allocation of Resources in Fisheries: An Economic Perspective

DENNIS M. CAUVIN

Fisheries and Marine Service
Freshwater Institute, 501 University Crescent
Winnipeg, Manitoba R3T 2N6

Abstract

The absence of a market system in the management of publicly owned fisheries resources has handicapped management agencies in measuring economic values, and, therefore, allocating resources efficiently. Hypothetical proxy values have been used to help estimate trade offs among conflicting objectives and to rationalize the conservation of fish. None has been very satisfactory. The adoption of the price system to allocate scarce resources in fisheries appears to hold the greatest promise of resolving many of our fisheries management problems.

Public resource managers agree that resources must be apportioned, distributed, assigned, or alloted to specific uses or to particular people. For example, land may be allocated to agriculture, forestry, mineral development, or urban development. Water may be allocated to fisheries production, domestic consumption, industrial cooling, irrigation, navigation, or domestic and industrial effluent disposal. Similarly, fisheries resources may be allocated to commercial and/or recreational use.

Resource managers also agree that resources which have alternative uses may be described as "scarce," that is, there is a cost (opportunity cost) associated with the use to which they are allocated. For example, a decision to allocate a body of water exclusively for commercial fisheries production may entail a sacrifice of alternative uses such as recreational fishing or effluent disposal. Where scarcity exists, conflicts arise and this gives rise to a need for criteria with which to make allocation decisions. The need for criteria on which to base allocation decisions is readily apparent in fisheries management where urban and industrial development has intruded on fisheries habitat and is likely to continue to displace fisheries production in the future.

Economics is a social science concerned with the allocation of scarce resources among alternative uses. From this point of view, I would like to discuss two divergent resource allocation systems, the price system and central planning and control, and make some cursory observations about the system which has been most commonly adopted by public resource management agencies. Secondly, I would like to look at some contemporary resource valuation techniques which are assumed by many as requirements for making allocation decisions. Examples will be drawn from recreational and commercial fisheries. Finally, I would like to discuss the contribution that the price system could make to resource allocation decision in fisheries.

Allocation Systems

In a free enterprise economy, an economy in which resources are privately owned, individual decision-making units allocate resources in accordance with price signals emitted in the market place. The objective of each decision-making unit is to obtain the maximum output from given inputs. Poor performance is rewarded by losses and possible bankruptcy. Good performance is rewarded by profits. Profit and loss phenomena have a cleansing action as enterprises enter and exit from the market place in response to changing preferences, production costs, technology, and prices. It was this type of system that led Adam Smith (1776) to assert that in a competitive market system, the activities of individual decision units are led, as if by an "invisible hand", to achieve the most efficient allocation of resources for society as a whole. Although a purely competitive system, as perceived by Adam Smith, is virtually impossible to attain, the success of North American society in terms of economic growth and development has largely been a result of a free enterprise market system.

In a centrally controlled economy, an economy in which resources are generally publicly owned, the allocation of resources is directed by the visible hand of a planning commission. In place of a price system, central governments generally coordinate the flow of materials and set production targets. Allocation efficiency is generally measured in terms of physical output and political goals. According to Sirkin (1968) prices of factors of production "bear no dependable relationship to scarcity of those inputs, and the prices of outputs bear no dependable relationship to their relative costs of production or to their utility". The lack of price signals, and of profit and loss measurements of performances, commonly leads to excess demands and shortages.

Free enterprise and centrally controlled economies exist only in theory. In North America, certainly in Canada, we have elements of both. We have a mixed economy. Government is involved in our economic system through its monetary and fiscal policies aimed at promoting economic growth and development and in attempting to provide a stabilizing influence. It is not always successful.

Government is also involved in the provision of goods and services. Although there is agreement that government should participate in the economic affairs of our society, the extent of the participation has been subject to debate. Galbraith (1958) has asserted that more resources should be allocated through the public sector to provide public needs. Others such as Keynes (1926) and Friedman (1962) have argued for a minimum of government intervention in the economy.

Generally speaking, the rationale for government involvement in economic activity will include four activities (Herber 1967).

(1) *The provision of non-market products* such as national defence, aids to navigation, police protection, and national parks. To the extent that it is not possible to exclude an individual from enjoying the benefits of such goods and services, if he chooses not to pay for them, such goods and services are regarded as public goods. They are "indivisible" and most efficiently provided by the public sector.

(2) *The correction of market defects* which result in a divergence between private and public costs. The use of a river as a zero-cost effluent disposal system, for example, can impose costs, known variously as "externalities" or "spillovers", on downstream users of the water resource.

(3) *The improvement in the functioning of the market* where market imperfections exist. Government participation will include stabilizing economic activity through the appropriate monetary and fiscal policy and improving or eliminating imperfect or defective competition through various incentives and disincentives and through anti-trust legislation.

(4) *The alleviation of defective supply characteristics* of resources such as the unique scarcity of certain productive resources, immobility of resources, and inadequate or imperfect information.

In North America it would appear that the justification of public ownership and management of fisheries resources has largely been rationalized on the basis of these criteria. We have selected a system of allocation which closely resembles that of a centrally controlled economy. In doing so we have chosen to ignore price signals in the market place. This situation, I believe, has created problems in the allocation of fisheries resources and has strained our credibility as resource managers.

Economically "accessible" commercial fisheries of Canada, such as the fisheries of the Great Lakes, have generally been characterized by excessive investment in fishing effort relative to the productive capacities of the resources. This situation has resulted in depleted fish stocks, depressed returns to labour and capital, economically depressed communities, perpetual government subsidies, high tax burdens, foregone returns to society (sacrificed rents) and, in short, we have misallocated our resources.

In my opinion, the fundamental reason for these problems can be attributed to our management of fisheries as common property resources. Public management agencies, as stewards of public fisheries resources, have failed to exercise proprietary rights to control fishing effort. They are reluctant to use prices to allocate scarce fisheries resources. Resources used to manage fisheries bear no dependable relationship to scarcity (the alternative use of those resources), and the prices of fish bear no dependable relationship to their costs of production. In foregoing the opportunity to utilize a price system for guiding resource allocation decisions, perhaps on the premise that the price system sends out bad signals, we have adopted a system resembling that of a central planning commission complete with the efficiency criteria of physical output (e.g., physical stocks and yields of fish). In the process, it is possible that this system has generated worse resource valuation signals than a price system would produce.

Resource Valuation

The ability to value a resource in alternative uses is considered to be an underlying requirement for making allocation decisions. Despite our reluctance to establish market values we have, in the absence of price signals, attempted to develop proxy values as a rationale for making investment and allocation decisions. Consider for example, efforts made to value recreational and commercial uses of fisheries resources.

Recreational Fisheries Valuation

Such recreational valuation techniques as the travel cost, value-added, expenditure, and willingness-to-pay are familiar. A method that appears to be favoured in western Canada combines the willingness-to-pay and expenditures made on recreational fishing activities. This method attempts to calculate the net social value accruing to society from recreational fisheries as a proxy for market value. Complex and costly survey techniques have been used.

A good example of this approach, in

TABLE 1.—*Net social benefits from recreational fisheries in Alberta.*

Direct social benefits	
Licence sales	$466,000
Extra-market benefits to residents	16,148,000
Indirect social benefits	
Expenditures by non-residents in pursuit of angling activity	1,269,000
Total social benefits	$17,883,000
Direct social costs	
Fishery management costs	954,000
Indirect social costs	
Business expenditures for non-resident anglers	266,000
Total social costs	$1,220,000
Net social benefits	$16,663,000

Alberta (Miller 1971), measured the total amount of recreational fishing in terms of the number of fishermen, the total number of days spent fishing, and the number of fish caught. This information then was combined with estimates of value obtained by surveying anglers in order to calculate total value.

In 1969, an estimated 153,000 anglers, including 5,500 nonresidents, fished 2.3 million angling days in Alberta. This approximates 15 days of angling per fisherman. The net social benefits emerging from this activity are presented in Table 1. The rationales underlying these benefits and costs follow.

Licence sales. This benefit should be obvious. It is that revenue generated from the sale of licences to resident and nonresident anglers.

Extra-market benefits. To the extent that licence sales do not reflect the value of the fishery, the extra-market (intangible or nonmarketed) benefits are an estimation of what fishermen would be willing to pay, over and above the nominal licence fee, for the privilege of fishing. It also represents what fishermen would have to be paid to forego their favourite sport. The figure of $16,148,000, derived from questionnaires, represents approximately $7.20 per resident angler-day, or approximately $110.00 per resident angler-year. Since the fisheries resource is being managed by Albertans, the extra market benefits to nonresidents are not included.

Expenditures by nonresidents in pursuit of fishing activity. These are regarded as an indirect social benefit. This information is also obtained from the surveys. Total expenditures of non-residents ($1,269,000) approximates $31.00 per angler. In this category only net gains to the Alberta economy are of importance when determining the value of the resource. Therefore, expenditures by Albertans are deleted on the premise that they represent transfers of money which would otherwise by spent on other activities within the province.

Fisheries management costs. These represent the government cost of providing recreational fisheries resources. They encompass such items as salaries and wages, enforcement, general administration, and operating expenses.

Indirect social costs. These represent the costs of the goods and services purchased by businesses in order to meet the purchases of fishermen. Again, these are costs attributed only to non-resident fishermen.

The net value is of considerable magnitude. These net social benefits are assumed to be generated annually and to extend over time. The resultant values should make people aware of the importance of recreational fisheries resources. They may even be used to promote more government spending on recreational fisheries. Recreational fishermen and managers have an incentive to try to do this.

There is no unanimity among resource managers as to the validity of this type of valuation. I am personally concerned with its validity in making public investment and allocation decisions. First, there is the implicit assumption that the recreational fishermen represents all of Alberta society. In fact recreational fishermen represent only a subset of society. Based on a total Alberta population of 1.75 million people, 147,500 resident anglers represent less than 10% of the population. During 1970–1971, the last year in which residents were required to purchase fishing licences, approximately 563,000 residents of Ontario purchased licences. This also represents less than 10% of a population of 7.4 million people (Environment Canada 1975). These facts suggest that some accounting might be desirable to determine who receives the benefits and who pays the bills. This is especially so if the objectives of management include the enhancement of welfare for all. To ignore such an accounting can result in an undesirable transfer of benefits to one group of society at a cost to the general population. This is a problem of income redistribution. Consider some of the benefit and cost components in the analysis.

Angling licence sales. This is a direct benefit. Much depends upon how receipts are used. The proceeds from licence sales can be invested back into fisheries resource management, used to develop other resources, build roads, invest in education, increase old age pensions and/or reduce taxes. This is a possible means of using the wealth of the province to generate internal growth and development.

Extra-market benefits. This benefit is also known as consumer surplus. It is a measure of the satisfaction experienced by the fisherman in excess of the price paid for the experience. My initial reaction to such benefits is that they do not mean too much in resource valuation and allocation until you have been required to step up to the counter and lay your money down. Sirkin (1968) has observed: "The . . . consumer's surplus arguments for projects that cannot cover total cost contains a fundamental fallacy. The cost of the product is taken as a measure of the value to consumers of the alternative products which might have been bought instead. The valuation of these alternative products does not include any consumer's surplus. A comparison is thus being made between the value of one product including consumer's surplus and the value of other products excluding consumer's surplus. A correct comparison between the value of alternative products must be based on the same valuation procedure for all products . . . . Unless all allocative decisions are to be based on calculations that include consumer's surplus (a problem which is rendered impractical by the problem of measuring consumer's surplus), the principle is not applicable." Scott (1965) has responded to the questionnaire method employed in estimating recreational values by remarking: "Ask a hypothetical question and you get a hypothetical answer; the results of this procedure have yielded some extremely fanciful valuations". Scott goes on to say that "When the frontier encroaches on the wilderness it is essential to know what recreational values are being destroyed, and when their value at the margin is sufficient to 'buy-off' further encroachment, . . . or indeed to buy out existing users in favour of extending the recreational area . . . to be brief, the only route to efficiency for this kind of problem is actually to levy a toll."

Indirect social benefits and costs. These two components can be combined to the extent that they generate net indirect social benefits. The proceding arguments opposing extra-market values has some relevance here. That is, if we include secondary benefits for recreational fisheries we must include them in alternative uses of the resource that we might wish to invest in.

Another of my concerns is related to our system of national or provincial accounts. Should the profits from liquor sales, which are included in the indirect social net benefits, be attributed as value added to the liquor industry or the fishing industry? Similarly, should the net proceeds from the sale of ice cream, bread, gasoline, etc., be attributed to their respective industry classifications or the fishing industry? There appears to be a danger of double counting, or at least a shuffling of accounts.

The question of attracting nonresidents and including as secondary benefits money spent in the province is also a bit puzzling. Apart from the question of double accounting, there appears to be some redistribution of income, where direct costs of management exceed the direct return to the people who pay the bill even if we include secondary benefits. Those secondary benefits accrue to the businesses serving them. There is little or no return to the resource or the people who pay the cost of management.

A basic fallacy here is the assumption that the resources used in providing secondary benefits would not otherwise be employed. Perhaps this would be true for a particular industry serving recreational interests. But given the scarce nature of resources, we should not assume they would not be used elsewhere in our economy. Consider, for example, petroleum products. Receipts from the sale of gasoline to anglers can be regarded as a benefit or a cost. We can hardly assume that gasoline would not otherwise be utilized.

Zivnuska (1961) handles this problem very nicely when he notes that: "The U.S. Fish and Wildlife Service estimated that in 1955 some 25 million American anglers and hunters spent nearly $3 billion for 500 million days of sport. But what does this mean about the value produced by allocating resources to provide hunting and fishing opportunities? Absolutely nothing. These expenditures were for food, lodging, travel clothes, guns, rods and similar items. These values account for the three billion dollars, and there is nothing left over as a return to the recreational use of the land.

"Certainly the particular groups serving the recreationists are benefiting from these expenditures, and data on expenditures may well be highly useful in gaining the support of such groups in lobbying for higher budgets for recreation. But from a broader public standpoint, all that this is achieving is a transfer of expenditures from one group to another: there is little if any net gain to society from this level of effects. The social case for public support of recreation must rest on the value to the users, not the increased profits of certain recreational service industries."

If you accept this point of view and net out extra-market benefits and indirect social benefits and costs from my previous example, it will be noted that the public sector is generating red ink (−$448,000). Quite possibly these net losses will also extend over time. If so, this situation indicates that the economic value of the resource to the resource stockholders (citizens of Alberta) is negative.

Commercial Fisheries Valuation

Since there is generally a zero or nominal price charged to the commercial industry for publicly owned fish, little or no revenue is generated by the public sector to offset the cost of management. According to the public accounts for Canada, the costs of fisheries management, research, and other Fisheries and Marine Service activities approximated $152 million during 1973–1974. Revenues generated during this same time period totaled $5 million dollars (Finance Canada 1974). These statistics clearly indicate that someone is being subsidized.

Commercial fisheries under federal control in Canada also generated red ink. As in the case of recreational fisheries, some interesting rationales have been developed for expenditures in excess of receipts. Valuation and allocation decisions have literally been predicted on the contribution of fisheries to value added in the Canadian economy! In addition, the conservation of fish, the generation of employment opportunities, increased incomes to fishermen, and the promotion of regional development and stabilization of communities have been used as rationale for government expenditures in excess of receipts. These calculations deserve comment.

CONTRIBUTION TO VALUE ADDED[1]

Value added is calculated by measuring the gross value of fisheries output less the costs of material and supplies, fuel, and electricity used in production. The deductions are made to avoid the problem of double counting the output of other industry groups.

During 1974, the contribution of the fish processing industry, in terms of value added was estimated at $248 million. The contribution of the primary level of production during 1974 was recorded at $291 million for a total contribution of $539 million (Fisheries and Environment Canada 1977). It should be noted here that the measure of the primary level of production was not value added but, instead, the gross value of production. Information on materials and supplies, fuel, and electricity was unavailable. Also, since fish are priced at zero, a major cost of materials to the primary industry is ignored.

Government's contribution to value added in the economy is also ignored. The values of government goods and services generally enter the accounts as being equal to the cost of their production (Branson 1972). While this may conform to accepted practices, I consider it to be dangerous. First, net revenue generated by the public sector is negative. If this is incorporated into the value-added accounts, the net contribution of Canada's fisheries resources to the economy would be reduced. As a matter of fact, if the value added by the primary sector was 50% of the gross value of production ($291 million), Fisheries and Marine Service expenditures would exceed the value added generated in the primary sector of the industry.

Second, if one is concerned with the distributional consequences of this accounting system, it is important to note that the Canadian taxpayer incurs tremendous costs of managing the fishery, the benefits of which accrue to certain segments of the industry. This is clearly a subsidy. It may be of interest here to note that the industry includes foreign ownership. Further, to the extent that the price of fish is lower because of a subsidy, the final consumer receives the benefit. This has interesting implications if we acknowledge that a large percentage of Canadian fish are exported (Fisheries and Environment Canada 1977).

Value added in the primary sector is irrelevant for making public investment decisions in public resource management. In my opinion, the reason for such inclusions is to disguise inefficient government expenditures. If one is concerned with efficiency in public resource management, it is important to maximize the return to the resource in question. Efficiency considerations in the management of a fisheries resource are independent of how many resources are used in processing and other secondary activities required to move the final product to market. We want all of these activities to be efficient.

To adopt the point of view that value added in the private sector is a measure of the value of public fisheries resources, it would seem necessary to demonstrate that the value of national products would decrease in the fisheries sector and not

---

[1] Value added is a measure of the contribution of the fishing industry to net national product.

increase in other sectors. No one has attempted to do this. J. A. Crutchfield has pointed out (in personal correspondence) that "Benefit cost analysis and national income accounting make it abundantly clear that net secondary benefits are not to be included unless a specific finding can be made that they are truly net." This is, in Crutchfield's experience, "roughly equivalent to finding a unicorn walking down the middle of the freeway."

CONSERVATION OF FISH POPULATIONS

Generally speaking, the natural scientist's interest in conservation is related to the maintenance, enhancement, and protection of physical stocks of resources. In this regard, I have been advised that I must learn to live in harmony with fish. Fish are an indicator of environmental quality. Fish have intangible psychic "values." The importance of fishing is beyond the capability of the dismal science of economics to measure.

Conservation, from an economic point of view, may be equated with investment in a resource which will produce a stream of benefits over time. In this sense, to conserve means to curtail present consumption in the interest of future consumption. To apply this concept to fisheries, it must be understood that the process of conservation entails a sacrifice of investments in other resources. Clearly, it does not make sense to sacrifice ten cents worth of one resource to conserve five cents worth of another. I do not mean to imply that fisheries resources do not justify maintenance or even enhancement measures. I do mean to say that to ignore prices and costs does not enhance the resource manager's competence in making such investment decisions.

GENERATION OF EMPLOYMENT

The provision of full or part-time employment is used continuously as justification for the management of fisheries resources. The provision of employment opportunities is an extremely important consideration in our society. While the promotion of these benefits may be valid for some fisheries, the indiscriminate application of this philosophy can result in the misallocation of resources.

In many remote northern communities, regional growth and development is dependent on the existence of economic activity associated with the primary level of production in resource-based industries (e.g. trapping, logging, fishing, mining). Unfortunately, many of these communities are characterized by limited opportunities for the development of resource-based industries. This situation is often compounded by the immobility of labour and its inadaptability to opportunities that do exist. Under these circumstances a case is often made to promote fisheries exploitation (even though such activity may be uneconomical) in the interest of engaging such unfortunately located communities in some form of production-oriented activity. To the extent that such activity provides the most efficient means of redistributing income and providing for the social well being of the communities, it is probably a legitimate program for government to promote. Note, however, that this is an argument for the redistribution of income to disadvantaged people. It cannot be construed as an argument for larger budgets for management agencies to spend public funds on fish. Perhaps a more appropriate role or opportunity for government would be to divert those funds to eliminate the immobility and inadaptability of human resources over time. Otherwise, there may be a danger of perpetuating the social and economic problems associated with remote communities. (See Adams 1978, this volume, for one overview of this problem.)

In other regions of Canadian society, such as the Great Lakes region, resources (labour, capital, land) are scarce. There are numerous opportunities for labour and public and private investment funds. Under these circumstances, resource development incentives designed to generate employment may simply result in a transfer of benefits from one region to another. As a matter of fact, such incentives could actually impose a cost on society as bidding for scarce resources increases prices. This should be of considerable concern in the

Canadian economy today with its present rate of inflation.

The creation of employment which will entail additional capital investments and increased fishing effort is in direct conflict with the objective of protecting fish stocks. If the current depletion of fish stocks is attributable to excess fishing effort, then the objective of increasing employment opportunities will further stress fish populations, and will further depress returns to the industry. The problem, it would seem, is that we already have too many fishermen.

INCREASED INCOMES TO FISHERMEN

Although many fishermen earn good incomes, many do not. As a result, there is much interest in increasing income and welfare to fishermen. There is necessarily a trade off between protecting fish stocks, generating employment benefits, and increasing incomes to fishermen. Without resorting to oversimplification, a reduction of labour and capital in fishing which would relieve pressure on fish populations would increase income to those who remained. Crutchfield (1962) has pointed out that the "unprecedented growth of economic welfare in the western world has been achieved by getting more output with less inputs. North American living standards reflect our continuing success in meeting basic needs (particularly for food) with fewer and fewer resources, thus providing the basis for expansion of both capital and consumer goods industries. The question of issue is fundamental: do we use the gains of the conservation program to support more marginal fisheries units, or do we provide better incomes for those actually needed and produce more of other things?"

PROMOTION OF REGIONAL DEVELOPMENT, STABILIZATION OF COMMUNITIES, AND MINIMIZATION OF SOCIAL DISRUPTION

Many communities associated with fisheries are characterized by limited opportunities for community growth and development. Understandably, therefore, there is an interest on the part of federal and provincial agencies to use fisheries resources as a mechanism for regional growth and development. Fisheries resources are an important element in development strategies for many regions of Canada.

Economic growth and development in North America has involved "breaking down the old and building the new." This has entailed deserting communities and mobilizing labour and capital to more profitable regions and productive activities. If we had adopted some of our present day philosophies concerning the stabilization of communities and the maintenance of a cultural way of life 100 years ago we might still be supporting a labour force of glass blowers or at the very minimum buffalo hunters. We live in a dynamic society with continual change. Given the continual adjustments in response to changing demands and costs of production, the question must be asked whether policies of goverment which constrain natural adjustment will alleviate social disruption or magnify social hardship by perpetuating weak and declining industries. There is a real danger of inhibiting natural adjustments if goverment support becomes institutionalized and perpetuated. This is not to ignore the fact that social hardships confronting declining industries can be quite severe.

Having expressed dissatisfaction with what appears to be the present rationale for managing fisheries resources, I should offer an alternative management stratagy. This will also entail a focus on a principle objective of management.

Resource Allocation

I will start with the proposition that the principle objective of fisheries management is to assure that fisheries resources contribute as much as possible to the welfare of society. This encompasses more than the welfare of the fish and the interests of fishermen. It encompasses the requirement of assuring that labour and capital engaged in exploiting the resource, and the resource itself, receive an economic return consistent with the productive potential of the resource. As noted previously, the economic return to the resource will provide an appropriate measure of value as an underlying requirment for making allocation and investment decisions.

The attainment of the principle objective requires that management agencies exercise proprietary rights over fisheries resources and utilize the price system to allocate scarce fisheries resources. In this regard, restricted entry and a royalty system on the landed value of commercial production appears to be the solution favoured by a number of economists working in fisheries. Restricted entry to promote the greatest efficiency in the use of vessels and fishermen, "as is our national purpose in every sector of our economy" (Bell 1977), will provide the fishing industry the opportunity to harvest the productive potential of the resource at the least possible cost. The initiation of a royalty system, set high enough to permit a competitive return to labour and capital required to harvest fish requires controlling fishing effort. The resultant measure of the resource value provides a firm foundation for allocation decisions. It provides a measure of the value foregone to society in the event that commercial fisheries are displaced in favour of alternative uses.

Obviously, decisions to allocate resources among alternative uses require consistent measurements of values. Thus, decisions to allocate resources between commercial and recreational fisheries uses require that the valuation of resources in recreational uses conform to that in commercial uses. In Canada, the opportunity to engage in recreational fishing, as in the case of commercial fishing, is provided free or at a nominal licence fee. The only solution to this dilemma, in my opinion, is to employ prices. If the objective of pricing is to raise revenue, it would be possible to increase licence fees in aggregate. If the price is set too high the number of people purchasing licences may decline to the point where total revenue would be less than that which would accrue from a smaller licence fee. We should beware of using estimates of what fishermen are willing to pay as determined by hypothetical surveys. If the objective is to ration use from one fishery which is undergoing severe fishing pressure to another area, the solution lies in using differential prices.

The question at issue is, quite simply, to allocate fisheries resources to the use that generates the greatest net value over time. Since values of alternative uses occur over time, comparisons are made easier by calculating net present values.

It should be pointed out that alternative uses will seldom be mutually exclusive. It will generally be found that the limited productive capability of fisheries resources necessitates a trade off between one use and another. Between the decision to allocate a fishery exclusively to commercial production or recreational activity, there is a range of production possibilities.

Conclusions

When industrial and urban growth and development displace fisheries production as a primary use of water, or when decisions must be made between alternative uses such as commercial fisheries production or recreational fisheries activity, it is important to know what values are being sacrificed in making allocation decisions. Despite various attempts at resource valuation, resource managers' hands are tied because of the absence of a price system that will provide tangible evidence of value. Actually, it is my belief that the current value of fisheries resources in commercial and/or recreational uses, in terms of the cost of providing fishing opportunities and the returns from providing those opportunities, is negative. Is it any wonder, therefore, that the blue pike has disappeared from Lake Erie and Lake Ontario, or that the frontier of civilization will continue to intrude on one's favourite fishing hole.

If we are really interested in a solution, we might consider pricing the opportunity to fish. We might be able to generate tangible evidence of the value of our favourite resource. I suspect this might conflict with the intangible physic values which are often attributed to fish, evidence of which I consider to be gratuitous and largely unsubstantiated. Some would call it political salesmanship. More important, given the opportunity to express our preferences through a price system, we might even be able to demonstrate that everybody wishes to live in harmony with fish.

References

ADAMS, G. F. 1978. An historical review of the commercial fisheries of the boreal lakes of central Canada: their development, management, and potential. Am. Fish. Soc. Spec. Publ. 11:347–360.

BELL, F. C. 1977. World wide economic aspects of extended fisheries jurisdiction management. Pages 3–28 in L. E. Anderson, ed. Economic impacts of extended fisheries jurisdiction. Ann Arbor Science, Ann Arbor, Mich.

BRANSON, W. H. 1972. Macro economic theory and policy. Harper and Row, New York.

CRUTCHFIELD, J. A. 1962. Regulation of the Pacific coast halibut fishery. Pages 353–392 in R. Hamlish, ed. Economic effects of fishery regulation, report of an FAO expert meeting in Ottawa, June 1961. F.A.O.U.N., Rome.

ENVIRONMENT CANADA. 1975. Statistics on sale of sport fishing licences in Canada. Fish. Mar. Serv. Recreat. Fish. Branch, Ottawa.

FINANCE CANADA, 1974. Public accounts of Canada. Information Canada, Ottawa.

FISHERIES AND ENVIRONMENT CANADA. 1977. Annual statistical review of Canadian fisheries. Fish. Mar. Serv. Market. Serv. Branch, Ottawa.

FRIEDMAN, M. 1962. Capitalism and freedom. Univ. Chicago Press, Chicago.

GALBRAITH, J. K. 1958. The affluent society. Houghton Mifflin, Boston.

HERBER, B. P. 1967. Modern public finance. Richard D. Irwin, Homewood, Ill.

KEYNES, J. M. 1926. The end of laissez-faire. Page 67 in Laissez-faire and capitalism. New Republic, New York.

MILLER, R. J. 1971. Alberta's hunting and fishing resources, an economic valuation. Alberta Dep. Agric. Econ. Div., Edmonton.

SCOTT, A. 1965. The valuation of game resources: some theoretical aspects. Pages 27–47 in Canadian Fisheries Reports. Dep. Fish. Can., Ottawa.

SIRKIN, G. 1968. The visible hand: the fundamentals of economic planning McGraw-Hill, New York.

SMITH, A. 1776. The wealth of nations. Routlege, London. (Reprinted 1913.)

ZIVNUSKA, J. A. 1961. The multiple problems of multiple use. J. For. 59(8):555–560.

The Concept of Balance for Coolwater Fish Populations

RICHARD O. ANDERSON AND A. STEPHEN WEITHMAN[1]

Missouri Cooperative Fishery Research Unit
University of Missouri, Columbia, Missouri 65211

Abstract

We propose a new approach to evaluate the state of balance or structure of coolwater fish populations and communities. All that is needed to calculate the index is a length-frequency distribution of a fish stock. The index, Proportional Stock Density (PSD) is calculated as the percentage of fish of a quality size (total length) that is longer than a minimum stock size. Minimum stock and quality sizes (20–26% and 36–41% of world-record length, respectively) are defined for: yellow perch (13 and 20 cm); walleye (25 and 38 cm); smallmouth bass (18 and 28 cm); northern pike (35 and 53 cm); and muskellunge (43 and 66 cm). An analysis of model stocks based on representative growth and mortality rates and analogies between largemouth bass and bluegill populations lead to the suggestions that the PSD for balanced populations of yellow perch may range from 30 to 50%; the PSD for balanced populations of coolwater game fish may range from 30 to 60%. Balanced fish communities are defined as those with the potential to provide a satisfactory catch and harvest of both game fish and panfish; i.e. both predators and prey. Overharvest is defined as harvest of more than the surplus of stock- or quality-size fish; the surplus is determined according to management objectives for favorable values of PSD. Minimum length limits at a longer than minimum quality size and protected size ranges (slot length limits) which protect some portion of quality-size fish are proposed as tactics to rebuild depleted stocks of game fish and to sustain balance in fish populations and communities. The goal is to achieve satisfactory sustained yield, benefits, and fishing quality.

The concept of balance is basic to the effective management of recreational fisheries and the goal of optimum yield. Swingle (1950) defined balanced populations as those that have the capacity to provide a satisfactory harvest of fish in proportion to the productivity of their habitat. Balance is a management term based on human values. At any given level of productivity or carrying capacity expressed as fish biomass, there is a broad range of potential densities and length-frequency distributions. Balanced populations have a structure that is intermediate between the extremes of a large number of small fish and a small number of large fish. For the structure of a fish population to be balanced, the rates of reproduction, growth, and mortality must be satisfactory.

Warmwater and coolwater fish communities are similar in that they include both game and panfish—i.e., both predators and prey. When a fish community of largemouth bass and bluegills is balanced, both populations can provide satisfactory yields. Balanced coolwater fish communities should also be able to provide satisfactory catch and harvest of both game fish and panfish.

A new index based on length-frequency distributions, Proportional Stock Density (PSD), has been developed to describe the structure of stocks of largemouth bass and bluegills (Anderson 1976). The objective of this paper is to extend the concept of PSD to populations of yellow perch, walleye, smallmouth bass, northern pike, and muskellunge.

The PSD models that follow incorporate some speculations, assumptions, and simplifications. Our intention is to establish the concept of PSD as a technique for the evaluation of fish population and community structure. Research will be necessary to define optimal values and to develop management strategies and tactics that will achieve objectives for various species and ecosystems.

Methods and Definitions

Proportional stock density is the percentage of the stock that is of quality size:

$$PSD\ (\%) = \frac{number \geq quality\ size \times 100}{number \geq stock\ size}$$

---

[1] Contribution from the Missouri Cooperative Fishery Research Unit, a cooperative program of the U.S. Fish and Wildlife Service, the University of Missouri, and the Missouri Department of Conservation.

TABLE 1.—*Proposed minimum stock and quality total lengths for selected coolwater fishes and the percentages of the record length.*

Species	Record length (cm)	Stock length		Quality length	
		cm	Percent of record length	cm	Percent of record length
Yellow perch	53.3	13	24	20	38
Walleye	104.1	25	24	38	37
Smallmouth bass	68.6	18	26	28	41
Northern pike	133.4	35	26	53	40
Muskellunge	163.8	43	26	66	40

All that is needed to calculate the index is a definition of minimum lengths for stock- and quality-size fish and a length-frequency distribution of the stock. We define the minimum length for the stock of all species as some length within 20 to 26% of the world record length; the minimum length for the quality size of all species is defined as some length within 36 and 41% of the world record length. For convenience the length is a whole number when expressed in centimeters or inches (Table 1). All lengths used herein are total lengths.

We developed models of PSD for stocks of coolwater species. Length-at-age data were selected for stocks with fast, moderate, and slow growth rates (Table 2). Survivorship curves that reflect annual mortality rates of actual or typical populations were developed when possible. Patterns of annual mortality in the PSD models include: a moderate constant rate; a moderate rate that varied with age; and a high rate for quality-size fish.

For all models we calculated the number in each age group by assuming a rate of reproduction of 100 age-I fish. The number of stock- and quality-size fish in each age group was estimated by assuming a range and a normal distribution about the mean length at a given age. The total number of stock- and quality-size fish for the populations was determined by summing each, for all age groups. Tables showing the actual numbers for all models are available upon request.

PSD Models

Yellow Perch

The growth patterns selected for yellow perch are: fast, Lake Erie in 1927–1937 (Jobes 1952); moderate, West Okoboji Lake in Iowa (Moen 1964); and slow, an average size in Michigan (Laarman 1963). Yellow perch in these populations are recruited to stock in their second or third growing season and reach the minimum quality length in the third to fifth growing season (Table 2).

When annual length increment is plotted

TABLE 2.—*Growth data used to develop models of proportional stock density for selected coolwater fishes.*

Species and growth rate	Total length at age (mm)									
	I	II	III	IV	V	VI	VII	VIII	IX	X
Yellow perch										
Fast	94	170	216	241	264	279				
Moderate	53	127	183	216	244	264				
Slow	117	155	178	203	229	251	272			
Walleye										
Fast	124	231	323	401	485	549	615	676		
Moderate	178	287	373	434	480	526	559	605		
Slow	142	211	267	310	345	376	396	419	442	
Smallmouth bass										
Fast	118	258	358	411	445	457	473			
Moderate	94	173	249	309	346	373	393	410	421	433
Slow	64	142	201	234	269	325	348	376	389	399
Northern pike										
Fast	152	330	483	635	762	864	965	1,041	1,092	
Moderate	287	417	508	577	635	686	754	833	876	
Slow	208	328	391	439	483	526	559	683	752	
Muskellunge										
Fast	198	437	622	754	861	958	1,036	1,105	1,133	1,163
Moderate	267	432	569	671	770	848	922	991	1,041	1,087
Slow	175	318	434	546	655	737	848	993	1,062	1,105

as a function of length at annulus formation, all three growth patterns exhibit similar length increments when fish are 20–25 cm long. The average Michigan yellow perch, however, exhibit a relatively low annual increment at lengths of 13-17 cm (Fig. 1).

The three survivorship patterns reflect: a constant 50% annual mortality; a concave curve with a high mortality rate of young fish and moderate rate for adults; and a convex curve with a low mortality rate for young fish and high rate for adults (Fig. 2). The second and third curves reflect survivorship patterns for bluegills from balanced and unbalanced populations, respectively (Anderson 1973). We have assumed that populations of these two important species of prey may exhibit similar dynamics.

The model PSD values for yellow perch range from 8 to 46%. The lowest value and the poorest structure result from slow growth and high mortality of quality-size fish (Table 3).

Walleye

Growth patterns selected for walleyes are: fast, Minnesota lakes (Eddy and Carlander 1939); moderate, Clear Lake in Iowa (Carlander and Whitney 1961); and slow, Red Lakes, Minnesota (Smith 1977). With these growth patterns walleyes are recruited to stock size in the second and third growing season and reach quality size in the

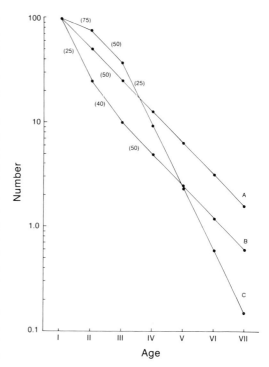

FIGURE 2.—Survivorship curves for model populations of yellow perch; numbers in parentheses are annual survival rates in percent. Lines A, B, and C represent constant, concave, and convex mortality patterns, respectively.

fourth to the seventh growing season (Table 2).

The three survivorship patterns reflect: constant 40% annual mortality; a concave curve with a mortality of 60% for age I, 50% for age II, and a low annual mortality of 35% for age III and older walleyes; and a convex curve with mortality of 20% for age I, 30% for age II, and a high annual mortality of 60% for age III and older walleyes. Carlander and Payne (1977) used an annual mortality of 42% in their analysis of the walleye population in Clear Lake, Iowa.

The model PSD values for the three growth and mortality patterns range from 3 to 55% (Table 3). The lowest value and the poorest structure are the result of slow growth and high mortality.

Smallmouth Bass

The selected smallmouth bass growth patterns are: fast, Norris Reservoir (Stroud 1948); moderate, Oneida Lake, New York

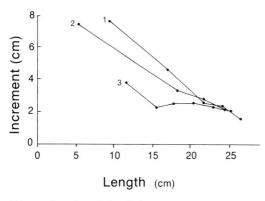

FIGURE 1.—Annual length increment as a function of total length at annulus for selected yellow perch populations. Lines 1, 2, and 3 represent fast, moderate, and slow growth, respectively.

TABLE 3.—*Proportional stock densities (%) of model populations of selected coolwater fishes. A concave mortality pattern represents high mortality of young fish and moderate mortality of adults; a convex mortality pattern represents low mortality of young fish and high mortality of adults.*

Species and mortality pattern	Growth rate		
	Fast	Moderate	Slow
Yellow perch			
Constant, 50%/yr	46	31	15
Concave	42	28	10
Convex	37	10	8
Walleye			
Constant, 40%/yr	53	44	18
Concave	55	40	19
Convex	34	35	3
Smallmouth bass			
Constant, 43%/yr	64	42	22
Constant, 65%/yr	48	20	4
Northern pike			
Constant, 50%/yr	41	32	5
Concave	50	40	12
Convex	16	19	<1
Muskellunge			
Constant, 30%/yr	62	49	47
Convex	54	35	23

(Forney 1972); and slow, the upper White River, Missouri (Purkett 1958). Smallmouth bass are recruited to stock size in the second and third growing season and reach quality size in the third to the sixth growing season in these populations (Table 2).

Two constant annual mortality rates were assumed—43 and 65%. A 43% rate is the average over a 14-year life span in Oneida Lake (Forney 1972); a 65% rate is estimated from the catch-curve analysis of smallmouth bass from ages I to VI in the Plover River, Wisconsin (Paragamian and Coble 1975).

The model PSD values for these growth and mortality patterns range from 4 to 64% (Table 3).

Northern Pike

The growth patterns selected for northern pike are: fast, Ohio (Roach 1948); moderate, a typical population in Michigan (Latta 1972); and slow, Bucks Lake, Wisconsin (Snow and Beard 1972). Northern pike in these populations are recruited to stock size in the second or third growing season and reach quality size in the fourth to the seventh growing season (Table 2).

The three survivorship patterns reflect: constant 50% annual mortality; 50% annual mortality for ages I and II and 40% annual mortality for ages III and older; and 50% mortality for ages I and II and 83% annual mortality for ages III and older. The third survivorship curve was proposed as typical for northern pike populations in Michigan (Latta 1972).

The model PSD values for the three mortality and growth patterns range from less than 1 to 50% (Table 3). The PSD values are low at all levels of mortality when growth is slow, and at all growth rates when mortality is high.

Muskellunge

The growth patterns selected for muskellunge are: fast, the Pennsylvania state average (Buss and Miller 1961); moderate, an average for the species (Karvelis 1964); and slow, an average for Minnesota (Carlander 1969). Muskellunge with these growth patterns are recruited to stock size in the second or third growing season and reach quality size in the fourth to the sixth growing season (Table 2).

The two survivorship patterns for the models are: a constant 30% annual mortality; and an annual mortality of 15% for ages I and II, 20% for age III, 50% for age IV, 64% for age V, and 80% for ages VI and older. A rate of 30% is similar to that of muskellunge age V–XVII in the life table for Lac Court Oreilles, Wisconsin developed by Johnson (1975); the second pattern is similar to the catch curve determined for the population in Nogies Creek, Ontario by Muir (1964).

The model PSD values for these growth and mortality patterns range from 23 to 62% (Table 3).

Young-Adult Ratios

A satisfactory annual rate of reproduction is necessary in order to sustain a favorable balance of fish stocks. We define the rate of reproduction as the number of age-I fish produced per unit area per year. Determining this rate is difficult, expensive, and impractical for management purposes.

Reynolds and Babb (1978) proposed a convenient index for evaluation of the success of reproduction—the Young-Adult Ratio (YAR). The favorable ratio for largemouth bass in the fall where populations and communities were balanced was 1–3:1 for the number of young-of-the-year fish to the number of quality-size fish.

The ratio was calculated for the populations developed in this paper by determining the ratio of age-I fish (100) to the number of quality-size fish. The average ratio for the populations of game fish with PSD values higher than 25% is 1.1–3.0:1 (Table 4). The ratios are higher for populations with PSD values less than 25% because of the relatively low density of quality-size fish. Ratios much less than 1:1 can obviously occur in real populations and indicate a weakness or a failure of a year class.

Centrarchid Analogies

Minimum stock and quality lengths for bluegills have been defined as 8 and 15 cm, respectively (Anderson 1976). These lengths are about 25 and 40% of the world-record length. The original balanced PSD suggested for bluegills was 25% with a range of 15–35% (Johnson and Anderson 1974; Anderson 1976). Subsequent analyses of bluegill populations in small midwestern impoundments led to the conclusion that a range from 20 to 40% provided the best number of fish for anglers and prey for bass; a range of 40–60% provided fewer bluegills as prey and for harvest but a higher average size of bluegills for anglers (Novinger and Legler 1978). Research is needed to determine whether similar relationships exist for populations of yellow perch. On the basis of our models, we suggest that stocks of yellow perch exhibit satisfactory or favorable structure, i.e. balance, when PSD is near or within a range of 30–50%.

Minimum stock and quality lengths for largemouth bass have been defined as 20 and 30 cm (Anderson 1976). These lengths are about 21 and 36% of the world record length. Largemouth bass and coolwater game fish serve the same dual functions, i.e. as a terminal link in aquatic food chains and

TABLE 4.—*Ratios of young to adults (YAR) for model populations of selected coolwater fishes as related to proportional stock densities (PSD).*

	PSD > 25%		PSD < 25%	
Species	Mean	Range	Mean	Range
Yellow perch	3.6	2.2–5.5	9.0	7.0–13.3
Walleye	2.4	1.6–3.2	20.0	7.1–41.7
Smallmouth bass	3.0	2.3–3.7	23.8	14.0–33.6
Northern pike	3.0	2.3–3.8	16.0[a]	6.7–31.2
Muskellunge	1.1	0.8–1.6	3.1	

[a] Model population with PSD < 1% not included.

as popular species for anglers. The original models of balance for largemouth bass populations resulted in a range of PSD from 45 to 65 (Anderson 1975, 1976). Reynolds and Babb (1978) proposed 40–60% as the best model for largemouth bass in small impoundments. We recommend a PSD from 50 to 70% for largemouth bass in Missouri impoundments where gizzard shad is a dominant species of prey (paper in preparation). Research will be needed to develop satisfactory ranges of PSD for coolwater game fish in different ecosystems. On the basis of our models, we suggest that stocks of walleyes, smallmouth bass, northern pike, and muskellunge exhibit satisfactory or favorable structure, i.e. balance, when PSD is near or within a range of 30–60%.

PSD of Coolwater Species

The literature contains only a few sources with length-frequency distributions of coolwater fish populations that suit the purposes of this paper. As an example, only 1 of the 58 papers in the Percid International Symposium (*Journal of the Fisheries Research Board of Canada* 34 (10), 1977) includes a length-frequency distribution (Kelso and Ward 1977). West Blue Lake, Manitoba, sustains populations of yellow perch and walleyes. The yellow perch population was relatively stable and balanced, with a PSD of 42%.

Yellow perch populations have a PSD of less than 10% if game fishes are scarce or absent (Eschmeyer 1937; Chadwick 1976). Low PSD of yellow perch populations may be caused by too few game fish preying on young yellow perch or too many large

TABLE 5.—*Proportional Stock Densities (%) of yellow perch and bass before and after stocking muskellunge.*

Lakes and species	Year									
	1953	1954	1955	1956[a]	1957	1958	1959	1960	1963	1964
Corrine										
Yellow perch		7		6	9	9	24	0	0	2
Largemouth bass		20		16	22	43	56	100		
George										
Yellow perch	12	10	3	2	28	0	0	0		
Smallmouth bass	79	13	30	17	17	52	63	97		
Largemouth bass	35	9	16	10	22	47	75	65		

[a] Year muskellunge stocked.

predators feeding on adult yellow perch. In Corrine and George Lakes, Wisconsin, yellow perch stocks had poor structure (Gammon and Hasler 1965); large young-of-the-year muskellunge were stocked in 1956 at densities of 23 and 27 per hectare. The PSD's of yellow perch changed from 7 to 12% prior to stocking to zero afterward (Table 5). The low PSD was evident even 8 years after the introduction (Schmitz and Hetfield 1965). The muskellunge also had a drastic effect on survival of young smallmouth and largemouth bass. The result was poor recruitment of bass to stocks, low bass density, and PSD values higher than those for favorable structure.

Clear Lake, Sawyer County, Wisconsin, was dominated by stunted bluegills (Snow 1968). The lake was stocked with walleyes (30 per hectare) in 1959 and with muskellunge (about 10 per hectare, 20–30 cm) in 1960 and 1961. A total of 5,188 bluegills were sampled by electrofishing in the fall from 1960 to 1967; PSD was always less than 1%. Yellow perch were second in abundance during the 8-year period; average PSD based on all the yellow perch collected (222) was 7%. The PSD of largemouth bass was within the range for a balanced population, 53%. The absolute density, however, must have been low as only 53 largemouth bass of stock size were collected in 8 years. Although the relative density of largemouth bass of stock and quality size was favorable, the absolute density was obviously not. The rates of growth and mortality of bass were satisfactory; however, the rate of reproduction was much too low. Stocking a high density of walleyes and muskellunge did not improve the balance of fish populations in Clear Lake, Wisconsin.

Data provided us on the length frequency of yellow perch in Oneida Lake, New York in the years 1970–1975 indicate an unstable population (J. L. Forney, personal communication). The successive PSD values of 90, 25, 38, 95, 27, and 77% were above the model PSD range of 30–50% in 3 of the 6 years. We hesitate to speculate on factors related to this instability but suspect that it is related to unfavorable structure and density of the yellow perch and walleye populations.

The walleyes in Clear Lake, Iowa, have been studied extensively by Carlander and his students. We made estimates of PSD in the absence of length-frequency data from calculated population densities by age groups and mean lengths at annulus for year classes 1948–1972 (Carlander and Payne 1977). The PSD estimates for the years 1948–1974 ranged from 15% in 1974 to 85% in 1959. The estimated PSD is within the suggested range of 30–60% for 16 of the 27 years. Low values were evident when strong year classes of walleyes entered the stock; high values were evident after successive years of a low rate of recruitment to the stock.

Problems and Solutions

Values of PSD that are too high or too low are symptoms of problems. The problems are unsatisfactory rates of reproduction, growth, or mortality. Management should aim to adjust unsatisfactory rates to solve the problem and eliminate the symptom.

Reproductive success is obviously a relatively difficult rate to manipulate. Studies have shown the importance of density-independent variables such as water temperature, water level fluctuations, and wind. Density-dependent relationships, however, have also been demonstrated (Anderson 1973; Reynolds and Babb 1978). The reproduction curves for bluegills and largemouth bass exhibit a pattern in which the maximum number of young are produced when adult density or biomass is relatively low. To the right of the peak there is an inverse relationship between the number of young and the density or biomass of adults.

Year-class weakness or failure for largemouth bass was always associated with exceptionally low densities of quality-size bass in small impoundments with favorable habitat (Reynolds and Babb 1978). More favorable rates of reproduction were observed when the density of quality-size fish was intermediate. Reproductive success may be influenced more by the density of quality-size fish which serve as predators to balance community structure rather than just contributing to fecundity and egg production. We suspect that a less than satisfactory density of quality-size adults and a resulting higher than satisfactory density of prey may adversely influence the rate of reproduction for most populations of game fish in favorable habitat. The rate of reproduction for the largemouth bass population was a problem in Clear Lake, Wisconsin. This supports the concept that poor reproductive success is associated with a low density of quality-size fish and unfavorable fish community structure.

The rate of reproduction for the walleye population in Clear Lake, Iowa is a problem. Stocking of fry has significantly increased the number of young of the year. Average population biomass is relatively low for this productive lake (Carlander and Payne 1977). We tentatively speculate that the low rate of natural reproduction is related to a relatively low biomass of adult walleyes and less than favorable fish community structure.

Slow growth rate, characterized by the attainment of quality size at a relatively advanced age—V and older—is a problem that can lead to low PSD. Slow growth rate of stock-size game fish and panfish in favorable habitat is the result of higher than satisfactory densities. Relatively high densities are caused by either a higher than satisfactory rate of reproduction or survival of too many fish less than a quality size. The solution is to adjust the mortality rate with increased predation.

A favorable pattern of annual mortality is probably the most important of the rates determining the structure and dynamics of a population. High annual mortality rates of quality-size game fish always result in unfavorable PSD's. Low relative density of quality-size fish also leads to unfavorable reproduction and growth rates for the game fish, as well as for their prey, the panfish species and other non-sport fishes, e.g., Catostomidae, Cyprinidae, or Clupeidae. The result is a fish community with unfavorable structure, and unsatisfactory yields and benefits.

Management Goals

The traditional goal of fishery management has been maximum sustained yield (MSY). A tactical approach to achieve MSY involves the concept of critical size (length). The critical size is reached when the instantaneous rate of mortality equals the instantaneous rate of growth. It can be demonstrated mathematically that MSY is achieved by high rates of exploitation of fish longer than the critical size. Critical size for populations with satisfactory growth and mortality rates is often about the same length as minimum quality size. We believe that the goal of MSY and the concept of critical size have contributed to the overharvest of quality-size game fish and poor structure of many fish populations and communities. Larkin (1977) suggested that MSY is dead—but what will take its place remains to be seen.

A new goal has been expressed as optimum sustained yield (Anderson 1974; Roedel 1975). The concept of optimality is important; however, better expression of a management goal might be satisfactory sustained benefits. For recreational fisheries, optimum or satisfactory yield and associated tangible and intangible benefits must obviously be related to the structure and dynamics of fish populations and communities and the resulting quality of fishing.

Manipulation of Population and Community Structure

If management is faced with the goal of satisfactory sustained benefits, the objectives must be satisfactory habitat, and balanced fish populations and communities. The latter are difficult objectives when anglers are increasing in number and knowledge, and are using more efficient and effective equipment. On the basis of evidence available, stocking of coolwater game fish as a general practice is often an ineffective tactic for improving the structure of stocks of stunted panfish. Guidelines or options have yet to be developed to determine what species, what number, what size, where, and when in order to most effectively use these hatchery products to improve the quality of fishing.

Minimum length limits on game fish have been struck from the code books for many species in many states. The results of studies that have evaluated this move are that the absence of length limits does not lead to extinction of species and loss of fishing opportunities. However, no study has demonstrated a positive effect on sustained harvest, population and community structure, or the quality of fishing after the size limits were removed.

Length limits have often been judged unfavorably, when evaluated. Snow and Beard (1972) concluded that a 46-cm minimum size limit for northern pike in Bucks Lake, Wisconsin, was not an effective management technique. Preliminary surveys before the initiation of the length limit revealed the presence of sparse and fluctuating populations of fast-growing panfish. Northern pike, the only predators present, were abundant and growing slowly. The results of the study demonstrate that it is illogical to expect to improve the rate of growth and structure of a population of slow-growing northern pike by protecting most of the fish in the population.

Based on the concept of critical size and equilibrium yield calculations, a size limit of 56 cm on northern pike was projected to achieve MSY in Escanaba Lake, Wisconsin (Kempinger and Carline 1977). However, the assumptions of no change in growth or mortality rates after the initiation of the minimum length limit proved invalid. This regulation resulted in a 73% reduction of harvest, a decline in growth rate by 40%, an increase in natural mortality from 14 to 76%, and a doubling of recruitment rates. After the length limit was established there was a collapse in the harvest of yellow perch, pumpkinseeds, rock bass, and black crappies—all of the panfish species. Fishing effort also decreased substantially. The application of the concept of critical size was unsuccessful.

Latta (1972) rejected the calculated critical size as a basis for establishing a length limit on northern pike in Michigan. On the basis of typical growth and mortality rates, critical size was 41 cm. Such a size limit was projected to have an adverse effect on the biomass of spawners. A 51-cm minimum size limit was recommended instead.

Results of the studies in Wisconsin lead us to believe that a minimum length limit of 51 cm may have an adverse effect on northern pike populations and fish community structures in lakes where: (1) northern pike are the dominant predators; (2) the rate of northern pike reproduction is moderate or higher; (3) there is a higher than optimal density of stock-size fish; (4) growth is slower than satisfactory; and (5) natural mortality of stock-size northern pike is high. A new approach to regulating harvest may be needed for northern pike populations with these characteristics.

The symptoms of unfavorable population and community structure, and fair to poor fishing quality, caused by the problems of high mortality rates and overharvest of quality-size fish are common in warmwater and coolwater populations alike. One step in solving the problem is defining the concept of overharvest. We define overharvest as the removal of more than the surplus of stock- or quality-size fish. The surplus of game fish or panfish should be related to the existing stock density and biomass, population and community structure, and management objectives. For example, if the PSD of game fish in a lake is less than the objective range set by management, there is no surplus of quality-size fish. When the PSD of panfish populations is less than the objective range set by management, it is likely that annual length increment of stock-size fish is less than satisfactory and the annual rate of natural mortality of quality-size fish is too high. If the habitat is favorable, these problems are usually associated with a low stock density of game fish. In such circumstances there is no surplus of game fish.

Responses of northern pike to length limits near minimum quality size appear to be similar to those of largemouth bass to length limits of 30 cm. With warmwater communities in small impoundments, better community and panfish structure results from a 30-cm limit than when no regulation is applied, but the resulting structure of the largemouth bass stock is less than optimal (Funk 1974). We conclude that minimum length limits set at or near the minimum quality size will not improve the balance of game fish stocks unless stock density is low, rate of growth is average or better, and the rate of natural mortality is low.

The symptom of low PSD for stocks of muskellunge may be relatively infrequent. Muskellunge are less vulnerable to angling than northern pike (Weithman and Anderson 1978a) and probably less vulnerable than smallmouth bass and walleyes. Growth and mortality rates of natural populations are usually satisfactory. Success of reproduction is usually low. The length limit of 76 or 81 cm is the only commonly applied state-wide length limit for a game fish that is longer than minimum quality size (66 cm). This tactic and the practice of catch and release of quality-size fish is apparently effective for a species with these dynamics.

Strategies and Tactics to Achieve Balance

Favorable balance or structure and dynamics of a fish population can be evaluated for management purposes by analysis of PSD and YAR. Balance in fish communities can be evaluated by a plot of the PSD of panfishes or prey as a function of PSD of the predators or game fish. Definition of the favorable ranges of PSD for panfishes and game fishes results in a tic-tac-toe graph (Fig. 3). The central window represents a balanced fish community. In communities with a diversity of panfishes or game fishes, the overall balance of each component could be determined by appropriate weighting factors according to the relative abundance (biomass) of each species in the community or the relative importance to the fishery. If the PSD of yellow perch or other panfish is consistently less than 20%, analysis of scale samples may reveal the problems of slow growth and high mortality. In such circumstances the PSD of game fish may range from 0 to 100%. Whatever the value, it may be inferred that there is no surplus of game fish. The first phase of management could be a minimum length limit above the minimum quality size. Stocking of large fingerling or subadult walleyes, smallmouth bass, or largemouth bass may facilitate rebuilding the stocks if the PSD is in the balanced or high range and rate of reproduction is a problem. Stocking an esocid as well may be appropriate if large, non-sport fishes make up a significant proportion of the fish community. After the stocks of game fish have recovered to a favorable density, biomass, and structure, a second phase of regulating harvest that protects some portion of quality-size fish may be appropriate.

When stock density of game fish is relatively high and PSD is low, analysis of scale samples may reveal the same problems as observed in panfish—slow growth and high mortality. If the PSD of panfish is high or variable over time, it may be inferred that there is a surplus density of game fish of stock size but not quality size. In this circumstance also, a regulation to protect quality-size game fish may be appropriate.

Largemouth bass harvest in Philips Lake, Missouri, was regulated with a 30-cm minimum length limit from 1967–1973 and a protected size range or slot length limit of 30 to 38 cm from 1974 to 1977 (Johnson and Anderson 1974; Anderson 1976). The protected size range has produced more

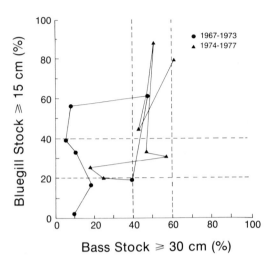

FIGURE 3.—*Proportional stock densities of largemouth bass and bluegills in Philips Lake, Missouri. The sequence of points reflect samples collected in (month and year): 3/67; 4/68; 3/69; 5/70; 5/71; 5/72; 6/73; 5/74; 6/75; 10/75; 5/76; 11/76; 5/77; and 10/77.*

favorable results than the minimum length limit (Fig. 3). The strategy is to: (1) protect quality-size bass for about 25% of their expected life span at a time when they exhibit a high absolute annual weight gain; (2) allow a harvest of a surplus of less than quality-size bass in the stocks; and (3) reduce the probability of excessive predation on young bluegills.

The initial results were also encouraging at Watkins Mill Lake, a state park lake in Missouri. In the fall of 1977, after the first fishing season a slot length limit of 30 to 38 cm was in effect, largemouth bass PSD increased from 1 to 25% (S. Eder, personal communication). In Missouri this approach appears promising because for most bass anglers, size is more important than numbers caught; catch and release contributes to fishing quality and personal satisfaction (Weithman and Anderson 1978b).

The same regulation is now being evaluated for several state fishing lakes in Kansas (D. Gabelhouse, personal communication) and two lakes in Illinois (T. Miller, personal communication); the Central States Pond Management Work Group is evaluating the regulation for several small impoundments in the Midwest (G. Novinger, personal communication). It is important to evaluate not only the biological responses and the resulting structure of the populations and the fish community but also the response of anglers to a regulation that calls for the catch and release of quality-size fish. We expect that this approach to regulating harvest will succeed if anglers are interested, and are willing to support an effort to improve the state of balance of fish communities, the size of fish caught, and the quality of fishing. Many anglers are developing this interest and ethic. We believe this approach to regulating harvest of game fish holds great promise in the future of recreational fishery management for sustaining the balance of fish populations and communities, maintaining fishing quality, and approaching optimum yield.

References

ANDERSON, R. O. 1973. Application of theory and research to management of warmwater fish populations. Trans. Am. Fish. Soc. 102(1):164–171.

―――. 1974. Problems and solutions, goals and objectives of fishery management. Proc. Annu. Conf. Southeast. Assoc. Game Fish Comm. 27(1973):391–401.

―――. 1975. Factors influencing the quality of largemouth bass fishing. Pages 183–194 *in* R. H. Stroud and H. Clepper, eds. Black bass biology and management. Sport Fishing Institute, Washington, D.C.

―――. 1976. Management of small warm water impoundments. Fisheries 1(6):5–7, 26–28.

BUSS, K., AND J. MILLER. 1961. The age and growth of the muskellunge in Pennsylvania. Pa. Angler, April, 1961.

CARLANDER, K. D. 1969. Handbook of freshwater fishery biology. Iowa State Univ. Press, Ames. 752 pp.

―――, AND P. M. PAYNE. 1977. Year-class abundance, population, and production of walleye (*Stizostedion vitreum vitreum*) in Clear Lake, Iowa, 1948–74, with varied fry stocking rates. J. Fish. Res. Board Can. 34(10):1792–1799.

―――, AND R. R. WHITNEY. 1961. Age and growth of walleyes in Clear Lake, Iowa, 1935–1957. Trans. Am. Fish. Soc. 90:130–138.

CHADWICK, E. M. P. 1976. Ecological fish production in a small Precambrian shield lake. Environ. Biol. Fish. 1(1):13–60.

EDDY, S., AND K. D. CARLANDER. 1939. Growth of Minnesota fishes. Minn. Conserv. 69:8–10.

ESCHMEYER, R. W. 1937. Some characteristics of a population of stunted perch. Pap. Mich. Acad. Sci. Arts Lett. 22(1936):613–628.

FORNEY, J. L. 1972. Biology and management of smallmouth bass in Oneida Lake, New York. N.Y. Fish Game J. 19(2):132–154.

FUNK, J. L., ED. 1974. Symposium on largemouth

bass overharvest and management in small impoundments. Am. Fish. Soc. N. Central Div. Spec. Publ. 3. 116 pp.

GAMMON, J. R., AND A. D. HASLER. 1965. Predation by introduced muskellunge on perch and bass, I.: years 1–5. Trans. Wis. Acad. Sci. Arts Lett. 54:249–272.

JOBES, F. W. 1952. Age, growth, and production of yellow perch in Lake Erie. U.S. Fish Wildl. Serv. Fish. Bull. 52(70):203–266.

JOHNSON, D. L., AND R. O. ANDERSON. 1974. Evaluation of a 12-inch length limit on largemouth bass in Philips Lake, 1966–1973. Am. Fish. Soc. N. Central Div. Spec. Publ. 3:106–116.

JOHNSON, L. D. 1975. How many muskies aren't there anymore? Wis. Conserv. Bull. Sept.–Oct.:20–21.

KARVELIS, E. G. 1964. The true pikes. U.S. Fish Wildl. Serv. Fish. Leaflet 569. 11 pp.

KELSO, J. R. M., AND F. J. WARD. 1977. Unexploited percid populations of West Blue Lake, Manitoba, and their interactions. J. Fish. Res. Board Can. 34(10):1655–1669.

KEMPINGER, J. J., AND R. F. CARLINE. 1977. Dynamics of the walleye (*Stizostedion vitreum vitreum*) population in Escanaba Lake, Wisconsin, 1955–72. J. Fish. Res. Board Can. 34(10):1800–1811.

LAARMAN, P. W. 1963. Average growth rates of fishes in Michigan. Mich. Dep. Conserv. Inst. Fish. Res. Rep. 1975. 9 pp.

LARKIN, P. A. 1977. An epitaph for the concept of maximum sustained yield. Trans. Am. Fish. Soc. 106(1):1–11.

LATTA, W. C. 1972. The northern pike in Michigan: a simulation of regulations for fishing. Mich. Acad. 5(2):153–170.

MOEN, T. 1964. Fish studies on Spirit and Okoboji Lakes. Dingell-Johnson Proj. F-68-R-3, Iowa Conserv. Comm. 9 pp. (Mimeo.)

MUIR, B. S. 1964. Vital statistics of *Esox masquinongy* in Nogies Creek, Ontario II. Population size, natural mortality, and the effect of fishing. J. Fish. Res. Board Can. 21(4):727–746.

NOVINGER, G. D., AND R. E. LEGLER. 1978. Bluegill population structure and dynamics. Am. Fish. Soc. N. Central Div. Spec. Publ. 5. (In press.)

PARAGAMIAN, V. L., AND D. W. COBLE. 1975. Vital statistics of smallmouth bass in two Wisconsin rivers, and other waters. J. Wildl. Manage. 39(1):201–210.

PURKETT, C. A., JR. 1958. Growth rates of Missouri stream fishes. Mo. Dep. Conserv. Dingle-Johnson Ser. 1. 46 pp.

REYNOLDS, J. B., AND L. BABB. 1978. Structure and dynamics of largemouth bass populations. Am. Fish. Soc. N. Central Div. Spec. Publ. 5. (In press.)

ROACH, L. S. 1948. In fishing circles. Ohio Conserv. Bull. 12(6):12–13.

ROEDEL, P. M. 1975. A summary and critique of the symposium on optimum sustainable yield. Am. Fish. Soc. Spec. Publ. 9:79–89.

SCHMITZ, W. R., AND R. E. HETFIELD. 1965. Predation by introduced muskellunge on perch and bass, II: years 8–9. Trans. Wis. Acad. Sci. Arts Lett. 54:273–282.

SMITH, L. L., JR. 1977. Walleye (*Stizostedion vitreum vitreum*) and yellow perch (*Perca flavescens*) populations and fisheries of the Red Lakes, Minnesota, 1930–75. J. Fish. Res. Board Can. 34(10):1774–1783.

SNOW, H. E. 1968. Stocking of muskellunge and walleye as a panfish control practice in Clear Lake, Sawyer County. Wis. Dep. Nat. Resour. Res. Rep. 38. 18 pp.

―――, AND T. D. BEARD. 1972. A ten-year study of native northern pike in Bucks Lake, Wisconsin. Wis. Dep. Nat. Resour. Tech. Bull. 56. 20 pp.

STROUD, R. H. 1948. Growth of the basses and black crappie in Norris Reservoir, Tennessee. J. Tenn. Acad. Sci. 23(1):31–99.

SWINGLE, H. S. 1950. Relationships and dynamics of balanced and unbalanced fish populations. Auburn Univ. Agric. Exp. Stn. Bull. 274. 74 pp.

WEITHMAN, A. S., AND R. O. ANDERSON. 1978a. Angling vulnerability of Esocidae. Proc. Annu. Conf. Southeast. Assoc. Fish Wildl. Agencies 30(1976):99–102.

―――, AND ―――. 1978b. A method of evaluating fishing quality. Fisheries 3(3):6–10.

Dynamics of the Northern Pike Population and Changes that Occurred with a Minimum Size Limit in Escanaba Lake, Wisconsin

JAMES J. KEMPINGER AND ROBERT F. CARLINE[1]

Department of Natural Resources
Box 440, Woodruff, Wisconsin 54568

Abstract

A 56-cm minimum length limit on catches of northern pike was applied to Escanaba Lake in 1964 and removed 9 years later. While the regulation was in effect, the density of northern pike shorter than 56 cm (sublegal size) doubled, and that of larger (legal) fish increased by 47%; total biomass increased by 43% to 7.9 kg/hectare. Growth rates decreased. Prior to 1964, northern pike reached 56 cm at age IV, but by 1972 they did not attain this size until age VIII. Total mortality rates rose from 60% to 82% during 1964–1972, but fishing mortality decreased from 46% to 6% and annual harvest of northern pike fell from 3.6 to 1.2 kg/hectare. Numbers of legal northern pike in the harvest remained about the same after the size limit was imposed, and angling success increased from 0.4 to 0.8 legal fish/100 hours, but mean weights decreased from 1.6 to 1.2 kg/fish. Panfish populations declined when the size limit on northern pike was implemented, and yellow perch increased in abundance after the regulation was abandoned in 1973.

Minimum size limits have been applied since 1909 to northern pike harvested by sport fishermen from Wisconsin waters. Length limits have ranged from 31 to 56 cm.

In recent years, length limits have been characterized by regional regulations. Some counties in southeast Wisconsin have a 56-cm limit, those in the northwest counties a 46-cm one, and the remainder have no size limit. Among lay people and fishery managers alike, there are considerable differences of opinion on the efficacy of size limits as a tool to increase harvests. The interest in improved northern pike harvests led to this size-limit study.

The objective of this study was to determine if the harvest of "large" northern pike could be increased through the application of a length limit. The study also was an opportunity to observe the effects of protecting northern pike on the population dynamics of that species and on predation of the coexisting panfish community.

Escanaba Lake was chosen for this study because a compulsory fishing permit system had been in effect since 1946, there was a long record of harvest and fish population data, and no size limits had ever been imposed. The 56-cm size limit on northern pike was implemented in 1964 and removed in 1973.

Study Area

Escanaba Lake, Wisconsin (latitude 46°04′N, longitude 89°35′W), has a surface area of 119 hectares, a shoreline of 8.2 km, a maximum depth of 8 m, and a mean depth of 4.3 m. The shoreline is irregular and there are several small islands with rocky bars. As the inlet and two outlets are intermittent, fish migration is considered inconsequential.

There are at least 24 species of fish in the lake (Kempinger and Carline 1978). The northern pike is not native to the lake, but approximately 547,000 northern pike fry were stocked between 1937 and 1941. The first natural reproduction that created a sizable year class occurred in 1956. Successful reproduction in subsequent years led to the establishment of northern pike as a major predator.

Methods

Northern pike were captured with fyke nets during the spring spawning season and were marked by fin removal or by affix-

---

[1] Present address: Ohio Cooperative Fishery Research Unit, The Ohio State University, Columbus, Ohio 43210.

ing an aluminum strap tag to the preopercular bone. Ages were determined from scales collected at time of marking. From 1957 to 1963 and in 1977, densities of age I and older males and age II and older females were estimated by the Petersen method. Proportions of marked fish in the population were determined from the sport fishing harvest. After the size limit went into effect, it was only possible to estimate the number of northern pike over 56 cm by the Petersen method. Densities of fish less than 56-cm were estimated by the Schnabel method. Netting periods ranged from 6 to 12 days. Prior to 1964, and in 1977, standing crops of northern pike were calculated by multiplying the average weight of harvested fish by the spring population estimate. While the size limit was in effect, mean lengths of fish caught in nets were converted to weights by a regression established previously; weights then were multiplied by population densities to obtain standing crops.

Total annual mortality rates for individual year classes were calculated from catch curves (Ricker 1975). Instantaneous total mortality rates (Z) were estimated from the slope of the regression of age on natural log of fish numbers. Mortality rates for the 1957 and 1958 year classes were calculated from population estimates at successive ages. For the 1962–1966 year classes, numbers of northern pike captured in fyke nets were used to construct catch curves. Exploitation rates were estimated from the proportion of marked fish caught by anglers. Exploitation rates were calculated for individual cohorts from 1957 to 1963 and for fish over 56 cm from 1964–1972. Total lengths (L, mm) and weights (W, kg) of northern pike caught throughout the fishing season were used to calculate condition factors (K): $K = 10^5 W/L^3$.

All anglers were required by law to obtain a permit without charge from the checking station located at the only landing on the lake. Upon completion of each fishing trip, fishing hours were recorded and catch was inspected. Fish were measured to the nearest 0.1 inch and weighed to the nearest 0.01 pound and then converted to metric measurements. Scale samples were collected for age determination. Annual fishing statistics were based on the fishing year beginning with disappearance of ice-out in spring, approximately April 20th, and therefore consists of a season of open-water fishing plus the ice fishing season.

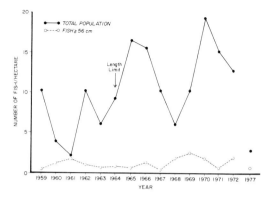

FIGURE 1.—*Northern pike population estimates in spring, Escanaba Lake. Although the 56-cm minimum size limit began in the spring of 1964, it would not have affected density that year.*

Results

Population Density

The first known natural reproduction by northern pike occurred in 1956, but few of this cohort were captured in fyke nets. There were large year classes in 1957, 1960, and 1962 which accounted for the population increases (Fig. 1). Population densities in years without a size limit ranged from 2.1 to 10.1/hectare (mean = 6.9) (Kempinger and Carline 1978). Population biomass closely paralleled density from 1959 to 1964, ranging from 1.9 to 9.9 kg/hectare annually (mean = 5.5). Mean population density of pike 56 cm and greater was 0.9/hectare, and biomass was 1.2 kg/hectare.

Spring population estimates from 1964 to 1972 (size limit in effect) were determined from multiple mark-recapture procedures. There were no year-class failures during this period as year classes from 1962–1966 were all well represented in net catches (Table 1). The population increased substantially, but most of the increase was due to fish less than 56 cm. Fyke net catches indicated northern pike were not fully vulnerable until age IV, whereas

TABLE 1.—*Mortality rates of northern pike calculated as the slope of age versus $\log_e$ of fish numbers. Numbers of fish shown for the 1957 and 1958 year classes are population estimates; 95% limits are in parentheses. Those for the 1962–1966 year classes are numbers of fish captured in fyke nets in spring.*

Year class	Number of fish at age						Ages used to calculate mortality	Correlation coefficient	Instantaneous rate of total mortality (Z)	Total annual mortality (%)
	II	III	IV	V	VI	VII				
1957	1,213 (±73)	511 (±72)	132 (±20)	90 (±22)			II–V	−0.98	−0.915	60
1958		128 (±180)	49 (±16)	21 (±22)			III–V	−0.99	−0.903	59
Means									−0.910	60
1962	370	466	451	163	44	1	IV–VII	−0.99	−1.163	69
1963	1	144	245	188	19	4	IV–VII	−0.96	−1.463	77
1964	394	112	163	161	28	1	IV–VII	−0.92	−1.703	82
1965	1	34	184	124	2	1	IV–VII	−0.94	−1.977	86
1966	6	194	339	108	3		IV–VI	−0.96	−2.363	91
Means									−1.734	82

during the early part of the study they were fully vulnerable to the nets at age II. Therefore, Schnabel estimates from 1964–1972 are useful for assessing numbers of mature fish, but underestimate total population size. The population ranged from 5.9 to 19.3/hectare (mean = 13.2) and biomass from 4.7 to 10.1 kg/hectare (mean = 7.9). Mean density of fish, 56 cm and greater, was 1.3/hectare or 1.6 kg/hectare.

Adult northern pike populations were not estimated 1973–1976. In spring 1977, the northern pike population was calculated by both Schnabel and Petersen methods. Estimates from both methods were similar, 2.9 northern pike/hectare.

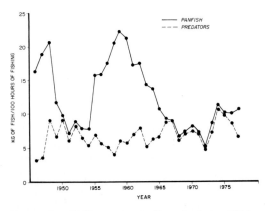

FIGURE 2.—*Fishing success on Escanaba Lake, 1946–1977. Solid line shows panfish (yellow perch, pumpkinseeds, rock bass, bluegills, and black crappies). Dashed line shows predators (walleyes, northern pike, muskellunge, largemouth bass, and smallmouth bass).*

Total Mortality

There was a substantial increase in total annual mortality of northern pike after the size limit went into effect. Mortality rates of fish from the 1957 and 1958 year classes were about 60%, but it should be noted that estimated densities of the 1958 year class were based on small sample sizes and 95% confidence limits were broad (Table 1). Mortality rates of the 1962–1966 year classes continually increased from 69 to 91%. Much of this was natural mortality, because after the size limit went into effect, only a small proportion (about 10%) of the population was subjected to fishing mortality. It is conceivable that total mortality would have increased, even if a size limit had not been implemented, because during early years of the study the population was in the process of building up.

Fishing Pressure and Harvest

Fishing pressure in Escanaba Lake was influenced by availability of panfish. From 1958 to 1963, pumpkinseed and yellow perch were abundant (Kempinger et al. 1975) and fishing pressure was high, 155–222 hours/hectare (mean = 185). By 1965, when the size limit was in effect on northern pike, panfish densities had declined precipitously and fishing pressure dropped to 83–137 hours/hectare (mean = 107). The panfish harvest (Fig. 2) again began rising in 1976 and anglers harvested 5,000

TABLE 2.—*Fishing success for, and catch of, northern pike from Escanaba Lake. Asterisks denote years when the 56-cm size limit was in effect. TL is total length.*

Year	Northern pike 56 cm and larger				Total catch			
	No.	Catch per hectare	Catch per 100 hours	Mean size TL (cm)	No.	Catch per hectare	Catch per 100 hours	Mean size TL (cm)
1958	44	0.4	0.2	59	935	7.9	3.5	41
1959	155	1.3	0.7	60	882	7.4	4.0	50
1960	97	0.8	0.4	61	152	1.3	0.7	58
1961	98	0.8	0.5	62	294	2.5	1.6	49
1962	122	1.0	0.5	64	691	5.8	3.1	50
1963	70	0.6	0.3	60	638	5.4	3.1	44
Mean, 1958–1963	98	0.8	0.4	61	599	5.0	2.7	47
1964*	62	0.5	0.5	59				
1965*	70	0.6	0.6	59				
1966*	70	0.6	0.5	60				
1967*	31	0.3	0.2	59				
1968*	81	0.7	0.8	58				
1969*	217	1.8	1.9	59				
1970*	148	1.2	1.0	59				
1971*	62	0.5	0.4	59				
1972*	103	0.9	0.9	59				
Mean, 1964–1972	93	0.8	0.8	59				
1973	253	2.1	1.9	59	325	2.7	2.4	49
1974	208	1.7	1.9	59	622	5.2	5.6	54
1975	98	0.8	0.5	61	214	1.8	1.2	55
1976	51	0.4	0.3	61	183	1.5	1.2	52
1977	69	0.6	0.4	61	174	1.4	1.0	53
Mean, 1973–1977	136	1.1	0.9	60	304	2.6	2.0	53

yellow perch in 1977. Fishing pressure after the size limit was removed (1973–1977) ranged from 93 to 150 hours/hectare (mean = 127).

Implementation of the 56-cm size limit severely reduced the number of legal size northern pike available to anglers and catches declined commensurately (Table 2). Mean angler catch 1958–1963, years without size limits, was 5.0 fish/hectare. Mean catch rates for all sizes was 2.7 fish/100 hours of fishing. Fifteen percent (0.8/hectare) of the catch rate was 56 cm and larger and mean catch rate of these was 0.4/100 hours. Mean yield (Table 3) was 3.6 kg/hectare. Approximately one-third (1.3 kg/hectare) of the total yield was 56 cm and larger.

During years when the size limit was in effect, mean angler harvest was 0.8 fish/hectare and yield was 1.0 kg/hectare. Mean catch rate was 0.8 fish/100 hours of fishing. Since the size limit was removed (1973–1977), mean catch was 2.6 fish per hectare. Catch rate for all size fish was 2.0/100 hours of fishing. Forty-five percent (1.1/hectare) of the catch was 56 cm and larger and mean catch rate of that size group was 0.9/100 hours. Mean yield of northern pike 56 cm and larger was 1.4 kg/hectare.

Growth Rates, Maturity, Condition Factors, and Exploitation Rates

After the size limit was implemented, population density of northern pike increased and growth rates declined. Prior to 1964, most fish reached 56 cm at age IV. In subsequent years, it was increasingly difficult to accurately read scales from fish age V and older; however, it appeared that after the size limit was in effect, most northern pike did not reach 56 cm until age VIII. Reduced growth rates delayed maturity. Before 1964, males first matured at age I and females at age II. By 1967, males matured at age III and females at age IV.

The best quantitative information we have on reduction in growth rates is from age I–III northern pike. Average lengths of these age groups in 1958–1964 were consistently larger than those from 1965–

TABLE 3.—Yield (kg) of northern pike from Escanaba Lake. Asterisks denote years when the 56-cm size limit was in effect.

Year	Northern pike 56 cm and larger			Total yield		
	Weight	Catch per hectare	Average weight	Weight	Catch per hectare	Average weight
1958	64	0.5	1.5	421	3.5	0.5
1959	233	2.0	1.5	763	6.4	0.9
1960	149	1.3	1.5	202	1.7	1.3
1961	154	1.3	1.6	264	2.2	0.9
1962	211	1.8	1.7	553	4.6	0.8
1963	99	0.8	1.4	379	3.2	0.6
Mean, 1958–1963	152	1.3	1.6	430	3.6	0.7
1964*	88	0.7	1.4			
1965*	89	0.7	1.3			
1966*	88	0.7	1.3			
1967*	38	0.3	1.2			
1968*	93	0.8	1.1			
1969*	262	2.2	1.2			
1970*	173	1.5	1.2			
1971*	70	0.6	1.1			
1972*	139	1.2	1.3			
Mean, 1964–1972	116	1.0	1.2			
1973	305	2.6	1.2	369	3.1	1.1
1974	253	2.1	1.2	580	4.9	0.9
1975	136	1.1	1.4	225	1.9	1.1
1976	71	0.6	1.4	163	1.4	0.9
1977	96	0.8	1.4	171	1.4	1.0
Mean, 1973–1977	172	1.4	1.3	302	2.5	1.0

1972 (Kempinger and Carline 1978). Mean lengths of age II and III males from 1962–1964 were 45 and 53 cm compared to 38 and 42 cm for those of 1969–1971; reduction in growth was 16 and 26%, respectively. When lengths were converted to weights, reduction in mean sizes were 38 and 55% for males age II and III, respectively.

Fishing mortality accounted for most of the total mortality prior to 1964. Exploitation rates (proportion of vulnerable stock actually caught) ranged from 27 to 64% (mean = 46) while total mortality was estimated at 60%. Because of their rapid growth, northern pike recruited to the fishery at an early age, males at age I and females at age II. For fish age II and older, exploitation rates remained approximately the same throughout their life. After the size limit was implemented, mean exploitation rate of legal size fish was 44%, nearly the same as prior to 1964. However, fish over 56 cm accounted for about 10% of the mature population, so that exploitation rate of all mature fish was about 6%. From 1964 to 1972 mean total mortality was 82%; hence, mean rate of natural mortality was 76%, substantially greater than the mean of 14% prior to 1964. The exploitation rate of the northern pike in 1977, without the size limit, was 39% of the 2.9 fish/hectare estimated to be in the lake.

Changes in condition factors of northern pike provided further evidence that growth rates declined substantially after the size limit was implemented. We compared mean condition factors of northern pike within a relatively narrow length range (57–59 cm) to avoid possible biases. From 1958 to 1973 mean condition factors declined greatly with only two major deviations in the trend, 1964 and 1972 (Fig. 3). It is noteworthy that condition factors had been steadily declining since the population first became established and in recent years (1974–1977) there is indication of improvement. The rate of decline in condition factors did not appear to increase after the size limit was implemented, even if data from 1964 and 1972 were omitted. Whether condition factors would have continued to decline if the size limit were not implemented is a matter of conjecture.

Discussion

Development of the northern pike population in Escanaba Lake had a marked effect on the sport fish community (Kempinger et al 1975; Kempinger and Carline 1977). Changes in abundance of prey species, in turn, had an influence on northern pike. Interpretation of data on effects of the size limit were attempted in light of these changes in the fish community.

Density of northern pike increased by at least 90% after the size limit went into effect. Because of the difficulty in estimating numbers of immature fish, we could not define the exact magnitude of this increase. However, it appears that the increased number of northern pike was not simply due to a reduction in fishing mortality. Changes in rates of natural mortality and recruitment also played a role.

Prior to 1964, fishing mortality (46%) accounted for much of the total annual mortality (60%). Although fishing mortality of legal size fish changed little after the size limit was implemented, the impact of fishing on the entire population was negligible; about 6% was harvested annually. Because the actual population was probably greater than the estimated population, actual fishing mortality must have been less than 6%.

The gradual increase in total mortality from 1964 to 1972 was due to natural mortality. Prior to 1964 natural mortality was about 14%. Based on the average exploitation of 6% and average total mortality of 82% (Table 1), natural mortality was about 76% from 1964 to 1972. Therefore, decreased fishing mortality was more than compensated for by an increase in natural mortality, from 14 to 76%.

We were unable to determine changes in recruitment rates because after the size limit went into effect, growth rates declined, age at maturity increased, and vulnerability to capture by fyke nets of fish aged I–III changed. However, if we assume the population was in a steady state, recruitment rates can be estimated from average total

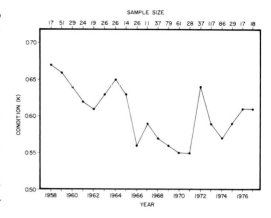

FIGURE 3.—*Mean condition factor of angler-caught northern pike in Escanaba Lake. Data are from fish 57 to 59 cm, inclusive.*

mortality rates and population sizes prior to and after 1964. Mean population density prior to 1964 was 6.9 fish/hectare and total annual mortality was 60%. In a steady state, annual recruits would equal annual deaths, so the expected number of recruits would be $6.9 \times 0.60 = 4.1$ fish/hectare. From 1964 to 1972 mean density was 13.2/hectare and mean total mortality was 82%, so expected recruitment would have been $13.2 \times 0.82 = 10.8$ fish/hectare; a 160% increase over pre-size limit recruitment. This should be a minimum estimate of recruitment, because population estimates represent mature fish and age at maturity increased by about 2 years after the size limit was implemented.

Based on mean lengths of 1–3-year-old northern pike, condition factors, and age at maturity, it is clear that growth rates declined substantially after the size limit went into effect. Reduced growth rates occurred as population density increased, suggesting that growth was density-dependent and that the food supply was being effectively cropped by northern pike. In recent summaries of changes in the sport fish community of Escanaba Lake, it had been suggested that establishment of the northern pike population was responsible for high mortality of juvenile centrarchids and yellow perch (Kempinger et al 1975; Kempinger and Carline 1977). From 1959 to 1972 only one year class of yellow perch survived in sufficient numbers to influence the sport fishery. There was negligible survival of centrarchid year classes so none was of sufficient magnitude to have an impact on the sport fishery. We have suggested that a small number of prey adults were able to produce enough juveniles to sustain walleye and northern pike populations, but mortality of juvenile prey was sufficiently high during years of the size limit to allow only a small number of prey to survive to adulthood.

Changes in condition factors of northern pike provided additional evidence that declining growth rates were related to reduced food supply. From 1958 to 1972 there was a continual decline in condition factors of northern pike with the exception of 1964 and 1972. Since size limits elapsed, condition factors generally improved, most likely due to a burgeoning food supply. The yellow perch population (estimated at 20,000 in 1977) was bolstered by 1974 and 1975 year classes that survived and recruited to the adult population. During 1964, 1972, and 1977, walleye fingerlings also exhibited above-average growth rates. Juvenile yellow perch were the predominant food of walleye fingerlings, and growth rates of walleye were directly related to perch density (Morsell 1970). Thus, the relatively high condition factors of northern pike in 1964, 1972, and 1977 was likely due to an improved food supply.

If the hypothesis that predation by northern pike significantly influenced survival of juvenile centrarchids and yellow perch is correct, then reductions in prey populations and concurrent declines in growth rates of northern pike would be likely outcomes. Aquatic vegetation in Escanaba Lake is confined to protected bays, and much of the shoreline offers little protective cover for prey species. This apparent lack of cover, coupled with increasing northern pike densities in the mid-1960's, could account for drastic declines in abundance of perch and centrarchids.

A decline in growth rates and an increase in mortality are phenomena that might be expected several years after a species has been introduced into a favorable environment. Recruitment of northern pike increased after high population densities were attained, the opposite of what one might predict. We doubt that increased recruitment was due to greater numbers of adults because the large 1957 year class was produced by a population of adults that was too sparse to estimate. If year-class production by northern pike was most influenced by abiotic factors as in some walleye populations (Busch et al. 1975), the increased recruitment after 1964 may have been due to the vagaries of climatic conditions.

Sport Fishery

Northern pike were the most heavily exploited predators in Escanaba Lake (Kempinger et al. 1975). Their preference for littoral habitats and voracious feeding habits probably account for their vulnerability to sport fishing. Within the range of observed fishing pressure (83–222 hours/hectare), exploitation rate (mean = 45%) of legal-size fish did not change appreciably (Kempinger and Carline 1978). Similarly, exploitation rates of walleyes in Escanaba Lake were not influenced by fishing pressure (Kempinger et al. 1975).

The 56-cm size limit resulted in an 84% reduction in average annual catch, and harvest of northern pike greater than 56 cm changed little. Prior to the size limit, average annual yield was 3.6 kg/hectare which was well within the range (1.0–8.5; mean = 4.3 kg/hectare) of northern pike yields reported from other waters (Groebner 1964; Kempinger 1965; Lawler 1961). High yields of 6.7 and 11.0 kg/hectare of northern pike from Murphy Flowage, Wisconsin (Snow 1978) and Bucks Lake, Wisconsin (Snow and Beard 1972) reflect the more favorable habitat of flowages for northern pike, probably because of their extensive littoral areas.

The only thorough study on size limits that can be compared to this report is that from Bucks Lake where a 46-cm size limit resulted in an 82% decrease in yield. In our study, yield declined by 72%. Northern pike in Bucks Lake grew more slowly than those in Escanaba Lake, but size limits at both lakes protected northern pike until age IV. It is noteworthy that in both populations, total annual mortality did not decline as a result of reduced fishing mortality. Snow and Beard (1972) suggested that population size and mortality rates were a function of the food supply. Similarly, in Escanaba Lake, total annual mortality increased as forage abundance declined. However, increased mortality may have been a compensatory response to increased recruitment. We have no way of separating out the effects of food supply and recruitment on total mortality. It was clear from Bucks Lake and Escanaba Lake studies that size limits on northern pike resulted in significant reduction in harvest.

We have suggested that protection of northern pike and the subsequent population buildup led to a decline in panfish populations. After abandonment of the size limit, the panfish population increased (Fig. 2). The limited littoral area in Escanaba Lake may have offered juvenile panfish little cover and made them vulnerable to predation. In lakes with extensive cover, protection of northern pike could conceivably lead to higher rates of predation on panfish without severely depleting their numbers.

Where northern pike populations are relatively dense, the probability of compensatory declines in growth would appear high, if a size limit were implemented. Even where populations are not dense, the most prudent approach to effect a decrease in fishing mortality might be to start with a modest size limit and increase it gradually as conditions dictate.

The questions of where size limits can be used to improve harvest, and how appropriate size limits should be determined, will be continuing problems in management of Wisconsin waters. The impact of size limits will vary, depending upon prevailing exploitation rates, growth rates, and structure of the fish community. It seems unreasonable to assume that a single length limit can produce desirable results over a wide range of lake types and fishing pressures. Even regionally, large differences among lakes can be anticipated, so that a uniform length limit may allow excessive harvest in some lakes and underutilization in others. We recognize that length limits can be an effective management tool, but we suggest that they be implemented only where data indicate they are necessary and even then, limits should be initially conservative.

References

Busch, W.-D.N., R. L. Scholl, and W. L. Hartman. 1975. Environmental factors affecting the strength of walleye (*Stizostedion vitreum vitreum*) year-classes in western Lake Erie, 1960–70. J. Fish. Res. Board Can. 32:1733–1743.

Groebner, J. F. 1964. Contributions to fishing harvest from known numbers of northern pike fingerlings. Minn. Dep. Conserv. Invest. Rep. 280. 16 pp.

Kempinger, J. J. 1965. Estimate and exploitation for fish populations. Wis. Conserv. Dep. Annu. Prog. Rep. 15 pp. (Mimeo.)

———, and R. F. Carline. 1977. Dynamics of the walleye (*Stizostedion vitreum vitreum*) population in Escanaba Lake, Wisconsin, 1955–72. J. Fish Res. Board Can. 34:1800–1811.

———, and ———. 1978. Changes in population density, growth, and harvest of northern pike in Escanaba Lake after implementation of a 22-inch size limit. Wis. Dep. Nat. Resour. Tech. Bull. 104. 15 pp.

———, W. S. Churchill, G. R. Priegel, and L. M. Christenson. 1975. Estimate of abundance, harvest and exploitation of the fish population of Escanaba Lake, Wisconsin 1946–69. Wis. Dep. Nat. Resour. Tech. Bull. 84. 30 pp.

Lawler, G. H. 1961. Heming Lake experiment. Fish Res. Board Can. Prog. Rep. Biol. Stn. Tech. Unit, London, Ontario. 2:48–50.

Morsell, J. W. 1970. Food habits and growth of young-of-the-year walleyes from Escanaba Lake. Wis. Dep. Nat. Resour. Res. Rep. 56. 14 pp.

Ricker, W. E. 1975. Computation and interpretation of biological statistics of fish populations. Bull. Fish. Res. Board Can. 191. 382 pp.

Snow, H. 1978. Fifteen year harvest, exploitation and mortality of fishes in Murphy Flowage, Wisconsin. Wis. Dep. Nat. Resour. Tech. Bull. 103. 22 pp.

———, and Thomas D. Beard. 1972. A ten-year study of native northern pike in Bucks Lake, Wisconsin. Wis. Dep. Nat. Resour. Tech. Bull. 56. 20 pp.

Effects of a Minimum Size Limit on the Walleye Population of a Northern Wisconsin Lake

STEVEN L. SERNS

Department of Natural Resources
Box 440, Woodruff, Wisconsin 54568

Abstract

After the establishment of a 381-mm minimum length limit on walleyes in Big Crooked Lake, Wisconsin, the angler catch and yield of walleyes 381 mm and larger decreased four-fold, while there was approximately a three-fold decline in the number of walleyes of this size in the lake. Mean length and weight of angler-caught walleyes declined as did growth and condition. The number of walleyes less than 381 mm increased during the study period, mainly due to large consecutive year classes. The total annual mortality rate of walleyes 381 mm and greater was directly related to their density but was not correlated with the density of walleyes less than 381 mm. There was a positive relationship between the population density of walleyes 381 mm and larger and angler catch. The exploitation rate ranged from 11.1 to 20.5% and was not related to walleye density.

The value of size limits has been a controversial topic in fish management in recent years. To gain more insight into this problem, a study was initiated in 1971 to evaluate the effects of a recently imposed size limit on the walleye population of Big Crooked Lake, Wisconsin. A 381-mm minimum length limit was established on walleyes for the 1970–1973 fishing seasons and then was reduced to 356 mm before the 1974 season. The data presented in this paper will primarily deal with the population of walleyes 381 mm and greater, as this was the original length limit and because these data should indicate whether the objective of increased angling quality was achieved with the size limit. There was no minimum length limit on Big Crooked Lake walleyes from 1956 to 1970 and the bag limit has been five/day since the mid-1950's.

Study Area

Big Crooked Lake, located in northcentral Wisconsin, is a soft-water lake with a surface area of 276 hectares and a maximum depth of 11.6 m. Total alkalinity and conductivity are approximately 13.0 mg/liter and 37.0 μmhos/cm at 25.0 C, respectively, while total nitrogen and phosphorous levels at turnover are approximately 0.69 and 0.03 mg/liter (Water Resources Research Section 1975). Profiles of dissolved oxygen and temperature obtained on eight different occasions from 1971 through 1974 indicated little stratification during the summer months and only a minor amount of dissolved oxygen depletion in the bottom waters during winter (Serns, unpublished data; Water Resources Research Section 1975).

Of the larger game and forage fishes in Big Crooked Lake, northern pike, burbot, largemouth bass, and rock bass are rare; muskellunge and smallmouth bass are common; yellow perch and walleye are abundant. Nearly all of the fishing pressure on this lake is directed at the walleye.

Methods

Walleyes were caught in fyke nets during the spring spawning runs (late April–early May) from 1971 through 1975. The nets were 25-mm square mesh in 1971 and 1972 and 25-mm and 19-mm square mesh during 1973–1975. All were 1.2 m in diameter and 4.9 m long and were fished with either a 30.5 m or 22.9 m lead net of 25-mm or 19-mm square mesh. Fish were measured (total length) and sex was recorded. There was no significant difference in mean lengths of walleyes caught in 25-mm vs. 19-mm square mesh fyke nets during 1973 ($t = 0.13$, $df = 1424$), so the net catch in the two mesh sizes was combined each year from 1973 through 1975. A chi-square test was used to test for differences in percent size distribution of walleyes <381 mm vs. those ≥381 mm in spring net catches from 1971 through 1975.

Walleyes 381 mm and longer were tagged in 1971 with aluminum preopercle strap tags and from 1972 through 1975 with monel butt-end jaw tags and all were fin-clipped; those walleyes less than 381 mm were given a temporary clip (top of caudal fin) in each year. From 1973 through 1975 all walleyes captured in nets were examined for *Lymphocystis* sp. disease.

Bailey's modification of the Petersen method (Ricker 1975) was used to estimate the number of walleyes 381 mm and greater in Big Crooked Lake from 1971 through 1975. Because the recapture data were obtained from a creel census extending over several months, an attempt was made to determine the effect of recruitment on the population estimates. Two estimates were calculated, one from walleyes caught in May and June only, and another from walleyes caught during the entire season. For each of the five years, the estimates from walleyes creeled during the entire season were slightly higher than those from walleyes caught during May and June. Because of this apparent effect of recruitment, only the estimates calculated from fish caught in May and June are presented.

The number of walleyes in the 267–380 mm size range was estimated by the Schnabel method (Ricker 1975) from capture-recapture data on successive days of netting during the spring spawning runs. An estimate of the standing crop of walleyes 381 mm and greater was obtained by multiplying the population estimate by the mean weight of angler-caught walleyes in May.

Survival rates of walleyes 381 mm and greater were calculated by a formula described by Ricker (1975) that can be applied if recapture data are available through two seasons of tagging. Exploitation rates were estimated each year by determining, through a complete creel census, the percentage of tagged fish recovered by anglers that season.

Complete creel census data were obtained for Big Crooked Lake because all of the fish caught by anglers were taken to one of two lodges for cleaning and freezing. Fish were measured, weighed, and examined for tags by dock attendants who recorded these data on printed creel census forms. During the study period, walleye angling on Big Crooked Lake extended from the first Saturday in May to mid October.

Angler catch rates were calculated in 1972 and 1973 from data supplied on voluntary fishing success forms given to the anglers fishing Big Crooked Lake. Fishing effort was back-calculated by multiplying the catch of walleyes ≥381 mm in the complete creel census by the reciprocal of the catch rate of walleyes of this size and dividing this value by lake area. The number of walleyes <381 mm caught by anglers was estimated by multiplying the catch rate for walleyes of this size by the estimated fishing effort.

Scale samples were taken from the left side of the fish below the lateral line and just behind the pectoral fin. The mean lengths at capture of sexed and scale-aged walleyes from the spring net catches in 1971 and 1972 were compared to those collected in 1974 and 1975 with a t-test. The age groups reported represent the number of growing seasons completed. Lengths at the various annuli were back-calculated for walleyes collected in the springs of 1971 and 1975 to allow construction of Walford (1946) lines for estimates of ultimate attainable lengths. Sexes were combined for calculation of the Walford lines because there was nearly the same percent sex distribution in both the 1971 and 1975 samples.

Condition factors of angler-caught walleyes were calculated from $K = 10^5 \, W/L^3$, where K = condition factor; W = weight in grams; and L = total length in mm. Condition factors were averaged monthly and those of 1971–1972 were compared with those of 1974–1975 by t-tests.

Results

Length Distribution and Disease

The number of large (≥457 mm) walleyes captured during spring netting periods decreased from 1971 through 1975. A

TABLE 1.—*Number of walleyes caught/10 net-days during spring fyke netting periods in Big Crooked Lake from 1971–1975.*

Total length (mm)	Year and net-days effort				
	1971 14	1972 8	1973 16	1974 24	1975 21
254–278	5	215	34	17	2
279–304	41	334	244	242	39
305–329	41	172	238	411	334
330–354	78	184	104	205	450
355–380	108	194	81	109	131
381–405	113	253	106	98	61
406–431	43	116	44	49	29
432–456	9	36	14	12	5
457–481	5	6	3	2	0
482–507	0	3	1	0	0
≥508	0	4	1	1	1

large group of walleyes in the 254–304-mm total length range were taken in the fyke nets in spring 1972; these fish were identified as the 1969 year class. In spring 1973, another relatively large group of walleyes in the 254–304 mm range were captured and most of these fish belonged to the 1970 year class; members of the 1969 year class were primarily in the 305–329 mm range. By 1975 most of the 1969 year class was still below 381 mm, and the majority ranged from 330 to 380 mm (Table 1). The percentage of walleyes less than 381 mm in the spring net catch increased from 1971 through 1975, while the percentage of walleyes 381 mm and greater declined. This change in size structure was significant at the 0.01 level between each year from 1971 through 1974 and significant at the 0.02 level between 1974 and 1975.

The incidence of *Lymphocystis* sp. infection on walleyes under 381 mm increased from 1973 to 1975, as did the percentage of walleyes of this size in the net catch. Similarly, as the percentage of walleyes 381 mm and greater declined from 1973 through 1975, the incidence of *Lymphocystis* sp. on walleyes of this size also decreased (Table 2).

Population Parameters

The number of walleyes 381 mm and greater in Big Crooked Lake declined from 19.8 and 23.8/hectare in 1971 and 1972 to 7.9 and 7.8/hectare in 1974 and 1975.

The standing crop estimates for walleyes 381 mm and greater also decreased from 11.6 and 14.2 kg/hectare in 1971 and 1972 to 4.2 and 4.1 kg/hectare in 1974 and 1975 (Table 3).

Exploitation rates for walleyes ≥381 mm ranged from a low of 0.111 in 1973 to a high of 0.205 in 1971, with an average from 1971 through 1975 of 0.140 (Table 3). Through the five-year period, there was a relatively constant rate of walleye exploitation in Big Crooked Lake with no apparent correlation with population density ($r = 0.50$; $P > 0.05$).

There was an increase in the number of walleyes in the 267–380 mm range from 1971 through 1973 and a slight decline in 1974 and 1975. The increase from 1971 through 1973 can mostly be attributed to the 1969 and 1970 year classes being recruited into this size range in 1972 and 1973. The ratio of 267–280 mm walleyes to those ≥381 mm increased from 0.8:1 in 1971 to 5.2:1 and 4.4:1 in 1974 and 1975, respectively (Table 3).

Total annual mortality rates for walleyes 381 mm and greater from 1971 through 1974 were 0.458, 0.478, 0.322, and 0.157, respectively, while natural mortality rates in the same years were 0.253, 0.340, 0.211, and 0.024, respectively. There was a significant positive correlation ($r = 0.98$; $P \leq 0.05$) between the population density of walleyes 381 mm and greater and the total annual mortality rate of walleyes of this size from 1971 through 1974 in Big Crooked Lake. There was no significant correlation ($r = 0.35$; $P > 0.05$) between the density of walleyes in the 267–380 mm range and the total mortality of walleyes 381 mm and greater.

TABLE 2.—*Incidence of* Lymphocystis *sp. disease on walleyes less than 381 mm and 381 mm and greater caught during spring fyke netting periods from 1973 through 1975.*

	Total length < 381 mm		Total length ≥ 381 mm	
Year	% in catch	% with *Lymphocystis* sp.	% in catch	% with *Lymphocystis* sp.
1973	81.2	4.5	18.8	5.0
1974	85.6	6.0	14.4	2.9
1975	87.7	16.8	12.3	1.8

TABLE 3.—*Population density, standing crop and exploitation rate estimates of Big Crooked Lake walleyes from 1971 through 1975. 95% confidence intervals are in parentheses.*

	Total length 267–380 mm	Total length ≥381 mm			
Year	Population density (no/hectare)	Population density (no/hectare)	Standing crop (kg/hectare)	Exploitation rate	Ratio of walleyes 267–380 mm:≥381 mm
1971	16.8 (10.4–43.9)	19.8 (13.8–25.8)	11.6 (8.0–15.2)	0.205	0.8:1
1972	44.6 (26.0–121.5)	23.8 (16.7–30.8)	14.2 (10.1–18.3)	0.138	1.9:1
1973	48.5 (32.3–110.0)	14.8 (8.5–21.1)	8.7 (5.0–12.4)	0.111	3.3:1
1974	40.7 (33.0–53.1)	7.9 (5.5–10.2)	4.2 (2.9–5.4)	0.133	5.2:1
1975	34.2 (30.4–39.2)	7.8 (4.1–11.5)	4.1 (2.3–6.0)	0.113	4.4:1

Angler Harvest

Angler harvest of walleyes 381 mm and greater in Big Crooked Lake decreased from 1971 through 1975. The catch declined from 4.6/hectare in 1971 to 1.2/hectare in 1975, while the yield decreased from 2.7 to 0.6 kg/hectare. The mean length of angler-caught walleyes dropped from 417 and 419 mm in 1971 and 1972 to 401 mm in 1975 and the mean weight decreased from 584 and 595 g in 1971 and 1972 to 533 and 527 g in 1974 and 1975 (Table 4). There appeared to be a positive correlation between the catch of walleyes 381 mm and larger and population size ($r = 0.86$; $P \leq 0.05$ at $r = 0.88$).

The yield of walleyes 381 mm and greater in Big Crooked Lake in 1971 and 1972 (2.7 and 2.1 kg/hectare) compares closely to an average annual walleye yield of about 2.2 kg/hectare in Leech Lake, Minnesota (no minimum size limit on walleyes). According to Schupp (1972), this is considered a typical summer harvest in Minnesota walleye lakes. The walleye yield in Escanaba Lake, Wisconsin, a 119-hectare lake with no size, bag, or season restrictions, averaged 9 kg/hectare from 1946 to 1969 (Kempinger et al. 1975).

In 1974 and 1975, after the size limit was reduced from 381 to 356 mm, more fish in the 356–380 mm range were caught than those 381 mm and greater. The yield of walleyes 381 mm and greater was slightly larger in 1974 but equal in 1975 to the yield of walleyes in the 356–380 mm range (Table 4). I do not feel that the size limit reduction from 381 mm in 1973 to 356 mm in 1974 caused a serious decrease in the number of walleyes 381 mm and greater caught during 1974 and 1975. In these two years, less than 5% of the anglers caught a daily bag limit of walleyes and less than 1% of the anglers caught a limit comprised only of fish less than 381 mm.

From 1971 through 1975 the number of large walleyes taken by anglers decreased. The percentage of walleyes 457 mm and longer in the catch of walleyes greater than or equal to 381 mm decreased from 14.6% in 1972 to 4.5% in 1975.

The catch rates for walleyes 381 mm and greater from May through July of 1972 and 1973 were comparable to those reported for walleyes by other investigators (Olson 1958; Forney 1967; Kempinger et al. 1975). The catch rate of sublegal (<381 mm) walleyes was 1.5 times higher than for legal walleyes in 1972 and 7.2 times higher in 1973 (Table 5). Many of the voluntary fishing success forms used to calculate the catch rates mentioned above were returned with unsuccessful fishing time recorded. Therefore, I do not feel that these data are seriously biased toward the successful angler. Not enough forms were returned from

TABLE 4.—*Angler harvest of walleyes from Big Crooked Lake (276 hectares).*

Year	Catch (no/hectare)	Mean total length (mm)	Mean weight (g)	Yield (kg/hectare)
Total length ≥381 mm				
1971	4.6	417	584	2.7
1972	3.5	419	595	2.1
1973	1.9	414	590	1.1
1974	1.4	411	533	0.7
1975	1.2	401	527	0.6
Total length 356–380 mm				
1974	1.7	363	366	0.6
1975	1.7	361	380	0.6

TABLE 5.—*Angler catch rates for Big Crooked Lake walleyes from May through July in 1972 and 1973, with back-calculated estimates of fishing effort.*

Year	Month	Catch rate of walleyes ≥381 mm (fish/hour)	Catch rate of walleyes <381 mm (fish/hour)	Number of walleyes ≥381 mm in catch	Estimated number of walleyes <381 mm in catch	Fishing effort (hours/hectare)
1972	May	0.33	0.42	484	616	5.3
	June	0.28	0.55	357	701	4.6
	July	0.30	0.53	64	113	0.8
	May–July	0.31	0.48	905	1,430	10.7
1973	May	0.35	0.86	128	314	1.3
	June	0.15	1.49	232	2,305	5.6
	July	0.03	0.18	53	318	6.4
	May–July	0.11	0.79	413	2,937	13.3

August to October of either 1972 or 1973 to allow for a satisfactory calculation of catch rates during these months.

Fishing effort estimates varied during the three months in each year, but for the period May–July of 1972 and 1973, effort was similar (Table 5). The dock attendants, club managers, and several anglers reported that they observed no obvious change in fishing effort on Big Crooked Lake through the five-year study.

The estimates of fishing effort on Big Crooked Lake were considerably less than those reported on similar sized lakes in the same area of Wisconsin. McKnight and Serns (1974) found an average effort of between 40 and 50 hours/hectare from June–August on three northeastern Wisconsin Lakes with surface areas between 200 and 250 hectares.

TABLE 6.—*Comparison of mean total lengths (mm) at capture of scale-aged walleyes during spring netting periods in 1971–1972 vs. those in 1974–1975. Asterisks indicate two-tailed t-test significance at $P \leq 0.05$*; $P \leq 0.01$**.*

		1971–1972		1974–1975		
Sex	Age	N	Length (a)	N	Length (b)	(a) − (b)
M	III	39	284.5	6	289.6	−5.1
	IV	54	335.3	54	315.0	20.3**
	V	74	373.4	181	335.3	38.1**
	VI	53	396.2	71	370.8	25.4**
	VII	11	414.0	20	408.9	5.1
F	IV	2	381.0	7	332.7	48.3*
	V	10	403.9	113	350.5	53.4**
	VI	11	424.2	39	375.9	48.3**
	VII	3	447.0	7	411.5	35.5

Age, Growth, and Condition

There was a decrease in the growth of walleyes in Big Crooked Lake as this study progressed. This is evidenced by a decrease in the total lengths at capture of scale-aged walleyes collected during spring netting operations between 1971–1972 and 1974–1975 (Table 6).

Walford (1946) lines for 1971 and 1975 indicated a decrease in ultimate attainable lengths ($l\infty$) in the years following the establishment of the size limit. Calculated ultimate attainable lengths were 525 mm and 1971 and 427 mm in 1975 (Fig. 1).

There was a significant decrease in the condition of walleyes 381 mm and larger during the five-year study period after the imposition of the size limit. Mean condition factors (K) for walleyes of the same size caught by anglers during the same month were almost always higher (15 of 18 comparisons) in 1971 and 1972 than in 1974 and 1975 (Table 7).

Mean condition factors increased from May through October for walleyes caught both in 1971–1972 and 1974–1975 (Table 7). This may be a natural change associated with post-spawning increases in weight through the summer. This phenomenon may also be related to a seasonal change in diet noted in 1971 and 1972. Walleyes ate mostly ephemeropteran and dipteran larvae and crayfish from May to July, but from August to October fishes were the major item found in walleyes stomachs (Serns, unpublished data).

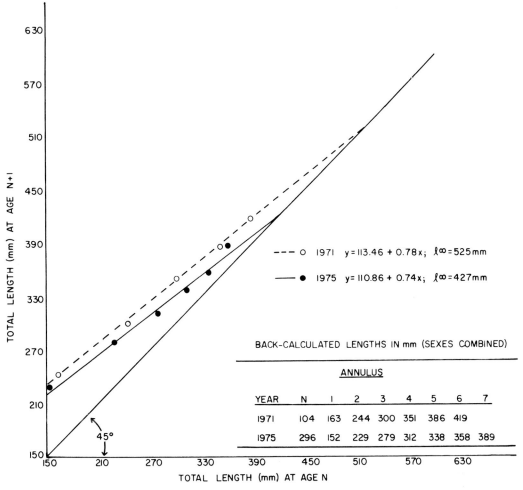

FIGURE 1.—*Walford lines for walleyes collected in the springs of 1971 and 1975 from Big Crooked Lake.*

Discussion

Some of the major objectives for establishing size limits are (1) to maximize the yield or weight harvested (Ricker 1945), (2) to increase the catch of large fish (Anderson 1974) and (3) to protect potential spawners. The predicted success of a minimum size limit presumes that the growth rate and natural mortality rate will remain unchanged (Hackney 1974).

In the present study, there was a drastic decline in the yield and catch of walleyes 381 mm and greater from 1971 through 1975. This decrease in angler harvest appeared to be directly related to a decrease in the number of walleyes of this size in the lake during these years. There was also a decrease in the mean length and mean weight of angler-caught walleyes during the five-year period.

After the establishment of a size limit in Big Crooked Lake, there was an increase in the number of walleyes in the 267–380 mm range, mainly due to the 1969 and 1970 year classes. Although the buildup of fish under 381 mm may have caused the decline in the growth and condition of walleyes 381 mm and greater, there was no apparent effect on the total mortality of the larger walleyes. An accumulation or stockpiling of largemouth bass below the

TABLE 7.—Comparison of mean condition factors of angler-caught walleyes 381 mm and greater in 1971–1972 vs. those in 1974–1975. Asterisks indicate two-tailed t-test significance at $P \leq 0.01$**; all other differences were non-significant at $P > 0.05$.

Total length range (mm)	Month	1971–1972		1974–1975		(a) − (b)
		N	Mean condition factor (a)	N	Mean condition factor (b)	
381–405	May	350	0.80	118	0.77	0.03**
	June	353	0.80	112	0.74	0.06**
	July	107	0.83	39	0.78	0.05**
	August	23	0.85	26	0.82	0.03**
	September	32	0.85	114	0.83	0.02
	October	42	0.85	17	0.84	0.01
406–431	May	293	0.79	52	0.79	0.00
	June	193	0.79	52	0.75	0.04**
	July	68	0.81	16	0.77	0.04
	August	7	0.59	8	0.90	−0.31
	September	12	0.84	31	0.84	0.00
	October	30	0.88	10	0.84	0.04
432–456	May	204	0.78	21	0.75	0.03
	June	123	0.78	15	0.71	0.07**
	July	40	0.79	9	0.75	0.04
	August	6	0.81	5	0.86	−0.05
	September	14	0.86	9	0.85	0.01
	October	14	0.88	3	0.85	0.03

size limit and reduction in growth has occurred in several Missouri impoundments after the establishment of minimum size limits (Rasmussen and Michealson 1974; Farabee 1974). A similar response occurred in Escanaba Lake, Wisconsin, after the imposition of a 559-mm minimum length limit on northern pike (Kempinger and Carline 1978, this volume).

In Big Crooked Lake, there was a direct relationship between the density of walleyes 381 mm and greater and the total annual mortality rate for walleyes of this size from 1971 through 1974. Although no data on mortality rates of walleyes less than 381 mm were obtained in this study, as the percentage of walleyes of this size in the spring net catches increased from 1973 through 1975, the incidence of *Lymphocystis* sp. on these fish also increased. This suggests that with the increase in the density of walleyes less than 381 mm after the imposition of the size limit, there may have been a resultant increase in stress and natural mortality rate.

With a decrease in growth and the same or higher annual mortality rate for fish under a size limit, there would be an expected drop in the recruitment of fish into the legal size range. This would eventually lead to a reduction in the number of fish above the size limit. If a significant increase in the condition of these fish above the size limit did not accompany a decline in their numbers, there would also be a decrease in standing crop and yield (assuming a constant exploitation rate, as indicated in the present study).

It appears that some or all of the factors mentioned above are responsible for the decline in the quality of the walleye sport fishery of Big Crooked Lake after the establishment of a minimum size limit. With no minimum size limit, fish may be harvested above a certain size which is satisfactory to the angler. In lakes where the possibility of the production of large year classes exists, a sport fishery with no size limit helps "thin out" large year classes or groups of year classes which might otherwise cause stunting.

The large 1969 and 1970 year classes in Big Crooked Lake complicated this study. Without them, the detrimental effects of the size limit may not have been as pronounced or the size limit may have been beneficial to the walleye fishery. It should be emphasized however, that these two year classes were produced by an adult population that had not been protected by a minimum size limit.

Knowledge of the population structure and rates of recruitment, mortality, and growth is necessary before a size limit can be intelligently assigned to a fishery. A size limit may increase the sport fishing yield where there is a combination of one or more of the following factors affecting the selected species: (1) limited natural reproduction; (2) good growth; (3) low natural mortality; and (4) high exploitation. One would have to assume, however, that after the size limit was imposed, the above-mentioned parameter(s) that influenced the application of the size limit would not change appreciably. In a lake where there is good natural reproduction, a size limit that protects fish in the medium size range may increase the quality of the sport fishery (Johnson and Anderson 1974). A size limit of this type would allow anglers fishing mainly for food to harvest the smaller fish thereby "thinning out" unusually large year classes. Protecting fish in the medium size ranges may provide greater numbers of large fish for the "trophy" angler.

Acknowledgments

I would like to thank the fish committee and the members of the country club where this study was conducted, and also the dock attendants who took special care in accurately recording the creel census data. I also appreciate the interest of the following who made helpful comments on the paper at various stages in its development: R. O. Anderson, R. F. Carline, L. M. Christenson, W. S. Churchill, D. W. Coble, J. L. Forney, W. C. Latta and T. L. Wirth.

References

ANDERSON, R. O. 1974. Influence of mortality rates on production and potential sustained harvest of largemouth bass populations. Am. Fish. Soc. N. Central Div. Spec. Publ. 3:54–68.

FARABEE, G. B. 1974. Effects of a 12-inch length limit on largemouth bass and bluegill populations in two northeast Missouri lakes. Am. Fish. Soc. N. Central Div. Spec. Publ. 3:95–99.

FORNEY, J. L. 1967. Estimates of biomass and mortality rates in a walleye population. N.Y. Fish Game J. 14:177–192.

HACKNEY, P. A. 1974. Largemouth bass harvest in the midwest, an overview. Am. Fish. Soc. N. Central Div. Spec. Publ. 3:114–116.

JOHNSON, D. L., AND R. O. ANDERSON. 1974. Evaluation of a 12-inch length limit on largemouth bass in Philips Lake, 1966–1973. Am. Fish. Soc. N. Central Div. Spec. Publ. 3:106–113.

KEMPINGER, J. J., AND R. F. CARLINE. 1978. Dynamics of the northern pike population and changes that occurred with a minimum size limit in Escanaba Lake, Wisconsin. Am. Fish. Soc. Spec. Publ. 11:382–389.

―――, W. S. CHURCHILL, G. R. PRIEGEL, AND L. M. CHRISTENSON. 1975. Estimate of abundance, harvest and exploitation of the fish population of Escanaba Lake, Wisconsin, 1946–1969. Wis. Dep. Nat. Resour. Tech. Bull. 84. 30 pp.

MCKNIGHT, T. C., AND S. L. SERNS. 1974. A summer creel census of Stormy, Black Oak, and Laura Lakes, Vilas County. Wis. Dep. Nat. Resour. Fish. Manage. Sect. Rep. 71. 27 pp.

OLSON, D. E. 1958. Statistics of a walleye sport fishery in a Minnesota Lake. Trans. Am. Fish. Soc. 87:52–72.

RASMUSSEN, J. L., AND S. M. MICHEALSON. 1974. Attempts to prevent largemouth bass overharvest in three northwest Missouri lakes. Am. Fish. Soc. N. Central Div. Spec. Publ. 3:69–83.

RICKER, W. E. 1945. A method of estimating minimum size limits for obtaining maximum yield. Copeia 1945(2):84–94.

―――. 1975. Computation and interpretation of biological statistics of fish populations. Bull. Fish. Res. Board. Can. 191. 382 pp.

SCHUPP, D. H. 1972. The walleye fishery of Leech Lake, Minnesota. Minn. Dep. Nat. Resour. Sect. Fish. Invest. Rep. 317. 11 pp.

WALFORD, L. A. 1946. A new graphic method of describing the growth of animals. Biol. Bull. (Woods Hole, Mass.) 90(2):141–147.

WATER RESOURCES RESEARCH SECTION. 1975. Water quality of selected Wisconsin inland lakes 1973–1974. Wis. Dep. Nat. Resour. Spec. Rep. 185 pp.

Selection of Minimum Size Limits for Walleye Fishing in Michigan[1]

JAMES C. SCHNEIDER

Michigan Department of Natural Resources
Ann Arbor, Michigan 48109

Abstract

Typical walleye stocks and fisheries were simulated and trade-offs between yield in weight, catch in numbers, stock size, and egg production were predicted. Under the most likely conditions of growth, natural mortality, and fishing, maximum yield per recruit occurred when walleyes recruited to the fishery during ages III or IV but near-maximum yields were obtained over a broad range of recruitment ages (and sizes). This allows the manager flexibility in managing for alternative goals such as recreational values or stock size. For average sport fisheries in Michigan, the recent change from a 330-mm to a 381-mm minimum size limit is expected to cause: (1) little change in yield; (2) a 20–25% increase in walleye egg production; (3) a 15–20% increase in the total numbers of walleyes caught (legal plus sublegal); (4) a 15–20% increase in the biomass of the population; and (5) a 10–25% decrease in the numbers of legal-size walleyes taken home.

Ever increasing fishing pressure on walleye stocks necessitates an ever increasing level of sophistication in walleye management. In Michigan, as elsewhere, there is concern that native stocks might be over exploited and that stocks maintained by planting might not be providing the highest possible return. Selection of the most appropriate management strategies must be based on a thorough quantitative understanding of the effects of exploitation on populations, and of the effects of fishing regulations and other management procedures on fisheries and stocks.

In this report the Ricker Equilibrium Yield Model is used to examine the effects of age of entry into fisheries and fishing rate on fisheries characteristics (number, weight, and size distribution of the catch) and on stock characteristics (density, reproductive potential, and size-age structure) for typical Michigan walleye populations. The methods and results are applicable to the management of either commercial or sport fisheries (in Michigan or elsewhere), but the sport fishery aspect will receive more emphasis because Michigan's commercial walleye fisheries have been closed since 1970 due to the low levels of certain Great Lakes stocks. A less complex version of these calculations was used as a guideline for increasing the minimum size limit for Michigan's sport fisheries from 330 mm to 381 mm in 1976.

Methods

The Ricker Equilibrium Yield Model (Ricker 1975) provides a simple technique for estimating yields and fish stocks under steady-state conditions given instantaneous rates of natural mortality (M), growth (G), and fishing mortality (F). Beginning with a cohort of 1,000 fish recruited to age II, and continuing to age XXIII, the arithmetic mean biomass of the cohort and its yield in weight were computed at five time intervals per year. Yield divided by the geometric mean weight of individual fish during the interval gave an estimate of the number of fish caught. Under equilibrium conditions the total catch from a cohort over its lifetime will be the same as the annual catch from the entire population. All results are expressed "per 1,000 recruits to age II."

Assumptions are that rates of recruitment, growth, natural mortality, and fishing mortality do not change as the density of the hypothetical population is altered; the validity of these assumptions will be considered later. The model is especially well suited to populations where recruitment can be controlled by stocking small walleyes.

---

[1] A contribution from Dingell-Johnson Project F-35-R.

The values for the variables were selected so as to span the range likely to be encountered in Michigan walleye populations. Where Michigan data were insufficient, I borrowed from elsewhere or made assumptions.

Growth

Growth is an important biological variable in the analysis for it is known to vary considerably among walleye populations (Fig. 1). An empirical average was compiled for Michigan's inland waters by Laarman (1963). This growth pattern is characteristic of the walleye stocks in the Inland Waterway (a chain of lakes including Burt, Black, Mullett, and Crooked lakes) in particular. I have fitted a smoothed step curve to Laarman's data (and the other data in Fig. 1) to simulate the seasonal growth pattern. Nearly all growth in length occurs from late May or June to October at Michigan latitudes (Kelso and Ward 1972; Forney 1977; Kempinger and Carline 1977).

Laarman (personal communication) also provided the following average length-weight relationship: log $W = -5.14176 + 3.03606$ log L, where W = weight in grams and L = total length in millimeters. This equation was used to convert lengths to weights for all growth types modeled.

A typical slow-growing walleye population is found in Lake Gogebic, Michigan. Eschmeyer (1950) reported the growth of Lake Gogebic walleyes up to age X; virtually no growth occurs thereafter. By way of comparison, a number of other walleye populations also grow at rates below the Michigan average: Oneida Lake, New York (Forney 1965); Lake Winnebago, Wisconsin (Priegel 1969); West Blue Lake, Manitoba (Kelso and Ward 1972); Escanaba Lake, Wisconsin (Kempinger and Carline 1977); and Lake of the Woods, Minnesota (Heyerdahl and Smith 1972).

A typical fast-growing Michigan walleye stock is found in Manistee Lake. P. W. Laarman (personal communication) has again provided the basic growth data. Fife Lake, Michigan (Schneider 1969) and Pike Lake, Wisconsin (Mraz 1968) also have fast-growing walleyes. All three populations are

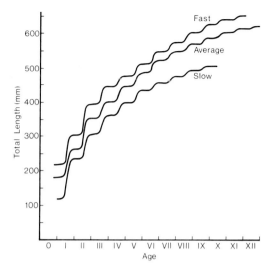

FIGURE 1.—*Typical growth patterns for slow-, average-, and fast-growing walleye stocks in Michigan.*

supplemented by plantings of hatchery fish and are typical of the kinds of situations where more intensive management of walleyes seems to hold promise.

Fishing Mortality

Estimates of total, fishing, and natural mortality of walleyes are summarized in Table 1. For Michigan populations, only rough estimates of fishing mortality can be made because studies depended on voluntary tag returns. I judge that exploitation is usually light to moderate (but it could exceed 50% in certain situations) and that it has been increasing. The minimum rate of exploitation (based on tag returns) increased at Lake Gogebic from 2% in 1947 (Schneider et al. 1977) to 6% in 1976 (R. Juetten, personal communication); at Fife Lake from 4% in 1964–1965 (Schneider and Crowe 1977) to 6% in 1974 (Pettengill 1975); and at the Inland Waterway (Burt Lake) from 7% in the 1950's (Schneider and Crowe 1977) to 18% in 1975 (M. Shouder, personal communication). For modeling purposes an instantaneous fishing mortality (F) of 0.25 (equivalent to an annual exploitation rate, u, of 21% by number) will be considered typical for present-day Michigan sport fisheries. (Mraz 1968 cites a similar level of fishing for Wisconsin

TABLE 1.—Annual mortality rates for walleye.

Location	Age or length and sex	Actual mortality rates			Conditional natural mortality[a]	Reference
		Total	Fishing	Natural		
Michigan						
Bay de Noc	adult	0.65				Schneider and Crowe (1977)
Eastern L. Michigan	adult	0.62				Schneider and Crowe (1977)
Inland Waterway	adult	0.37				Schneider and Crowe (1977)
Fife L.	III–IV	0.22	0.05	0.17	0.17[b]	Schneider (1969)
L. Gogebic	III+ ♂	0.20				Schneider et al. (1977)
	III+ ♀	0.35				Schneider et al. (1977)
Ontario						
Nipigon Bay	360+ mm	0.55	0.07–0.34	0.21–0.48	0.27–0.50	Ryder (1968)
Mississagi R.	III–VI	0.56	0.27	0.30	0.36	Spangler et al. (1977)
Ground Hog Is.	IV–IX	0.60				Spangler et al. (1977)
Shawanaga R.	IV–IX	0.32				Spangler et al. (1977)
Moon R.	V–XV	0.34				Spangler et al. (1977)
Dexter L.	IV–XVII	0.48				Moenig (1975)
Minnesota						
L. Winnibigoshish	IV+	0.44	0.11–0.22	0.28[b]	0.31[b]	Johnson and Johnson (1971)
Leech L.	IV+	0.37				Schupp (1972)
Many Pt. L.	III+	0.31–0.36	0.21–0.33	0.03–0.05	0.05[b]	Olson (1958)
Red Lakes	adult	0.65–0.70				Smith (1977)
Lake of the Woods	VI–IX	0.54				Schupp (1974)
Wisconsin						
Escanaba L.	III+	0.31–0.68	0.13–0.42	0.0–0.33	0.22[b]	Kempinger and Carline (1977)
Pike L.	III+	0.34	0.25	0.09[b]	0.10[b]	Mraz (1968)
L. Winnebago	250+ mm		0.18			Priegel (1968)
South Dakota						
Lewis and Clark L.	II–VIII	0.49				Nelson and Walburg (1977)
L. Sharpe	III–IX ♂	0.49				Nelson and Walburg (1977)
	III–IX ♀	0.48				Nelson and Walburg (1977)
L. Oahe	III–IX ♂	0.56				Nelson and Walburg (1977)
	III–IX ♀	0.38				Nelson and Walburg (1977)
Elsewhere						
L. Erie	II+	0.80				Regier et al. (1969)
West Blue L., Manit.	II+	0.80[b]	0.0	0.80[b]	0.80[b]	Kelso and Ward (1972)
Oneida L., N.Y.	IV+	0.11–0.54	0.10–0.47	0.01–0.07	0.01–0.10	Forney (1967)
Clear L., Iowa	IV+	0.35	0.16	0.19	0.21[b]	Whitney (1958)
Spirit L., Iowa			0.11–0.29			Rose (1949, 1955)

[a] The estimated rate of natural mortality if there was no fishing.
[b] Calculated from data in reference.

walleyes.) For some of the simulations, fishing mortality was varied from 0 to 3 times the typical level, but only the range 0.5 to 2.0 is likely to be realistic with respect to likelihood of occurrence and the response predicted by the model.

As a rule, walleyes in Michigan waters do not become vulnerable to fishing until age II because of their small size. For this analysis it was assumed that insignificant numbers of walleyes would be captured by anglers prior to age II and that during age II they would be half as vulnerable as older and larger walleyes. The rate of exploitation was held constant among older ages in accord with field observations (Johnson and Johnson 1971; Mraz 1968).

A typical seasonal distribution of the walleye sport catch was derived from unpublished catch records (Michigan Department of Natural Resources) for Lake Gogebic, Houghton Lake, and Bear Lake and is as follows: 15% from December 15 to February 28; 0% from March 1 to May 14 (closed season statewide); 45% from May 15 to June 30; 32% from July 1 to September 30; 8% from October 1 to December 14.

Natural Mortality

A number of estimates of conditional natural mortality appear in the literature or can be calculated from data in the literature (Table 1). They range from 0.01 to

0.80 but they are not clearly related to growth or to fishing mortality. Both Oneida Lake, with a very dense, slow-growing population, and Pike Lake, with a sparser, fast-growing stock, have low rates of natural mortality—5–10%. Limited data from Michigan suggest that natural mortality may be on the order of 20% for the lightly fished populations in Lake Gogebic (slow growth) and Fife Lake (fast growth) and that is assumed to be a typical figure for the state.

Whether natural mortality of adult (age II+) walleyes changes with age is not clear (Table 1). Mortality reportedly decreases with age in the Shawanaga stock. Evidence that M is constant comes from tag return data for Lake Gogebic and the Inland Waterway, and the low values of M at Oneida and Many Point lakes. Evidence that M increases with age comes from catch curves for Lake Winnibigoshish, Leech Lake, Lewis and Clark Lake, Lake Sharp, Lake Oahe, and also for the Inland Waterway. If $F = 0.25$ for each age group of walleye in the Inland Waterway, then M increased from 0.01 at age III to 0.83 at age X (Table 2), and was 0.09 overall.

In the simulations, the age-specific values for M from Table 2 were used as well as constant values of 0.10, 0.20, and 0.40. For lack of good information, M was assumed to be evenly distributed through the year.

Selecting Minimum Size Limits

Selection of minimum size limits for walleye fisheries should be based on the biological characteristics of the stocks and on goals established by the manager. Possible goals include maximum yield per recruit (MY/R), increased recreational benefits, or an increase in the walleye population itself. The simulations provide quantitative estimates of the trade-offs accompanying these goals.

Managing for Yield

Yields from various combinations of age of entry into the fishery and multiples of the typical fishing mortality rate were computed by means of Paulik and Bayliff's (1967) computer program of Ricker's equa-

TABLE 2.—*Estimated mortality rates of walleyes in the Inland Waterway based on age-frequency of 852 walleyes in net catches, 1954–1975. ($M = Z - F$, where F is set equal to 0.25).*

Age	Number collected	S[a]	Z	M
I	24			
II	102			
III	210	0.77	0.26	0.01
IV	166	0.73	0.31	0.06
V	116	0.68	0.39	0.14
VI	77	0.64	0.45	0.20
VII	87	0.60	0.51	0.26
VIII	48	0.54	0.62	0.37
IX	11	0.41	0.89	0.64
X	7	0.34	1.08	0.83
XI	2	0.34	1.08	0.83
XII	2	0.34	1.08	0.83

[a] Survival (S) is derived from a smoothed catch curve. Survival of XI and older is set equal to that of age X.

tion. A summary of these results will be presented.

For the rates of growth (state average) and natural mortality (age-specific) believed to be typical of the Inland Waterway, yield isopleths were plotted as a function of age at entry into the hypothetical fishery and fishing mortality rate (Fig. 2). The eumetric fishing line denotes the combinations of age of entry and F which produce the maximum yields per recruit. The dashed isopleths, in the areas of the graph where $0.12 > F > 0.5$, indicate that these predicted yields are less likely to be accurate because the assumptions that growth and mortality

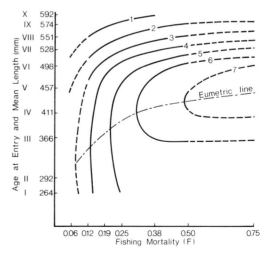

FIGURE 2.—*Predicted yields of walleyes (kg × 100) as a function of age and size (mm) of entry into the fishery, and fishing rate (F). (G = average, M = 0.09, age-specific.)*

TABLE 3.—Predicted maximum yields per 1,000 recruits (MY/R) and corresponding mean lengths at entry in relation to growth and natural mortality (M) at typical (F = 0.25) and twice typical (F = 0.50) rates of fishing mortality.

Growth	M[a]	F = 0.25		F = 0.50	
		MY/R (kg)	Entry lengths[b]	MY/R (kg)	Entry lengths[b]
Slow	0.09	385	310–380	484	365–405
	0.10	448	375–445	490	405–470
	0.20	224	310–365	258	320–405
	0.40	87	[c] –265	113	[c] –310
Average	0.09	559	360–450	701	405–465
	0.10	701	450–525	763	485–550
	0.20	328	300–420	373	400–460
	0.40	128	[c] –307	168	250–350
Fast	0.09	678	400–450	825	430–480
	0.10	824	480–570	905	525–605
	0.20	406	320–445	458	395–450
	0.40	169	[c] –325	225	260–375

[a] M = 0.09 is the average for a series of age-specific rates which increase with age (Table 2); M = 0.10, 0.20, and 0.40 are constant rates independent of age.
[b] Mean lengths (mm) at ages of entry which produce the MY/R within 2%.
[c] Less than 230 mm, the size at which walleyes first become partially vulnerable to angling (in effect, no size limit).

do not change in response to stock density changes may be invalid.

For the typical fishing mortality rate, $F = 0.25$, the largest yield, 559 kg per 1,000 recruits, was achieved by beginning harvest in October of age III when the fish in the cohort averaged 400 mm long. Thus under the present statewide minimum size limit of 381 mm, the Inland Waterway fisheries are operating at close to the ideal level; however we do not have sufficient field data on stock size and catch to verify the yield prediction. The model predicts that yield would drop to 500 kg if entry began at a mean length of 264 mm and the stock would be overexploited in the sense that potential biological production was not being realized. Conversely, delaying fishing until the fish in the cohort reached a length of 470 mm would also decrease harvest to 500 kg and the stock would be underexploited from the standpoint of yield. Note, however, that yield has varied only 11% over that broad range of entry.

Changing growth, or changing the magnitude or age distribution of natural mortality, had a pronounced effect on MY/R and optimum size of entry (Table 3). Varying growth caused up to a 2-fold change in MY/R; varying natural mortality caused up to a 5.5-fold change in MY/R. Combinations of fast growth and low natural mortality resulted in higher yields and necessitated higher minimum size limits. The best minimum size limits, at the typical fishing mortality, ranged from about 230 mm (virtually no limit for fisheries using conventional gear) to 525 mm; however near-maximum yields were achieved over a fairly broad range of minimum size limits. For the combinations of variables closest to the norm, the highest yields were obtained when harvest began in ages III or IV.

The simulations also predicted that as F increases (the trend of the future) the highest yields would be obtained by delaying entry to progressively larger sizes (up to a maximum of about 500 mm for the example shown, in part, in Fig. 2), and the selection of the optimum size limit becomes increasingly critical.

So far as is known, under the present minimum size limit none of the Michigan fisheries are operating in the area below the eumetric fishing line (see Fig. 2), the area where potential biological production is not being realized and where there is a higher risk of failure in natural recruitment.

Managing for Recreation

The above assessment is more useful for commercial than sport fisheries because in commercial fisheries entry can be more readily controlled by restrictions on fishing gear. In sport fisheries significant numbers of sublegal walleyes could be inadvertently caught (and presumably released), thereby providing additional recreation, but also causing some added mortality. Consequently, the predicted increases in yield produced by higher size limits must be tempered by this hooking mortality. Hooking mortality has not been reported for walleye; consequently, an assumed figure of 10% was used to recalculate yield, the catch of legal and sublegal fish, and the biomass of the hypothetical stocks (Table 4). The data presented are for stocks with average growth, but identical numbers of fish would be caught from populations with other growth rates which experience the same

TABLE 4.—Predicted catches (yields and numbers of legal, sublegal, and total fish), biomass of the population (age II and older) in the spring, and egg production in relation to natural mortality (M) and age of entry into the fishery. Given: average growth; F = 0.25 (0.12 for age II); 10% hooking mortality for sublegal fish.

M^a	Entry age	Yield (kg)	Catch Number Legal	Catch Number Sublegal	Catch Number Total	Biomass age II+ (kg)	Eggs (millions)
0.09	II	508	818	0	818	2,097	52
	III	543	767	173	940	2,426	60
	IV	551	691	414	1,105	2,902	75
	V	522	614	636	1,250	3,365	92
0.10	II	545	676	0	676	2,440	61
	III	589	615	163	778	2,638	71
	IV	628	540	369	909	3,181	89
	V	651	475	549	1,024	3,753	110
	VI	654	416	707	1,123	4,336	136
0.20	II	313	476	0	476	1,298	29
	III	322	397	149	546	1,494	34
	IV	311	304	316	620	1,766	43
	V	293	236	445	681	2,026	52

[a] $M = 0.09$ is the average for a series of age-specific rates which increase with age (Table 2); $M = 0.10$, and 0.20 are constant rates independent of age.

mortality. To obtain corresponding estimates of yields for slow-growing stocks, multiply the yields in Table 4 by 0.68 (ages II–V); for fast-growing stocks multiply by 1.26 (age II), 1.24 (age III), 1.21 (age IV), or 1.17 (age V). To obtain corresponding estimates of stock biomass for slow-growing stocks multiply the figures in Table 4 by 0.68 (ages II–V); for fast-growing stocks multiply by 1.28 (ages II–III), 1.27 (age IV), and 1.25 (age V).

These calculations quantify some of the indices of "recreational benefits." For example, for populations with average growth and $M = 0.20$, increasing the age of entry from III to IV would increase the total number of fish caught by 74 (14%). However, this gain would be in exchange for 93 walleyes of legal size which could have been taken home under the age-III entry (reduction of 23%). Yield would not decrease appreciably (to 363 kg from 370 kg) because the legal catch would now be comprised of fish 26% larger. In addition, the residual walleye population would increase by 18% and that might improve catch rate as well as total catch.

Whether or not a walleye fisherman would elect this trade-off is problematical since walleyes are noted for their eating qualities and not their sporting qualities. Certainly this is an area where attitude surveys can help the manager form goals and select options. Of course catch and release strategies become more attractive as exploitation rates increase. The ages or sizes at which walleyes become partially and completely vulnerable to the fishery can have a large effect on the numbers caught and on the choice of restrictions.

The simulations which take into account hooking mortality may give more realistic predictions of yields in sport fisheries for walleyes (Table 4). The optimal size of entry was slightly smaller, but still indicated that yield was not sensitive to size limits.

Managing for Stock Size and Reproduction

The simulations show that fishing dramatically alters the size and age structure of populations given that growth, mortality, and recruitment are not density dependent. For example, with the G and M rates for the Inland Waterway, increasing F from 0 to the typical level of 0.25 decreased standing crop 57% and shifted the predominant age from VI–VII to IV–V (Fig. 3). In general, the initial decrease in stock results from relatively little fishing effort and if feedback mechanisms (such as reduced catchability) did not come into play, fish stocks could often be seriously depleted (Schneider 1973). Beyond that, there is concern that piscivorous species, such as walleye, should have the greater protection

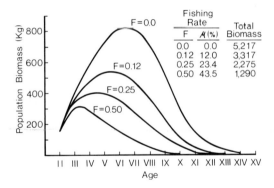

FIGURE 3.—Predicted effects of fishing rate on the total biomass and age structure of a walleye population. (G = average, M = 0.09, age-specific.)

afforded by a higher size limit because they play a key role in maintaining a desirable structure in fish communities (Anderson 1975).

Another possible advantage of a higher size limit on walleye is the assurance of a higher reproductive potential as a result of the accumulation of a larger spawning stock. The increase in reproductive potential was estimated from the following conclusions reached by study of literature. First, the increase in fecundity would be roughly equal to the increase in biomass of mature females (Wolfert 1969). Second, about 72,000 eggs would be produced per kilogram irregardless of growth rate (Eschmeyer 1950; Forney 1976; Johnson 1971; Wolfert 1969). Third, the relationship between growth and maturation of females is approximately: slow growth—50% of the fish mature at age V, 100% at VI+; average growth—50% at IV, 100% at V+; fast growth—50% at III, 100% at IV+. A 1:1 sex ratio was assumed for all ages. Results for average growth conditions are listed in Table 4. To obtain estimates for slow growth multiply the figures in Table 4 by 0.5; for fast growth multiply the figures in Table 4 by 1.58 (ages II and III), 1.54 (age IV), or 1.47 (age V).

Even with slow growth and early entry into the fishery (age II—no size limit) seemingly excess numbers of eggs would be produced. For example with $M = 0.20$, the stock is maintained if survival from egg to age II is only 0.007%. Yet recruitment in some walleye stocks–including dense ones (Forney 1976; Kempinger and Churchill 1972)–has been enhanced by the addition of hatchery-reared fry or fingerlings. Consequently, the possibility that natural reproduction can be enhanced by means of fishing regulations must be considered carefully.

Delaying the onset of harvest from age III to IV increased egg production by about 25%. Whether or not this would result in higher recruitment in a specific walleye population cannot be predicted without a careful analysis of the factors limiting that population. Walleye recruitment is notoriously variable due to weather and a number of poorly understood causes (Koonce et al. 1977). In many lakes a lack of coarse substrate suitable for the incubation of eggs is limiting (Newburg 1975; Prentice and Clark 1978, this volume) so increasing the number of eggs could be either ineffectual or perhaps even detrimental by causing overcrowding of eggs on the spawning grounds. However, in other habitats the increase in eggs will likely increase the probability of higher and more stable recruitment in the long term. Fast-growing and highly exploited stocks seem the most likely to respond favorably.

Discussion

The above analysis has used general data to produce a quantitative assessment of the compromises involved when minimum size limits for different types of walleye populations and fisheries are selected. Naturally, more specific data are desirable when applying the methodology to the management of a specific stock. In the modeling process a number of gaps in our knowledge became apparent.

Results of the analysis showed that walleye stocks with low rates of natural mortality, high fishing mortality, or fast growth should be managed with higher minimum size limits if the objective is MY/R. Under typical conditions, maximum yield on a weight basis will occur when harvest begins during ages III or IV. Fortunately, near-maximal yields are obtained over a relatively broad range of ages and sizes of entry for the walleye as for many other species of fish (Ricker 1975, p. 247). Thus, it is possible that alternative strategies such as managing for recreational benefits, stock size, or reproduction can be pursued with little sacrifice in yield. It also means that few stocks are likely to be grossly mismanaged, from a yield point of view, by a single, statewide regulation tailored to average conditions. Nevertheless, walleye stocks and fisheries with unusual characteristics (especially those with fishing mortality rates less than 0.1 or greater than 0.5) should be identified, studied, and given special consideration.

For some typical Michigan stocks and sport fisheries, such as are found in the Inland Waterway, delaying the onset of harvest from age III to IV (roughly equivalent to the change in minimum harvest size limits from 330 to 381 mm) seems wise because a 10% decrease in number of legal-sized walleyes is exchanged for increases in yield (2%), total numbers caught (18%), population biomass (20%), and reproductive potential (25%). For these fisheries the trade-off is very attractive because natural mortality is low and growth is high between ages III and IV. For fast-growing stocks, such as those found in Manistee Lake and some other waters supplemented by hatchery fish, entry at either age III or IV (approximately 381- or 430-mm minimum size limits) seems best; the choice depends on whether or not it might be worthwhile to increase stock size and reproductive potential. The recent change in minimum size limit should effect less than a 7% change in the characteristics considered. For slow-growing populations, at average levels of exploitation, a somewhat lower (280–330 mm) size limit is appropriate unless there is specific concern about reproductive success. For Lake Gogebic, where walleyes grow slowly and fishing rate is probably about one-fourth the norm, no minimum size limit (equivalent to entry at age II at about 250 mm) would optimize yield and, as compared to the statewide limit of 381 mm, would double the catch in numbers. Reproductive potential might decline by 13% but that is not likely to harm recruitment there.

The assumptions that recruitment, growth, and mortality are not sensitive to the changes in walleye density generated by the manipulation of fishing are crucial to the validity of model predictions. Alterations in fishing mortality rate could have a relatively large effect: changing F from 0.25 to 0.12 or from 0.25 to 0.50 could cause about a 50% change in stock biomass.

While density-dependent processes must place ultimate bounds on walleye stocks, the published field data on the dynamics of walleye populations are sparse, conflicting, and difficult to interpret. Overall, they suggest that among average or fast-growing walleye populations modest (less than 30%?) changes in adult density are not likely to be counterbalanced through compensation of recruitment, growth, or natural mortality. The clearest evidence comes from pond studies on sub-adults where, for yearlings, natural mortality was not related to density (Schneider 1975), and where for yearlings and young of the year, changes in growth were less than proportional to changes in density (Dobie 1956; Schneider 1975). For adult walleyes in natural lakes, some information is available from slow-growing stocks. In one, growth and natural mortality responded to changes in density (Moenig 1975); in another, no such dependence could be demonstrated (Kempinger and Carline 1977); and in a third stock, density-dependent processes appeared to be buffered by food supply (Forney 1977).

Solid information with which to test the model predictions is not available. After the minimum size limit was raised from 0 to 381 mm at Big Crooked Lake, Wisconsin, Serns (1978, this volume) noted an increase in natural mortality and decreases in yield, catch, growth, and numbers of large walleyes in the stock. However, interpretation of these data is confounded by a lack of baseline information and by the chance occurrence of two large year classes which may have triggered those responses in that slow-growing population. Results of the equilibrium yield calculations by Kempinger and Carline (1977) for the slow-growing walleyes in Escanaba Lake were similar to those obtained here but they felt that, in practice, establishing a size limit to optimize yield might be offset by a compensatory response—as occurred in Escanaba Lake with northern pike (Kempinger and Carline 1978, this volume).

A rough idea of the impact that raising the minimum size limit from 330 to 381 mm has had to date on the 15 largest and best Michigan walleye sport fisheries can be obtained by comparing catch estimates for 1975 with those for 1976. These estimates are based on mail surveys of 1% of the state's anglers and are adequate for monitoring major trends. Overall, walleye harvest declined 13% by number, which is about as expected. Harvest should be 10–25% lower once a steady state has been achieved. Actually, the reported catches failed to decline in two-thirds of the fisheries, but this is not surprising considering the high year-to-year variability frequently observed in walleye recruitment and catch. It is the model's ability to average out such variability and reveal the most likely long-term trends that makes it a useful tool.

References

ANDERSON, R. O. 1975. Optimum sustainable yield in inland recreational fisheries management. Am. Fish. Soc. Spec. Publ. 9:29–38.

DOBIE, J. 1956. Walleye pond management in Minnesota. Prog. Fish-Cult. 18(2):51–57.

ESCHMEYER, P. 1950. The life history of the walleye in Michigan. Mich. Dep. Conserv. Inst. Fish. Res. Bull. 3. 99 pp.

FORNEY, J. L. 1965. Factors affecting growth and maturity in a walleye population. N. Y. Fish Game J. 12(2):217–232.

———. 1967. Estimates of biomass and mortality rates in a walleye population. N.Y. Fish Game J. 14(2):176–192.

———. 1976. Year-class formation in the walleye (*Stizostedion vitreum vitreum*) population of Oneida Lake, New York, 1966–73. J. Fish. Res. Board Can. 33(4):783–792.

———. 1977. Evidence of inter- and intraspecific competition as factors regulating walleye (*Stizostedion vitreum vitreum*) biomass in Oneida Lake, New York. J. Fish. Res. Board Can. 34(10):1812–1820.

HEYERDAHL, E. G., AND L. L. SMITH, JR. 1972. Fishery resources of Lake of the Woods, Minnesota. Univ. Minn. Agric. Exp. Stn. Tech. Bull. 288. 145 pp.

JOHNSON, F. H. 1971. Numerical abundance, sex ratios, and size-age composition of the walleye spawning run at Little Cutfoot Sioux Lake, Minnesota 1942–1969, with data on fecundity and incidence of lymphocystis. Minn. Dep. Nat. Resour. Invest. Rep. 315. 9 pp.

———, AND M. W. JOHNSON. 1971. Characteristics of the 1957–1958 and 1939 sport fishery at Lake Winnibigoshish and connecting waters with special emphasis on the walleye population and catch. Minn. Dep. Nat. Resour. Invest. Rep. 312. 31 pp.

KELSO, J. R. M., AND F. J. WARD. 1972. Vital statistics, biomass, and seasonal production of an unexploited walleye (*Stizostedion vitreum vitreum*) population in West Blue Lake, Manitoba. J. Fish. Res. Board Can. 29(7):1043–1052.

KEMPINGER, J. J., AND R. F. CARLINE. 1977. Dynamics of the walleye (*Stizostedion vitreum vitreum*) population in Escanaba Lake, Wisconsin, 1955–72. J. Fish. Res. Board Can. 34(10):1800–1811.

———, AND ———. 1978. Dynamics of the northern pike population and changes that occurred with a minimum size limit in Escanaba Lake, Wisconsin. Am. Fish. Soc. Spec. Publ. 11:382–389.

———, AND W. S. CHURCHILL. 1972. Contribution of native and stocked walleye fingerlings to the anglers' catch, Escanaba Lake, Wisconsin. Trans. Am. Fish. Soc. 101:644–648.

KOONCE, J. F., T. B. BAGENAL, R. F. CARLINE, K. E. F. HOKANSON, AND M. NAGIEC. 1977. Factors influencing year-class strength of percids: a summary and a model of temperature effects. J. Fish. Res. Board Can. 34(10):1900–1909.

LAARMAN, P. W. 1963. Average growth rates of fishes in Michigan. Mich. Dep. Conserv. Inst. Fish. Res. Rep. 1675. 9 pp.

MOENIG, J. T. 1975. Dynamics of an experimentally exploited walleye population in Dexter Lake, Ontario. M.S. Thesis. Univ. Toronto, Toronto. 145 pp.

MRAZ, D. 1968. Recruitment, growth, exploitation, and management of walleyes in a southeastern Wisconsin lake. Wis. Dep. Nat. Resour. Tech. Bull. 40. 38 pp.

NELSON, W. R., AND C. H. WALBURG. 1977. Population dynamics of yellow perch (*Perca flavescens*), sauger (*Stizostedion canadense*), and walleye (*S. vitreum vitreum*) in four main stem Missouri River reservoirs. J. Fish. Res. Board Can. 34(10):1748–1763.

NEWBURG, H. J. 1975. Evaluation of an improved walleye (*Stizostedion vitreum*) spawning shoal with criteria for design and placement. Minn. Dep. Nat. Resour. Invest. Rep. 340. 39 pp.

OLSON, D. E. 1958. Statistics of a walleye sport fishery in a Minnesota lake. Trans. Am. Fish. Soc. 87:52–72.

PAULIK, G. J., AND W. H. BAYLIFF. 1967. A generalized computer program for the Ricker model of equilibrium yield for recruitment. J. Fish. Res. Board Can. 24(2):249–259.

PETTENGILL, T. D. 1975. Evaluation of a walleye, *Stizostedion vitreum vitreum* (Mitchill), stocking program, Fife Lake, Michigan. M.S. Thesis, Central Mich. Univ., Mt. Pleasant. 36 pp.

PRENTICE, J., AND R. CLARK, JR. 1978. Walleye fishery management program in Texas: a systems approach. Am. Fish. Soc. Spec. Publ. 11:408–416.

PRIEGEL, G. R. 1968. The movement, rate of exploitation and homing behavior of walleyes in Lake Winnebago and connecting waters, Wisconsin, as determined by tagging. Trans. Wis. Acad. Sci. Arts Lett. 56:207–223.

———. 1969. Age and growth of the walleye in Lake Winnebago. Trans. Wis. Acad. Sci. Arts Lett. 57:121–139.

REGIER, H. A., V. C. APPLEGATE, AND R. A. RYDER. 1969. The ecology and management of the walleye in western Lake Erie. Great Lakes Fish. Comm. Tech. Rep. 15. 101 pp.

RICKER, W. E. 1975. Computation and interpretation of biological statistics of fish populations. Bull. Fish. Res. Board Can. 191. 382 pp.

ROSE, E. T. 1949. The population of yellow pikeperch (*Stizostedion v. vitreum*) in Spirit Lake, Iowa. Trans. Am. Fish. Soc. 77:32–42.

———. 1955. The fluctuations in abundance of walleyes in Spirit Lake, Iowa. Proc. Iowa Acad. Sci. 62:567–575.

RYDER, R. A. 1968. Dynamics and exploitation of mature walleyes, *Stizostedion vitreum vitreum*, in the Nipigon Bay region of Lake Superior. J. Fish. Res. Board Can. 25(7):1347–1376.

SCHNEIDER, J. C. 1969. Results of experimental stocking of walleye fingerlings, 1951–1963. Mich. Dep. Nat. Resour. Res. Dev. Rep 161. 31 pp.

———. 1973. Angling on Mill Lake, Michigan, after a five-year closed season. Mich. Acad. 5(3):349–355.

———. 1975. Survival, growth and food of 4-inch walleyes in ponds with invertebrates, sunfishes or minnows. Mich. Dep. Nat. Resour. Fish. Res. Rep. 1833. 18 pp.

———, AND W. R. CROWE. 1977. A synopsis of walleye tagging experiments in Michigan, 1929–1965. Mich. Dep. Nat. Resour. Fish. Res. Rep. 1844. 29 pp.

———, R. H. ESCHMEYER, AND W. R. CROWE. 1977. Longevity, survival, and harvest of tagged walleyes in Lake Gogebic, Michigan. Trans. Am. Fish. Soc. 106(6):566–568.

SCHUPP, D. H. 1972. The walleye fishery of Leech Lake, Minnesota. Minn. Dep. Nat. Resour. Invest. Rep. 317. 11 pp.

———. 1974. The fish population structure and angling harvest of Lake of the Woods, Minnesota, 1968–70. Minn. Dep. Nat. Resour. Invest. Rep. 324. 16 pp.

SERNS, S. L. 1978. Effects of a minimum size limit on the walleye population of a northern Wisconsin lake. Am. Fish. Soc. Spec. Publ. 11:390–397.

SMITH, L. L., JR. 1977. Walleye (*Stizostedion vitreum vitreum*) and yellow perch (*Perca flavescens*) populations and fisheries of the Red Lakes, Minnesota, 1930–75. J. Fish. Res. Board Can. 34(10): 1774–1783.

SPANGLER, G. R., N. R. PAYNE, AND G. K. WINTERTON. 1977. Percids in the Canadian waters of Lake Huron. J. Fish. Res. Board Can. 34(10):1839–1848.

WHITNEY, R. R. 1958. Numbers of mature walleyes in Clear Lake, Iowa, 1952–3, as estimated by tagging. Iowa State Coll. J. Sci. 33(1):55–79.

WOLFERT, D. R. 1969. Maturity and fecundity of walleyes from the eastern and western basins of Lake Erie. J. Fish. Res. Board Can. 26(7):1877–1888.

Walleye Fishery Management Program in Texas—
A Systems Approach

JOHN A. PRENTICE

Texas Parks and Wildlife Department
Ingram, Texas 78025

RICHARD D. CLARK, JR.[1]

Texas Parks and Wildlife Department
Austin, Texas 78744

Abstract

This report recounts progress made in Texas walleye fishery management, results of walleye research, and use of a systems approach to walleye management. Seventeen reservoirs of varying features were stocked with walleyes and sampled to determine water quality, standing crops of fishes, and survival, spawning success, and age and growth of walleyes. Results indicated three major factors affected success of walleye introductions: (1) water temperature during spawning time; (2) amount of potential walleye spawning area; and (3) standing crops of potential walleye predators. These factors were used to develop a walleye population dynamics similator model, WALLEYE for walleye management. Rapid assimilation of data and a general systems approach to aid in fishery management decisions were the main contributions of WALLEYE.

Freshwater resources of Texas have undergone major changes in the past 50 years. Once, most of the resource consisted of streams and rivers, but now large reservoirs dominate. Texas has approximately 729,000 hectares of public reservoirs ranging in size from 1.6 to 60,750 hectares and existing under a variety of physical and climatological conditions. Projected reservoir construction indicates an additional 364,500 hectares of water will be impounded in the next 25 years.

Most fishes inhabiting Texas reservoirs originated from stream and river environments. Differences between ecological adaptations inherent to stream fishes and those required by reservoir life may be responsible for many of the problems associated with reservoir fisheries. Reservoirs have often been characterized by initial high fish production and harvest followed by rapid declines (Ellis 1937; Dalquest and Peters 1966; Redmond 1974). Bennett (1947) attributed declining production to changes in fish population structure. Game fishes, generally restricted to littoral zones, are subject to competition for space from forage fishes which are better able to occupy the entire reservoir.

One possible solution to prolonging or improving quality reservoir fishing is to introduce open-water predator fishes to effectively exploit problematic forage fishes. Walleye has been one of the first species chosen for introduction into Texas reservoirs for this purpose.

This report recounts progress made in Texas walleye fishery management, results of walleye research, and use of a systems approach to walleye management.

History of Walleye in Texas

Walleyes were first stocked into Texas waters in 1953 (Toole 1953), but these stockings were not successful. A second walleye stocking attempt was made during 1965 and 1966 in Canyon, Possum Kingdom, and Meredith Reservoirs. Walleye stockings in Canyon and Possum Kingdom failed (Smith 1965, 1967; White 1969), but a significant walleye fishery was established in Meredith (Peters 1965; Crabtree 1967). Walleyes in Meredith provided substantial benefits to anglers and showed little evidence of competition with native game fishes (Crabtree 1967; Kraai and Prentice 1974).

---

[1] Present address: Michigan Department of Natural Resources, Ann Arbor 48109.

TABLE 1.—*Descriptive data for walleye study reservoirs in Texas. Potential walleye spawning area is defined as number of water surface hectares less than 1.5 m deep over clean gravel or rubble bottom.*

	Surface area (hectares)	Drainage area (km^2)	Year impounded	Surface elevation (m)	Mean depth (m)	Maximum depth (m)	Outlet depth (m)	Shore length (km)	Growing season (days)	Potential spawning area (hectares)
Belton	4,982	9,220	1954	180	15	34	26	219	260	27.9
Twin Buttes	3,677	9,645	1962	592	6	17	17	72	233	20.2
Somerville	4,641	2,605	1967	73	4	12	10	137	275	4.0
Conroe	8,499	1,153	1972	61	6	23	18	242	270	20.2
Garza-Little Elm	9,428	4,299	1953	159	8	24	20	295	226	3.2
Eagle Mountain	3,442	5,102	1934	198	7	16	15	319	230	6.5
Diversion	1,385	606	1924	320	4	11	4	45	220	6.1
Canyon	3,337	3,691	1964	278	14	49	38	129	270	16.6
L. B. Johnson	2,582	93,991	1951	253	7	27	10	322	260	2.4
Medina	2,258	1,642	1913	323	14	38	32	97	263	11.3
Casa Blanca	672	303	1966	137	4	11	0	27	322	4.0
J. B. Thomas	3,167	9,127	1952	689	8	25	18	72	217	1.2
Ft. Phantom Hill	1,720	1,238	1938	500	5	20	16	47	223	30.4
Cypress Springs	1,397	194	1970	116	6	18	15	69	234	0.8
Possum Kingdom	8,019	34,473	1941	305	11	44	38	499	221	4.0
Blundell	810	93	1972	104	6	14			234	0.4
Corpus Christi	8,870	43,139	1958	27	4	12	11	322	305	6.1

Observations at Meredith located spawning walleye concentrations along the dam during March and April in 1968 (Crabtree 1969). Artificial propagation of walleyes was initiated in 1969 to supply fry and fingerlings for a statewide introduction program (Clark 1972). Hatch rates initially were low, but experience in walleye capture and stripping techniques and development of adequate egg incubation facilities significantly increased hatch rates. Meredith walleye egg procurement combined with walleyes received from other states from 1969 to 1975 provided 52,712,325 fry and 8,741,780 fingerlings for stocking in study reservoirs throughout Texas. Statewide walleye research began in 1972 to determine reservoir criteria for successful establishment of walleye fisheries.

Methods

Seventeen reservoirs representing various climatological and physical features in Texas were selected for walleye stocking and study (Table 1). Literature was reviewed to compile physical and biological factors affecting the walleye life cycle. Walleye fry and/or fingerlings were stocked in these reservoirs at rates of 72 to 2,863 fry (<2 cm) per hectare and 10 to 408 fingerlings (5–12.5 cm) per hectare (Table 2). Data were collected in each reservoir for at least two years by standardized collection and preparation procedures.

Reservoir descriptive data (Table 1) were compiled from engineering reports (Dowell and Petty 1971, 1973, 1974) for each reservoir at the beginning of the study. Potential walleye spawning area (water less than 1.5 m deep over clean gravel or rubble substrate) was estimated by biologists familiar with each reservoir. At least two stations (upper, lower and/or mid-reservoir) were selected on each reservoir for determining monthly water quality profiles. Water temperature, dissolved oxygen, conductivity, pH, turbidity, total alkalinity, and total dissolved solids were measured from surface to bottom.

One to three coves representing different habitats were sampled with rotenone to estimate standing crop and species composition of fishes in each reservoir. Procedures used were those described by Crandall et al. (1978) and Hayne et al. (1967). Potential walleye predators were catalogued by species and size, and were grouped (Table 3) according to their ability to feed on walleye eggs, fry (<5 cm), fingerlings (5–12.5 cm), and yearlings (12.5–30 cm) as described by Prentice (1977).

Permanent seining stations were established in each reservoir to monitor walleye stocking success and determine relative

TABLE 2.—Stocking rates in number per hectare of walleye fry (A) and fingerlings (B) in Texas study reservoirs, 1969–1975.

Reservoir	1969	1970	1971	1972	1973	1974	1975
Belton	0	0	0	0	99 (B)	67 (B)	0
Twin Buttes	0	0	20 (B)	96 (A) 124 (B)	272 (A) 131 (B)	30 (B)	0
Somerville	0	0	0	0	141 (B)	37 (B)	54 (B)
Conroe	0	0	0	0	694 (A)	529 (A)	0
Garza-Little Elm	0	0	0	30 (B)	49 (B)	49 (B)	0
Eagle Mountain	0	0	0	0	408 (A)	902 (A)	625 (A)
Diversion	0	72 (A)	37 (B)	314 (B)	721 (A) 242 (B)	49 (B)	0
Canyon	0	0	0	0	225 (B)	69 (B)	0
L. B. Johnson	0	0	0	0	2,169 (A)	620 (A)	0
Medina	0	0	0	0	284 (B)	10 (B)	0
Casa Blanca	0	0	0	0	54 (B)	0	0
J. B. Thomas	158 (A)	430 (A)	0	0	190 (B)	0	0
Ft. Phantom Hill	0	0	0	0	408 (A) 40 (B)	408 (B)	464 (A) 59 (B)
Cypress Springs	0	250 (A)	2,863 (A)	32 (B)	0	0	0
Possum Kingdom	0	0	0	0	748 (A)	798 (A)	375 (A)
Blundell	0	0	0	0	1,235 (A)	49 (B)	0
Corpus Christi	0	0	0	0	225 (A)	0	0

strength of year classes. Five stations were established for reservoirs less than 4,050 hectares and 7–10 stations for reservoirs greater than 4,050 hectares. Stations were sampled with 7.6-m bag seines (0.6 cm ace weave mesh) near the middle of each month on an altering day-night schedule from April through October each year.

Fishes were collected over suitable walleye spawning areas during February, March, and April each year to determine gonadal condition and spawning success of walleye. Sampling gears were modified Wisconsin trap nets (1.25 cm mesh) and multi- or monofilament gill nets (46–53 m long × 1.8 m deep with 1.25–8.75-cm mesh).

Total lengths, weights, and scale samples were collected from walleyes captured during spawning observations and during gill net sampling. At least three gill nets such as those used in spawning observations were set for two consecutive nights during September, October and November. This effort was made to observe year-class strengths and to compare growth of walleyes from different reservoirs. Age determinations followed methods similar to those presented by Prentice and Whiteside (1975).

A two-year creel survey was conducted on Cypress Springs Reservoir during 1973 and 1974 (Toole and Ryan 1975). Walleye harvest information on other study reservoirs was estimated by biologists familiar with each reservoir from reported catches.

TABLE 3.—Sizes (cm) of potential predators on various life stages of walleyes in Texas reservoirs.

Predator taxon	Walleye life stage			
	Egg	Fry	Fingerling	Yearling
Lepisosteidae		30–46	>30	>30
Amiidae		23–38	>23	>23
Anguillidae		30–46		
Esocidae			>30	>30
Carassius	>20			
Cyprinus	>20			
Notemigonus	>10			
Catostomidae	>20			
Ictaluridae	15–23	15–23	>23	>38
Noturus	>8	>10		
Aphredoderidae	>8	>10		
Percichthyidae	15–23	>23	>30	>30
Morone		>30	>30	
Ambloplites		15–23	>23	>30
Centrarchidae	>8	>8	>15	
Micropterus		15–23	>23	>25
Percina	>8	>8		
Stizostedion			>30	>30
Cichlasoma	>8	>8	>15	
Tilapia	>8	>8		

Predictive Modeling

Monte Carlo simulation techniques (Schmidt and Taylor 1970; Taha 1971, Clark and Lackey 1975) were used to develop a walleye population simulation model which is discussed in "Evaluating Texas reservoirs for walleye introductions with the aid of a simulation model," by R. D. Clark, Jr. and J. A. Prentice, 1978 (unpublished), Texas Parks and Wildlife Department. Weibull probability density functions were used exclusively to describe random variables in the model as described by Clark and Lackey (1975). Model structure was an adaptation of the Leslie matrix model (Leslie 1945, 1948) which assigns elements of the population matrix to critical stages in the animal's life history. The model was written in FORTRAN IV and developed for IBM 360/50 computer use.

Data input needed to evaluate walleye stocking in a particular reservoir were (1) year of reservoir impoundment, (2) standing crop estimates for walleye predator groups, (3) year of cove rotenone sample, (4) amount of good walleye spawning substrate, (5) reservoir surface area, (6) average water temperature on April 1, and (7) year and number per hectare of each walleye fry and/or fingerling stocking. Model output consisted of a series of walleye life tables for each year of simulation. Projected abundance of all life stages were shown in tables as median number per hectare. Medians were from frequency distributions of possible outcomes produced by Monte Carlo techniques. Output also provided a Monte Carlo probability for each life stage existing in each year of simulation. The projected number of catchable walleye (sum of median numbers per hectare of all adult year classes) for each year of simulation was included in model output and can be graphed against year to monitor catchable population size through time.

Walleye populations were simulated for all study reservoirs, except Corpus Christi. Simulations were also made for two extensively studied walleye populations in Meredith Reservoir, Texas and Canton Reservoir, Oklahoma. Comparisons between simulated and actual walleye populations were made to test model accuracy.

Habitat Conditions

Factors affecting walleye reproductive success and survival of young were compiled by literature review and summarized by Machniak (1975) as (1) quality of spawning sites, (2) water temperature, (3) dissolved oxygen levels, (4) siltation, (5) flow rates, (6) water level fluctuations, (7) pollution, and (8) predation. Water quality, forage supply, and predation factors affect survival and growth of adult walleyes.

The amount of potential spawning area ranged from 0.4 to 30.4 hectares (Table 1). Walleyes spawn in water less than 1.5 m deep over gravel and rubble substrates in most reservoirs (Grinstead 1971; Machniak 1975). Spawns on other bottom types survive poorly (Johnson 1961).

Smith and Koenst (1975) reported the optimum temperature ranges for walleye survival to be 6–12 C during egg fertilization, 9–15 C during egg incubation, 9–21 C for hatched fry through yolk sac absorption, and 21 C for juveniles. Water temperatures were at or near optimum levels for fry, juvenile, and adult walleye survival in all study reservoirs, but optimum temperatures for egg fertilization and incubation were exceeded in many study reservoirs by March. Medina, Casa Blanca, Blundell, and Corpus Christi reservoirs had yearly low water temperatures that were above 10 C at all depths throughout the study.

Dissolved oxygen, conductivity, pH, turbidity, total alkalinity, and total dissolved solids were satisfactory for walleye survival and growth throughout the study. Annual water level fluctuations in study reservoirs ranged from 0 to 3.7 m, but fluctuations during spawning and egg incubation periods each year ranged from 0 to 0.3 m.

Fish Sampling

Cove rotenone sampling was not a satisfactory walleye collection method. A total

TABLE 4.—Adjusted standing crops (kg/hectare) of potential predators of walleye eggs, fry (<5 cm), fingerlings (5–12.5 cm), and yearlings (12.5–30 cm) in Texas study reservoirs, 1972–1976. Year of cove samples used for estimates are in parentheses.

Reservoir	Walleye life stage			
	Egg	Fry	Fingerling	Yearling
Belton	436 (1974)	146 (1974)	184 (1974)	89 (1974)
	114 (1976)	42 (1976)	48 (1976)	33 (1976)
Twin Buttes	147 (1974)	56 (1974)	46 (1974)	12 (1974)
	154 (1975)	70 (1975)	84 (1975)	46 (1975)
Somerville	313 (1976)	84 (1976)	126 (1976)	64 (1976)
Conroe	184 (1974)	147 (1974)	126 (1974)	45 (1974)
	236 (1976)	126 (1976)	145 (1976)	68 (1976)
Garza-Little Elm	105 (1974)	34 (1974)	27 (1974)	17 (1974)
	164 (1975)	36 (1975)	19 (1975)	14 (1975)
Eagle Mountain	338 (1974)	46 (1974)	66 (1974)	48 (1974)
	207 (1976)	56 (1976)	61 (1976)	28 (1976)
Diversion	668 (1975)	37 (1975)	54 (1975)	29 (1975)
Canyon	225 (1974)	140 (1974)	114 (1974)	42 (1974)
	179 (1976)	81 (1976)	68 (1976)	40 (1976)
L. B. Johnson	325 (1974)	30 (1974)	40 (1974)	16 (1974)
	221 (1976)	39 (1976)	46 (1976)	27 (1976)
Medina	132 (1974)	21 (1974)	19 (1974)	9 (1974)
	89 (1975)	34 (1975)	40 (1975)	12 (1975)
Casa Blanca	374 (1975)	151 (1975)	130 (1975)	33 (1975)
	308 (1976)	101 (1976)	96 (1976)	49 (1976)
J. B. Thomas	57 (1974)	57 (1974)	35 (1974)	21 (1974)
	217 (1975)	35 (1975)	43 (1975)	30 (1975)
Ft. Phantom Hill	339 (1974)	20 (1974)	35 (1974)	22 (1974)
	269 (1975)	53 (1975)	73 (1975)	27 (1975)
Cypress Springs	84 (1972)	109 (1972)	135 (1972)	53 (1972)
	129 (1973)	132 (1973)	101 (1973)	53 (1973)
	147 (1974)	128 (1974)	120 (1974)	59 (1974)
Possum Kingdom	371 (1974)	35 (1974)	70 (1974)	34 (1974)
	160 (1976)	72 (1976)	95 (1976)	46 (1976)
Blundell	47 (1974)	83 (1974)	137 (1974)	68 (1974)
	72 (1976)	78 (1976)	329 (1976)	268 (1976)
Corpus Christi	347 (1976)	232 (1976)	231 (1976)	77 (1976)

of 14 walleyes were collected by this method during the study. Standing crop estimates of walleye predators showed highest levels of potential walleye predation occurred during the walleye egg developmental stage (Table 4). Levels of predation generally decreased from egg to fry, fingerling, and yearling developmental stages.

Seine sampling was not an adequate method of collecting young walleyes or evaluating stocking success. Only three walleyes were collected by this method. Low densities of recently stocked walleyes could explain the absence of walleyes in seine collections, but more likely, habitat selection of walleyes placed them in water too deep to be sampled with seines. Seine collections of native fishes during this study were comparable to seine collection records in previous years in all study reservoirs, which indicated sufficient forage to support walleye growth.

Spawning Observations

Spawning activity in study reservoirs generally occurred from mid-March through April, which was earlier than that reported in northern areas (Rawson 1957; Johnson 1961). However, even with early spawning, walleyes in some reservoirs were subjected to water temperatures above the optima indicated by Smith and Koenst (1975) for egg fertilization and, in fewer cases, incubation. Walleye egg hatch rate data (Prentice and Dean 1978) for selected study reservoirs agreed with those of Smith and Koenst (1975); egg hatch rate declined when optimum temperatures were exceeded.

Collections of walleyes over potential spawning areas showed stocking survival and sexual development in 14 study reservoirs. Optimum water temperatures at spawning time (6–12 C) occurred in four study reservoirs: Twin Buttes; Garza-Little Elm; J. B. Thomas; and Ft. Phantom Hill. Walleye eggs were collected in Twin Buttes and Ft. Phantom Hill, confirming reproduction in these reservoirs (Prentice and Dean 1978). Limited spawning areas when compared to reservoir surface area (Table 1) may have caused reproduction failure in Garza-Little Elm and J. B. Thomas; no evidence of reproduction was found in these two reservoirs during the study. Marginal water temperatures (12–15 C) at spawning time may have allowed limited spawning success in seven study reservoirs: Belton; Conroe; Diversion; Canyon; L. B. Johnson; Cypress Spring; and Possum Kingdom. Walleye egg collections at Canyon showed a mean hatch rate of 2.6% under conditions similar to those existing in the reservoir (Prentice and Dean 1978), but conclusive evidence of reproduction was not found for the other reservoirs in this group. Belton, Conroe, Diversion, and Canyon had large amount of spawning area which improved the possibility of successful spawning (Table 1). Walleye reproduction in Conroe was probably limited by rapid water temperature increases from March to April which exceeded the optimum for egg incubation. Walleye reproduction in Diversion may have been reduced by a high level of egg predation (Table 4). L. B. Johnson, Cypress Springs, and Possum Kingdom had relatively small spawning areas which limited walleye spawning success (Table 1). Water temperatures in the remaining six study reservoirs were high enough to cause spawning failure.

Age and Growth

Walleye growth in Texas was similar to that reported in more northern locations, but generally more rapid in young age groups (Prentice 1977). Growth also appeared to decline more rapidly with age in Texas than in northern walleye populations.

Naturally reproduced walleyes were recovered from Canyon, Ft. Phantom Hill, and Cypress Springs. However, extensive studies at Cypress Springs concluded that the walleye fishery did not succeed in spite of this (Bonn and Moczygemba 1975). It should be noted that walleyes less than two years old were difficult to collect by any method. But, evidence of recruitment from age analyses generally corresponded with reservoirs indicated by spawning observations to have successful spawns.

Walleye Harvest

Walleyes were harvested by sport anglers from all study reservoirs except Corpus Christi. Walleye fishing in Somerville, Conroe, Garza-Little Elm, Eagle Mountain, L. B. Johnson, Medina, Casa Blanca, J. B. Thomas, Possum Kingdom, and Blundell exhibited the same pattern as described for Cypress Springs by Toole and Ryan (1975). Walleye fishing in these reservoirs was fair to good for one month to one year after stocked fish reached harvestable size. Fishing success dropped markedly after this period which indicated depletion of those walleye populations. Walleye fishing in Belton, Diversion, and Canyon Reservoirs was low in yield but generally consistent, which indicated possible recruitment at low levels. Walleye fishing in Twin Buttes and Ft. Phantom Hill Reservoirs was good and improved with time, which indicated recruitment and expanding populations.

Predictive Modeling

Study results indicated successful establishment of walleye in Belton, Twin Buttes, Diversion, Canyon, and Ft. Phantom Hill, but unlikely or unsuccessful walleye establishment in the other study reservoirs. Model simulations (Fig. 1) for these reservoirs showed good agreement with walleye populations indicated by study results. Predictions for Medina and Eagle Mountain were the only deviations. Reasons for these deviations are not yet clear, but probably involve the mathematical interpretation of walleye mortality due to high water temperature during egg and fry developmental stages. Simulated walleye populations in Meredith Reservoir, Texas and Canton Reservoir, Oklahoma were also in good agreement with walleye populations de-

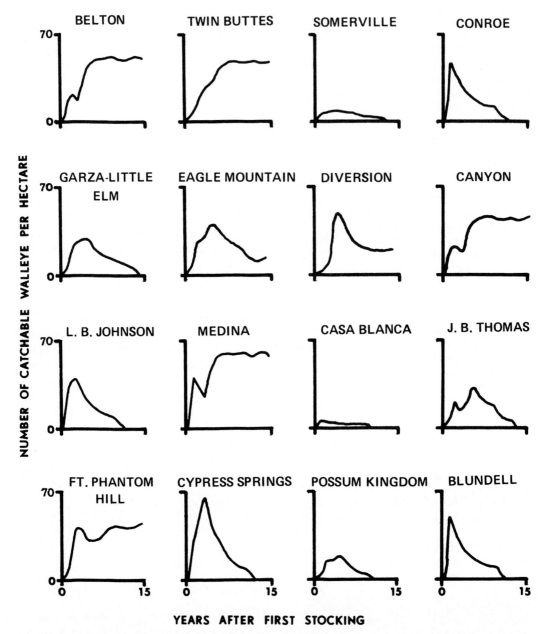

FIGURE 1.—Simulated walleye populations in Texas study reservoirs with most recent predator standing crop information and actual stocking rates as model input.

scribed for these reservoirs by Kraai and Prentice (1974) and Grinstead (1971), respectively. Model predictions were accurate within the ranges of precision of field techniques employed to assess actual populations in 17 of the 19 reservoirs modeled (89% accuracy). Agreement between modeled and expected walleye populations over the wide range of reservoir characteristics sampled showed strong evidence that major factors affecting Texas reservoir walleye stocking success had been identified and that relationships between these factors were accurately described in the population dynamics simulation model, WALLEYE.

The goal of the Texas walleye management program is to maximize angling

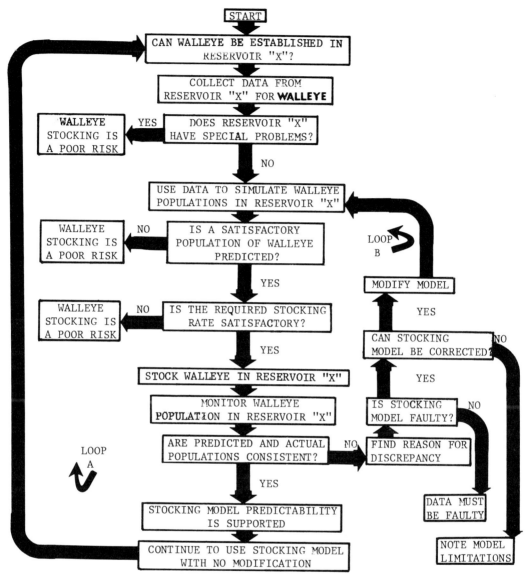

FIGURE 2.—Decision-making system used for evaluating potential of Texas reservoirs for supporting walleye fisheries.

benefit while minimizing cost. WALLEYE was used to devise a decision-making system to help attain this goal (Fig. 2). Fishery managers can formulate a statewide strategy for stocking limited numbers of walleyes with high probability of establishing successful fisheries. WALLEYE aids in rapid assimilation of relevant data from candidate reservoirs and in decisions as to which reservoirs to stock and how many walleyes to stock in each reservoir. Walleye management strategies, such as supplemental stockings or construction of spawning areas, can also be evaluated before they are implemented.

The major limitation of WALLEYE is its status as a single-species model. It cannot predict effects of stocking other fishes along with walleyes, nor are effects of walleyes on native fishes projected.

Acknowledgments

The authors gratefully acknowledge the efforts of most biologists and technical staff of the Texas Parks and Wildlife Department Inland Fisheries Branch. The Texas walleye management program would not exist without the many hours of dedicated field work conducted by these people. Thanks also go to various Department personnel for reviewing the manuscript. Partial funding for this study was provided by the U.S. Fish and Wildlife Service under the Federal Aid in Fish and Wildlife Restoration Program.

References

BENNETT, G. W. 1947. Fish management—a substitute for natural predation. Trans. N. Am. Wildl. Nat. Resour. Conf. 12:276–285.

BONN, E. W., AND J. H. MOCZYGEMBA. 1975. Cypress Springs walleye study. Tex. Parks Wildl. Dep. Dingell-Johnson Fed. Aid Proj. F-31-R-1 Final Rep. Job IV.

CLARK, R. D. JR., AND R. T. LACKEY. 1975. Computer-implemented simulation as a planning aid for state fisheries management agencies. Va. Polytech. Inst. State Univ. Pub. FWS-75-03. 179 pp.

CLARK, S. G. 1972. Statewide walleye program for Texas, section 1: egg procurement. Pages 12–18 in Inland Fisheries staff meeting, Feb. 29–March 2, 1972 (Austin, Texas).

CRABTREE, J. E. 1967. Fish population control recommendations. Tex. Parks Wildl. Dep. Dingell-Johnson Fed. Aid Proj. F-7-R-15 Segment Rep. Job 4.

———. 1969. Walleye and northern pike study. Tex. Parks Wildl. Dep. Dingell-Johnson Fed. Aid Proj. F-7-R-17 Prog. Rep. Job 17.

CRANDALL, P. S., B. W. LYONS, AND R. W. LUEBKE. 1978. Evaluation of gill net and rotenone sampling techniques with total reservoir drawdown. Proc. Annu. Conf. Southeast. Assoc. Fish Wildl. Agencies 30:222–229.

DALQUEST, W. W., AND L. J. PETERS. 1966. A life history study of four problematic fish in Lake Diversion, Archer and Baylor Counties, Texas. Tex. Parks Wildl. Dep. Inland Fish. Rep. 6.

DOWELL, C. L., AND R. G. PETTY. 1971. Engineering data on dams and reservoirs in Texas: Part III. Tex. Water Dev. Board Rep. 126.

———, and ———. 1973. Engineering data on dams and reservoirs in Texas: Part II. Tex. Water Dev. Board Rep. 126.

———, and ———. 1974. Engineering data on dams and reservoirs in Texas: Part I. Tex. Water Dev. Board Rep. 126.

ELLIS, M. M. 1937. Some fishery problems in impounded waters. Trans. Am. Fish. Soc. 66:63–75.

GRINSTEAD, B. G. 1971. Reproduction and some aspects of the early life history of walleyes, Stizostedion vitreum (Mitchill) in Canton Reservoir, Oklahoma. Am. Fish. Soc. Spec. Publ. 8:41–51.

HAYNE, D. W., G. E. HALL, AND H. M. NICHOLS. 1967. An evaluation of cove sampling of fish populations in Douglas Reservoir, Tennessee. Pages 244–297 in Reservoir fishery resources symposium. Am. Fish. Soc. South. Div. Athens, Ga.

JOHNSON, F. H. 1961. Walleye egg survival during incubation on several types of bottom in lake Winnibigoshish, Minnesota, and connecting waters. Trans. Am. Fish. Soc. 90(3):312–322.

KRAAI, J. E., AND J. A. PRENTICE. 1974. Walleye life history study. Tex. Parks Wildl. Dep. Dingell-Johnson Fed. Aid Proj. F-7-R-22 Final Rep. Job 17a.

LESLIE, P. H. 1945. On the use of certain population mathematics. Biometrika 33:183–222.

———. 1948. Some further notes on the use of matrices in population mathematics. Biometrika 35:213–245.

MACHNIAK, K. 1975. The effects of hydroelectric development on the biology of northern fishes (reproduction and population dynamics) III. Yellow walleye, Stizostedion vitreum vitreum (Mitchill): a literature review and bibliography. Environ. Can. Fish. Mar. Serv. Tech. Rep. 529.

PETERS, L. J. 1965. Stocking recommendations. Tex. Parks Wildl. Dep. Dingell-Johnson Fed. Aid Proj. F-7-R-13 Segment Rep. Job 5.

PRENTICE, J. A. 1977. Statewide walleye stocking evaluation. Tex. Parks Wildl. Dep. Dingell-Johnson Fed. Aid Proj. F-31-R-2 Final Rep. Job 12.

———, AND W. J. DEAN. 1978. Effect of temperature on walleye egg hatch rate. Proc. Annu. Conf. Southeast Assoc. Game Fish Comm. 31 (In press.)

———, AND B. G. WHITESIDE. 1975. Validation of aging techniques for largemouth bass and channel catfish in central Texas farm ponds. Proc. Annu. Conf. Southeast. Assoc. Game Fish Comm. 28:414–428.

RAWSON, D. S. 1957. The life history and ecology of the yellow walleye, Stizostedion vitreum, in Lac la Ronge, Saskatchewan. Trans. Am. Fish. Soc. 86:15–37.

REDMOND L. C. 1974. Prevention of overharvest of largemouth bass in Missouri impoundments. Pages 54–68 in J. L. Funk, ed. Symposium on overharvest and management of largemouth bass in small impoundments. Am. Fish. Soc. N. Central Div. Spec. Publ. 3.

SCHMIDT, J. W., AND R. E. TAYLOR. 1970. Simulation and analysis of industrial systems. R. D. Irwin, Homewood, Ill. 644 pp.

SMITH, D. Q. 1965. Stocking of walleye in Possum Kingdom Lake. Tex. Parks Wildl. Dep. Dingell-Johnson Fed. Aid Proj. F-4-R-11 Prog. Rep. Job 3.

———. 1967. Evaluation of walleye fry stocked in Possum Kingdom Lake. Tex. Parks Wildl. Dep. Dingell-Johnson Fed. Aid Proj. F-4-R-13 Prog. Rep. Job 35.

SMITH, L. L. JR., AND W. M. KOENST. 1975. Temperature effects of eggs and fry of percoid fishes. U.S. Environ. Prot. Agency Rep. EPA-660/3-75-0017.

TAHA, H. A. 1971. Operations research: an introduction. MacMillan, New York. 703 pp.

TOOLE, J. E., AND M. J. RYAN. 1975. Creel survey of Lake Cypress Springs. Tex Parks Wildl. Dep. Dingell-Johnson Fed. Aid Proj. F-31-R-1 Final Rep. Job VI.

TOOLE, M. 1953. Walleye pike are coming to Texas. Tex. Game Fish 9:16–17.

WHITE, R. L. 1969. Fishery management recommendations. Tex. Parks Wildl. Dep. Dingell-Johnson Fed. Aid Proj. F-2-R-16 Prog. Rep. Job B-26.

Fisheries Management Theory

ROBERT T. LACKEY

Department of Fisheries and Wildlife Sciences
Virginia Polytechnic Institute and State University
Blacksburg, Virginia 24061

Abstract

This paper places general management theory in a fisheries management framework. Fisheries management is the practice of analyzing, making, and implementing decisions to maintain or alter the structure, dynamics, and interactions of habitat, aquatic biota, and man to achieve specified human goals and objectives through the aquatic resource. Managers usually predict the consequences of a proposed fisheries management decision in a number of ways, including rules of thumb, past experience, formal models, experimentation, trial and error, and of course, pure guess. A key and obvious problem in making accurate predictions of the consequences of a proposed management decision is the complexity of most fisheries. Numerical analysis or lack of data have often been identified as the major problems with using formal models in fisheries management, but lack of basic management theory is even more critical. A fundamental premise in all fisheries management is that all benefits derivable from fisheries management are accruable solely to man. Given this premise, a simple general theory of fisheries management can be developed in which most of the controversy surrounding fisheries management decisions revolves around which goals and objectives are selected and *who* selects them. In such a "general theory of fisheries management" biological factors are largely constraints and are only rarely major decision variables.

To analyze fisheries management and develop management theory, it is necessary to define the system of concern. A *fishery* (either recreational or commercial) is a system composed of habitat, aquatic animal and plant populations (biota), and man. In a broad sense, *fisheries science* is the study of the structure, dynamics, and interactions of habitat, aquatic biota, and man, and the achievement of specified human goals and objectives through use of the aquatic resource. *Fisheries management* is the analysis of alternative decisions, and implementation of a decision or decisions to meet human goals and objectives through use of the aquatic resource. When one considers the number and diversity of components which constitute a fishery (i.e., fishes, plankton, benthos, rooted plants, chemical and physical water characteristics, various sorts of anglers, and related recreational and commercial activities), the true complexity of a fishery and fisheries management becomes apparent. A slight change in part of the fishery may result in substantial change in another, seemingly unrelated part.

Although their size is usually relatively small, even recreational fisheries are complex. There are usually many game fish populations to consider and angler diversity is also great. Some anglers exclusively pursue a single fish species, while many exhibit little species preference. Management strategies for each type of fishery differ greatly.

Prediction is the essence of all management, including fisheries management. Managers usually predict consequences of a proposed decision in a number of ways, including rules of thumb, past experience, formal models, experimentation, trial and error, and pure guess. None of these ways is totally acceptable as a predictive tool, but all have a place in fisheries management. Depending on the circumstances, any method of prediction may be appropriate.

A key problem in accurately predicting the consequences of a proposed management decision is the inherent complexity of recreational fisheries. Even if some components of a fishery are well understood, the number of interrelationships is staggering. Further, the dynamic aspects of all components of a fishery are important because *rates* of change are as important as the magnitude of the components themselves. For example, growth rate of an individual fish is affected by many components of a fishery, even though some of those linkages may be unknown or obscure.

The definition of a *model* must be clarified for the purpose of subsequent discussion: in a general sense, a model is simply an abstraction of a system. Models may be verbal, graphical, physical, or mathematical (including computer-implemented models). However, fisheries modeling nowadays usually connotes modeling of a mathematical nature and typically involves computer-implemented simulators.

Management Goals and Objectives

Any realistic theory of fisheries management must have a "bottom-line," or what and how much is produced through management efforts—i.e., goals, objectives, and desired performance. A management "objective" is a statement of the desired result of a decision or set of decisions. In classical management terminology, an objective is not equated with a "goal," which is defined as the end toward which a strategy tends; that is, a goal is an ideal or aim which is usually expressed in general and abstract terms. A few examples of goals in recreational fisheries management are: "best" or "wisest" use of resources; conservation, protection, and enhancement of the resource; and realization of the greatest amount of recreational opportunity for the greatest number of people. Whereas goals provide general direction to agency programs and are useful in public relations, clear, sound objectives are vital to developing sound management strategies.

Objectives can be described from many vantage points: aspirations for preferred or desirable conditions; end-points to be reached which are attainable and measurable; or simply fairly specific goals. However defined, objectives have some very important properties which affect their use (as contrasted to goals) in fisheries management: (1) objectives are clearly stated; (2) objectives are specific and not filled with broad, general terms; (3) objectives are quantifiable by some means, if not empirically, then subjectively; and (4) objectives have a performance measure which can be used to evaluate management progress and effectiveness.

Effective management of any natural resource system is based upon clear and formally stated objectives. Most fisheries managers have recognized the inherent difficulties of operating without functional objectives and have tried to institute measurable objectives (Roedel 1975). Historically, the most common management objective has been to maximize the sustained yield in weight or numbers of fish (MSY) (Larkin 1977). Some common variants are maximizing yield of certain species or certain sizes of fish. MSY is desirable because it is conceptually simple and is an objective-oriented approach to management. However, maximum sustained yield is restrictive because most anglers regard catch as only one of several measures of output from a fishery (McFadden 1969).

Most anglers agree that their interest is not solely in the fish they catch, but also in *fishing* itself (Martin 1976; Kennedy and Brown 1976). Aspects of fishing important to the angler may be the outdoor experience, environmental aesthetics, and the sporting challenge, as well as the species and sizes caught and the method by which they are pursued (Driver and Knopf 1976).

Among efforts to incorporate other criteria into objectives have been attempts to measure quantities such as man-days of use. The assumption is that measuring the number of angler-days of recreation on a particular fishery is a realistic index of benefits received. Some may also go further and assume that this approach could be used to maximize recreational benefit. However, maximizing angler-days may, in fact, reduce both individual and total societal benefits. Neither potential fish yield (all species considered) nor intolerance to angler crowding constitute foreseeable limits on potential recreational fishing in North America. The angler-day concept does not incorporate a quality aspect, but is at least measurement of management performance in human-oriented terms rather than biological ones (Radovich 1975).

Other possible objectives in recreational fisheries management are maximizing aesthetics and diversity of angling opportunity. While these are altruistic approaches, they are not readily quantifiable. Further, without a functional pricing system, the value of various recreational factors cannot be easily determined by a market survey of the angling public. Quality is an extremely vague and variable parameter to measure, but many factors which contribute to the quality of the recreational fishing experience can be presumably be delineated and measured. The number of potential variables is large, but specific fisheries may have only a few aspects which determine quality. Recent work with "proportional stock density" represents an indirect approach to quantifying the benefits of fishing through a measure of catch (Anderson 1976; Anderson and Weithman 1978, this volume).

Beyond the issue of *what* goals and objectives are appropriate in recreational fisheries management is the question of how to *select* those goals and objectives. Identifying, selecting, articulating, and ranking of goals and objectives is never easily achieved. In particular, there are many problems concerning quantifying and measuring aesthetic and environmental factors. Further, managers may be unwilling to formulate objectives because of fear that some of the "real" objectives would be disapproved under scrutiny by the public, and that some might not be approved by all interested parties. Even given a willingness to set objectives, managers may be unable to formulate objectives because of three additional major difficulties: incomplete problem awareness; incomplete knowledge of the intracacies of the problem; and inability, due to time, money, or manpower constraints, to devote sufficient thinking to the effort. Further, most classical objective-setting methodology stresses only the importance of objectives without providing practical means for determining or detailing them. However, several techniques are available and, when used in combination, could provide a reasonably sound framework for determining objectives. The strawman/discussion technique, tree structures, relevance trees, the brainstorming technique, Delphi method, and attitude surveys are just six objective and goal determination procedures. These are discussed elsewhere (Lackey 1975).

An issue for resolution is: *Who* should set goals and objectives: agency personnel, the public, or a combination of the two? Historically, fisheries management goals and objectives have been arrived at by professionals in organizational positions, a planning process which theoretically allows those who are best qualified and most knowledgeable to set goals. Public participation is often minimal (Scheffer 1976). An informed and concerned public is essential for fisheries management decision-making in the current social and political climate. Fisheries managers cannot rely solely on public opinion in formulating goals, but public opinion is valuable input.

Decision-making in fisheries management is characterized by a division of labor around functional specialists capable of responding to narrow problems efficiently and competently. Advantages of this type of decision-making process are that it usually employs professional ethics and standards, and it often uses rational decision-making processes in which goals and objectives are often clearly defined, pertinent data collected, and alternatives surveyed and selected. This approach is intuitively appealing in principle but difficult to apply in practice. Goal setting primarily involves *value* judgments concerning desirable or undesirable consequences of alternative fisheries management programs. Scientifically trained personnel are no more qualified than the general public to make these value-based decisions.

Procedures for establishing broad goals permit citizen representatives or panels to collaborate with professionals in decision-making. Attitude and opinion survey techniques offer promising opportunities for agencies to procure direct public input. Sampling techniques based on fishing license records can be used in mail, telephone, or personal interviews. No single procedure should be emphasized, but combinations of the various techniques used as supplements to one another will likely prove most useful. The disproportionate influence of relatively small but well organized special interests groups is something all fisheries managers should recognize and consider with whatever goal setting procedure is used (Belusz 1978, this volume).

Modeling and Fisheries Management

Management theory and modeling are often closely related because the modeling process forces the modeler to state formally his theory; in fact a model can be looked upon as a formalization of a theory. Most fisheries-related models, even those seemingly unrelated, are quite similar in philosophy and approach, but there is substantial variation among models when their intended use is studied. Models in fisheries management can be categorized into families which include one or more fisheries components (habitat, aquatic biota, and/or man). The evolution of fisheries models has not followed a discrete path, but rather has taken a disjointed and often circuitous route. Major trends in development apply equally to recreational or commercial fisheries and marine or freshwater fisheries, but different evolutionary trends are of great importance when evaluated by the scientific effort expended.

Modeling in fisheries management may be justified in many ways, some of which result in benefit/cost ratios much greater than unity and others which do not. The first and perhaps most obvious potential benefit of modeling in fisheries management is organizational. Fisheries are highly complex systems and modeling (graphical or mathematical) does provide a medium for clarification and organization. In this context, a model is a theory about the structure, dynamics, and function of a fishery or a fisheries component.

A second potential benefit of modeling in fisheries management is as a self-teaching device to the builder or user. There is probably no better way to develop a "feel" for a fishery than to model it formally. Some fisheries models, particularly computer-implemented models, can serve as useful management exercises in agencies and universities.

Identifying gaps in understanding of a fishery is a third potential benefit from modeling in fisheries management. The modeler may become painfully aware of areas of missing or inadequate data. Acquisition of these data may well be top priority for improving management and the model. "Sensitivity analysis" identifies the parameters of most importance in determining model output, and data acquisition and research efforts may be allocated accordingly.

Models used as research tools may be considered as a fourth category of potential benefits. Manipulation of the model itself may generate "data" which are unattainable from the real system. For example, rainfall and water temperature may each have an impact on certain biotic components, and certain combinations of rainfall and temperature have been observed in the field to quantify the impact. Exercising a model may permit a reasonable assessment of the general relation by interpolation (based on existing data combinations) and prediction of what impact would occur if a new combination of rain and water temperature were to occur.

The fifth and most discussed potential benefit of modeling in fisheries management is predicting the impact of alternative management decisions or external influences. Historically, fisheries managers have been interested in predicting the impact of a proposed fishing or exploitation rate expressed in the form of a season, gear type, and size or creel limit. Managers wish to estimate the impact of decisions on the number of realized angler-days, catch, or some other measure.

Ecosystem Models

Ecosystem models address either or both of the habitat and biotic components of a fishery. Ecosystem models of the habitat type include those developed to predict aquatic temperature regimes, toxicant dispersal, and sediment transport. For example, one problem which exists in recreational fisheries management is predicting the structure and function of environments in proposed reservoirs, but managers (and modelers) must first address and solve the problem of predicting future habitat characteristics, including physical and chemical parameters, before, first, ecosystem and, then, fisheries models can be accurately predictive.

Models of the biotic component of a fishery include classical fish population dynamics models and models of single- and multiple-population systems. In this category we find the Schaefer, Beverton and Holt, and Ricker models. Nearly all of the extensive literature on population dynamics as applied in fisheries science falls into this category (e.g., Schneider 1978, this volume).

Ecosystem models which address both habitat and biotic components are becoming increasingly common in fisheries science and other areas of renewable natural resources management (e.g., Bovee 1978, this volume; Prentice and Clark 1978, this volume). Accounting for component interaction is a key point in ecosystem models and much of the profuse literature deals with interaction characteristics and mechanisms to describe them. Freshwater systems have been modeled more frequently than marine systems, due in part to the rather discrete nature of lakes and, to a lesser extent, streams. The next step in ecosystem model development may well be an effort to solve the problem of managing an evolving or unstable system.

Bioeconomic Models

Models which address solely the third fisheries component, man, may be termed social models. In commercial fisheries, managers have tended to measure fisheries output as weight of fish or perhaps gross income. In recreational fisheries, output is composed of many factors, including aesthetics as well as catch. From a management and modeling standpoint, the relevant questions are: How do people respond to changes in renewable natural resources? How can human behavior be predicted, or at least the behavior of part of the human population?

Bioeconomic models, as the name implies, include the biotic and human components of a fishery. Bioeconomic models are integral to management of commercial fisheries but relatively neglected in recreational fisheries. Managing trends in use of aquatic renewable natural resources may prove to be of much greater importance as human recreational and commercial demands continue to increase.

Fisheries Models

Fisheries models, in the broadest sense, combine the major fisheries components of habitat, biota, and human use. At such a comprehensive level of analysis, detailed modeling borders on the impossible. However, if certain realistic constraints (i.e., economic, political, and social realities) are added to a comprehensive fisheries model, a relatively complete decision-making system can result.

Few management decisions are explicitly based on these kinds of models. Most strategies, however, are implicitly based on two widely known single population models which must assume that the main or perhaps sole measure of human use should be biomass of fish harvested: the dynamic pool model (Beverton and Holt) and the logistic model (Schaefer). The dynamic pool model describes a stock in terms of the vital statistics of recruitment, growth, and mortality. Each statistic is assumed to be a continuous deterministic function of time. Implementing the dynamic pool model requires a large amount of data. The logistic model, also called the surplus yield model, combines the effects of recruitment, growth, and natural mortality into a single differential equation for change in population biomass. The logistic model, usually employed when information is relatively scanty, requires only catch and effort data.

Large-scale, computer-implemented simulations of fisheries are becoming increasingly common. While these kinds of models may be useful in improving management in a study area, their generality is usually limited.

General Theory of Fisheries Management

If, as a basic premise in recreational fisheries management, we assume that all benefits derivable from aquatic renewable natural resources are accruable exclusively to man, then it follows that a general theory of fisheries management may be written as

$$Q_{MAX} = f(X_1, X_2, \ldots X_n | Y_1, Y_2, \ldots Y_m);$$

Q = some numerical value of societal benefit;
X_n = management decision (n = the number of all possible decisions); and
Y_m = management constraint (m = the number of all possible of constraints).

The vertical line is short-hand for "given that." The theory reads *the greatest societal benefit* (Q) *derivable from a fishery can be realized by manipulating a series of decision variables* (X's) *given a set of constraints* (Y's). Controlled or partially controlled decision variables (X's) are those regarded as management prerogatives (stocking, habitat improvement, etc.). Noncontrollable variables (Y's) are random or dependent on other factors (weather, highway development, recreational attitudes, etc.). Variables may, however, overlap both categories. Within these constraints (Y's), the manager selects decisions which maximize Q.

Clearly the "bottom line" in the general theory is societal benefit (Q). Practically, the management problem facing all recreational fisheries agencies is evaluating how best to allocate limited financial resources to meet particular goals and objectives which, in total, constitute Q. Given the quality ranked angler-day (or some other quantity) as a measure of output (Q) from a fisheries management program, how can an agency allocate its resources to increase angler-day production within a relatively fixed budget (one of the Y's)? For example, how many angler-days accrue from: (1) building additional state-owned lakes; (2) improving support facilities at existing state-owned lakes; (3) stocking various species and numbers of fish; (4) managing intensively with lake fertilization and adjustment of fish populations; (5) educating the angling public; (6) law enforcement; or (7) improving angler or boat access to fisheries? Some agencies have additional methods of increasing the number of angler-days, while others have few alternatives.

Management problems, such as evaluating the "mix" of decision alternatives (the X's), are not unique to fisheries management. Planning in business is integrally involved in allocating resources toward maximizing specific objectives. In fact, many current business management decisions are made within a simulation of alternative marketing strategies. With computer assistance, many past, present, and future technological conditions can be analyzed, performance measured under each, and the best course of action selected.

Angler use of aquatic resources is one of the major interactions of man with aquatic biota and habitat. Angler use may be treated as a decision variable (X), but it is typically viewed as a decision constraint (Y). Thus, the level and type of resource use is a major concern of management agencies, but use trends in recreational fisheries are generally uncontrolled (Clark and Lackey 1975). In practice, angler use trends are nearly always viewed as phenomena extrinsic to fisheries management, but in reality, they are only partially extrinsic. Virtually all management agency programs and activities have an effect on the location and intensity of angler use. Land acquisition, dam construction, pollution control, fish stocking, and access development are common examples.

Management policies in fisheries have typically been designed to respond to angler use trends but rarely to shape or influence them (McFadden 1969). If fisheries management policies were explicitly designed to regulate angler use, greater benefits might be accrued to society from fisheries. Regulation of angler use could be achieved by limiting licenses, but such a tactic is often neither politically nor culturally acceptable. A less dictatorial approach, based on subtle relations between individual management activities and angler use and preferences, might also be effective and perhaps more politically palatable.

Angling regulations, information transfer, and educational programs address human components in fisheries management, but such efforts alone cannot be relied upon to direct angler use in a desirable direction (Clark and Lackey 1975). One or two actions in a complex management system are invariably inadequate to achieve a desired change. For example, while information and education efforts are working to direct angler use along a particular course, other agency activities may be working subtly to contravene that course.

The general theory of fisheries management is not difficult to accept if one accepts the premise that all benefits from fisheries are accruable to man. Unfortunately, this premise often gets lost in semantic jargon and emotionalism. Further, many "objectives" in recreational fisheries management revolve around vague terms such as maximizing recreational "benefit" or making "best" use of the resource. These "objectives" have a strong emotional appeal and are philosophically valid, but they are too ambiguous for developing meaningful management strategies. These are really goals. Quantifiable objectives (even though they might be subjective in origin) which are defined at all organizational levels, from the local manager to the director, will inevitably make management more efficient—although not necessarily "better."

References

ANDERSON, R. O. 1976. Management of small warmwater impoundments. Fisheries 1(6):5–7, 26–27.

———, AND A. S. WEITHMAN. 1978. The concept of balance for coolwater fish populations. Am. Fish. Soc. Spec. Publ. 11:371–381.

BELUSZ, L. C. 1978. The role of private organizations in management of coolwater fisheries resources—a panel. Am. Fish. Soc. Spec. Publ. 11:426–427.

BOVEE, K. D. 1978. The incremental method of assessing habitat potential for coolwater species, with management implications. Am. Fish. Soc. Spec. Publ. 11:340–346.

CLARK, R. D., JR., AND R. T. LACKEY. 1975. Managing trends in angler consumption in freshwater recreational fisheries. Proc. Annu. Conf. Southeast. Assoc. Game Fish Comm. 28:367–377.

DRIVER, B. L., AND R. C. KNOPF. 1976. Temporary escape: one product of sport fisheries management. Fisheries 1(2):21, 24–29.

KENNEDY, J. J., AND P. J. BROWN. 1976. Attitudes and behavior of fishermen in Utah's Uinta Primitive Area. Fisheries 1(6):15–17, 30–31.

LACKEY, R. T. 1975. Recreational fisheries management and ecosystem modeling. Publ. VPI-FWS-4-75, Va. Polytech. Inst. State Univ. 44 pp.

LARKIN, P. A. 1977. An epitaph for the concept of maximum sustained yield. Trans. Am. Fish. Soc. 106(1):1–11.

MARTIN, R. G. 1976. Philosophy of sport fisheries management. Fisheries 1(6):8–10, 29–30.

MCFADDEN, J. T. 1969. Trends in freshwater sport fisheries of North America. Trans. Am. Fish. Soc. 98(1):136–150.

PRENTICE, J. A., AND R. D. CLARK, JR. 1978. Walleye fishery management program in Texas—a systems approach. Am. Fish. Soc. Spec. Publ. 11:408–416.

RADOVICH, J. 1975. Application of optimum sustainable yield theory to marine fisheries. Am. Fish. Soc. Spec. Publ. 9:21–28.

ROEDEL, P. M. 1975. A summary and critique of the symposium on optimum sustainable yield. Am. Fish. Soc. Spec. Publ. 9:79–89.

SCHEFFER, V. B. 1976. The future of wildlife management. Wildl. Soc. Bull. 4(2):51–54.

SCHNEIDER, J. C. 1978. Selection of minimum size limits for walleye fishing in Michigan. Am. Fish. Soc. Spec. Publ. 11:398–407.

PANELS

Hatchery Design for Coolwater Species—A Panel[1]

JOHN G. NICKUM

New York Cooperative Fishery Research Unit
Cornell University
Ithaca, New York 14853

Demands for more and larger coolwater fishes for both stocking and food production purposes have exceeded the production capabilities of traditional hatchery and pond systems. A strong interest in the design of hatcheries for intensive culture of coolwater fishes has resulted. The panel described the principles which underlie the design of successful coolwater fish hatcheries, including planning processes, engineering considerations, and biological functions.

In pond culture, the pond provides both the place of residence and the source of food for the fish. In flow-through systems, whether pond, tank, or trough, the water serves as a residence and delivers oxygen, removes metabolic wastes, and maintains temperature, but food is supplied from outside sources. Flow-through systems require less space and can provide greater control of environmental conditions than static systems. Production capacity depends largely on the flow rates, or water exchange rates, in the rearing units, but generally is much higher in flow-through systems than in pond systems.

Hatchery Design and Development

A hatchery must be designed to meet the nutritional, physiological, physical, and behavioral needs of the fishes which will be reared in it. At this time there is a serious lack of reliable information on many such factors for coolwater fishes. Those few studies which have been done, as well as the more extensive work with salmonid fishes, provide some guidance relative to rates of oxygen consumption, minimum allowable levels of dissolved oxygen, rates of ammonia production, and maximum allowable levels of ammonia. It appears that some coolwater fishes may tolerate lower oxygen levels and higher ammonia levels than are permissible for salmonids. Feeding behavior, as well as nutritional requirements, and other behavioral reactions must be considered for each species. Harry Westers demonstrated that feeding rates, fish metabolism, density levels, behavioral characteristics, and flow requirements can be accommodated by hatchery design in a quantitative manner.

Flexibility. This word occurred to me repeatedly as I considered the remarks of Cecil Fox, Edward Miller, and Maurice Moore. The best hatcheries make it possible to change rearing units, water-supply piping, and other physical features as opportunities arise to improve fish culture. Needs change; technology changes; the biological information base grows; and a hatchery must have flexible capabilities if the persons operating it are to respond to these changes.

The hatchery development process should provide for continuing input from the various groups involved: owner; management; user; consultants; architects; engineers; biologists. It must be remembered that a considerable period of time may elapse between the identification of the need for a new hatchery and full-scale production. The process of hatchery development is essentially the same in most cases; however, the hatchery which results

---

[1] Panel members: Jack R. Hammond, Michigan Department of Natural Resources (Moderator); John G. Nickum, U.S. Fish and Wildlife Service (Rapporteur); Cecil L. Fox, Kramer, Chin & Mayo, Inc.; Edward R. Miller, Pennsylvania Fish Commission; Maurice H. Moore, Michael Baker, Jr., Inc.; Harry Westers, Michigan Department of Natural Resources.

from the process is unique. Fish hatcheries generally are one of a kind. The hatchery management manual must consider the unique capabilities and limitations of each facility and must be continually updated.

Hatchery Facilities

A large supply of high quality water is as basic to the culture of coolwater fishes as it is to raising salmonids. The ability to heat or cool water for an acceptable cost appears to be an additional problem for coolwater culture since water of desirable temperature (ca. 20 C) for growth of coolwater fishes is seldom available for more than a few weeks each year. Recirculation units or heated effluents may help reduce the cost.

The specific size and design of hatch buildings and rearing units is generally determined by water supply, projected production, and individual experiences. Fiberglass troughs and tanks provide flexibility in hatchhouse operations, while double-concrete raceways seem to be effective outdoor rearing units. Silos appear to have potential for some species, but have not been thoroughly tested for coolwater culture. Experience in Pennsylvania indicates that each rearing unit should be accessible to an electric outlet, have paved access, be equipped with automatic feeders, have capabilities for aeration and water blending, and have an individual water supply and drain. The Linesville (Pennsylvania) Fish Culture Station not only provides a working example of many of these features, but also incorporates a capability for public education within an aesthetically pleasing facility.

The design and operation of hatcheries for coolwater fishes seems to involve the same principles as those for coldwater fishes. Unique biological considerations, based upon the nutritional, physiological, physical, and behavioral requirements of coolwater fishes must be included; however the process and the basic principles remain the same. A good water supply and flexible capabilities are essential.

The Role of Private Organizations in Management of Coolwater Fisheries Resources—A Panel[1]

LAWRENCE C. BELUSZ

Missouri Department of Conservation
1104 S. Grand, Sedalia, Missouri 65301

Special-interest organizations are the fastest growing, most exciting feature in the recent history of sporting groups. Increased fishing pressure has coincided with the establishment and growth of national and state fishing organizations. New fishing techniques as well as sophisticated equipment have been developed through the efforts of these groups. The overall result has been highly efficient fishermen capable of harvesting increased numbers of sportfish.

In his opening remarks, Mr. Radonski challenged both the state agencies and private fishing groups by asserting that the future of fisheries resources was closely tied to the degree of cooperation or antagonism between the two groups. He further stated that private organizations have a challenging role to play in directing fisherman opinions, but to function properly they need open lines of communication with state agencies.

Meeting the Challenge

Representatives of the private fishing organizations concurred with Mr. Radonski's remarks. They felt that state agencies should no longer characterize fishing groups as "chronic complainers" but rather as sources of manpower, support, and feedback for agency fishing programs. Members of private organizations realize, perhaps for the first time because of their organization, that they are the source of pressure which can deplete a delicately balanced resource. Therefore, private organizations today do not function to provide bigger and better catches of fish. Rather, they are deeply interested in the resource and wish to insure its future value for quality recreation.

The stated goal of today's fishing organization is to become involved in "what's happening" at local, state, and national levels. They stand ready to offer manpower, money, and encouragement to agencies willing to provide guidance and direction for their efforts.

Agency Response

Responding to the offer of assistance and involvement by private fishing organizations, panel members representing state resource agencies welcomed the cooperation of the fishing groups. They stressed the importance of communication, and agreed that state agencies could not be completely at odds with organized fishing groups and still maintain their effectiveness in dealing with the fishing public. The opportunity to present information to organized groups rather than to individuals effectively magnifies the agencies' potential audience. More importantly, the private group is a receptive audience able to support agency decisions, particularly when it has assisted with the decision-making process.

The involvement by organization members, who may not have scientific training, in the collection of biological data for use by the state agency requires careful planning. However, the overall view was expressed that well-directed efforts could supply useful information when it is obtained by informed and enthusiastic participants.

---

[1] Panel members: Gilbert C. Radonski, Sport Fishing Institute (Moderator); Lawrence C. Belusz, Missouri Department of Conservation (Rapporteur); Eugene Harbage, Ohio Huskie Muskies Club; Dale Henegar, North Dakota Game and Fish Department; Larry Ramsell, Muskies, Inc.; Charles Shaw, in' Fisherman Society; Duane Shodeen, Minnesota Department of Natural Resources.

Future Roles

The future functions of private organizations cannot be assessed at this time because of their continuing evolution. Many are still maturing, and attempting to determine those areas in which to best direct their efforts. A word of caution, then, may be appropriate at this time. Do not direct all efforts toward one species or that once-in-a-lifetime trophy. Instead, direct efforts to protect all fishing resources. Promote the sport, not just sport fish. Realize that many other fishermen prefer other fish species and while not organized still deserve to be heard. A proliferation of special-interest groups at odds with each other and with the state agency would surely be counterproductive.

To help guide this maturation into future involvement, the state agency plays a most important role. Efforts of private organizations are often directed toward short-range goals and limited solutions. If the agency advises the organizations of its long-term planning policy and goals, it can eliminate misunderstandings and direct all efforts toward a common goal. Many agencies need to reconfirm that they are responsible to their constituents. Open cooperation with the fishing public will assure that the common goal of improving fisheries resources will be reached.

Allocating Percid Resources in the Great Lakes: Biological, Institutional, Political, Social, and Economic Ramifications — A Panel[1]

Kenneth M. Muth

*U.S. Fish and Wildlife Service
2022 Cleveland Road
Sandusky, Ohio 44870*

The Great Lakes, shared between Canada and United States and spanning an area nearly 1,300 km from east to west and 800 km from north to south, represent the greatest reservoir of fresh water in the world. The fishery resources contained within these lakes assumed ever increasing importance as regional settlement expanded and commercial fisheries became established. Coldwater fish species such as lake trout, whitefish, cisco, and chubs from Lakes Superior, Michigan, Huron, and Ontario were the species most highly prized and widely acclaimed during the early history of the fisheries. Coolwater species such as yellow perch, walleye, sauger, and blue pike from Lake Erie and the shallower waters of some of the other Great Lakes were also harvested but the importance of these species developed more slowly and with less fanfare.

Since the 1930's, environmental, cultural, and economic events have greatly changed the character of the fisheries. With the decimation of many populations of coldwater species that resulted from the sea lamprey invasion, the increased market demand for fish that resulted from increased population levels, the expansion of recreational fishing, and the deterioration of aquatic habitat as the result of increased pollution, exploitation of coolwater species has increased to such a degree that some stocks have been severely depleted or, in the case of blue pike, perhaps driven to biological extinction in Great Lakes waters.

In response to these fishery problems, an international body known as the Great Lakes Fishery Commission (GLFC) was formed in 1956 with the initial responsibility of dealing with research and control of the parasitic sea lamprey. Other responsibilities under the terms of the convention are: "1) . . . to formulate a research program or programs designed to determine the need for measures to make possible the maximum sustained productivity of any stock of fish in the convention area which in the opinion of the commission is of common concern to the fisheries of the United States of America and Canada to determine what measures are best adapted for such purposes; 2) coordinate research pursuant to such programs and if necessary, undertake such research; and 3) recommend measures based on these findings."

Although the GLFC has no jurisdictional authority or regulatory power over the fisheries, which are the direct responsibilities of the Province of Ontario and the states bordering the Great Lakes, it provides a cohesive force both through leadership and as a focus for interagency coordination and negotiation. The Commission has created a lake committee for each of the five Great Lakes; each committee is comprised of senior fishery management personnel from provincial and state resource agencies bordering the respective lake. When the committees are confronted with technical problems that need resolution, they often appoint ad hoc or standing technical committees composed of scientific and technical specialists from within the cooperating agencies, or from other agencies

---

[1] Panel members: Carlos Fetterolf, Great Lakes Fishery Commission, (Moderator); Kenneth Muth, U.S. Fish and Wildlife Service, (Rapporteur); Dennis Cauvin, Freshwater Institute; Robert Haas, Michigan Department of Natural Resources; Art Holder, Ontario Ministry of Natural Resources.

and universities, and charge these specialists with devising the most scientifically justifiable recommendations. These recommendations are incorporated in the lake committees[1] fishery management strategy recommendations which are provided to the GLFC for endorsement and transmittal to the responsible management agencies for action.

The development and acceptance of a management strategy for a fishery resource which is under the jurisdiction of several political entities having varied philosophies, socioeconomic conditions, and political constraints is never an easy task. This is illustrated by the allocation of percid resources in the Great Lakes, particularly by the recently developed walleye quota management system in western Lake Erie.

Sport Fishery Considerations

The allocation of Great Lakes percid resources implies that different and competing uses of the resource exists, namely the sport and commercial fisheries. Three types of values can be derived from these fisheries which include direct value to the angler and fish purchasers, indirect values to local economies through economic impacts, and the more humanistic values such as traditional occupations (Talhelm 1973). However, the decision to harvest percid resources by sport and/or commercial fisheries must be based on more than tradition and the direct values must be of greatest concern.

The commercial fishing value may be estimated from such data as dockside price of the catch, wholesale price in the market, or some other index value; but what is the value of the sport fishery? Obviously, the sport fishery value depends on the ability and willingness of people to spend their time and money on fishing. These attributes, in turn, are dependent on such factors as abundance and catchability of fish; the size of fish; the geographic location of the fish resource with respect to user population centers; the age, fishing experience, economic status, education, and leisure time of the fishermen; and many other factors. The influence of these factors is poorly understood but if the public is going to derive maximum value from a fishery resource, and if a sport fishery harvest is part of that value, then managers must determine what the sport fishermen want from their fishing experience.

A survey of Great Lakes state and provincial management agencies indicates that only a few percid stocks are presently important to both commercial and sport fisheries. When both fisheries compete for a percid resource, management agencies tend to give harvest priority to the sport fishery. This priority has little to do with the relative values of each fishery, but instead seems to be related to the ever increasing number of sport fishermen who voice their objections to the presence of the commercial fishery. A more reasonable management strategy for apportioning the resource should include value factors as well as social and political considerations, and such determinations can only be possible when data on the sport fishing effort and catch are known.

In recent years, Ontario, Ohio, Michigan, New York, and perhaps other state resource agencies have begun collecting data on sport fishing effort and catch. A mail survey of Michigan licensed anglers engaged in yellow perch fishing in Michigan's Great Lakes waters revealed that an estimated 2.6 million angler days in 1976 resulted in a catch of 27.6 million yellow perch. About 20% of this effort and 29% of the catch were from Saginaw Bay (Michigan Department of Natural Resources, unpublished) where there are simultaneous sport and commercial yellow perch fisheries. Ontario and Ohio are engaged in vigorous sport angler surveys in western Lake Erie to determine effort and catch for walleye and other species. Data on sport catch of walleyes are particularly important because they are incorporated in walleye quota allocations under the current interagency cooperative management program. No commercial fishery for walleyes is permitted in Ohio waters but the sport catch increased from 47,000 fish in 1975 to 476,000 fish in 1977. The 1977 sport catch represented approximately 91% of the Ohio quota (Ohio Department of Natural Resources 1978). Ontario also reported an increased catch in the sport fishery from 19,000 fish in 1975 to 92,000 in 1977. Both sport and commercial walleye fisheries occur in Ontario waters and the 1977 sport catch represents 5% of the provincial quota (Ontario Ministry of Natural Resources 1978).

In summary, the sport fishery for percid resources in the Great Lakes is becoming more important in terms of management allocations between competing sport and commercial fisheries. Additional data on sport fishery effort and catch are needed but we have at least one example of an extensive sport fishery for walleyes which could harvest the entire surplus production if that were the selected use of the resource.

Commercial Fishery Considerations

The allocation of fishery resources is predicated on assumptions that the total quantity of product available for distribution is known, the numbers of "shares" or members of society who wish to harvest the resource are identified, and the value system which will be used to divide the resource among the users is defined. Identification of these components in past management efforts has been deficient but some progress is being made to correct these deficiencies. The current fishery management practices and philosophy of the Province of Ontario might serve as an example of some of this progress as well as point out other areas for improvement.

Resource Availability

Historically, the harvest availability of fish in Ontario waters of the Great Lakes has not been adequately determined. The early emphasis on life history studies of Great Lakes fishes did not provide data on the harvest availability and the more recent population dynamic studies have been used principally for postmortem explanations of the cause of changes in population levels. The theoretical concepts of maximum sustained yield (MSY) and optimum sustained yield (OSY) have not proven to be adequate for fishery management in the practical sense. It is obvious that fishing pressure can be a major contributing factor to the commercial extinction of important fish populations and it is necessary to control this pressure within projected limits of available harvest if a fishery resource is to be maintained. An initial attempt to control harvest within projected total allowable catch (TAC) of western Lake Erie walleyes is in progress under the interagency walleye quota management system, and this program will be discussed later.

User Participation

Data on the second factor, participation in the Ontario fishery, date back to 1869 when commercial fishermen were first licensed (Adams and Kolenosky 1974). Commercial access to the resource has been controlled to a varying degree ever since that time. The early practice of impounding fish limited access to sites which had space and depth suitable for the gear, but with the advent of gill nets, and the greater fishing mobility associated with this gear, an open-access fishery existed. Since about 1960, following the collapse of some fish stocks such as the blue pike in Lake Erie, the issuance of commercial fishing licenses has been more strictly controlled. On the other hand, the recreational angler participation in the fishery is not as well known or controlled. Licensing of nonresident anglers, established in 1887, indicated adult participation in the Ontario fishery in 1975 totalled 647,848. The current nonresident license fee of $10.75 is not considered to be a restrictive factor for angler use of the resource. Licensing of resident anglers only occurred in 1969 and 1970, and was abandoned after that time, but the 1975 estimate of Ontario adult resident anglers was 1,967,000. About 28% of the resident and 19% of the nonresident anglers participated in the Great Lakes sport fishery alongside 68% of the 1,121 licensed commercial fishermen.

Fishing Values

Finally, a value system for the Ontario fishery resources based on some type of a pricing system has not been established. The concept of free and equal access to fisheries resources is a strong and integral part of the heritage of Ontario residents and probably explains part of the reluctance to restrict usage of a resource which clearly needs protection. The free-access concept has resulted in little or no value being attached to the fishery resources which, in turn, has contributed to the degradation of this resource.

Management Initiatives

A unique opportunity to deal with these resource management deficiencies became apparent when the walleye commercial fishery in western Lake Erie was closed after 1969 because of mercury contamination. The walleye stock had been severely stressed during the years prior to 1969 and commercial harvest had seriously declined (Regier et al. 1969). The Great Lakes Fishery Commission recognized in this unfortunate incident an opportunity to develop a more rational plan for future management of the walleye resource when the fishery reopened. In 1973, the GLFC formed a Special Protocol Committee (SPC) on Interagency Management of Walleye Resources of Western Lake Erie, "for the purpose of developing the technical data needed for determining the total allowable catch, fair quota shares, criteria for future adjustment, and specifications for the mechanism for implementing the sharing system". The membership of the SPC included representatives from provincial, state, and federal governments. The first walleye quota allocations were recommended in 1976 and Ontario divided their allocation between sport and commercial fisheries by first meeting the existing angler demand, as determined via creel census, and then assigning the balance of the allocations to commercial licenses. In addition, two open fishing seasons for the commercial fishery were established from May 1 to June 30 and from the first Tuesday in September to December 31. The rational for commercial fishing seasons included such considerations as protection of walleye spawning concentrations, encouragement of product quality, and minimized conflict with recreational boaters and anglers. An additional subdivision of Ontario's commercial fishery allocation was made on the basis of individual fisherman shares within each of the two counties involved in the fishery. A similar procedure was followed in 1977 but a review date of October 15 was established to redistribute quotas between counties if existing quotas were not being achieved.

After two years of walleye management under the quota allocation system, there is some indication that the TAC concept may be at least partially successful as a strategy for maintaining a healthy fishery. However, many questions and problems relating to this concept have been evident during these two years. It is obvious that there is a need for improvement in the quantity, quality, and distribution of walleye data that are used to predict TAC's. Although current sport fishery needs are being accommodated, any future demands by the sport fishery will likely be met at the expense of the commercial fishery share. Because no benefit-cost analysis for the harvest has been done, the merits of this course of action remain unproven. The economic questions become even more complex when one considers that nearly 85% of the Ontario sport anglers are from the United States; they consume a portion of Ontario's walleye allocation while contributing only the cost of a low nonresident license fee to Ontario's economy. The current number of commercial fishermen engaged in catching the allowable harvest of walleyes may be inappropriate and may result in excessive fishing capability and over-capitalization, but no definite study of this problem has been conducted for the Lake Erie fishery. If this is determined to be an actual problem, what fair method can be devised to adjust the number of fishermen? These issues, along with many others, must be solved before individual quota control can be accepted for long-term management.

Ontario's future application of resource allocation as a means of effective management is predicated on the following recommendations (Ontario Ministry of Natural

Resources 1978):

"1. increased emphasis on resource protection (or rehabilitation), and the adjustment of harvests accordingly;

"2. explicit allocation of allowable yields among competing users based on adequate criteria reflecting an ongoing appraisal of society's values; and

"3. the adoption of a 'user pays' policy to meet increased costs of management and to better reflect the true value of the fishery."

The key factors which must be recognized with respect to fishery resource allocation are (1) that a large portion of each annual year class must be dedicated to the continuing health of the stock if the fishery resource is to be maintained, and (2) that trade-offs made at the expense of the resource for immediate economic or social gains can have lasting and sometimes irreversible effects on the resource.

Walleye Quota Procedures

As previously indicated, the SPC developed the analysis procedures for projecting the annual total allowable catch (TAC) that is harvestable from the entire walleye stock of western Lake Erie. This TAC is expressed in numbers, not weight, of fish that can be harvested, and by definition includes fish which are two years old and older.

An analysis of historic walleye reference data from the 1963–1969 period provides a ratio of average yearling standing stock to average young-of-the-year (YOY) index of abundance and this ratio is used to estimate recruitment. The YOY abundance index, expressed as the number of YOY walleye caught per hour of bottom trawling, is determined each year and applied to the reference ratio to generate the new estimate of recruitment.

The current characteristics of the walleye stock are determined from an analysis of pooled data provided by Ontario, Michigan, Ohio, and the U.S. Fish and Wildlife Service. Mortality coefficients, growth rate, and age group composition of the stock are determined from these data. The application of mortality coefficients to each age group provides an estimate of the standing stock of each age group. Following each age group in a sequential manner throughout its existence in the fishery provides an estimate of the individual age group contributions to the harvest, and summing these contributions provides a projected annual TAC. Mean biomass estimates of the stock in any given year are derived by applying age-specific growth rates to the standing stock of individual age groups and summing these values.

The recommended TAC is applicable to the entire walleye stock of western Lake Erie but this stock is subjected to the fisheries of three political entities: Ontario, Michigan, and Ohio. Subdivisions of the TAC follow the percentage of western basin lake surface area that is under the jurisdiction of each political unit (Ontario, 38.8%; Michigan, 8.8%; Ohio, 52.4%). The manner in which each individual quota is harvested, whether it be by sport fishermen, commercial fishermen, or a combination of the two, is determined by the respective governments.

The responsibilities of the SPC were completed in 1977 and the committee was disbanded. In its place, a Standing Technical Committee (STC) of the Lake Erie Committee was formed to annually provide the walleye quota recommendations for interagency management and to seek improvements in the analysis which would enhance the reliability of the quotas. Although only Ontario, Michigan, and Ohio fishermen harvest the walleye stock in western Lake Erie, the management of this resource can have lakewide implications and it is essential that all states bordering on Lake Erie, the Province of Ontario, and federal government agencies participate in the development of resource management strategies. Representatives from these governments contributed to the development of the quota management system under the SPC and are currently involved in the continuation of this program under the STC.

The current quota allocations recommended for walleye harvest in 1978 and projected for 1979 are 827,000 fish and 2,564,000 fish, respectively. This threefold increase for 1979 is possible because the walleye spawning success in 1977, as determined from the YOY index of abundance, was phenomenal and these fish will begin entering the fishery as catchable two-year-olds in 1979.

The quota management system did not achieve complete harvest of the TAC in either 1976 or 1977. The 1976 catch of 490,000 fish and the 1977 catch of 918,000 fish represented 53.6% and 92.3%, respectively, of the recommended allocations. Recommended TAC values are intentionally conservative to allow for walleye stock rehabilitation, and actual harvests which are less than recommended quotas ensure stock rehabilitation at the expense of achieving the estimated best utilization of the resource. Rehabilitation of the walleye stock appears to be successful as indicated by the current high levels of abundance and the large brood stock available for future production.

Economic Considerations

While many fishery managers might agree that protection of resources by means of an allocation process is desirable, agreement on appropriate criteria for these allocations is not so readily achieved. This is perhaps most evident with regard to economics, where we have chosen to ignore allocation based on a price system or else have attempted to measure proxy values and use these substitutions to justify allocation decisions (Cauvin 1978, this volume).

An overall value for the recreational use of fisheries resources in western Canada is established by combining a value associated with the willingness of the users to pay with an estimate of expenditures related to recreational fishing activities. The estimated 1969 recreational fishery in Alberta was 2.3 million angler days for 153,000 resident anglers and 5,500 nonresident anglers (Miller 1971). Although the net social value of this fishery derived from the valuation process is an impressive $16.7 million, the realism of many of the arbitrarily derived values is questionable. A comparison of the two known monetary values, fishery management costs and license sales, provides a net deficit for the recreational fishery of $488,000.

The commercial fishery valuation process in Canada is also based on some interesting rationale. A concept of "value added" to the Canadian economy is included, based on the objectives of maximizing stocks and yield of fish, maximizing employment, increasing incomes to fishermen, and promoting regional growth and development. Taken individually, these objectives in relation to a commercial fishery may be achievable but in combination, they are in conflict. It is mathematically impossible to simultaneously maximize two or more conflicting objectives. In an industry which is notorious for excess labor and capital investment, any valuation which is based on an expansion of the industry and an associated increase in regional development is questionable.

Alternate methods for valuating fisheries resources as criteria for allocation decisions are favored by resource economists. It will be necessary to consider a market system which generates price signals. For commercial fisheries, a restricted entry and a royalty system on landed value of fishery products will permit a competitive return to the minimum amount of labor and capital required to harvest the resource while controlling aggregate fishing effort and providing tangible evidence of resource value. For recreational fisheries, an increase in licensee fees in aggregate would be required if the pricing objective is to raise revenue. If the objective is to divert fishing pressure from an over-fished area to one with less use, it could be possible to use differential pricing.

The resultant measure of the resource values will provide a firm foundation for making allocation decisions on the basis of that combination of uses that generates the greatest net value. Since values of alternative uses occur over time, comparisons are made easier by calculating net present values. These values represent the opportunities foregone by allocating resources to one use at a sacrifice to another. Since alternative uses are seldom mutually exclusive, there will necessarily be a trade-off between one use and the other.

Discussion and Comments

Following the presentations of the panel members, open discussion with the audience generated the following questions and comments (C) and panel responses (R):

(C) Is the aggregate net value of the resource more important or are politics the dictating factor when resource allocation decisions are made?

(R) An accounting is required as to who receives the benefits and who pays the bill, particularly if the objective of management is to enhance the welfare of all of society. Currently taxpayers are providing for the management and maintenance of the fishery resources while the real resource value is being reaped by other user groups at little or no cost to them.

(C) Why do not resource agencies devote more management effort and development to the onshore yellow perch fishery which can provide recreation for people who cannot afford boats or who live in urban areas and have limited access to the fishery?

(R) Michigan has initiated a metro fishing project which is still in the preliminary stages of development and the success of this program remains unknown at this time. Fishery managers generally fail to understand the wants and needs of this segment of recreational fishermen. They tend to rely on older programs and management concepts that have been successfully used elsewhere but, in reality, these may not be applicable to such problems as the onshore recreational fishery for yellow perch.

(C) What did the economist mean when he suggested that the resources used to manage a fishery bear no relationship to the scarcity of these resources?

(R) Dollars are usually the standard used when talking about the cost of managing a fishery resource and there are always alternate uses for this money such as welfare programs, reduction in taxation, development of new social programs, etc. The alternate uses of money are never considered when fishery management costs are analyzed so there is no relationship between the management costs and the real value obtained from this expenditure.

(C) Fishery management costs are often attributed to the sport anglers and commercial fisheries are ignored. Sport anglers may actually be paying for the fishery management costs through taxation on fishing equipment.

(R) The Miller study used as an example for determination of direct and indirect costs and benefits deals only with a sport fishery. An indirect benefit such as the tax on fishing equipment used to support fishery management may help alleviate costs but it does not help in solving such management problems as controlling access to the fishery resource or rationing a valuable resource which is undergoing stress. When restrictive management measures are required, the expenditure on recreational equipment is decreased and the availability of tax money for management is decreased.

(C) The commercial fishing industry is characterized by excessive labor and capitalization and the first step in fishery economics is to bring these two factors down to a rational level with respect to economic gains that can be obtained from the industry. How do we do that?

(R) A pricing system will regulate these two factors. As the operational costs increase, the tendency toward excessive labor and capitalization will decrease.

(C) When the value of the resource is less than the cost of managing that resource, such as is the case of some depleted fishery resources in the Great Lakes, what do we do? Can raw economic theory be implemented in a political situation?

(R) When a pricing system is applied to the resource, you may discover that no one wants it enough to pay the cost. Water resources are not priced in terms of the quantity used or the quality of water returned to the system. The economic theory for pricing these resources is available and, when it is applied so that realistic values can be established, the results are convincing data that can be taken to your politicians to try and stop abuse and misuse of the resources.

(C) If it is alright to harvest fish for recreation, human consumption, industrial fishery products, etc. and we allocate fishery resources on this basis, why is it not equally justifiable to allocate a segment of the fishery resource to power plants and other industries which are going to destroy fish through entrainment and impingement processes associated with water withdrawal for cooling purposes?

(R) Mortality factors caused by entrainment and impingement are a continuing concern and the STC is attempting to improve the estimates of mortality rates used in determining walleye quotas. Although any adjustment made in this regard would not be in the form of an allocation to nonfishery industries, we do recognize that these industries influence the walleye resource. Adjustments in the quotas, because of this influence, are necessary.

(C) With increasing development of nuclear power plants along the shores of Lake Erie, can the STC make recommendations concerning the location of future plant sites so as to minimize the adverse effects on fish?

(R) No. The Ecological Services Division of the U.S. Fish and Wildlife Service is responsible for analyzing potential environmental effects on aquatic resources and makes recommendations in regard to power plant sitings, locations of water intakes, etc. However, these recommended alternatives are not easily accepted by the industry. Industry tends to minimize the severity of adverse effects by ignoring the long-term effects or by considering only single-site effects rather than the cumulative effects on fishery resources of all power plants.

(C) Can intangible values associated with natural resources, such as esthetics, ever be assigned a real value under the pricing system advocated for reforming the use and abuse of natural resources? Is not the pricing system only applicable to tangible and exchangeable goods of the market place?

(R) Numerous cases exist where intangibles are assigned monetary values and included in a pricing system.

(C) Is not money spent on recreational fisheries discretionary money that would simply be spent for other forms of recreation if the Great Lakes fisheries resources were not available? Is that money really an economic value of the fisheries resources?

(R) Recreational expenditures often quoted are those monies spent for services and goods such as food, lodging, clothing, equipment, etc. used by the recreational fishermen. They do not relate to the actual fishery resource cost/value.

References

ADAMS, G. F., AND D. P. KOLENOSKY. 1974. Out of the water: Ontario's freshwater fish industry. Ont. Minist. Nat. Resour., Toronto. 68 pp.

CAUVIN, D. M. 1978. The allocation of resources in fisheries: an economic perspective. Am. Fish. Soc. Spec. Publ. 11:361–370.

MILLER, R. J. 1971. Alberta's hunting and fishing resources, an economic valuation. Alberta Dep. Agric., Edmonton.

OHIO DEPARTMENT OF NATURAL RESOURCES. 1978. Status of Ohio's Lake Erie fisheries. 19 pp.

ONTARIO MINISTRY OF NATURAL RESOURCES. 1978. Lake Erie and Lake St. Clair fisheries report. Presented to the Great Lakes Fish. Comm., Lake Erie Comm., Feb. 28–March 1, 1978. 28 pp.

REGIER, H. A., V. C. APPLEGATE, AND R. A. RYDER. 1969. The ecology and management of the walleye in western Lake Erie. Great Lakes Fish. Comm. Tech. Rep. 15. 101 pp.

TALHELM, D. R. 1973. Evaluation of the demands for Michigan's salmon and steelhead sport fishery of 1970. Mich. Dep. Nat. Resour. Fish. Res. Rep. 1797. 69 pp.

SUMMARY

Review and a Look Ahead

G. H. LAWLER

Freshwater Institute, Fisheries and Environment Canada
501 University Crescent, Winnipeg, Manitoba R3T 2N6

I would like to preface my brief remarks by pointing out that in Canada, the management of the fishery resource in inland waters is carried out by provincial governments with one exception—the Northwest Territories—where the federal government has the responsibility for both management and research. The region under my direction for federal programs encompasses the three prairie provinces and the NWT, which collectively represents a tremendous volume of cool water and of fish species associated with cool water.

During the past 2½ days, I was impressed by the number and quality of the papers presented on various aspects of the biology and management of these species. Back in 1961 when I assisted in producing "A Synopsis of Pike" for FAO distribution, I thought there was very little more knowledge to be gained. This symposium has certainly opened my eyes to the expansion of our knowledge base in such a short time period. Who would have thought coolwater species would be tolerant of southern waters? Research and experimental plantings in what one would have thought were unsuitable environments have proven that educated gambles can pay off. The fields of genetics and genetic engineering in fisheries are young sciences and it would appear that in future years, some outstanding achievements will occur. We will be building fish to suit the environment, whatever it may be!

I was struck, however, during the discussions by the absence of attention given the impact of contaminants such as PCBs, heavy metals such as mercury and cadmium, etc., particularly their absorption by fishes which serve as human food items. Also lacking were papers dealing with the impacts of toxic chemicals and eutrophication on natural populations and their continuing success. The major emphasis appeared to be geared to the management of harvest and culture; these areas are of concern to both Canadians and Americans. The question of resource allocation was debated strongly and vociferously by resource managers and economists, and, although the results may have been declared a draw, the resource economists proposed a hypothesis which I feel will stimulate further such debate.

Although I have a strong interest in the Ontario/Great Lakes situation, my knowledge in recent years is not as current as I would like it to be. As an administrator, one's major concerns are budgets and balance sheets and attempts to keep the ship afloat in times of shrinking dollars. It is only when one has the opportunity to meet as we are doing today that one's training and experience come back into perspective. I must admit even my knowledge of the prairie situation is probably dated. With this in mind, I will now make some general observations on the future.

(1) With respect to commercial fishing, walleye, sauger, and northern pike are especially valuable species in the market place, and increasing emphasis must be placed on effective management of these species. There may be occasions where enhancement may be justified but the feasibility and cost benefit must be carefully examined. Some of the major lakes in Manitoba such as Lake Winnipegosis and Lake Manitoba come to mind. During the early days of experimental planting of rainbow trout in prairie potholes, we decided

to see how walleyes would survive and grow in one season. The results clearly indicated that both survival and growth were excellent. Newly hatched fry reached lengths of up to 13 cm during the growing season but harvesting presented a problem. We have been advocating a large scale pilot plant operation, utilizing shallow, productive waters adjacent to large lakes that would permit flushing the fish to the large lakes at the end of the growing season. This, we feel, would result in major economic benefits to the fishermen—hopefully at minimum cost to governments!

(2) The species under discussion in this symposium are utilized by the greatest number of fishermen across the prairies. *Except* in the vicinity of major population centres, the supply generally exceeds the demand. Trophy-sized fish, particularly northern pike, are still to be found—certainly in northern Alberta, Saskatchewan, and Manitoba. The lakes containing these fish should be managed for the trophy fishery as long as this is feasible. Walleyes are highly prized and pressure on this species will continue to increase. Utilization of more intensive management techniques will become increasingly necessary.

(3) The need for more intensive management is a direct result of increasing human populations (or redistribution of populations to more remote areas). I believe the area of major growth in Canada in the next few years will be in Alberta. The massive tar sands program will result in the creation of several towns with populations of 25–30,000 people. Thus the capability of fish populations close to large urban centres to withstand increasing pressures must be constantly reassessed. At the present time or in the relatively near future, the need for culture of these species is not foreseen.

(4) More hydro power, nuclear generating stations, and flood control projects may be anticipated in the coming years. We shall have to look critically at the potential impact of these projects, in order to determine appropriate mitigating measures.

There has been increasing evidence in recent times that a seminar/symposium should be held on goldeye to assess the current knowledge of this species. It is truly a coolwater species whose range has now been considerably extended. I was a little surprised that we had not included some discussion of the goldeye during our stay here. Perhaps we can include a session during future North Central meetings.

I would like to extend my personal congratulations and thanks for the excellent job done by the organizers of the symposium: the steering committee; the program chairman, Bud Griswold; the session chairmen and rapporteurs; and particularly I would like to single out Dick Sternberg, our local arrangements chairman, for the dedication he has shown.